32

AF781161

Nuclear Particle Correlations and Cluster Physics

Nuclear Particle Correlations and Cluster Physics

Editor

Wolf-Udo Schröder
University of Rochester, USA

World Scientific

NEW JERSEY • LONDON • SINGAPORE • BEIJING • SHANGHAI • HONG KONG • TAIPEI • CHENNAI • TOKYO

Published by

World Scientific Publishing Co. Pte. Ltd.
5 Toh Tuck Link, Singapore 596224
USA office: 27 Warren Street, Suite 401-402, Hackensack, NJ 07601
UK office: 57 Shelton Street, Covent Garden, London WC2H 9HE

British Library Cataloguing-in-Publication Data
A catalogue record for this book is available from the British Library.

NUCLEAR PARTICLE CORRELATIONS AND CLUSTER PHYSICS

Copyright © 2017 by World Scientific Publishing Co. Pte. Ltd.

All rights reserved. This book, or parts thereof, may not be reproduced in any form or by any means, electronic or mechanical, including photocopying, recording or any information storage and retrieval system now known or to be invented, without written permission from the publisher.

For photocopying of material in this volume, please pay a copying fee through the Copyright Clearance Center, Inc., 222 Rosewood Drive, Danvers, MA 01923, USA. In this case permission to photocopy is not required from the publisher.

ISBN 978-981-3209-34-3

Desk Editor: Rhaimie Wahap

Typeset by Stallion Press
Email: enquiries@stallionpress.com

Printed in Singapore

PREFACE

Much of the 80-year old story of nuclear science can be told, framed in one of two extreme theoretical model views of the atomic nucleus. Initially, the nucleus is viewed as a classical, essentially structure-less, charged liquid drop, which is governed by short-ranged, nearest-neighbor nucleon-nucleon interactions. Much later, this simple model is surpassed in predictive power by the picture of the nucleus as a hollow shell containing a swarm of independent particles, the nucleons, which self-organize in the beautiful patterns imposed by quantum mechanics. In practice, often hybrids of nuclear liquid-drop and independent-particle (shell) models are devised to explain observations in terms of subtle details in nuclear structure and reaction dynamics.

Already early investigations of extended nuclear matter, conducted to serve as foundation to a unified theory of light and heavy nuclei, demonstrate that a liquid phase of independent nucleons loses stability at subnormal matter densities. At mean densities below about one third of the equilibrium saturation density, the most stable phase is shown to be inhomogeneous, made of conglomerates of most strongly bound nuclear clusters, the alpha particles. The general phenomenon is reflected in the charge and matter density distributions of light nuclei, which are systematically more diffuse than those of their heavier counterparts. Unlike the latter, light nuclei below iron, specifically a suite of so-called alpha-nuclei, have an internal consistency prone to developing configurational cluster sub-structure.

Such configurational substructure is revealed in characteristic macroscopic matter distributions and in the characteristic energy level schemes of specific nuclei. It is expressed as well, in terms of molecular entrance or exit channel configurations, in reaction and decay modes of such nuclei at moderate excitations or angular momenta. The importance of cluster structure of light nuclei, for nuclear astrophysics and beyond, has been demonstrated in a spectacular and well-publicized fashion for the 3-alpha Hoyle state in carbon-12, a long-lasting challenge to nuclear theory. However, while propensity for particle correlations sustaining nuclear cluster structure is well established for sub-normal matter density, the development of large-scale correlations in systems at supra-normal densities is not obvious at first sight. Yet, such correlations appear to be theoretically inescapable in dense cosmological systems such as neutron stars, defining a new frontier in nuclear structure theory.

Contemporary research can take advantage of new opportunities to study cluster sub-structure in nuclei at the subnormal densities that develop at excitations close to the limits of stability, or with unusual mass-to-charge densities, e.g., for nuclei far off the beta-stable valley. Challenging theoretical problems persist in understanding the dynamic generation and decay of clusters within the nuclear medium, as experienced perhaps in heavy-ion reactions and in fission. New experimental work reporting unexpected patterns of nuclear clusters, ejected in sudden nuclear breakup and fission-like disruptions, awaits theoretical analysis.

The 20 or so invited articles appearing in this volume on nuclear particle correlations and cluster physics have been authored by nuclear scientists currently working actively on these topics in some 35 different laboratories around the world. Invitations to join the workforce on the volume have been issued after much appreciated consultations with expert colleagues, and following their thoughtful recommendations. It is in the nature of conducting research in modern physical sciences and, specifically, in experimental nuclear science, which are often executed in the frameworks of large international collaborations, that themes of research, as well as memberships of research teams overlap. From reading the articles in this volume, it should also be evident that the present collection will not provide a final judgement on the intriguing subject of nuclear particle correlations and cluster physics. Many fundamental questions and issues remain unresolved, hopefully instigating vigorous new research ventures in this field.

W. Udo Schröder
Rochester, NY 14627, USA
schroeder@chem.rochester.edu

February 2017

CONTENTS

Part 1

General Cluster Properties

CHAPTER 1

SIMILARITIES AND DIFFERENCES BETWEEN NUCLEI AND METAL CLUSTERS

P. M. Dinh

Laboratoire de Physique Théorique,
Université de Toulouse, CNRS, UPS, France

P.-G. Reinhard

Institut für Theoretische Physik,
Universität Erlangen, Staudtstraße 7,
D-91058 Erlangen, Germany

E. Suraud*

Laboratoire de Physique Théorique,
Université de Toulouse, CNRS, UPS, France

Besides the topic of this volume, namely clustering in nuclei, the keyword "cluster" addresses a special form of large molecules consisting out of the same building blocks (atoms or small molecules) piled up to a large compound, so to say a small piece of a solid. A particular species are metal clusters whose valence electrons can be viewed as Fermi liquid. This establishes the similarity to nuclei which also consist of the Fermi liquid of protons and neutrons. In this article, we discuss similarities and differences of nuclei and metal clusters with respect to structure, resonance excitations, and fission. The latter process is closely related, again, to clustering, the emergence of sub-units in a larger compound.

1. Introduction

Clustering in nuclei is associated with more or less identifiable substructures within a nucleus which consist themselves of small nuclei, particularly of α particles. Molecular or atomic clusters are, at first glance, totally different

*Corresponding author: suraud@irsamc.ups-tlse.fr

objects. These are large composites of molecules or atoms of the same sort piled up to a large molecule or, looking at it the other way round, small pieces of bulk material.[1] According to the huge variety of chemical materials, there is a huge variety of cluster types with different bonding types.[2] Unlike nuclei which consist out of one sort of material with one type of binding ("metallic"), all sorts of relations between the substructures (atoms or small molecules) and the total system can be found for atomic/molecular clusters. The task to discriminate substructures is thus also very relevant in molecular or cluster physics and several tools have been developed for this. A useful tool is, for example, the electron localization criterion which allows to identify areas where one electron wave function dominates.[3] In fact, this localization criterion can also be used to identify α clustering in light nuclei.[4] This is only one example out of many more tools which are common to nuclear and cluster physics. Already the basic theoretical description in terms of energy-density functionals proceeds very similar.[5,6] But in addition to the similar tools, nuclei and molecules cover also systems with comparable properties. From the many conceivable bonding types in molecules, metallic binding comes closest to nuclear binding. In both cases, the system's fermions (valence electrons, or nucleons respectively) have a long mean-free path and thus belong to the whole system. This suggests that metal clusters should have something in common with nuclei. This is indeed the case, particularly for simple metals (Li, Na, K, Cs, Rb) where the valence shell is well decoupled from ionic cores (that is the nucleus and the core electrons). Thus a comparison between structural and dynamical properties of nuclei and metal clusters has been a much discussed issue, for examples see Refs. 7–9. Many similarities have been worked out concerning spatial profiles, shell structure (magic numbers, deformation) and resonance excitations (giant resonances in nuclei, plasmons in clusters). We will address them in sections 2 and 3. In section 2, we show that, by introducing appropriate natural units, even a similarity at quantitative level can be found for density distributions and potentials. Section 3 takes a look at spectral properties in the stationary state as well as in excitations.

Clustering phenomena as they are discussed in nuclear physics can appear for metal clusters in decay channels. The dominant decay channels for excited clusters are electron or monomer (atom) emission. At sufficient excitation energy, multiple electron emission can lift the cluster charge above the stability limit. The strong repulsive Coulomb force thus generated drives dissociation of the system, Coulomb expansion with subsequent fission as moderate process near threshold[10–12] and Coulomb explosion producing several smaller fragments for higher charging.[13,14] We will address cluster fission in section 4 taking numerical simulations for Coulomb induced fission of Na_{14} as an example.

Before carrying on with the detailed comparison of structural and (low-energy) dynamical properties at the level of a mean-field description, we summarize here briefly a couple of differences and similarities which are not addressed later on. At the experimental side, the dramatically different dimensions of nuclei and clusters require, of course, much different machines. It requires much higher efforts and patience to elicit information from the awfully small nuclei. Nonetheless, nuclei are probably the best studied objects, both what concerns structural and dynamical properties. In particular the long developments of detailed experimental investigations have allowed to reach a remarkable degree of detail, both in terms of spectroscopic properties (for structure and low energy dynamics) and in terms of analysis of (possibly violent) dynamical scenarios with the highly sophisticated 4π multi-detectors even able to work on an event-by-event basis. Metal clusters have not yet been scrutinized at that level of resolution and details. Development is here still going on, e.g., in detailed measurement of angular distributions and kinetic energies of electrons emitted in connection with laser excitation. On the other hand, clusters are better to handle and have time scale in reach of present-days clocks. This allows experiments which nuclear physicists can only dream of (pathway to huge systems, time resolved dynamics, tunable temperature). Thus cluster physics supplies worthwhile complementing information on finite fermion systems.

At the side of theoretical description, nuclei are much more demanding than clusters. Molecular physics is governed by the well understood electro-magnetic interaction. Modern methods of many-body theory allow a description at each level refinement and to derive systematically effective approaches for efficient lower-level treatment, e.g., density functional theory. The nuclear many-body problem, on the other hand, has not yet fully converged. The case is much more involved because nucleons are composite objects whose intrinsic energy scales are too close to nuclear scales. The solution lies in a cumbersome treatment at the elementary level of quantum-chromodynamics and we witness presently breath-taking progress in that direction, see e.g. Refs. 15–17. In spite of the huge differences at deeply microscopic level, the effective energy-density functionals for a mean-field treatment of structure and low-energy dynamics come out at the end very similar. Finally, there is a regime intermediate between fully microscopic treatment and mean-field approximation. This concerns the extension of effective mean-field theory by dynamical correlations which become important with increasing excitation energy. This can still be treated with effective interactions (in-medium scattering cross section) and it can be attacked much the same way for nuclei and electronic systems. The method of choice so far are kinetic equations at the semi-classical level of the Vlasov-Uehling-Uhlenbeck approach.[18–20] These powerful kinetic equations to not yet include quantum shell effects. This is still in a development stage.[21] Although much could be said about similarities between metal clusters and nuclei concerning kinetic equations, we will not address this topic here for reasons of space.

2. Basic properties: nuclei versus metal clusters

2.1. *Orders of magnitudes and scales*

At first glance, nuclei and metal clusters are extremely different objects. While nuclei are compact at fm scale and relate to energies at MeV scale, clusters are 5 orders of magnitude larger and 6 orders of magnitude less bound. However, looking at shear size can be deceiving. Rescaling observables in natural units reveals a great deal of similarities between these two systems. Both share a basic feature: they can be viewed as finite pieces of a Fermi fluid. Nuclei are composed as a two component system out of a proton and a neutron fluid. The valence electrons of simple metal clusters can be considered as an electron fluid immersed in a background of positively charges ions where the latter are treated as classical particles. Both, nucleon and electron fluids, have in bulk a well defined equilibrium density ρ_{eq} which is conveniently characterized by the Wigner-Seitz radius r_s, the radius of a reference sphere carrying at equilibrium density exactly one fermion (denoted in nuclear physics often as r_0). From r_s, we can deduce, for a system composed of N_{syst} fermions, the global radius R, the bulk equilibrium density ρ_{eq}, and the Fermi energy ε_{F} as

$$R = r_s N_{\text{syst}}^{1/3}, \quad \rho_{\text{eq}} = \frac{3}{4\pi r_s^3}, \quad \varepsilon_{\text{F}} = \frac{\hbar^2}{2m}\frac{3.7}{r_s^2}. \tag{1}$$

Mind that ρ_{eq} in the case of nucleons holds for neutron and proton density separately. The total nucleon density at equilibrium of symmetric nuclear

Table 1. Typical scales of finite fermion systems. To be specific, we have taken Na clusters as examples of simple metal clusters. The Wigner-Seitz radius for a quantum dot and an atom trap can be tuned by design; the table quotes only typical values.

system	*fermion*	r_s	ε_F	*interact.*	*bulk limit*
		self-binding fermion systems			
nucleus	n, p	1.2 fm	35 MeV	strong	limited
^{3}He droplet	^{3}He	6Å	2.7 K	atom-atom	liquid
		binding supported by ions			
Na cluster	e^-	2 Å	3 eV	Coulomb	solid
		binding via design through external fields			
quantum dot	e^-	≈10 nm	1.4 meV	Coulomb	?
atom trap	atom	≈10 nm	nK	atom-atom	—

matter is twice that value. Note also the particular notations for system size. In nuclei, one uses conventionally N for neutron number, Z for proton number, and A for total nucleon number which means that we have to identify in this case $N_{\text{syst}} \equiv A$. In metal clusters, we distinguish electron number N_{el} and number of ions N_{ion} and both could be identified with N_{syst}. However, this does not make much of a difference because clusters are mostly neutral or singly charged. We mentioned above that a comparison of finite fermion systems is done best when expressing observables in terms of natural units. The natural units take r_s as unit of length and ε_F as unit energy.

Table 1 compares the basic properties for a couple of finite fermion systems, restricted to saturating systems, i.e., those which can be characterized by an equilibrium density or Wigner-Seitz radius (nucleons, atoms and molecules are also finite fermion system, but not of saturating type). Nuclei and ^{3}He droplets consist exclusively in fermions. They are self-binding system in that they form bound finite drops without support from external fields. Metal clusters consist in valence electrons as Fermi fluid together with some ionic background. The electron cloud alone would not be stable due to the repulsive Coulomb force. A stable configuration is obtained only together with the charge-compensating ionic background. The equilibrium density depends on the material resulting from the interplay between the ions (described by a pseudo-potential[22]) and the electron cloud. In fact, the picture is that simple only for simple metals, i.e., metals whose valence electrons are well decoupled from deeper lying bands.[23] These are typically the alkaline metals Li, Na, K, Rb, and Cs. The bound electrons in the valence band have rather long wavelengths such that they will not be sensitive to the details of the ionic structure. As a consequence, binding and transport properties can be rather well described by the jellium model where the ionic background is replaced by as a homogeneous positive background charge.[23] Amongst the alkaline metals, Na is the ideal representative because its equilibrium density is practically identical with the density of the free electron cloud (stabilized by a jellium background). Those systems which are dominated by the properties of the electron cloud can then be compared with the self-binding fermion fluids in Table 1 in spite of the basically different binding mechanism. There is one important difference to be noted already here. This concerns the maximum conceivable system size. Clusters, being close to neutral, have no upper limit and can connect steadily to bulk material while nuclear size is limited due to the strong repulsive Coulomb force from the protons. This limit will be around $A \approx 300$. There is a second chance for extremely large nuclear systems. These are neutron stars where gravitation helps to keep neutron matter stable against β decay. But these do not fit into the present comparison because neutron stars are more bulk than finite systems.

The table has two more entries which are quantum systems stabilized by particularly designed external fields. The leading agent in quantum dots is, again, an electron could, here formed from the electronic quasi-particles in semi-conducting bands. Whole atoms, as in the ^{3}He fluid, are the constituents in atomic traps and the atom-atom interaction has much in common with the interaction between nuclei or ^{3}He atoms and the working point can be tuned by design. This is why atom clouds in traps are often used as model for many-body systems.[24] Design systems are distinguished from the other three systems in the table in that their fermion cloud is confined. They are not suited to study effects of particle emission, fragmentation, or fission. Thus we will not consider them in the following. We will also disregard ^{3}He droplets although they have much in common with nuclei and metal clusters. But they are extremely hard to assess experimentally because the leading agents, the ^{3}He atoms, are neutral. He droplets are mainly used as laboratory for high-precision spectroscopy of small molecules.[25] Thus we will confine ourselves in the following to a comparison between nuclei and Na clusters.

2.2. *Mean-field properties*

There is one more crucial property common to nuclear and electron fluid which is not listed in Table 1. This is the mean free path $\lambda_{\rm free}$ of the fermions in bulk matter. It is much larger than the Wigner-Seitz radius $\lambda_{\rm free} \gg r_s$ and in most cases also much larger than the system radius, i.e. $\lambda_{\rm free} \gg R$. This feature allows a treatment in terms of single-particle wave functions moving in a common mean field. State of the art is a self-consistent description of the mean field in terms of density functional theory (DFT). It is interesting to note that the development of DFT in nuclei and in electronic systems went much in parallel at comparable dates. Electronic DFT was conceptually formed with the seminal papers[26,27] while a most fruitful proposal for a nuclear effective interaction[28] appeared at about similar time. Big steps toward practical and reliable DFT was made in the seventies, e.g., for electrons with[29] and for nucleons with.[30,31] Since then, the energy functionals have been steadily developed further such that we now dispose of high-quality DFT in both areas, for reviews see Ref. 5 for electronic DFT and[32] for nuclear DFT. For the numerical calculations in the following, we use at the electronic site the energy-density functional of[33] augmented with an averaged self-interaction correction.[34] Clusters are generally computed with spherical jellium background (Wigner-Seitz radius $r_s = 3.9$ a$_0$, surface width $\sigma = 0.9$ a$_0$). For the fission example, we also employ detailed ionic background and use for the ions the soft local pseudo-potentials from.[35] Nuclear calculations are performed with the Skyrme-energy-density functional using parametrization SV-bas from a recent optimization.[36]

The leading quantities in a mean-field description are the one-body density and the mean-field potential. Figure 1 compares density and potentials for the nucleus ^{208}Pb and the neutral cluster Na$_{92}$. First, it is to be emphasized that drawing the quantities in terms of natural units allows to use exactly the same frames for nuclei and clusters. This already demonstrates a great deal of similarity. Next, we see that density and potential (for electrons in case of jellium) share the typical profile of saturating fermion systems, namely a more or less flat plateau in the volume of the system with some smooth transition in the surface. Even the transition width of about 1 r_s is comparable. The flat region of density keeps close to equilibrium density. This is obvious for the Na cluster. For the nucleus, we have drawn the density of symmetric matter as reference. However, ^{208}Pb has rather large neutron excess ($N = 126$ neutrons versus $Z = 82$ protons). Thus proton density stays below the equilibrium line and neutron density above

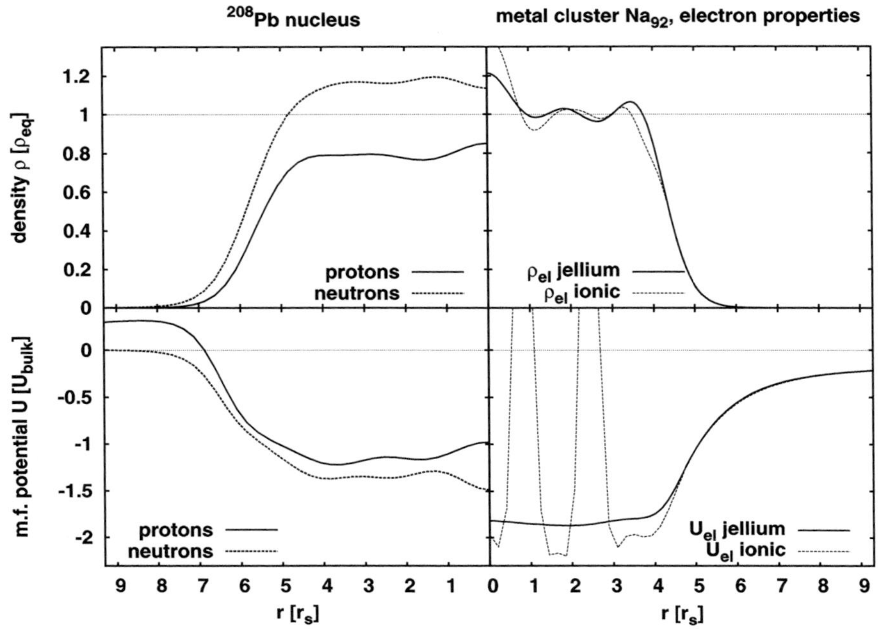

Fig. 1. Comparison between nuclei (left panels and reverse horizontal scale) and neutral Na clusters (right panels) for radial density distributions (upper panels) and mean-field potentials (lower panels). Energies, densities and lengths are given in natural units, that is, lengths in units of Wigner-Seitz radius r_s (see Table 1), energies in units of Fermi energy ε_F, and densities in units of bulk equilibrium density ρ_{eq} as given in Eq. (1).

this average. The mean-field potentials, although similar in general, show differences in detail. The proton potential (left lower panel) includes a large contribution from the (repulsive) Coulomb potential of the 82 protons. The figure just shows the thus generated Coulomb barrier. For larger r (not shown here), the proton potential decreases to zero as Ze^2/r. The electron potential of Na_{92} (right lower panel) decreases asymptotically as $-e^2/r^2$. This is the correct behavior for a neutral cluster because one electron, if it is departing, leaves a positively charge system beyond. This correct asymptotics is achieved by properly implementing a self-interaction correction.[34] Note that such a self-interaction correction is not yet included in nuclear calculations such that we see the asymptotic trend Ze^2/r rather than the correct $(Z-1)e^2/r$. But this is minor effect in case of nuclei and thus neglected.

The right panels of Figure 1 compare for the Na_{92} cluster results from the jellium model (full lines) with those from more realistic calculations including detailed ionic structure (faint dotted lines). The potential shows clearly the impact of ions. This, however, remains well localized and the average potential depth, even more so the surface potential, stay close to the jellium case. The densities (upper right panels) are even closer to each other. As argued above, the rather large wavelength of the valence electrons is unable to resolve the well confined ionic structures. Thus we see that the jellium model suffices for many purposes. Ionic structure is needed only if we are interested on fine details or explicitly on ionic motion.

2.3. *Shell closures, magic numbers and deformation*

Quantum shell structure in finite droplets leads to shell closures and thus particularly strong binding for certain fermion numbers, called magic numbers. Both, nuclei and clusters, shows a remarkable shell structure, however with different sequences of magic numbers. They are for Na clusters $N_{\rm el} = 2, 8, 20, 40, 58, 92, \ldots$ but $N, Z = 2, 8, 20, 28, 50, 82, \ldots$ for nuclei. This happens because there is a major difference in the mean field. Nuclei experience a strong spin-orbit interaction while this is negligible for Na clusters. The strong nuclear spin-orbit interaction is explained by the particular combination of relativistic effects[37] which are absent in electronic systems. There is another crucial difference, mentioned already above: clusters are not limited in size. This has given way to track the chain of magic numbers up to huge system sizes which revealed the occurrence of super-shells[38,39] as predicted much earlier from semi-classical quantization.[40]

Quantum shell structure has also a strong influence on deformation. In between the magic numbers, there is a large density of single-particle levels at the Fermi energy of spherical configurations. Many-body systems try to avoid such a situation and move to deformed configurations in order to reduce the level density, ideally finding some shell gap between the HOMO and the LUMO. This is the much celebrated Jahn-Teller effect which is abundant in molecular and solid-state physics.[41] It is also active in metal clusters and explains very well the regions of spherical or deformed clusters, even allowing to predict the sign of deformation (prolate or oblate).[42,43] The systematic sequence of nuclear deformations can also be explained in terms of the Jahn-Teller effect.[44] The great similarity between clusters and nuclei with respect to deformation was nicely worked out in.[45]

A word of caution is in order concerning quantum shell effects in metal clusters. Visibility and dominance of shell effects depends on cluster temperature.[46] They are particularly prominent at temperatures just above the melting point of the material. At low temperatures geometric shell effects from ions compete and often overlay electronic quantum shells. At too high temperatures, quantum effects fade away and shell beating is much reduced. For example, the ideal working point for Na clusters is around $T = 400$ K. This is safely above melting (in bulk at 370 K) and far below the 2000 K where quantum effects disappear.[38] Many experiments on plasmons and cluster deformation happen to be done in that temperature range.[47,48]

3. Spectral properties

In this section, we compare the systematics of spectra of nuclei and clusters. The comparison of mean-field potentials in Figure 1 indicates that the properties of the single-particle (s.p.) spectra generated from these potentials should be rather similar (besides difference in details as, e.g., spin-orbit splitting). Figure 2 compares the density of s.p. states for nuclei (lower panel) and neutral Na clusters (upper panel). There is one decision to be made for nuclei. They are not specified by system size A alone. One needs to choose additionally proton number Z or neutron number N. In order to define a unique series of nuclei with steadily increasing size A, we have followed the nuclei along bottom of the valley of stability, i.e. the best bound even-even nuclei in each isotopic chain. This way we have arranged comparable conditions and can now compare the two cases. Although the scales differ by order of magnitude, the trends are very much the same. There is a pronounced shell bunching, particularly visible for smaller system (toward right sides of the plots). The span between lowest level and Fermi energy ε_F remains constant for the metal clusters. More and more shells are diving below ε_F thus making the spectrum even denser until it can hardly be distinguished from a continuum. The pattern are similar for nuclei. Here, however, the spectral span is somewhat opening up with increasing system size and the Fermi energy moves toward continuum threshold. This is due to the fact that the neutron excess increases with nuclear size A which produces a deeper neutron potential well (see Figure 1) and a weaker Fermi energy. Moreover, the nuclear series terminates abruptly at about $A \approx 300$ due to insurmountable fission instability for heavier nuclei. Clusters, on the other hand, can proceed steadily to arbitrarily large system up to bulk.

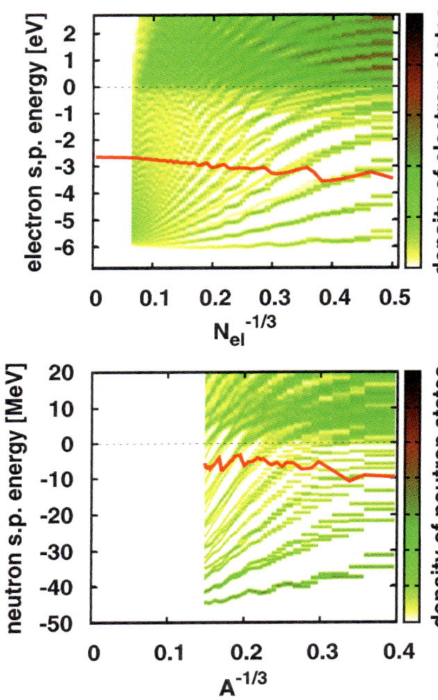

Fig. 2. Density of s.p. states for nuclei (lower panel) and neutral Na clusters (upper panel) drawn versus system size in terms of $A^{-1/3}$, or $N_{\rm el}^{-1/3}$ respectively. In case of nuclei, the density of neutron s.p. states is shown and the chain of nuclei is taken along the bottom of the valley of stability. The densities are scaled with one over system size and one over square root of energy (relative to bottom of potential) in order to compress the density scale for better plotting. The Fermi energy is indicated by a heavy line and the continuum threshold by a faint dotted horizontal line.

This is indicated in the upper panel of Figure 2 by carrying the heavy line for the Fermi energy through to $N_{\rm el}^{-1/3} = 0$. Technical reasons hinder us from doing the same for the density of states.

The s.p. levels are the ingredients for computing excitation spectra. The latter are described as coherent superposition of one-particle-one-hole ($1ph$) states in the framework of linearized time-dependent DFT, also known as random-phase approximation (RPA).[49] Of particular interest are here the remarked resonances which appear in the dipole channel, the isovector giant dipole resonance (GDR) in nuclei[50] and Mie surface plasmon resonance in clusters.[42,51] These dipole resonances show up in both systems as prominent peaks in the spectra with some fragmentation due to coupling

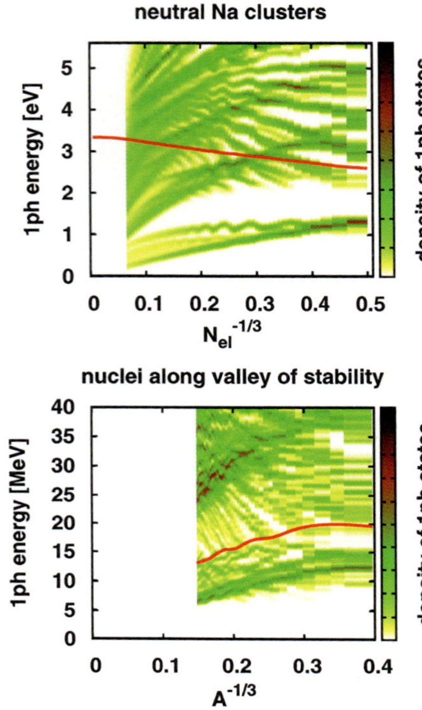

Fig. 3. Density of dipole $1ph$ states for nuclei (lower panel) and neutral Na clusters (upper panel) drawn versus system size in terms of $A^{-1/3}$ or $N_{\rm el}^{-1/3}$ respectively. The heavy line indicate the peak position of the dipole resonance (GDR for nuclei, Mie plasmon for clusters). As in Figure 2 for s.p. states, we take for nuclei a chain along the bottom of the valley of stability. The density of $1ph$ states is scaled with one over system size and one over excitation energy to fit the color scale better into one plot.

to energetically close $1ph$ states. Figure 3 shows the trend of the dipole resonance with system size together with the density of $1ph$ states. The evolution of $1ph$ energies follows in general a trend $A^{-1/3}$ or $N_{\rm el}^{-1/3}$ which is understandable because the level spacing shrinks inversely proportional to the radius which is $R \propto A^{1/3}$, or $R \propto N_{\rm el}^{1/3}$ respectively. Overlaid to this general trend is a slight increase of diffuseness with system size. Thus small systems (right side of the panels) still show quantum shell structure which produces a spectral gap above the first bunch of dipole $1ph$ states. The Mie plasmon resonance for very small clusters falls into such a gap, which allows to see rather sharp resonances in these systems. The GDR for small nuclei resides in a region of at least low density of states which delivers fragmentation into a couple of isolated peaks, particularly pronounced for ^{16}O.[52] The density of 1*ph* increases dramatically with system size such that spectral fragmentation produces one broad and smooth peak for heavy nuclei and clusters. For example, the fragmentation width becomes 2 MeV for ^{208}Pb and 0.3 eV for Na_{92}. The evolution of fragmentation width with cluster size is worth a comment. As said above, it increases when going from small clusters to larger ones because the density of 1*ph* states increases steadily. However density of states is not the only ingredient. Coupling matrix elements also play a role. And it turns out that coupling from the dipole resonance to 1*ph* states fades away for very large clusters such that we see at around $N_{\rm el} \approx 1000$ a maximum of width and decreasing width farther above.[53,54]

Note the major difference concerning the trend of the dipole resonance energy. The Mie plasmon energy stays almost constant throughout all cluster sizes while the nuclear GDR follows a trend $A^{-1/3} \propto R^{-1}$. This is an effect of the residual interaction. The structure of the dipole resonance as such is the same in both systems. It can be characterized as a collective oscillation of proton versus neutron cloud in nuclei and of the electron cloud against positive ionic background in clusters. However, the residual interaction is of short range in nuclei, but long-range Coulomb in clusters. The resonance is formed by a coherent superposition of 1*ph* states which thus set a trend R^{-1}. The nuclear residual interaction is of short-range, thus confined to the surface. Together with spectral density and trend of matrix element it amounts also to a trend $\propto R^{-1}$ which coincides with the trend to 1*ph* energies such that both move practically in parallel. The Coulomb interaction, however, has no spatial limitations and acquires the more grip the more charges are involved globally. This counterweights the trend of 1*ph* states and eventually yields the constant Mie plasmon energy.

4. Fission

4.1. *Fission in metal clusters*

Similar as in nuclei, fission is one important decay channel of metal clusters. Thus there exist several investigations on cluster fission see, e.g., Refs. 11 and 56 for experimental work and Refs. 10, 11, 57 and 58 for theory. The crucial results of this early period of fission studies have been gathered in the review.[12] Among the major issues of these studies was the global

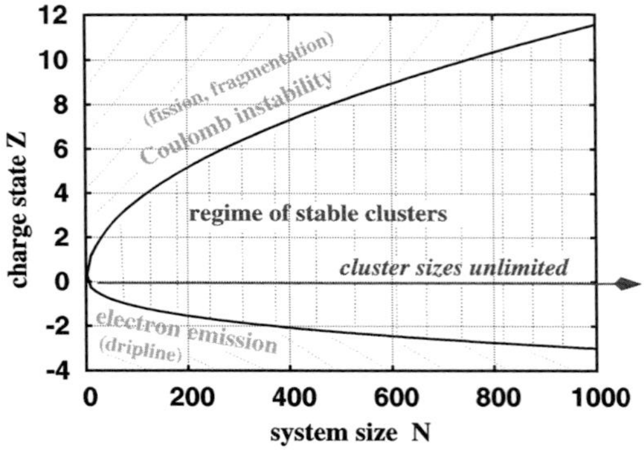

Fig. 4. Schematic plot of the region stable clusters. Preferred decay modes in unstable regions are indicated.

stability of the cluster. And this means here stability against destabilizing Coulomb pressure which acts against the cohesive forces. The situation is comparable to fission stability of (super-)heavy nuclei which emerges from the balance between Coulomb repulsion and binding from surface tension and shell effects.[59] The situation is also different to the extend that the nuclear charge is tied to the nuclear size because a certain fraction of protons to neutrons has to be maintained while clusters are near to neutral and can grow arbitrarily large. Cluster charge is an independent parameter which can be driven deliberately by ionization or electron attachment. This allows to explore systematically the limits of (Coulomb) stability. Figure 4 illustrates the stability landscape of Na clusters. Around net charge $Q = 0$ extends stability to arbitrary sizes. The stable region is limited toward negative and positive Q by what is called appearance size $N_{\rm app}$, i.e., the cluster size from which on a cluster with certain Q is found stable. Unstable anionic clusters ($Q < 0$) decay preferably by emission of an electron or monomer (neutral Na atom). Unstable cationic clusters ($Q > 0$) decay by fission, multi-fragmentation, or Na^+ emission depending on the dynamical regime. The division line is approximately $Q = \sqrt{0.13\,N_{\rm app}}$.[12] It is to be noted that this estimate is applicable to hot Na clusters at temperature around $T \approx 800$K. The actual appearance size depends sensitively on temperature. For example, the above estimate yields $N_{\rm app}(Q{=}2) \approx 30$. Lower temperature yield much lower $N_{\rm app}$ going down to about 8 for $T = 0$.[60,61] However, the shape of the boundary $N_{\rm app} \propto Q^2$, as illustrated in the figure, remains the same. It may be, of course, modified in detail by electronic and ionic shell effects, particularly for small clusters. This is also well known from nuclear physics where quantum shell effects shift the limits of stability high up to the world of super-heavy elements.[59]

There are marked differences to nuclear fission concerning the dynamical regimes and processes. Cluster fission proceeds through (slow) ionic motion which is a classical process. Spontaneous fission through tunneling below the Coulomb barrier does not exist for clusters. It is always a classical over-barrier dynamics. Moreover, asymmetric fission usually dominates and the most asymmetric case releasing just the small Na_3^+ fragment prevails. The latter case rather corresponds to α-decay nuclei which is, in fact, also a competing process in the decay chains of super-heavy elements. Another difference was mentioned already: temperature plays a great role in cluster dynamics while nuclear dynamics in the laboratory stays at comparatively low temperatures. Sizable temperatures become relevant only for nuclei in astro-physical environment.[62] Also unlike in the nuclear case, the charge state can be tuned deliberately. This allows to explore all regimes of Coulomb expansion. Fission happens just above stability threshold. Stepping to slightly larger charge states (one charge unit often suffices) leads to multi-fragmentation because the higher Coulomb pressure excites also higher order surface deformations.[12] Even higher charges then trigger straightforward Coulomb explosion producing a broad mix of fragments with high kinetic energies.[14]

As pointed out above, cluster fission proceeds at the slow time scale of ionic motion. This allows time resolved analysis by the well established pump and probe strategies.[63] This will be discussed more extensively in the next two sections.

4.2. *Pump and probe dynamics*

Pump and probe (P&P) dynamics has been developed to resolve in detail ionic dynamics in molecules and solids[63,64] and extensively used in these areas. More recently, the availability of attosecond pulses[65] has given way to track electron dynamics at sub-fs scale. First investigations focused still on simple systems, atoms[66] and very small molecules.[67] The field is quite active and more complex systems will possibly be considered in the near future. Metal clusters, however, are much more complex objects where it is already some effort to render P&P analysis at fs scale feasible. Nonetheless, it has been used successfully also in several cluster experiments see, e.g., Refs. 68–71. Many of these experiments deal already with Coulomb driven processes (expansion, explosion) to which cluster fission also belongs. The handle for analysis in these experiments on metal clusters is the unique dependence of the most prominent dipole mode, the Mie plasmon, on the cluster shape. In the next subsection, we will illustrate by simulating a model system how one could measure the time scales involved in the fission dynamics of a metal cluster by P&P techniques.

There are various tools for the theoretical description of fission of metal clusters. Early theoretical investigations stayed at the level of adiabatic approximation, mapping the fission-energy surface with simple jellium background, e.g., Ref. 57, or detailed ionic structure, e.g., Ref. 58, or coupling that to Born-Oppenheimer molecular dynamics,[10,11] and even adding quantum features to that.[72] All these approaches are not fully suited to simulate laser induced charging and fission because they do not include the details of the laser excitation and ignore the fact that the subsequent dynamics proceeds far above the Born-Oppenheimer surface. A time-resolved simulation of cluster fission requires to account simultaneously for both, electronic and ionic degrees of freedom (the latter being the target for fission studies). The same requirement exists for any laser induced cluster dynamics and a most robust tool to perform such investigations has been developed for that: the Time Dependent Local Density Approximation (TDLDA) for electrons, coupled to Molecular Dynamics (MD) for ions.[73,74] The setup has been shown to provide a versatile tool for many studies involving a detailed analysis of various decay channels as, e.g., electron emission following in particular laser irradiation scenarios.[75,76] The TDLDA part is handled as described in sections 2 and 3. Ionization (the doorway to fission) is computed via absorbing boundary conditions.[73] Ionic motion is propagated simultaneously together with electron dynamics solving the classical equations of motion for the ions. The laser fields are described in the dipole approximation as time-dependent external potentials

$$V_{\rm las} = E_0 \mathbf{e}_{\rm pol} \cdot \hat{\mathbf{d}} f_{\rm las}(t) \cos(\omega t) \tag{2}$$

where $\mathbf{e}_{\rm pol}$ is the laser polarization, $\hat{\mathbf{d}}$ the dipole operator, E_0 the field amplitude, and $f_{\rm las}(t)$ the envelope of the laser pulse. Pulses are short (Full Width at Half Maximum, FWHM = 48 fs for pump and 24 fs for probe pulses) which guarantees, first, well defined time slots, and second a most gentle charging without too much unwanted heating.[77] In order to analyze fission dynamics on the basis of electronic signals (see below) we record both electronic and ionic observables as a function of time. The two electronic key observables will thus be the dipole response explored by scanning the frequency of probe pulses and net ionization after probe.

4.3. *An illustrative example*

We choose as an example the cluster Na_{14}, which exhibits a prolate shape (with dimensionless quadrupole deformation $\beta_2 = 0.6$), and excite it with a pump pulse to a charge state of about 3 electrons. The pump pulse parameters are $\omega_{\text{pump}} = 2.31$ eV, FWHM = 96 fs, $I = 4.8 \times 10^{10}$ W/cm^2. The resulting charged cluster Na_{14}^{+++} thus becomes Coulomb unstable and undergoes fission. Key observables of the resulting dynamics are shown in Figure 5. The lower panel shows net ionization as function of time together with the envelope of the pump pulse. Ionization saturates quickly after the pump pulse is over. The upper panel sketches the time evolution of ionic positions. The net charge immediately leads to an expansion of the system which, after some time, provokes a breaking of the cluster into two pieces. Breaking seems to occur around 800 fs which is a rather moderate time at ionic pace. For comparison, ionic vibration periods in Na clusters are of order 200 fs. The breaking leads to two fragments of respectively 6 and 8 atoms. The total net charge (actually 3.2) is split in a balanced way between the two fragments about 1.7 for the Na_8 like fragment and 1.5 for the Na_6 like fragment. Mind that TDLDA-MD used here is a mean-field theory at the electronic side which explains that ionization is not quantized. The actual ionic evolution thus takes place in an average electronic potential according to average (non-integer) ionization. The identified fragments represent an average situation, close to the most probable fission channel.

The question is now how one can track fission dynamics. This will be achieved by probing the fissioning system with a second laser pulse, the probe pulse, with a delay time Δt after the pump pulse. The probe pulses are scanned with various Δt. This is illustrated schematically in Figure 6. As discussed in section 3 the dipole response usually exhibits a dominant peak whose energy is directly linked to the cluster mass, charge, and shape. The optical response of the system should thus vary in time depending on the fact the geometry of the system. A time resolved measurement of the full dipole response may not be simple experimentally. On the other hand the response of the system to a laser is strongly influenced by the closeness of the laser frequency to eigenfrequencies of the systems,[78] especially to the closeness to the dominating peaks, in metal clusters the Mie plasmon. Thus, if one uses probe pulses close to strong eigenmodes of the

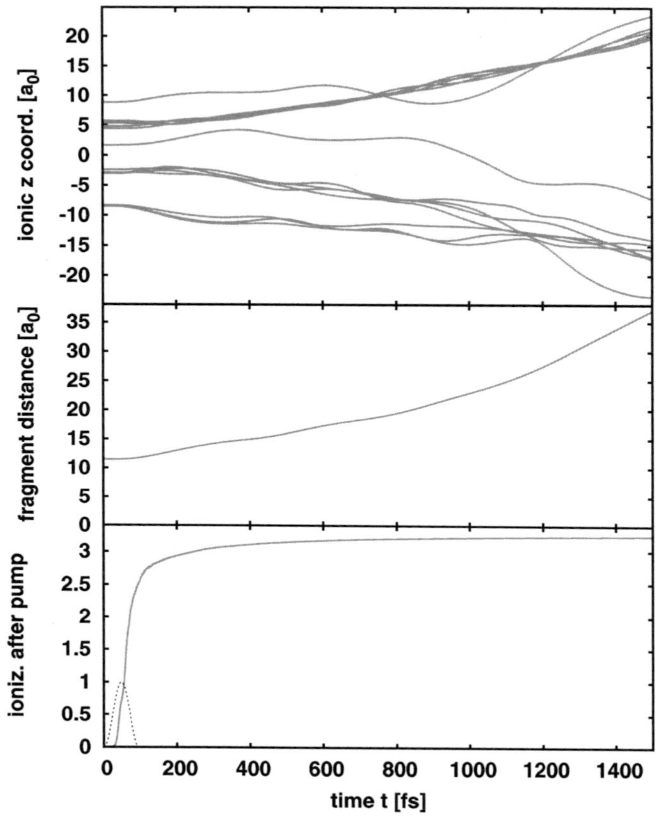

Fig. 5. Time evolution of key observables during Coulomb induced fission of Na_{14}. Lower panel: Ionization from pump pulse only, tuned to extract $\simeq 3.2$ electrons, just above reaction threshold. The faint dotted line indicates the envelope of the pump pulse. The pump pulse parameters are $\omega_{\rm pump} = 2.31$ eV, FWHM$=96$ fs, $I = 4.8 \times 10^{10}$ W/cm^2. Middle panel: distance between ionic centers of the upper fragment (6 ions) and the lower fragment (8 ions), computed as $\sqrt{\langle z^2 \rangle / N_{\rm el}}$. Upper panel: ionic z coordinates (along elongated direction of the prolate Na_{14} cluster).

system one will observe ionization varying with the spectral distance of the probe pulse to the mode. As the dominant dipole response is usually concentrated in a well defined frequency domain, and because it is directly linked to the ionic structure, probe ionization thus almost directly reflects the closeness of the laser frequency to the strong modes. The extra ionization provided by the probe pulse thus provides a strong clue to the ionic structure, and, analyzed as a function of time, to dynamics. Such a signal is

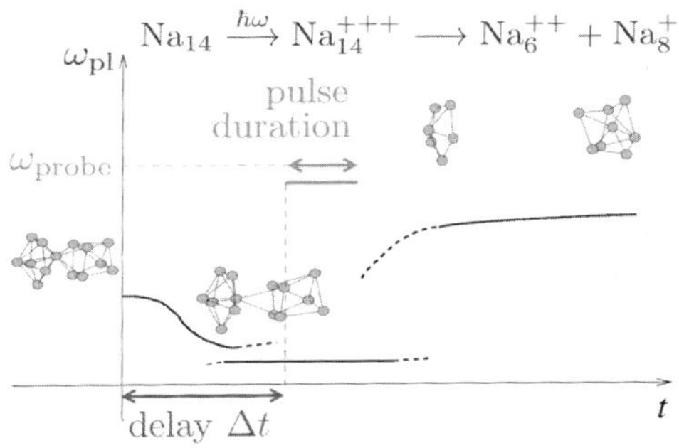

Fig. 6. Schematic drawing of the evolution of the plasmon response ω_{pl} of an irradiated Na_{14}^{+++} metal cluster along its fission path. Typical ionic structures are indicated for completeness. The charge state of the final fragments is indicated only schematically. In practice it comes out non-integer. The pump and probe analysis discussed here is to be considered in terms of both delay and probe frequency ω_{probe} as sketched on the figure. Adapted from Ref. 55.

easy to reach experimentally and this is what we shall indeed record. The expected scenario can thus be summarized as follows: The energy of the Mie plasmon decrease with increasing elongation of the cluster. The fuzzy shapes near scission lead to spectral fragmentation with a strong low lying mode, the latter terminates as soon as the electron cloud looses does not connect any more both fragments, and the Mie plasmons of each fragment shows up. Due to the occurrence of spectral fragmentation around scission one has to scan a few different frequencies in addition to scanning delay times.

There is one crucial detail contained in the initial ionic configuration shown in Figure 6. The strongly deformed Na_{14} already indicates a clustering into Na_6 and Na_8 sub-units. This is the starting point for the fission into two comparable units. Else-wise the strongly asymmetric fission with one Na_3+ would probably dominate.[58] This example shows that the detailed outcome of cluster fission varies strongly from system to system depending on shell effects and pre-configured fragments.

The practical outcome of such a program is illustrated in Figure 7. To establish connection with the dynamics after pump shown in Figure 5, we replot in the lower panel here the time evolution of fragment separation. The upper panel complements that by showing the time-evolution of dipole spectra along the stages of the fission dynamics. But in the middle panel, the time scale represents the delay time Δt between pump and probe pulses, and not the physical time. Let us concentrate first at the upper panel, the

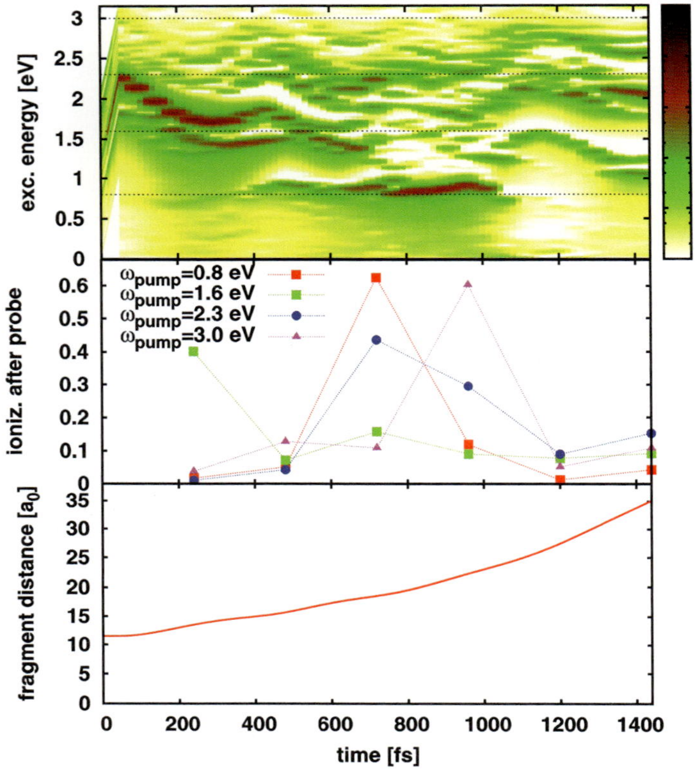

Fig. 7. Time evolution of probe observables during Coulomb induced fission of Na_{14}. Lower panel: distance between ionic centers of the upper fragment (6 ions) and the lower fragment (8 ions), the same as middle panel of Figure 5. Middle panel: additional ionization after probe pulse after delay given by the time axis. The probe pulse parameters are FWHM = 48 fs, $I = 4.8 \times 10^{10}$ W/cm^2 and various frequencies as indicated. The signal for $\omega_{\mathrm{probe}} = 0.8$ eV is multiplied by a factor of five for better visibility. Upper panel: dipole power spectrum for the ionic configurations at given times along dynamics after pump pulse. The faint horizontal lines indicate the frequencies of the four different probe lasers (see middle panel).

time-evolution of dipole power spectrum. During pump pulse up to 50 fs, we see the blue shift through ionization. Then, during the following 300 fs, a redshift occurs due to the cluster's stretching. The further expansion does not yield further red shift but a separate low lying dipole state (oscillations through the neck of the electron cloud between the forming fragments). This low-energy mode is suddenly terminated after scission and the plasmon of the fragments shows up at their typical energies. The middle panel demonstrates the effect of the probe pulse in terms of extra ionization through probe pulse piling up onto the original ionization delivered by the pump pulse. Peaks and minima can be well understood from the spectra in the upper panel following the expected scenario. This last signal clearly shows that the scission occurs around 800 fs, as was qualitatively visible from the original time evolution of ionic positions. But mind that the interest of the complicated pump probe setup is that it allows to see in a simply measurable quantity (total ionization) this scission time, otherwise impossible to grasp experimentally.

5. Conclusions

In this paper, we have compared side by side nuclei and metal clusters. They share several properties due to their common nature being a finite drop of a Fermi liquid. They possess a more or less well defined saturation point and particles with low momenta exhibit a long mean free path. Natural units for length and energy scales are set by the Wigner-Seitz radius r_s (related to saturation density, analogue of nuclear r_0 parameter of nuclear radius systematics) and Fermi energy. As finite systems they share several interesting features such as pronounced shell effects and resonance excitations. The properties of these different materials look very similar when expressed in the natural units of the material. There are, of course, also some differences between these systems, mostly in terms of availability and experimental access. For example, while nuclei are limited in size, metal clusters can be grown up to arbitrarily large size. This allows to study the path to bulk matter and to see how shell effects evolve for large systems giving rise to the phenomenon of super-shells. The strong coupling of coherent laser light to metal clusters allows dedicated driving of dynamical scenarios and so opens up an enormously rich field of research. We thus see that differences are highly welcome as they deliver complementing information which will favor a common understanding of finite fermion systems.

Beyond structural properties, which can be remarkably unified by a proper scaling, we have focused our analysis on spectral properties and on fission dynamics. Concerning spectral properties, both systems show pronounced quantum shell effects giving to rise to a well defined sequence of magic numbers and deformation. The trends of single-particle spectra with system size are practically the same in both species. There is also a great similarity in the appearance of a pronounced resonance excitations, the Mie surface plasmon in metal clusters and the giant dipole resonance in nuclei. Both modes have the same underlying mechanism as a collective motion of electron cloud versus ionic background, or proton versus neutrons respectively. A great difference is seen here in the trends, nearly constant peak position for clusters against position decreasing with system size in nuclei. This is due to a much different range of the residual interaction, long-range Coulomb in electronic systems versus short range interaction in nuclei.

Fission is also a phenomenon common to both systems. There are great similarities concerning the mechanism, namely Coulomb pressure acting against binding forces (surface tension, quantum shell effects), and simple modeling (liquid drop model) can be applied to both cases equally well. At variance with nuclei which can cannot avoid to undergo fission when sufficiently heavy, metal clusters need to be artificially charged (by a laser pulse or a collision with a charged particle) to trigger the Coulomb instability driving fission. The widely tunable charge state in clusters does also allow to explore the various stages of Coulomb driven dynamics from fission at charges just above barrier over multi-fragmentation at somewhat higher ionization up to violent Coulomb explosion. Furthermore, the dynamics of cluster fission can be tracked in time resolved manner. We have illustrated this possibility by considering a typical pump and probe scenario in which the pump pulse delivers the initial net charge to the system triggering its fission while the probe pulse allows to analyze the state of the system at various instants by recording the extra ionization its provokes. This extra ionization depends whether we are before or after scission. At variance with standard pump and probe scenarios, the above setup requires to scan both, delay time (between pump and probe pulses) *and* probe frequency in order to deal with the fragmentation of excitation spectra around the scission point. Three factors make an analysis then possible: the amount of ionization depends on the closeness to a resonance, the dipole spectrum is governed by strong resonance modes, the dipole spectrum depends sensitively on cluster shape. In a next step, one could go further and learn from nuclear physics where more information on fission dynamics can drawn from the distribution of emitted neutrons. in the same spirit, one should analyze the angular distribution electrons emitted electrons after the probe pulse. Work along that line is in progress.

Acknowledgments

This work was supported by the Institut Universitaire de France. It was granted access to the HPC resources of CalMiP (Calcul en Midi-Pyrénées) under the allocation P1238.

References

1. P. M. Dinh, P.-G. Reinhard and E. Suraud, *Introduction to Cluster Science* (Wiley, Berlin, 2013).
2. Y. Yamaguchi, Y. Osamure, J. D. Goddard and H. E. S. III, *A New Dimension to Quantum Chemistry* (Oxford University Press, Oxford, 1994).
3. A. Becke and K. Edgecombe, *J. Chem. Phys.* **92**, 5397 (1990).
4. P.-G. Reinhard, J. A. Maruhn, A. S. Umar and V. E. Oberacker, *Phys. Rev. C* **83**, 034312 (2011).
5. R. M. Dreizler and E. K. U. Gross, *Density Functional Theory: An Approach to the Quantum Many-Body Problem* (Springer-Verlag, Berlin, 1990).
6. I. Z. Petkov and M. V. Stoitsov, *Nuclear Density Functional Theory* (Clarendon Press, Oxford, 1991).
7. R. Schmidt, H. O. Lutz and R. Dreizler (eds.), *Nuclear Physics Concepts in the Study of Atomic Cluster Physics*, Lecture Notes in Physics, Vol. 404 (Springer, Heidelberg, 1992).
8. C. Bréchignac and P. Cahuzac, *Comm. At. Mol. Phys.* **31** (1995).
9. P.-G. Reinhard and E. Suraud, Resonance dynamics in metal clusters and nuclei, in *Metal Clusters*, ed. W. Ekardt (Wiley, New York, 1999), New York, p. 211.
10. R. N. Barnett, U. Landman and G. Rajagopal, *Phys. Rev. Lett.* **67**, 3058 (1991).
11. C. Bréchignac, P. Cahuzac, F. Carlier, M. de Frutos, R. N. Barnett and U. Landman, *Phys. Rev. Lett.* **72**, 1636 (1994).
12. U. Näher, S. Björnholm, S. Frauendorf, F. Garcias and C. Guet, *Phys. Rep.* **285**, 245 (1997).
13. T. Ditmire, T. Donnelly, A. M. Rubenchik, R. W. Falcone and M. D. Perry, *Phys. Rev. A* **53**, 3379 (1996).
14. T. Ditmire, J. W. G. Tisch, E. Springate, M. B. Mason, N. Hay, J. P. Marangos and M. H. R. Hutchinson, *Phys. Rev. Lett.* **78**, 2732 (1997).
15. E. Epelbaum, H.-W. Hammer and U.-G. Meiner, *Rev. Mod. Phys.* **81**, 1773 (2009).
16. H.-W. Hammer, A. Nogga and A. Schwenk, *Rev. Mod. Phys.* **85**, 197 (2013).
17. G. Hagen, A. Ekström, C. Forssén, G. R. Jansen, W. Nazarewicz, T. Papenbrock, K. A. Wendt, S. Bacca, N. Barnea, B. Carlsson, C. Drischler, K. Hebeler, M. Hjorth-Jensen, M. Miorelli, G. Orlandini, A. Schwenk and J. Simonis, *Nat. Phys.* **12**, 186 (2016).

18. G. F. Bertsch and S. Das Gupta, *Phys. Rep.* **160**, 190 (1988).
19. A. Domps, P.-G. Reinhard and E. Suraud, *Phys. Rev. Lett.* **81**, 5524 (1998).
20. T. Fennel, G. F. Bertsch and K.-H. Meiwes-Broer, *Eur. Phys. J. D* **29**, 367 (2004).
21. E. Suraud and P.-G. Reinhard, *New J. Phys.* **16**, 063066 (2014).
22. W. A. Harrison, *Pseudopotentials in the Theory of Metals* (W A. Benjamin, New York, 1976).
23. N. W. Ashcroft and N. D. Mermin, *Solid State Physics* (Saunders College, Philadelphia, 1976).
24. I. Bloch, J. Dalibard and W. Zwerger, *Rev. Mod. Phys.* **80**, 885 (2008).
25. F. Stiekenmeier and A. F. Vilesov, *J. Chem. Phys.* **115**, 10119 (2001).
26. P. Hohenberg and W. Kohn, *Phys. Rev.* **136**, 864 (1964).
27. W. Kohn and L. J. Sham, *Phys. Rev.* **140**, 1133 (1965).
28. T. H. R. Skyrme, *Nucl. Phys.* **9**, 615 (1959).
29. O. Gunnarsson and B. I. Lundqvist, *Phys. Rev. B* **13**, 4274 (1976).
30. D. Vautherin and D. M. Brink, *Phys. Lett. B* **32**, 149 (1970).
31. D. Vautherin and D. M. Brink, *Phys. Rev. C* **5**, 626 (1972).
32. M. Bender, P.-H. Heenen and P.-G. Reinhard, *Rev. Mod. Phys.* **75**, 121 (2003).
33. J. P. Perdew and Y. Wang, *Phys. Rev. B* **45**, 13244 (1992).
34. C. Legrand, E. Suraud and P.-G. Reinhard, *J. Phys. B* **35**, 1115 (2002).
35. S. Kümmel, M. Brack and P.-G. Reinhard, *Eur. Phys. J. D* **9**, 149 (1999).
36. P. Klüpfel, P. G. Reinhard, T. J. Bürvenich and J. A. Maruhn, *Phys. Rev. C* **79**, 034310 (2009).
37. H.-P. Duerr, *Phys. Rev.* **103**, 469 (1956).
38. M. Brack, *Rev. Mod. Phys.* **65**, 677 (1993).
39. S. Bjornholm and J. Borggreen, *Phil. Mag.* **79**, 1321 (1999).
40. R. Balian and C. Bloch, *Ann. Phys. (N.Y.)* **85**, 514 (1974).
41. R. Englman, *The Jahn-Teller Effect in Molecules and Crystals* (Wiley, London, 1972).
42. W. A. de Heer, *Rev. Mod. Phys.* **65**, 611 (1993).
43. H. Haberland (ed.), *Clusters of Atoms and Molecules 1- Theory, Experiment, and Clusters of Atoms* (Springer Series in Chemical Physics, Berlin, 1994).
44. P.-G. Reinhard and E. W. Otten, *Nucl. Phys. A* **420**, 173 (1984).
45. M. Koskinen, P. O. Lipas and M. Manninen, *Nucl. Phys. A* **591**, 421 (1995).
46. T. P. Martin, *Phys. Rep.* **273**, 199 (1993).
47. H. Haberland and M. Schmidt, *Eur. Phys. J. D* **6**, 109 (1999).
48. H. Haberland, Optical and thermal properties of sodium clusters, in *Metal Clusters*, ed. W. Ekardt (Wiley, New York, 1999), New York.
49. J. A. Maruhn, P.-G. Reinhard and E. Suraud, *Simple Models of Many-Fermion Systems* (Springer, Heidelberg, 2010).
50. A. van der Woude, The electric giant resonances, in *Electric and Magnetic Giant Resonances in Nuclei*, ed. J. Speth (World Scientific, 1991) pp. 99–232.
51. U. Kreibig and M. Vollmer, *Optical Properties of Metal Clusters* (Springer Series in Materials Science, 1993).

52. S. S. Dietrich and B. L. Bermann, *At. Data Nucl. Data Table* **38**, 199 (1988).
53. C. Yannouleas, J. M. Pacheco and R. A. Broglia, *Phys. Rev. B* **41**, 41 (1990).
54. J. Babst and P.-G. Reinhard, *Z. Phys. D* **42**, 209 (1997).
55. P. M. Dinh, P.-G. Reinhard and E. Suraud, *J. Phys. B* **38**, 1637 (2005).
56. F. Chandezon, C. Guet, B. A. Huber, D. Jalabert, M. Maurel, E. Monnand, C. Ristori and J. C. Rocco, **74**, 3784 (1995).
57. F. Garcias, J. A. Alonso, M. Barranco, J. M. Lopez, A. Mananes and J. Nemeth, *Z. Phys. D* **31**, 275 (1994).
58. B. Montag, T. Hirschmann, J. Meyer, P.-G. Reinhard and M. Brack, **52**, 4775 (1995).
59. S. Hofmann and G. Münzenberg, *Rev. Mod. Phys.* **72**, 733 (2000).
60. Y. Li, E. Blaisten-Barojas and D. A. Papaconstantopoulos, *Phys. Rev. B* **57**, 15519 (1998).
61. P. Blaise, S. A. Blundell, C. Guet and R. R. Zope, *Phys. Rev. Lett.* **87**, 063401 (2001).
62. C. Reiss, M. Bender and P.-G. Reinhard, *Eur. Phys. J A* **6**, 157 (1999).
63. A. H. Zewail, *Femtochemistry, Vol. I & II* (World Scientific, Singapore, 1994).
64. B. M. Garraway and K.-A. Suominen, *Rep. Prog. Phys.* **58**, 365 (1995).
65. F. Krausz and M. Ivanov, *Rev. Mod. Phys.* **81**, 163 (2009).
66. P. Agostini and L. F. DiMauro, *Rep. Prog. Phys.* **67**, 813 (2004).
67. G. Sansone *et al.*, *Nature* **465**, 763 (2010).
68. R. Heinicke and J. Grotemeyer, *Appl. Phys. B* **71**, 419 (2000).
69. G. Seifert, M. Kaempfe, K.-J. Berg and H. Graener, *Appl. Phys. B* **71**, 795 (2000).
70. T. Döppner, S. Teuber, T. Diederich, T. Fennel, P. Radcliffe, J. Tiggesbäumker and K.-H. Meiwes-Broer, **24**, 157 (2003).
71. T. Döppner, T. Fennel, T. Diederich, J. Tiggesbäumker and K. Meiwes-Broer, **94**, 013401 (2005).
72. M. Hartmann, J. Pittner, V. Bonacic-Koutecky, A. Heidenreich and J. Jortner, *J. Chem. Phys.* **108**, 3096 (1998).
73. F. Calvayrac, P.-G. Reinhard, E. Suraud and C. A. Ullrich, *Phys. Rep.* **337**, 493 (2000).
74. P.-G. Reinhard and E. Suraud, *Introduction to Cluster Dynamics* (Wiley, New York, 2004).
75. M. A. L. Marques, C. A. Ullrich and F. Nogueira, *Cluster Dynamics in Strong Laser Fields*, Lecture Notes in Physics, Vol. 706 (Springer, Berlin, 2006).
76. P. Wopperer, P. M. Dinh, P.-G. Reinhard and E. Suraud, *Phys. Rep.* **562**, 1 (2015).
77. P.-G. Reinhard and E. Suraud, *Int. J. Mol. Sci.* **1**, 92 (2000).
78. H. G. Muller and M. V. Fedorov (eds.), *Super-Intense Laser-Atom Physics IV* (Kluwer, Dordrecht, 1996).

Part 2

Particle Correlations in Dilute and Dense Matter

CHAPTER 2

CORRELATIONS AND CLUSTERING IN DILUTE MATTER

G. Röpke
Institut für Physik, Universität Rostock,
D-18051 Rostock, Germany,
gerd.roepke@uni-rostock.de

Nuclear systems are treated within a quantum statistical approach. Correlations and cluster formation are relevant for the properties of warm dense matter, but the description is challenging and different approximations are discussed. The equation of state, the composition, Bose condensation of bound fermions, and the disappearance of bound states at increasing density because of Pauli blocking are of relevance for different applications in astrophysics, heavy ion collisions, and nuclear structure.

1. Bound states in fermion systems: The chemical picture

1.1. *Structure of matter*

Our investigation refers to a fundamental property of matter: We can define elementary constituents (indivisible particles, Greek "$\alpha\tau o\mu o\varsigma$) that interact. Then, the macroscopic properties are calculated within a quantum statistical approach. The interaction may lead to bound states, which can fundamentally change the properties of matter, and new properties may emerge. A well known example is the formation of atoms from charged particles, electrons and atomic nuclei. While unbound electrons and nuclei form a plasma which conducts electricity, the formation of charge-neutral atoms results in non-conducting dielectric matter. Another example is Bose-Einstein condensation which may occur when bosonic bound states are formed in a fermionic system. A well-known bosonic bound state is Helium-4, consisting of two electrons, two protons, and two neutrons. It shows superfluidity at low temperatures. Nuclear systems where Bose-Einstein condensation of α particles may occur are discussed below.

Here, we consider nuclear systems consisting of neutrons and protons (and possibly electrons for charge neutrality) as constituting "elementary" particles. As a consequence of the nucleon-nucleon (N-N) interaction, bound states, i.e., nuclei, are formed. Examples are the light nuclei ^{2}H (deuteron, d), ^{3}H (triton, t), ^{3}He (helion, h), and ^{4}He (α particle). The fundamental interaction is the strong interaction, as described by QCD. Because the deduction of the N-N interaction from QCD is not solved yet, effective interactions are introduced which reproduce measured properties such as binding energies and scattering phase shifts. Examples are the Yukawa, Reid, Bonn, Paris, Argonne, etc., interactions, but also separable (Yamaguchi, PEST, BEST, etc.) parametrizations of the N-N interaction are available. The optimal form of the effective N-N interaction is widely discussed in nuclear physics but will not addressed in the present work.

The appearance of bound states is a general feature of the structure of matter. For example, quarks form hadrons, hadrons form nuclei, nuclei together with electrons form atoms and ions, atoms form molecules. The new bound states may be considered as the elements of a statistical description of matter, which is possible as long as the energies considered are small enough so that the internal structure of the bound states is not excited. A description where the "elementary" particles and the composed bound states are considered on the same footing, is denoted as chemical picture. In a simple form it is applicable in the low-density limit where we have an ideal mixture of the constituents and the composed particles. They are considered as non interacting, but accidentally reacting so that chemical equilibrium (mass action law) between the different components of the system is established. In the special case of nuclear systems considered here, bound states (nuclei) are expected to appear at low densities and low temperatures according to the chemical equilibrium. Such states may occur for expanding matter produced in heavy ion collisions (HIC), but also in astrophysical objects and in excited nuclei.

A schematic overview of the structure of matter is given in Table 1. "Elementary" particles are introduced on different levels, which prove to be composed particles (bound states) if the scale of energy is changed. An interesting aspect is the behavior at increasing density. As a general feature, the bound states are dissolved, and a phase transition to a condensed phase may occur. Different effects are responsible for the dissolution of bound states: In Coulomb systems (electrons, ions, and atoms in a plasma; electrons, holes, and excitons in excited semiconductors) screening

Table 1. Bound state formation and dissolution.

energy scale	*fermions*	*interaction*	*bound states*	*density effects*	*condensed phase*
1...10 meV	electrons, holes	Coulomb	excitons	screening	electron-hole liquid
1...10 eV	electrons, nuclei	Coulomb	ions, atoms	screening	liquid metal
1...10 MeV	protons, neutrous	$N-N$ int.	nuclei	Pauli blocking	nuclear matter
0.1...1 GeV	quarks	QCD	hadrons	deconfinement	quark-gluon plasma

of the long-range Coulomb interaction leads to the Mott transition, where an atomic insulator transforms into a metallic conductor; see Refs. 1, 2. Another effect which leads to the destruction of bound states is the Pauli principle. If the constituents of the bound states are fermions, the phase space which is available to form bound states is reduced at increasing density (Pauli blocking) so that the formation of bound states is suppressed; see Figure 1. This phenomenon will be discussed in detail below. It is responsible for the transition from a gas of nucleons and nuclei to nuclear matter (nuclear liquid), as observed at saturation density $n_{\rm sat} \approx 0.16$ fm^{-3}. Also at the level of the quark substructure, with increasing density a deconfinement transition from the hadronic phase to a quark-gluon plasma is expected where the hadrons are dissolved.

Starting from a microscopic description of matter, e.g., a Hamiltonian with an effective N-N interaction, a systematic treatment is given by the quantum statistical approach where correlation functions are evaluated for the equilibrium distribution. Infinite matter is characterized by the temperature T and the chemical potentials μ_c of the different components c. These thermodynamic variables, which define the often used grand canonical ensemble, are related to the average of the internal energy U and the average particle numbers $N_c = \Omega n_c$, where Ω denotes a normalization volume (volume of the system) and n_c the density of species c. Relations between these variables, such as $N_c/\Omega = n_c(T, \mu_{c'})$, are denoted as equations of state and will be considered in detail in this work.

Different equations of state are possible. In particular, thermodynamic potentials, such as the free energy F as function of T and n_c, have the property that all other thermodynamic relations can be obtained by taking the derivation with respect to the corresponding thermodynamic quantities.

In the case considered here, i.e., $\mu_c(T, n_{c'})$, the free energy is found by integration, $F(T,N,\Omega) = \Omega \int_0^n \mu(T,n')dn'$ (one component). In addition to the thermodynamic properties, it also describes phase transitions when the stability condition $\partial\mu/\partial n|_T \geq 0$ is violated, see also Ref. 3 for the multi-component case. A schematic overview of the phase diagram of symmetric nuclear matter is given in Figure 1.

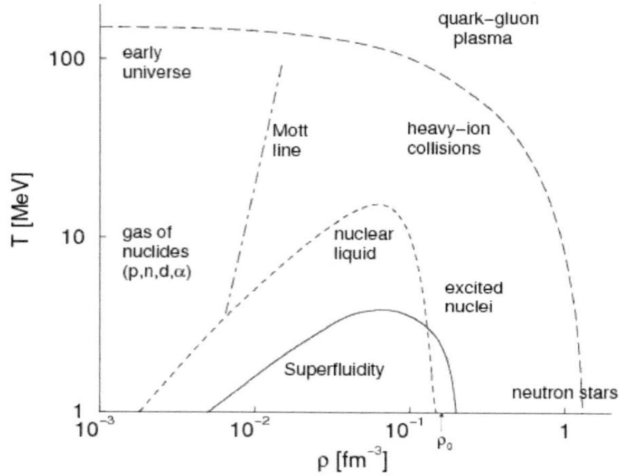

Fig. 1. Phase diagram of symmetric nuclear matter (schematically).[4] Baryon density $n_B = \rho$ with saturation density $n_{\text{sat}} = \rho_0$. The Mott line indicates the region where the formation of bound states is suppressed because of Pauli blocking.

1.2. *Quantum statistical calculation of cluster abundances in asymmetric hot dense matter*

Let us start with a brief historical review.[5] The thermodynamic properties of hot nuclear matter are of interest in connection with the theory of heavy-ion collisions as well as with astrophysical and cosmological problems. Of course, the behavior of dense matter under the conditions of star evolution, the expanding universe or deep inelastic relativistic ion collisions should be described by kinetic theory which describes the detailed nuclear processes during the time evolution of hot dense matter in non-equilibrium (see, e.g., Ref. 6). However, it is reasonable to compare the result of these non-equilibrium processes with results for the thermal equilibrium in the sense of an estimate for the most probable state the system is likely to attain.[7] Although we are fully aware of the problem of using equilibrium results, we suggest that a correctly formulated theory of thermal equilibrium leads to facts as the distribution of cluster abundances or possible phase instabilities which are also of interest for the non-equilibrium behavior of hot dense matter. For instance, a thermal model[8,9] and the concept of a freeze-out baryon density n_B and a freeze-out temperature T[10] were successfully employed in the theory of heavy ion collisions to determine the production of composite fragments.

In the present work, nuclear matter in thermodynamic equilibrium is investigated, confined in the volume Ω at temperature T, and consisting of N_n neutrons (total neutron density $n_n^{\rm tot} = N_n/\Omega$) and N_p protons (total proton density $n_p^{\rm tot} = N_n/\Omega$). In the thermodynamic limit, the state is given by the parameter set $\{T, n_n^{\rm tot}, n_p^{\rm tot}\}$, the dependence on Ω is trivial. The subsaturation region will be considered where the baryon density $n_B = n_n^{\rm tot} + n_p^{\rm tot} \leq n_{\rm sat}$ (with the saturation density $n_{\rm sat} \approx 0.155$ fm^{-3}), the temperature $T \leq 20$ MeV, and the proton fraction $Y_p = n_p^{\rm tot}/n_B$ between 0 and 1. This region of warm dense matter is of interest not only for nuclear structure calculations and heavy ion collisions explored in laboratory experiments,[11,12] but also in astrophysical applications, see Ref. 13. For instance, core-collapse supernovae at post-bounce stage evolve in this region of phase space,[14,15] see Figure 2. Different processes such as neutrino emission and absorption, which strongly depend on the composition of warm dense matter, influence the evolution of core-collapse supernovae.

Let us denote by an ideal nuclide gas an approximation where nuclear matter is considered as an ideal mixture of free particles and clusters which move relatively freely, except for occasional collisions and nuclear reactions. According to the law of mass action, the equilibrium abundances of clusters are determined by the temperature, the chemical potentials of the nucleons, and the internal energy of the clusters.[8–10] No phase transition is obtained within this simple approximation. At given temperature the part of nucleons which is bound in clusters increases with increasing density.

Recently, this simple chemical picture, also called nuclear statistical equilibrium (NSE),[16] has been applied for the nuclear matter EoS in the low-density limit. Fragmentation as observed in HIC has been treated within a micro-canonical approach which takes interaction into account via the restriction of available space because of non-overlapping clusters; for references see.[17]

A simple chemical equilibrium of free nuclei is not applicable up to saturation density because medium modifications by self-energy shifts and

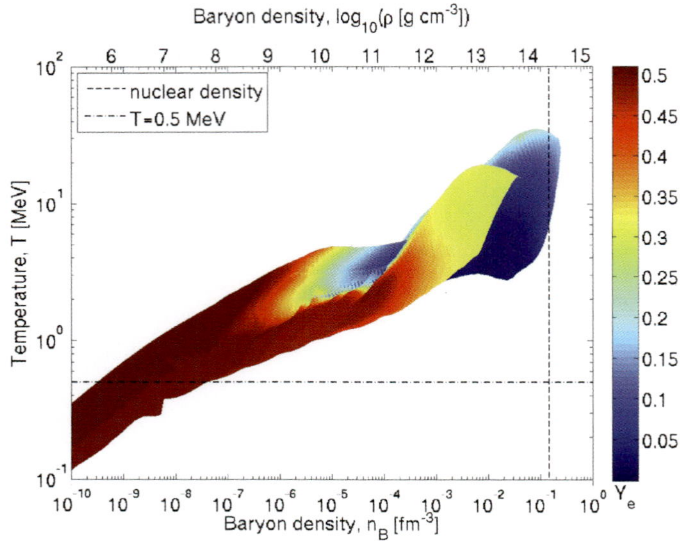

Fig. 2. Exploding supernova. States in the phase diagram which are met during the SN explosion. for core collapse SN the region on the phase diagram is slightly reduced. From Ref. 14.

Pauli blocking become relevant. Alternatively, standard versions[18,19] of the nuclear matter equation of state (EoS) considering single nucleons in mean-field (Skyrme or relativistic mean field (RMF)) approximation, as well as a representative heavy nucleus and α-particles, are available for astrophysical simulations such as supernova collapses. They have been improved elaborating concepts such as the heuristic excluded-volume approach or in-medium nuclear cluster energies within the extended Thomas–Fermi approach, see Refs. 16, 20–40. These concepts may be applied to heavier clusters but are not satisfactory to describe light clusters that require a more fundamental quantum statistical approach. A generalization of the RMF approach (gRMF) which includes the light nuclei ($A \leq 4$) as additional degrees of freedom, where the quasiparticle properties are taken from the quantum statistical approach described below, has been proposed in Refs. 23 and 41.

A rigorous quantum statistical (QS) approach to the thermodynamic properties of hot nuclear matter is formulated within the framework of thermodynamic Green functions.[42–44] The total number of neutrons N_n and protons N_p are introduced as conserved quantities, weak interaction processes leading to β equilibrium are not considered. Starting from the grand-canonical ensemble defined by the temperature T and the chemical potentials μ_n, μ_p of neutrons and protons, respectively, the chemical potentials μ_n, μ_p are fixed by the relations

$$\frac{1}{\Omega} N_n = n_n^{\text{tot}}(T, \mu_n, \mu_p), \qquad \frac{1}{\Omega} N_p = n_p^{\text{tot}}(T, \mu_n, \mu_p) \tag{1}$$

which are equations of state (EoS) that relate the set of thermodynamic quantities $\{T, \mu_n, \mu_p\}$ to the set $\{T, n_n^{\text{tot}}, n_p^{\text{tot}}\}$.

The QS approach considers correlation functions and its Fourier transform, the spectral function $S_\tau(1, \omega; T, \mu_n, \mu_p)$. The single-nucleon quantum state $|1\rangle$ can be chosen as $1 = \{\mathbf{p}_1, \sigma_1, \tau_1\}$ which denotes wave number, spin, and isospin, respectively. A rigorous expression for the nuclear matter EoS is found provided that the spectral function is known,

$$n_\tau^{\text{tot}}(T, \mu_n, \mu_p) = \frac{1}{\Omega} \sum_{p_1, \sigma_1} \int \frac{d\omega}{2\pi} \frac{1}{e^{(\omega - \mu_\tau)/T} + 1} S_\tau(1, \omega) \tag{2}$$

(Ω is the system volume, $\tau = \{n, p\}$; we take $k_B = 1$). The spectral function $S_\tau(1, \omega; T, \mu_n, \mu_p)$ is related to the self-energy $\Sigma(1, z)$ for which a systematic approach is possible using diagram techniques, see[45,46]:

$$S_\tau(1, \omega) = \frac{2\text{Im}\Sigma(1, \omega - i0)}{(\omega - E(1) - \text{Re}\Sigma(1, \omega))^2 + (\text{Im}\Sigma(1, \omega - i0))^2}; \tag{3}$$

$E(1) = \hbar^2 p_1^2 / 2m_1$.

The EoS (2) relates the total nucleon numbers N_τ^{tot} or the particle densities n_τ^{tot} to the chemical potentials μ_τ of neutrons/protons so that one can switch from the densities to the chemical potentials characterizing thermodynamic equilibrium of warm dense matter. If this EoS is known in some approximation, all other thermodynamic quantities are obtained consistently after calculating a thermodynamic potential as shown in Sec. 1.1. In the following sections, different approximations for $\Sigma(1, \omega)$ are discussed.

1.3. *Cluster decomposition of the equation of state and quasiparticle concept*

Within a Hamiltonian approach to the many-particle system, the self-energy $\Sigma(1, z)$ may be presented by a series of diagrams which are constructed from the free nucleon propagator $G_0^{-1}(1, z) = z - E(1)$ and the nucleon-nucleon interaction potential $V(12, 1'2')$. In order to obtain approximations for the equation of state (2) of nuclear matter we can proceed in a number of different ways.[42,43]

The spectral function $S_\tau(1,\omega;T,\mu_n,\mu_p)$ and the corresponding two-point correlation functions (density matrix) are quantities, well-defined in the grand canonical ensemble characterized by $\{T,\mu_n,\mu_p\}$. The self-energy $\Sigma(1,z;T,\mu_n,\mu_p)$ depends, in addition to the single-nucleon quantum state $|1\rangle$, on the complex frequency z. It is calculated at the Matsubara frequencies, the analytical continuation to the z plane must be performed. Within a perturbative approach it can be represented by Feynman diagrams. A cluster decomposition of the self-energy with respect to different few-body channels (c) is possible,[42–44,47] characterized, for instance, by the nucleon number A, as well as spin and isospin variables.

The cluster contributions to the self-energy are derived from an in-medium A-particle Schrödinger equation which describes the propagation of the A-particle cluster (the Fourier transform of the A-particle correlation function gives the corresponding spectral function). The Green function approach provides us with consistent approximations for these few-nucleon propagators. In particular, we introduce the quasiparticle concept to describe the propagation of few-nucleon clusters (including $A = 1$) in warm dense matter if the A-particle spectral function $S_{A,\nu_c}(\mathbf{P},\omega;T,\mu_n,\mu_p)$ shows a peak structure at the energy $\omega = E_{A,\nu_c}(\mathbf{P};T,\mu_n,\mu_p)$. The dispersion relation $E^0_{A,\nu_c}(\mathbf{P})$ of the free nucleon cluster $\{A,\nu_c\}$ is modified at finite densities.

The Green function approach describes the propagation of a single nucleon by a Dyson equation governed by the self-energy, $E_\tau(\mathbf{p};T,\mu_n,\mu_p) = E^0_\tau(\mathbf{p}) + \Delta E^{\rm SE}_\tau(\mathbf{p};T,\mu_n,\mu_p)$. It also describes the propagation of few-particle states by a Bethe-Salpeter equation containing the effective interaction kernel. Both quantities, the effective interaction kernel and the single-particle self-energy, should be approximated consistently. Approximations which take cluster formation into account have been worked out,[44,48] where within the cluster mean-field (CMF) approximation correlations in the surrounding medium are consistently taken into account.

For the A-nucleon cluster, the in-medium Schrödinger equation

$$
\begin{aligned}
&[E_{\tau_1}(\mathbf{p}_1;T,\mu_n,\mu_p) + \cdots + E_{\tau_A}(\mathbf{p}_A;T,\mu_n,\mu_p) \\
&\quad - E_{A\nu}(\mathbf{P};T,\mu_n,\mu_p)]\psi_{A\nu\mathbf{P}}(1\ldots A) \\
&\quad + \sum_{1'\ldots A'}\sum_{i<j}[1 - n(i;T,\mu_n,\mu_p) - n(j;T,\mu_n,\mu_p)]V(ij,i'j') \\
&\quad \times \prod_{k\neq i,j}\delta_{kk'}\psi_{A\nu\mathbf{P}}(1'\ldots i'\ldots j'\ldots A') = 0 \qquad (4)
\end{aligned}
$$

is derived from the Green function approach after the effective occupation numbers $n(i;T,\mu_n,\mu_p)$

$$n(1) = f_1(1) + \sum_{\bar{A}=2}^{\infty} \sum_{\bar{\nu}\bar{\mathbf{P}}} \sum_{2\ldots\bar{A}} \bar{A}\, f_{\bar{A},\bar{\nu}}[E_{\bar{A},\bar{\nu}}(\bar{\mathbf{P}};T,\mu_n,\mu_p)]|\psi_{\bar{A}\bar{\nu}\bar{\mathbf{P}}}(1\ldots\bar{A})|^2\,, \tag{5}$$

are introduced and exchange terms are neglected.

$$f_{A,Z}(\omega;T,\mu_n,\mu_p) = \frac{1}{\exp[(\omega - N\mu_n - Z\mu_p)/T] - (-1)^A} \tag{6}$$

is the Bose or Fermi distribution function for even or odd A, respectively, which is depending on $\{T,\mu_n,\mu_p\}$.

The in-medium Schrödinger equation (4) contains the effects of the medium via the single-nucleon quasiparticle shift

$$\begin{aligned} \Delta E_\tau^{\mathrm{SE}}(\mathbf{p};T,\mu_n,\mu_p) &= E_\tau(\mathbf{p};T,\mu_n,\mu_p) - \sqrt{m^2c^4 + \hbar^2c^2p^2} + mc^2 \\ &\approx E_\tau(\mathbf{p};T,\mu_n,\mu_p) - \frac{\hbar^2p^2}{2m} \end{aligned} \tag{7}$$

(nonrelativistic case), as well as in the Pauli blocking terms given by the occupation numbers $n(1;T,\mu_n,\mu_p)$ in the phase space of single-nucleon states $|1\rangle \equiv |\mathbf{p}_1,\sigma_1,\tau_1\rangle$. Thus, two effects have to be considered, the quasiparticle energy shift and the Pauli blocking.

As example, consider the two-nucleon ($A = 2$) in-medium Schrödinger equation

$$\begin{aligned} &[E_{\tau_1}(\mathbf{p}_1;T,\mu_n,\mu_p) + E_{\tau_2}(\mathbf{p}_2;T,\mu_n,\mu_p) - E_{2\nu}(\mathbf{P};T,\mu_n,\mu_p)]\psi_{2\nu\mathbf{P}}(12) \\ &\quad + \sum_{1'2'}[1 - f_1(1) - f_1(2)]V(12,1'2')\psi_{2\nu\mathbf{P}}(1'2') = 0 \end{aligned} \tag{8}$$

where in Eqs. (4) and (5), the Fermi distribution function is taken for the Pauli blocking term. As shown in Figure 3, the phase space to form a bound state is reduced owing to the Pauli principle so that the interaction is effectively reduced and the binding energy is reduced. It should be mentioned that the same equation (8) describes the Bose quantum condensation when $E_{\tau_2}(\mathbf{p}_2;T,\mu_n,\mu_p) = \mu_n + \mu_p$, as well as the cross-over from BEC to BCS.[49–53]

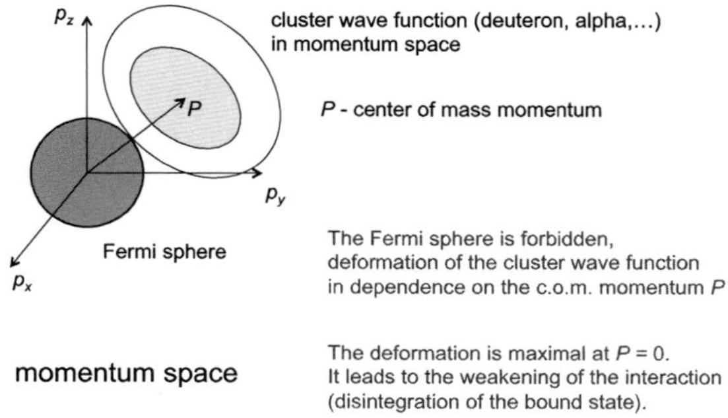

Fig. 3. Bound state wave function in phase space (momentum space). The Fermi sphere is already occupied by the matter and cannot be used to form a bound state (Pauli principle).

Using the cluster decomposition of the self-energy which takes into account, in particular, cluster formation, one obtains

$$n_n^{\rm tot}(T,\mu_n,\mu_p) = \frac{1}{\Omega}\sum_{A,\nu,\mathbf{P}} N f_{A,Z}\left(E_{A,\nu}(\mathbf{P};T,\mu_n,\mu_p)\right),$$

$$n_p^{\rm tot}(T,\mu_n,\mu_p) = \frac{1}{\Omega}\sum_{A,\nu,\mathbf{P}} Z f_{A,Z}\left(E_{A,\nu}(\mathbf{P};T,\mu_n,\mu_p)\right), \qquad (9)$$

where $\mathbf{P}$ denotes the c.o.m. momentum of the cluster (or, for $A = 1$, the momentum of the nucleon). The internal quantum state ν contains the proton number Z and neutron number N=A-Z of the cluster. The integral over ω is performed within the quasiparticle approach, the $\mathbf{P}$-dependent quasiparticle energies $E_{A,\nu}(\mathbf{P};T,\mu_n,\mu_p)$ are depending on the medium characterized by $\{T,\mu_n,\mu_p\}$. These in-medium modifications will be detailed in the following section 1.4.

1.4. *Different approximations*

As it is well known,[54] the self-energy occurring in Eq. (3) may be represented by an infinite series of irreducible diagrams. We give the physical ideas how to construct the approximation for the self-energy used in different approaches. Numerical estimates of these contributions are given in Sec. 2.

In nuclear systems, we are mainly concerned with the strong interaction. Consequences of the N-N interaction are the formation of correlations and bound clusters, but also self-energy shifts and the Pauli exclusion principle. A free nucleon feels a mean Hartree field due to the surrounding nuclear matter consisting of free nucleons and clusters, which is modified by the Pauli exclusion principle ("Fermi hole") so that the Hartree-Fock single particle energy shift $\Delta^{\rm HF}(1)$ is obtained,

$$\Sigma^{\rm HF}(1,z) = \Delta^{\rm HF}(1) = \sum_2 [V(12,12) - V(12.21)] f_{1,\tau_2}(2), \qquad (10)$$

where $f_{1,\tau_1}(1) = [\exp(E_{\tau_1}(\mathbf{p}_1) - \mu_{\tau_1})/T + 1]^{-1}$ is the Fermi distribution function.

Similarly, a bound state (cluster) of A nucleons with total momentum $\mathbf{P}$ and internal quantum state ν (including the proton number Z) feels a self-energy shift $\Delta E^{\rm SE}_{A\nu P}$ due to the surrounding matter. In addition, the binding energy is lowered by $\Delta E^{\rm Pauli}_{A\nu P}$ as a consequence of the Pauli exclusion principle, because phase space is occupied by the surrounding correlated matter and is not available to form a bound state. The lowering of the binding energy leads to the destruction of bound states in dense matter.

Notice that also the other forces (weak, Coulomb, gravitation) are of relevance describing matter in astrophysics of compact objects; they are not considered here. β equilibrium is not achieved in HIC because the time scales are short. Coulomb interaction contributions are of importance especially for heavy clusters because of its long range character. In the region of thermodynamic instability of nuclear matter, it is essential for the formation of pasta structures. The QS treatment of the Coulomb interaction is well elaborated, see Ref. 2.

Coming back to the strong interaction, we consider Eq. (9) together with (4) as the basic result for the EoS. They are treated at different levels of sophistication as will be explained from a very general point of view, see Table 2. This allows us to compare different approximations presently used to describe properties of dense matter. The consistent treatment of different effects is clearly demonstrated.

(1) The simplest approximation is obtained if cluster formation and mean-field effects, i.e., any effects of interaction, are neglected (zero self-energy). We obtain the ideal Fermi gas consisting of protons, neutrons, (electrons, neutrinos,...) well known from text books. In Eq. (9), we have only $A = 1$ and the free fermion energies $E^0_\tau(\mathbf{p})$.

This simple approximation can be improved in two directions:

(1 medium) At one hand by including mean-field effects when going to higher densities. We obtain a quasiparticle quantum liquid. Expanding $\Sigma(1,z)$ in a power series with respect to $V(12,1'2')$, we obtain in lowest order the Hartree-Fock approximation (10). Investigations of hot nuclear

Table 2. Different approximations.

low density limit	*high density*
(1) *Ideal Fermi gas:* neutrons, protons (electrons, neutrinos,...)	(1 medium) **medium effects** *Quasiparticle quantum liquid:* mean-field approximation Skyrme, Gogny, RMF
(2) **bound state formation** *Nuclear statistical equilibrium:* ideal mixture of all bound states chemical equilibrium, mass action law	(2 medium) *Chemical equilibrium of quasiparticle clusters:* medium modified bound state energies self-energy and Pauli blocking
(3) **continuum contributions** *Second virial coefficient:* account of continuum correlations ($A = 2$) scattering phase shifts, Beth-Uhlenbeck equation	(3 medium) *Generalized Beth-Uhlenbeck formula:* medium modified binding energies, medium modified scattering phase shifts
(4) **chemical & physical picture** *Cluster virial approach:* all bound states (clusters) scattering phase shifts of all pairs	(4 medium) *Correlated medium:* phase space occupation by all bound states in-medium correlations, quantum condensates

matter within the frame of such a Hartree-Fock approximation have been performed, see, for instance, Refs. 55–57. A region of thermodynamic instability was found, and the occurrence of a first order phase transition below a critical point $T_c \approx 20$ MeV has been discussed there. No clusters can be found within a Hartree-Fock approximation.

(2) On the other hand, expanding $\Sigma(1, z)$ with respect to the nucleon density, i.e., for $n_c \Lambda_c^3/2 \ll 1$ with $\Lambda_c = (2\pi\hbar^2/m_c T)^{1/2}$ being the thermal wavelength, in the lowest order the Beth-Uhlenbeck formula is obtained for the second virial coefficient. For hot dense matter, this approximation has been discussed, e.g., in Refs. 10, 58. The Beth-Uhlenbeck formula takes into account the formation of two-particle clusters. However, no in-medium corrections for these clusters are obtained.

To obtain the ideal nuclide gas approximation (NSE), all bound states should be taken into account on the same footing. This may be done within a cluster decomposition of $\Sigma(1, z)$ using the t-matrices of the isolated

A-particle system.[42–44] This approximation is the basis of the ordinary thermal model[8–10,60,61] widely used in describing the occurrence of clusters in hot dense matter.

(2 medium) In the high density region, the ideal nuclide gas approximation becomes not applicable owing to density corrections caused by the interaction of the clusters with the surrounding matter. A systematic approach to the in-medium corrections for the free particle and bound state energies as well as the wave functions can be given within the framework of many-body theory. A self-consistent ladder Hartree-Fock approximation[44] has been applied to the quantum statistical calculation of the abundances of deuterons, tritons and alpha particles[62] and to the description of a nuclear matter phase transition.[63,64] As discussed in these papers, the in-medium corrections to the energy and wave functions of the clusters lead to an interesting effect: At high densities, bound states are destroyed because of the Pauli quenching. Beyond a certain density (we will denote it as Mott density $n_A^{\rm Mott}(T)$) the abundances of the corresponding clusters decrease, and in the high density limit all bound states are dissolved so that a degenerate Fermi liquid of quasiparticles remains.

(3) In Eq. (9), the summation refers to the mass number A, the internal quantum number ν and the c. o. m. momentum $\mathbf{P}$. The internal quantum number ν includes in addition to the proton number Z also excited bound states with medium dependent energies $E_{A,\nu}(\mathbf{P}; T, \mu_n, \mu_p)$ as well as continuum states. In the two-particle case, these continuum contributions are expressed by the scattering phase shifts as function of the energy of relative motion. Only after including the continuum contributions, the correct expression for a virial expansion is obtained. The corresponding relation is known as Beth-Uhlenbeck equation (see Eq. (30) below).

(3 medium) To pass to higher densities, in addition to medium modified bound state quasiparticle energies, also the medium modified scattering phase shifts have to be calculated. For $A = 2$, the generalized Beth-Uhlenbeck formula[49] is obtained. As a peculiarity, when introducing the quasiparticle description one has to subtract contributions of the continuum scattering phase shifts to avoid double counting.

(4) This concept of inclusion of scattering state contributions is given in the cluster Beth-Uhlenbeck approach where arbitrary mass numbers A are considered.[48] Only estimates for the continuum correlations are known at present,[3] and the in-medium modifications of the corresponding cluster-virial coefficients are relevant for the composition and the EoS.

(4 medium) A consistent description of the medium effects should also take into account correlations in the medium. Extending the Hartree-Fock approximation to arbitrary clusters, the CMF approximation has been derived.[3,44] With respect to the Pauli blocking (4), the phase space occupation numbers $n(1; T, \mu_n, \mu_p)$ are not given by the free particle fermion distribution $f_{\tau_1}(p_1)$ but the effective occupation numbers (5) which considers also the phase space occupation owing to correlations and to the formation of clusters. An estimate is given in Ref. 3. For a more systematic approach, the self-consistent determination of in-medium correlations and cluster formation has to be treated which is not solved yet. For instance, the unified description of α matter and nuclear matter is not known at present.

2. Model calculations for in-medium corrections of cluster quasiparticle energies

Model calculations are performed to describe correlations and cluster formation in nuclear systems. We have to go beyond single-particle descriptions, which have been proven as very successful not only in nuclear structure calculations where the shell model has been worked out, but also in the thermodynamic properties of nuclear matter where mean-field approaches are very popular. Nuclear systems at low density and low excitation energy are dominated by correlations and cluster formation, if the kinetic energy characterized by the temperature or the Fermi energy is small compared with the potential energy.

Starting from the interaction potential $V(12, 1'2')$ which contains the long range Coulomb interaction as well as the short range nucleon-nucleon interaction, *ab initio* calculations of correlations, in particular the structure of bound states $\{A\nu\mathbf{P}\}$ are rather involved. In addition, in contrast to the fundamental Coulomb interaction which is known, the N-N interaction is introduced in an empirical way, fitted to measured properties of nuclear systems. Consequently, QS calculations for nuclear systems should implement, as much as possible, measured data, for instance the empirical binding energies of isolated nuclei.[65]

We focus on the contributions of the N-N interaction to the structure of nuclear systems. The Coulomb interaction is treated using approaches known from plasma physics;[1,2] see Sec. 2.6. It has to be taken into account for nuclear structure and nuclear reactions. In particular, in context with the EoS and the composition of nuclear matter, it becomes relevant for heavy nuclei and pasta-like structures.

2.1. *Single-quasiparticle approximation*

Within the QS approach, the influence of the medium is given by the self-energy $\Sigma(1, z)$. It determines the spectral function, Eq. (3) which then allows to calculate the EoS (9). For the self-energy, a systematic approach is possible using diagram techniques, see Refs. 45 and 46. The ideal Fermi gas is a trivial approximation of the EoS, Eq. (9), neglecting any self-energy contribution in Eq. (3) so that only $A = 1$ with the free dispersion relation $E(1) = \hbar^2 p_1^2/2m_1$ remains.

We discuss the approximation (1, medium) of Table 2, considering single-nucleon quasiparticle states in dense nuclear matter. In lowest order with respect to the interaction, neglecting correlations in the surrounding matter the Hartree-Fock (HF) result (10) is obtained. The approximation $\Sigma^{\rm HF}(1, z) = \Delta^{\rm HF}(1)$ is real and not depending on the frequency z. This is in correspondence to the Kramers-Kronig relation that any frequency dependence of $\Sigma(1, z)$ would produce an imaginary contribution Im $\Sigma(1, z)$. Because Im $\Sigma^{\rm HF}(1, z) = 0$, the spectral function (3) depends on ω as

$$S_\tau(1, \omega) = 2\pi\delta[\omega - E(1) - \Delta^{\rm HF}(1)]. \tag{11}$$

In HF approximation, we have a sharp quasiparticle with the shifted energy $E^{\rm qu,HF}(1) = E(1) + \Delta^{\rm HF}(1)$. This modification of the single-particle energy $E(1)$ is denoted as mean field.

In higher order with respect to the interaction, contributions are expected for Im $\Sigma(1, z)$ describing collisions in the system. The δ-like spectral function obtained from the quasiparticle pole $\omega = E^{\rm qu}(1)$, with the self-consistent solution of

$$E^{\rm qu}(1) = E(1) + {\rm Re}\,\Sigma(1, \omega)|_{\omega = E^{\rm qu}(1)}, \tag{12}$$

is broadened by Im $\Sigma(1, z)$ which describes the finite life-time of the quasiparticle.

The Hartree-Fock approximation has been improved taking into account two-particle correlations within the Brueckner approach, see Refs. 49 and 66. From the self-energy the spectral function is calculated; see Eq. (3). The position of the peak, i.e., the self-consistent solution of Eq. (12) gives the quasi-particle self energy $E_\tau(\mathbf{p}; T, \mu_n, \mu_p)$. However, the peak structure is not always clearly seen from the spectral function, for instance at low densities, where bound states appear. Near the saturation density phenomenological values for different properties such as the saturation density, binding energy, compressibility are known which are not correctly obtained within the Brueckner theory using a simple form of N-N interaction.

Instead of microscopic calculations of the quasiparticle energies, based on a semi-empirical N-N interaction, we can also directly parametrize the quasiparticle energies using the known properties of nuclear matter. A simple parametrization is given by Skyrme; see.[67,68] The Hartree-Fock shift (10) is estimated using, for instance, a zero range effective interaction $V(n_B)\delta(\mathbf{r}_1 - \mathbf{r}_2)$. Then, $\Delta E_\tau^{\rm SE}(1)$ is independent of momentum and temperature.

A more advanced parametrization is the relativistic mean-field theory (RMF) where the quasiparticle energies are parametrized by a scalar $S(T, n_B, Y_p)$ and a vector potential $V_\tau(T, n_B, Y_p)$; see Refs. 69,70, adapted to empirical data for nuclear systems:

$$\begin{aligned} &E_\tau(\mathbf{p}; T, n_B, Y_p) \\ &\quad = \sqrt{[m_\tau c^2 - S(T, n_B, Y_p)]^2 + \hbar^2 c^2 p^2} + V_\tau(T, n_B, Y_p) - m_\tau c^2. \end{aligned} \tag{13}$$

In the limit $p \to 0$, the quasiparticle dispersion relation leads to the effective mass approximation

$$E_\tau(\mathbf{p}; T, n_B, Y_p) \approx \Delta E_\tau^{\rm SE}(0; T, n_B, Y_p) + \frac{\hbar^2 p^2}{2m_\tau^*(T, n_B, Y_p)} \tag{14}$$

with the shift $\Delta E_\tau^{\rm SE}(0) = -S(T, n_B, Y_p) + V_\tau(T, n_B, Y_p)$ and the effective mass ratio $m_\tau^*/m_\tau = 1 - S(T, n_B, Y_p)/(m_\tau c^2)$. Explicit expressions for $S(T, n_B, Y_p)$ and $V_\tau(T, n_B, Y_p)$ in form of Padé approximations which are suitable for numerical applications are found in Refs. 3, 23 and 48.

As a mean-field approach, in the low-density limit a linear dependence $S, V_\tau \propto n_B$ is assumed. A linear shift of the quasiparticle energy as solution of Eq. (12) results. No cluster formation is described in the single-quasiparticle approximation. Therefore, the RMF approximation is not sufficient in the low-density, low-temperature region where correlations and bound state formation are of relevance.

In contrast, the EoS (9) obtained taking only $A = 1$ in quasiparticle approximation gives already a reasonable description near the saturation density. In addition, a region of thermodynamic instability $\partial\mu/\partial n_B < 0$

is obtained below a critical temperature (symmetric matter) $T_{\rm crit}^{\rm mf} \approx 13.72$ MeV[23] indicating the occurrence of a phase transition.

The concept to parametrize the single-nucleon quasiparticle energies is related to the density functional theory used in condensed matter theory. A problem is the exact treatment of correlations. Part of the interaction is already implemented in the parametrization of the energy density functional. If the correlations are considered separately one has to avoid double counting. We come back to this issue below.

2.2. *Cluster-quasiparticle approximation*

To obtain the formation of clusters and correlations, the term $\mathrm{Im}\,\Sigma$ in the spectral function (3) has to be analyzed. A cluster decomposition of the self-energy gives the contribution of the A-nucleon correlation, in particular the bound states. Neglecting all medium effects in the few-nucleon wave equation (4), the ideal nuclide gas approximation (NSE) for the EoS (9) is obtained. Instead solving the isolated A-nucleon Schrödinger equation with an appropriate N-N interaction to obtain the energy eigenvalues $E^{(0)}_{A,\nu}(\mathbf{P})$, we can directly use the empirical values for the masses of nuclei as given, e.g., by Ref. 65.

To describe medium effects consistently with the single-quasiparticle approximation, the few-nucleon wave equation (4) has to be solved. Within a perturbative approach, the cluster-quasiparticle shift has two contributions ($A > 1$),

$$E_{A,\nu}(\mathbf{P};T,\mu_n,\mu_p) = E^{(0)}_{A,\nu}(\mathbf{P}) + \Delta E^{\rm SE}_{A,\nu}(\mathbf{P};T,\mu_n,\mu_p) + \Delta E^{\rm Pauli}_{A,\nu}(\mathbf{P};T,\mu_n,\mu_p). \tag{15}$$

In addition to the single-quasiparticle energy shift $\langle\psi_{A\nu\mathbf{P}}|\sum_i \Delta E_{\tau_i}(\mathbf{p}_i)|\psi_{A\nu\mathbf{P}}\rangle$, the Pauli blocking $-\langle\psi_{A\nu\mathbf{P}}|\sum_{i\neq j} n(i)V(ij,i'j')|\psi_{A\nu\mathbf{P}}\rangle$ must be considered to obtain the approximation (2, medium) of Table 2.

Within a simple estimate for the medium-modified cluster-quasiparticle energies, the cluster self-energy shift is easily calculated in the rigid shift approximation

$$\Delta E^{\rm SE}_{A\nu P} = \sum_i^A \Delta E^{\rm SE}_{\tau_i}(0;T,n_B,Y_p) \tag{16}$$

neglecting the effective mass corrections. To improve it, the effective mass approximation (14) can be performed using empirical values for m^*_τ, see.[3]

In the Pauli quenching term

$$\Delta E^{\rm Pauli}_{A\nu P} = -\sum_{1\dots A}\sum_{1'\dots A'} \psi^{0*}_{A\nu P}(1\dots A)\psi^{0}_{A\nu P}(1'\dots A')$$
$$\times \sum_{ij}^{A}{}' f_{1,\tau_i}(\mathbf{p}_i - \mathbf{P}/A)V(ij,i'j')\delta_{kk'} \qquad (17)$$

the interaction potential $V(12,1'2')$ may be approximately eliminated if the cluster wave function is represented by the antisymmetrized product of single particle wave functions $\phi_n(i)$.[5,59] In the case $A = 2$ this can be done rigorously. After separating the center-of-mass motion $\mathbf{P}$, the relative motion (relative momentum $\mathbf{p}$) of the free deuteron obeys a Schrödinger equation $\sum_{\mathbf{p}'} V(\mathbf{p},\mathbf{p}')\psi_d^{(0)}(\mathbf{p}') = (E_d^{(0)} - \hbar^2 p^2/m)\psi_d^{(0)}(\mathbf{p})$ so that

$$\Delta E^{\rm Pauli}_{d,P} = \int \frac{d^3p}{(2\pi)^3}\left[f_n\left(\frac{\mathbf{P}}{2}+\mathbf{p}\right) + f_p\left(\frac{\mathbf{P}}{2}-\mathbf{p}\right)\right]$$
$$\times \left(-E_d^{(0)} + \hbar^2 p^2/m\right)|\psi_d^{(0)}(\mathbf{p})|^2. \qquad (18)$$

With the bound state energy $E_d^{(0)} = -2.225$ MeV, the Gaussian wave function (which can be replaced by a better wave function corresponding to the empirical density distribution) $\psi_d^{(0)}(\mathbf{p}) = (8^{1/2}\pi^{3/4}/k_d^{3/2})\exp(-p^2/2k_d^2)$ the evaluation of Eq. (18) is simple and straight forward. The parameter $k_d = 0.312\ \text{fm}^{-1}$ is fixed by the nucleonic point rms radius.[72] In the low-density limit where $f_\tau(p) = n_\tau^{\rm tot}\Lambda^3/2\exp(-\hbar^2p^2/2mT)$ with the thermal wave length $\Lambda = (2\pi\hbar^2/mT)^{1/2}$, the result reads

$$\Delta E^{\rm Pauli}_{d,P} = \frac{n_B\Lambda^3T^{3/2}}{2(2\epsilon_d+T)^{3/2}} e^{-X_d^2\frac{\epsilon_d+T}{2\epsilon_d+T}}\left(-E_d^{(0)} + \frac{3\epsilon_d T}{2\epsilon_d+T} + X_d^2\frac{2\epsilon_d^2T}{(2\epsilon_d+T)^2}\right) \qquad (19)$$

with $X_d^2 = \hbar^2P^2/8mT$ and $\epsilon_d = \hbar^2k_d^2/2m = 2.02$ MeV. In addition to the linear dependence on the nucleon density n_B as a consequence of the perturbation approach, the strong variation of the Pauli blocking shift with P and T is remarkable, in contrast with the single-nucleon mean-field shifts. For a rigorous result, the empirical form factor $|\psi_d^{(0)}(\mathbf{p})|^2$ has to be used instead of a Gaussian form factor, fitted to the rms radius. A more sophisticated evaluation of $\Delta E^{\rm Pauli}_{d,P}$ is found in Refs. 3,71 where also Padé approximations for the dependence on $\{\mathbf{P}, T, n_B, Y_p\}$ are given, see Figure 4. (Note that the shift given in Ref. 23, Eq. (72), contains an empirical factor to suppress the abundance at higher densities.)

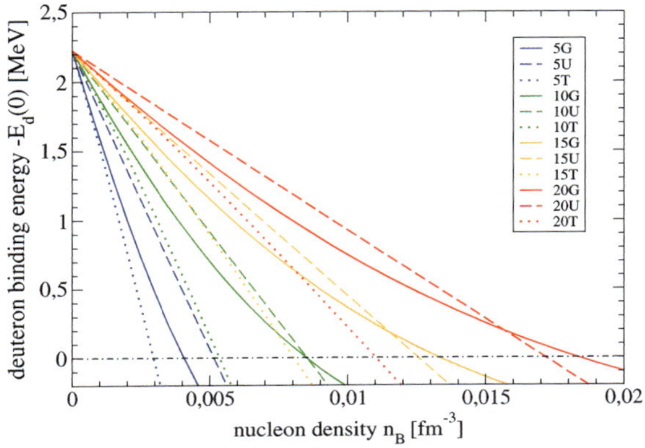

Fig. 4. Shift of the deuteron binding energy because of Pauli blocking. Temperatures $T = 5, 10, 15, 20$ MeV (from left to right), $P = 0$. The rigorous result[3,71] (G, full) is compared with the simple estimation (19) (U, dashed) and the expression given in Ref. 23 (T, dotted).

For the evaluation of the Pauli quenching term of the other light elements $A \leq 4$, i.e., t, h, α, assumptions for the interaction potential are necessary. Even for the product ansatz, the binding energy $E^{(0)}_{A\nu P}$ is not given by the sum of the single nucleon state energies E_n so that shell model calculations have to be used,

$$\Delta E^{\text{Pauli}}_{A\nu P} = -\sum_{n}^{\text{occ.}}\sum_{1} |\phi_n(1)|^2(E_n - E_1) f_{1,\tau_1}(\mathbf{p}_1 - \mathbf{P}/A). \tag{20}$$

Estimates based on the harmonic oscillator model are given in Refs. 5 and 59. A more sophisticated calculation where the c.o.m. motion is eliminated, has been performed[3,71] using a separable potential. Padé approximations for $\Delta E^{\text{Pauli}}_{A0P}(\mathbf{P}, T, n_B, Y_p)$ are given there which are suited for the evaluation of the EoS (9) and the corresponding thermodynamic potentials. A more simple parametrization of the quasiparticle shifts of the light elements has been considered in Ref. 23 where also various thermodynamic functions have been presented. The quasiparticle shifts ΔE_{A0P} of light clusters, depending on $\{\mathbf{P}, T, n_B, Y_p\}$, have been investigated recently.[3,71,72] In-medium bound state energies are calculated for the light clusters (d, t, h, α), and fit formulae are presented there.

Clusters with $5 \leq A \leq 11$ (light metals) are weakly bound and therefore very sensitive to the medium effects, similar to the deuterons. These elements are weakly bound (note that ^{8}Be is unbound) and experience a strong influence of the medium. A strong depletion because of the Pauli blocking is expected. Estimates of $\Delta E^{\mathrm{Pauli}}_{A0P}$ for $4 < A \leq 16$ are given in Refs. 5,44,59,73.

Different approaches are used to calculate the effect of medium modification for large nuclei. The heavy clusters $A \geq 12$ are usually considered as droplets of a dense phase with $n_B \approx n_{\mathrm{sat}}$. As a semiempirical treatment of medium effects, in particular Pauli blocking, the excluded-volume model[25,74] may be introduced. As an alternative, local density approaches such as the Thomas-Fermi model can be applied to calculate the modification of the cluster in a dense medium[18,19,26,30,34,36,37] which provide us with a microscopic treatment of large nuclei in a dense medium.

We give here an approach[5,44,59] which is closely related to the treatment of light clusters. The total energy shift of a cluster $(A\nu P)$

$$\Delta E_{A\nu P}(T, n_B, Y_p) = \Delta E^{\mathrm{SE}}_{A\nu P} + \Delta E^{\mathrm{Pauli}}_{A\nu P} = \sum_{\alpha}^{\mathrm{occ.}} \sum_{22'} \phi^*_\alpha(2)\phi_\alpha(2')$$
$$\times \left\{ \sum_{11'} f_{\tau_1}\left(\mathbf{p}_1 - \frac{\mathbf{P}}{A}\right) \left[\delta_{11'}\delta_{22'} - \sum_{\alpha'}^{\mathrm{occ.}} \phi^*_{\alpha'}(1)\phi_{\alpha'}(1')\right] \right.$$
$$\left. \times \left[V(12,1'2') - V(12,2'1')\right] \right\} \tag{21}$$

has a simple structure if within a (locally homogeneous) Fermi gas model the wave functions $\phi_\alpha(i)$ are taken as momentum eigenfunctions. At not too high temperatures, a strong compensation between the self-energy and the Pauli quenching term results. Thus large clusters remain nearly unshifted. Applying the homogeneous Fermi gas model to a cluster (A) with densities $n^A_\tau(r)$ for the neutrons and protons, respectively, we obtain

$$\Delta E_{A\nu P}(n_B) = \sum_{\tau=n,p} \int d^3r \Delta E^{\mathrm{SE}}_\tau(n^A_n(r), n^A_p(r)) n_\tau$$
$$\times \int_{\Lambda_\tau p_F(n^A_\tau(r))}^{\infty} \frac{y dy}{2\pi x_\tau} \left[e^{-(y-x_\tau)^2/4\pi} - e^{-(y+x_\tau)^2/4\pi}\right] \tag{22}$$

with

$$p_F(n_\tau) = (3\pi^2 n_\tau)^{1/3}, \qquad x_\tau = m_\tau v \Lambda_\tau/\hbar \approx P\Lambda/A. \tag{23}$$

An analysis of experimental data was performed (see, e.g., Ref. 75) to fit the following symmetrical density distribution function

$$n^A_B(r) = \frac{3A}{4\pi R^3} \frac{1}{1+(\pi b/R)^2} \left[\frac{1}{1+e^{(r-R)/b}} + \frac{1}{1+e^{(-r-R)/b}}\right]. \tag{24}$$

Appropriate values for $A > 16$ are $R = 1.05$ fm $A^{1/3}$, $b = 0.57$ fm. Furthermore we take $n_p^A(r) = n_B^A(r)Z/A$.

2.3. *Mott points and Mott momenta*

Of special interest are the binding energies

$$B_{A,\nu}^{\rm bind}(\mathbf{P};T,n_B,Y_p) = -[E_{A,\nu}(\mathbf{P};T,n_B,Y_p) - E_{A,\nu}^{\rm cont}(\mathbf{P};T,n_B,Y_p)] \quad (25)$$

with $E_{A,\nu}^{\rm cont}(\mathbf{P}) = NE_n(\mathbf{P}/A) + ZE_p(\mathbf{P}/A)$, that indicate the energy difference between the bound state and the continuum of free (scattering) states at the same total momentum $\mathbf{P}$. This binding energy determines the yield of the different nuclei according to Eq. (9), where the summation over $\mathbf{P}$ is restricted to that region where bound states exist, i.e., $B_{A,\nu}^{\rm bind}(\mathbf{P};T,n_B,Y_p) \geq 0$.

We denoted the density $n_{A,\nu}^{\rm Mott}(T,Y_p)$ as Mott density[42–44,47] where the binding energy of a cluster $\{A,\nu\}$ with c.o.m. momentum $\mathbf{P} = 0$ vanishes,

$$B_{A,\nu}^{\rm bind}(0;T,n_{A,\nu}^{\rm Mott},Y_p) = 0\,, \quad (26)$$

see Figure 4. For baryon densities $n_B > n_{A,\nu}^{\rm Mott}(T,Y_p)$ we introduced the Mott momentum $P_{A,\nu}^{\rm Mott}(T,n_B,Y_p)$, where the bound state disappears,

$$B_{A,\nu}^{\rm bind}(P_{A,\nu}^{\rm Mott};T,n_B,Y_p) = 0\,. \quad (27)$$

At $n_B > n_{A,\nu}^{\rm Mott}(T,Y_p)$, the summation over the momentum to calculate the bound state contribution to the composition (9) is restricted to the region $|\mathbf{P}| > P_{A,\nu}^{\rm Mott}(T,n_B,Y_p)$.

The condition (27) may be replaced by further restrictions in the $\mathbf{P}$-space if further decay modes are considered. Especially, the decay into α quasi-particles is of interest. This effect takes place, e.g., in the region $6 \leq A \leq 11$ where also the stability of the clusters with respect to the decay into other fragments such as deuterons and tritons/^{3}He must be checked in order to determine $P_{A\nu}^{\rm Mott}$.

The Mott point where the binding energy vanishes is determined by the Pauli blocking term, $E_{A,\nu}^0 + \Delta E_c^{\rm Pauli}(\mathbf{P};T,n_{A,\nu}^{\rm Mott},Y_p) = 0$. The self-energies $\Delta E_{A,\nu}^{\rm SE}$ for the bound state and the continuum states compensate if their momentum dependence is neglected. Crossing the Mott point by increasing the baryon density, part of correlations survive as continuum correlations so that the EoS changes smoothly. Therefore, the inclusion of correlations in the continuum is of relevance.

2.4. *Excited states, continuum correlations and virial expansions*

The chemical equilibrium in Eq. (9) contains the sum over all components. This includes not only the A-nucleon clusters in the ground state but also the ν-summation over all excited cluster states. This can be replaced by an integral after introducing the density of states[76]

$$D_A(E) = \frac{1}{12}\left(\frac{\pi^2}{a}\right)^{1/4} E^{-5/4}\exp(2\sqrt{aE}) \tag{28}$$

with $a = A/15\ \mathrm{MeV}^{-1}$ for the homogeneous Fermi gas model and arbitrary values Z. Then the abundance of the clusters with mass number A is given by

$$X_A = \frac{n_A}{n_B} = \frac{A^3}{2\pi^2 n_B \Lambda^3}\int_{P^{\mathrm{Mott}}_{A,0}\Lambda/A}^{\infty} x^2 dx \int_{E^A_{\min}}^{E^A_{\max}} dE D_A(E)$$
$$\times \exp\left\{-\frac{1}{k_BT}\left[E^0_{A,0}(n_B) + E + \Delta E_{A,0}(n_B,x)\right.\right.$$
$$\left.\left. + A\frac{k_BTx^2}{4\pi} - (A-Z)\mu_n - Z\mu_p\right]\right\}. \tag{29}$$

In this formula, it was assumed that Coulomb corrections, self-energy and Pauli blocking corrections to the cluster energy do not strongly depend on the excitation state. A lower bond $E^A_{\min} + E_F/A$ (E_F — nuclear matter Fermi energy) was introduced to take into account that below this energy the density of states (28) is not applicable, and the discrete structure of the excitation spectrum of the clusters should be considered instead. This lower limit can be flexible depending on the number of states which are explicitly taken into account. The upper bond $E^A_{\max}(x)$ is introduced into (29) to take into account that excited clusters may become instable with respect to the decay into smaller fragments as given, e.g., by $E^A_{\max} = E^A_{\mathrm{Mott}} = A\Delta E^{\mathrm{SE}}_\tau - E^0_{A,0}$.[77]

The summation over all excited states ν is not restricted to the bound states but includes also scattering states. Only taking into account the contribution of scattering states, the correct low-density limit of the EoS and related thermodynamic quantities are obtained. Expanding the EoS in powers of n_B, the lowest order gives the result for ideal quantum gases $n_B^{(0)} = 2\Lambda^3[\exp(\mu_n/T)+\exp(\mu_p/T)]$ and for the pressure $p = n_BT$. The second order of n_B is denoted as second virial coefficient. An exact expression is given by the Beth-Uhlenbeck equation

$$n_{2,d} = \frac{2^{3/2}}{\Lambda^3} e^{(\mu_n+\mu_p)/T} \left[3e^{-E_d^0/T} + \int_0^\infty \frac{dE}{\pi} e^{-E/T} \frac{d}{dE} \delta_{2,d}^{\rm tot}(E) \right] \quad (30)$$

with $\Lambda = (2\pi\hbar^2/mT)^{1/2}$ being the baryon thermal wavelength (the neutron and proton masses are approximated by $m_\tau \approx m = 939.17$ MeV$/c^2$), and $\delta_{2,T_I=0}^{\rm tot}(E) = \sum_{S,L,J}(2J+1)\delta_{^{2S+1}L_J}(E)$ the isospin-singlet ($T_I = 0$) scattering phase shifts with angular momentum L as function of the energy E of relative motion. A similar expression can also be derived for the isospin-triplet channel (e.g., two neutrons) where, however, no bound state occurs. The relation (30) gives an exact relation for the second virial coefficient in the low-density limit where in-medium effects are absent. For data see[78] where detailed numbers are given. Note that the second viral coefficient is expressed in terms of measured data, the binding energies and scattering phase shifts $\delta_2(E)$.

These second virial coefficients cannot directly be used within a quasi-particle approach. Because part of the interaction is already taken into account when introducing the quasi-particle energy, one has to subtract this contribution from the second virial coefficient to avoid double counting, see.[48,49] Instead of Eq. (30) we obtain for the d channel

$$n_{2,d}^{\rm qu} = \frac{2^{3/2}}{\Lambda^3} e^{(\mu_n+\mu_p)/T} \times \left[3(e^{-E_d^0/T} - 1) + \int_0^\infty \frac{dE}{\pi T} e^{-E/T} \left\{ \delta_{2,d}^{\rm tot}(E) - \frac{1}{2}\sin[2\delta_{2,d}^{\rm tot}(E)\right] \right\}. \quad (31)$$

Comparing (31) with the ordinary Beth-Uhlenbeck formula (30), we observe two differences:

(i) After integration by parts, the derivative of the scattering phase shift is replaced by the phase shift, and according to the Levinson theorem for each bound state the contribution -1 appears.
(ii) The contribution $-\frac{1}{2}\sin[2\delta_c(E)]$ appears to avoid double counting[48,49] when introducing the quasiparticle picture.

The EOS (9) is not free of ambiguities with respect to the subdivision into bound state contributions and continuum contributions, compare (31) and (30). The continuum correlations in the second virial coefficient are reduced if the quasiparticle picture is introduced. The remaining part of the continuum contribution in Eq. (30) is absorbed in the quasiparticle shift. This has been discussed in detail in Refs. 41, 48 and 49.

At finite densities, we can introduce also the quasiparticle picture for the $A = 2$ channel so that the deuteron energy E_d^0 is replaced by the in-medium (quasiparticle) deuteron energy, and the phase shifts $\delta_c(E)$ contain also the medium modifications, see.[49] The approximation based on the solution of the in-medium two-particle problem (8) leading to the generalized Beth-Uhlenbeck formula is denoted as (3, medium) in Table 2.

2.5. *Cluster virial approach and correlated medium*

A more advanced approach [(4, medium) in Table 2] to the nuclear matter EoS would include cluster with arbitrary A. A cluster Beth-Uhlenbeck approach is discussed recently[48] to include also higher-order ($A > 2$) correlations.[3]

For the A-nucleon cluster, the in-medium Schrödinger equation (4) is derived, depending on the occupation numbers $n(1; T, \mu_n, \mu_p)$ of the single-nucleon states $|1\rangle$. It is obvious that the nucleons found in clusters contribute to the mean field leading to the self-energy, but occupy also phase space and contribute to the Pauli blocking. The cluster mean-field (CMF) approximation[3,44,48] considers also the few-body t-matrices in the self-energy and in the kernel of the Bethe-Salpeter equation. The CMF approximation leads to similar expressions (10) but the free-nucleon Fermi distribution $f_{1,\tau_1}(1)$ replaced by the effective occupation number (5).

Because the self-consistent determination of $n(1; T, \mu_n, \mu_p)$ for given $\{T, \mu_n, \mu_p\}$ is very cumbersome, as an approximation the Fermi distribution with new parameters $\{T_{\rm eff}, \mu_n^{\rm eff}, \mu_p^{\rm eff}\}$ (effective temperature and chemical potentials) has been introduced,[3]

$$n(1; T, \mu_n, \mu_p) \approx f_{1,\tau_1}(1; T_{\rm eff}, \mu_n^{\rm eff}, \mu_p^{\rm eff}) \,. \tag{32}$$

The effective chemical potentials $\mu_\tau^{\rm eff}$ realize the normalization to the given baryon densities $n_\tau^{\rm tot}$. A simple relation

$$T_{\rm eff} \approx 5.5\,{\rm MeV} + 0.5\;T + 60\;n_B\;{\rm MeV\;fm}^3 \tag{33}$$

was given in Ref. 3 as an approximation for the region 5 MeV $< T <$ 15 MeV and densities $n_B < n_{\rm sat}/2$ of the parameter space. More detailed investigations are necessary to derive a more general expression for the effective temperature as function of T, n_B, Y_p including, for instance α matter where the medium consists of α nuclei. The present simple fit formula (33) may be considered as a first step in this direction.

2.6. *Coulomb correlations and Debye-Thomas-Fermi screening*

In Refs. 5,47,73 the Debye-Thomas-Fermi theory known from plasma physics was adapted to calculate the Coulomb corrections to the cluster energies $E^0_{A\nu P}(T, n_B, Y_p)$ and the nucleon density variations (pair distribution function) within the screening cloud around a cluster. A more sophisticated approach to these Coulomb corrections can be formulated by considering the cluster self-energy due to the Coulomb part of the interaction in the dynamically screened potential approximation, see, for instance, Refs. 1, 2.

The screened potential $V_D(r)$ is obtained from the Poisson equation, and the screened density follows from the self-consistent solution of the linearized Poisson-Boltzmann equation. A typical parameter is the screening length $1/\kappa$ where

$$\kappa^2 = 2\pi e^2 \frac{\partial}{\partial \mu_p} n_p(\beta, \mu_n, \mu_p). \tag{34}$$

Two effects are obtained from the Coulomb interaction:

(i) The density of the surrounding charged particles is reduced in the vicinity of the cluster. The total proton density n_p^* at the surface of a cluster is smaller than the average density $n_p^{\rm tot}$ because of the Coulomb repulsion.

This effect will diminish the in-medium corrections due to the short range nucleon-nucleon interaction especially for large clusters.

(ii) In addition to the Coulomb energy of the isolated nucleus, i.e., $n_B = 0$, a self-energy shift due to the finite charged nucleon density $n_p^{\rm tot}$ is obtained. For globally charge-neutral matter, in particular in high-density astrophysical objects, the Coulomb term in the Bethe-Weizsäcker formula is reduced. The Coulomb field of the charged nucleus extends for $n_B = 0$ over the entire space, but is confined to the compensating screening cloud at finite matter density. This means a reduction of the electric field energy. To give an estimate, the simple Debye theory gives the energy shift $-Z^2e^2\kappa/2$ of a charged particle. This effect makes the large clusters more stable. With increasing density, the valley of stable nuclei is shifted towards the symmetry line $Z = N$. Larger clusters can be formed because the destabilizing influence of the Coulomb term is reduced.

Expressions to calculate both effects (i) and (ii) are given in the literature.[5,47,73]

Note that a simplified description of the effects of Coulomb correlations is given by the Wigner-Seitz cell method as used, for instance, in Ref. 56. In the spirit of the Wigner-Seitz model, all protons are removed from a sphere with the radius $R_A^{\rm WS} = (3Z/4\pi n_p^{\rm tot})^{1/3}$ so that $n_p^{\rm WS}(r) = n_p^* = 0, r < R^{\rm WS}$, and $n_p^{\rm WS}(r) = n_p^{\rm tot}$ elsewhere. Instead of the Debye shift which reads in the classical limit (uncorrelated medium)

$$\Delta E_{A\nu P}^{\rm Coul,D}(T, n_B, Y_p) = -\frac{Z^2e^2}{2}\left(\frac{4\pi e^2 n_p^{\rm tot}}{mT}\right)^{1/2} \tag{35}$$

we have ($R_A = 1.25 A^{1/3} = (3/4\pi n_{\rm sat})^{1/3}$ fm)

$$\Delta E_{A\nu P}^{\rm Coul,WS}(T, n_B, Y_p) = -\frac{3}{5}\frac{Z^2e^2}{R_A}\left[\frac{3}{2}\frac{R_A}{R_A^{\rm WS}} - \frac{1}{2}\left(\frac{R_A}{R_A^{\rm WS}}\right)^3\right]. \tag{36}$$

Light clusters $A \leq 4$ are not significantly influenced by Coulomb interaction. In contrast, Coulomb interaction is of fundamental interest for the stability of large clusters, and it is very important in the phase transition region determining the pasta-like structures. The simple Wigner-Seitz model can give only a rough estimate of the effects of Coulomb interaction. It has to be replaced by more accurate calculations which treat the Coulomb correlations within a QS approach. The solution of the Poisson-Boltzmann equation with the consistent account of normalization gives an adequate treatment of Coulomb effects.

3. Experimental evidences and relevance

We discuss several in-medium effects which can be observed. We give not an exhaustive description of these properties, but intend only to point out the relevance of clustering in nuclear systems and the necessity to describe medium effects.

3.1. *HIC: EoS and chemical constants*

Quantum statistical (QS) calculations, Eq. (9), of the composition of symmetric ($Y_p = 0.5$) matter are shown in Figure 5.[3] Only light elements ($A \leq 4$) are considered. Deviations from the simple NSE at low densities are caused by the account of scattering states (in particular d). For densities

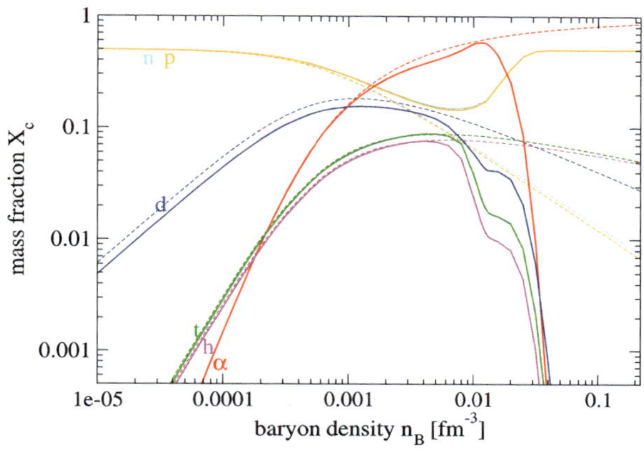

Fig. 5. Composition of symmetric matter in dependence on the baryon density n_B, $T = 5$ MeV. Quantum statistical calculation of light element ($A \leq 4$) mass fractions[3] compared with NSE.

$n_B > 0.001$ fm^{-3} medium effects become of relevance. Pauli blocking leads to the disappearance of bound states around $n_B = 0.03$ fm^{-3} so that a Fermi liquid of single nucleon quasiparticles remains.

The experimental verification of clustering in nuclear systems is not simple. Firstly, in laboratory experiments we do not have infinite nuclear matter, but always finite systems. Secondly, experiments like heavy ion collisions (HIC) are non-equilibrium processes and are only approximately described by equilibrium properties, for instance within a freeze-out approach. Despite the consistent quantum statistical description of cluster formation in HIC may be solved by future transport codes, the local thermodynamic equilibrium is a prerequisite for such a nonequilibrium approach, not only as a test case for the quality of the equilibrium limit solution, but also for the formulation of the kinetic theory beyond the Boltzmann single-nucleon description; see Sec. 5 below.

The cluster yields and respective energy spectra have been discussed intensely to derive the thermodynamic properties of nuclear matter at finite temperature. Recently investigations have been published[12] which clearly rule out the NSE, but demonstrate the relevance of medium corrections. Within a QS approach including quasiparticle shifts and correlations in the continuum,[3] it was possible to reproduce the data for the chemical constants of the light elements d, t, h, α obtained from the cluster yields, see Figure 5.

More simple models for medium corrections such as the excluded volume concept[25] can be adapted to reproduce the data.[38]

So-called Mott points have also been discussed when comparing the theory of cluster formation in dense matter with experiments.[79] Note that for $n_B > n_{A,\nu}^{\rm Mott}$ the abundance of the cluster $\{A, \nu\}$ is not vanishing. Contributions from high momenta $|\mathbf{P}| > P_{A,\nu}^{\rm Mott}(T, n_B, Y_p)$ as well as the continuum remain.

3.2. *EoS and gas-liquid nuclear matter phase transition*

The evaluation of the EoS (9) for symmetric matter ($Y_p = 0.5$) is shown in Figure 6 for different T. Further EoS such as the Free energy $F(T, N_n, N_p, \Omega)$ are obtained by integration; see Sec. 1.1. We will not present results for all thermodynamic quantities, see Ref. 23, but discuss only the influence of clustering. RMF is applicable at high densities where clusters disappear. The ideal Fermi gas, valid at very low densities, is improved by the NSE which becomes invalid around $n_B = 0.001$ fm^{-3}.

An interesting point is the stability of the EoS for homogeneous matter with respect to phase transitions. For instance, if $\partial\mu/\partial n_{|_T} < 0$, a separation of phases with different densities will occur. We would like to point out that some of the parameter values of T and n_B considered here lie within that region of the temperature-density plane of matter where a nuclear matter

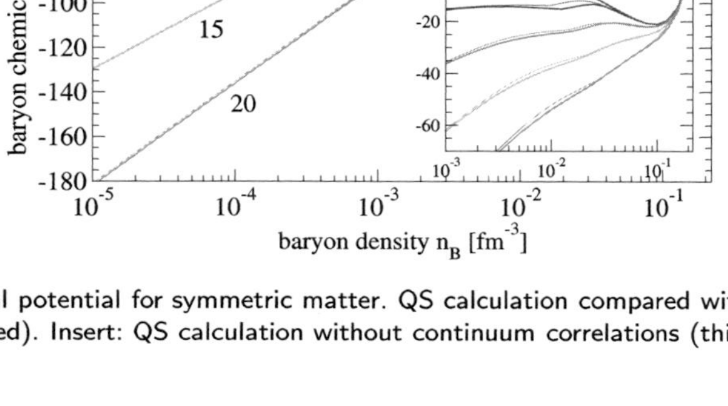

Fig. 6. Chemical potential for symmetric matter. QS calculation compared with RMF (thin) and NSE (dashed). Insert: QS calculation without continuum correlations (thin lines). From Ref. 3.

phase transition is possible.[42,43,55–57,63,80] Of interest is the reduction of the region of phase instability obtained from RMF, if correlations and cluster formation are taken into account; see Figure 6. The critical temperature $T_{\rm cr}^{\rm mf} = 13.72$ MeV (Sec. 2.1) is reduced to $T_{\rm cr}^{\rm QS} = 12.42$ MeV;[3,42,44] see also Refs. 28, 33.

The spinodal instability has been considered[81,82] and a limiting temperature for large clusters has been discussed as limit for thermalization owing to spinodal vaporization. This may be improved by considering the intrinsic partition function, in particular the continuum contributions. For homogeneous matter, phase separation is suppressed because of the long-range Coulomb interaction. Structures (droplets, rods, sheets, etc.) are formed separating high-density regions from low-density regions. Such so-called pasta phases give lower values for the free energy. Nuclear pasta phases, which may have complex structures, are discussed to derive the EOS also within the region of thermodynamic instability, see Refs. 26, 30, 33 and 34.

Note that cluster formation in hot dense matter has also been treated as phase separation and droplet formation near the phase transition region. In Ref. 55, the dense phase is assumed to be distributed as droplets (identical nuclei) immersed in the low dense phase which consists of free nucleons and ideal α particles. Within a finite temperature Thomas-Fermi approach, abundances of clusters and probabilities of fluctuations of the droplet size were considered.[56,83] A mass formula approach which takes into account excited nuclei and Coulomb effects was given in Ref. 61. These approaches, however, do not seem to be adequate to describe the abundance of small clusters where a more rigorous quantum statistical approach is necessary. Confusion arises when a nucleus which is a bound quantum state is treated as droplet of a dense phase which is a classical object.

3.3. *Symmetry energy*

Solving the EoS (9), after calculating the free energy as discussed above, further thermodynamic quantities are obtained such as the internal energy per nucleon $U(T, n_B, Y_p)$. The symmetry energy $U_{\rm sym}(T, n_B)$ describes the dependence of the internal energy on the asymmetry parameter Y_p and is related to the difference of the internal energy of neutron matter and symmetric matter, see Refs. 84 and 85.

The symmetry energy is sensitive to cluster formation, see Figure 7. Whereas quasiparticle approaches such as Skyrme Hartree-Fock and relativistic mean-field (RMF) models predict $U_{\rm sym}(T, n_B) \propto n_B$ in the

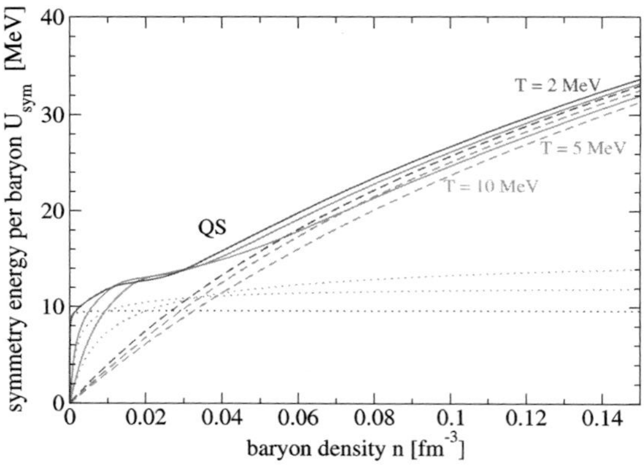

Fig. 7. Symmetry energy for different $T = 2, 5, 10$ MeV. QS calculation (full line) compared with RMF (dashed) and NSE (dotted). From Ref. 85.

low-density limit,[84] the QS calculations[3,85] show strong deviations at low densities compatible with the NSE. Cluster formation is strongly temperature-dependent, and the low-density, low-temperature limit is dominated by the binding energy per nucleon in nuclei which is ≈ 8 MeV.

Such a finite value of $U_{\mathrm{sym}}(T, n_B)$ in the low-density region in contradiction to the mean-field approaches has been confirmed experimentally.[11,86,87] At low density the symmetry energy changes mainly because additional binding is gained in symmetric matter due to formation of clusters and pasta structures.[88]

3.4. *Low temperatures and quantum condensates*

A special feature of correlations and bound state formation are quantum condensates. According to Eqs. (6), (9), the Bose distribution function exhibits a singularity when the energy eigenvalue $E_{A\nu}(\mathbf{P}; T, \mu_n, \mu_p)$ coincides with the cluster chemical potential $\mu_A = (A\text{-}Z)\mu_n + Z\mu_p$ (Thouless criterion). The case $A = 2$ is well investigated. Whereas in the low-density limit, depending on the asymmetry, Bose-Einstein condensation (BEC) of deuterons is expected to occur, at high densities Bardeen-Schrieffer-Cooper pairing (BCS) of continuum states happens. The dissolution of bound states is connected with the crossover from BEC to BCS.[51–53] These effects are described solving the two-nucleon wave equation (8).

However, in symmetric matter the formation of a BEC of deuterons interferes with the BEC of α particles (quartetting) which are strongly bound (7.1 MeV/A for α in contrast to 1.1 MeV/A for d) so that, at finite temperature, with increasing chemical potential (increasing density) the BEC of α particles occurs prior to the BEC of deuterons.[89,90] The BEC of α particles disappears abruptly when the bound states are dissolved owing to the Pauli blocking, which follows from the solution of the in-medium wave equation (4) for the special case $A = 4$, $E_\alpha = \mu_4$.

Quantum condensates are not rigorously incorporated in present standard EoS. Whereas pairing is rather well described, the transition from α matter to nuclear matter at low temperatures is not clearly described until now. Experimentally, signatures of pairing (rigorously defined for infinite matter) are seen in the even-odd staggering of the binding energies of nuclei. Signatures of quartetting have been identified for the Hoyle state of ^{12}C,[92] see the following section 3.5.

3.5. *Nuclear structure*

Results derived for homogeneous nuclear matter cannot directly be applied to finite nuclei. However, within a density functional approach, the local density approximation uses these results. Neglecting any correlations, the mean-field approximation [(1, medium) of Table 1] has been applied very successful as shell model of nuclei. The many-nucleon wave function is approximated by the antisymmetrized product of single-nucleon states which are not momentum eigenstates as in homogeneous matter but have to be determined self-consistently.

Inclusion of correlations is possible, for instance, by superposition of shell-model states which needs some effort to realize a cluster structure. Other approaches implement the formation of clusters from the beginning, for instance the resonating group method and related approaches, for a review see Ref. 91. Numerical simulations such as fermionic/antisymmetrized molecular dynamics (FMD/AMD) are at present restricted to small numbers ($A \leq 12$) of nucleons.

There is clear evidence for clustering in nuclei if the density is low. We give some examples for α clustering in nuclei. Symmetric $4n$ nuclei (^{8}Be, ^{12}C, ^{16}O, etc.) in dilute gas states near the $n\alpha$ threshold have been investigated, experimentally and by theory. A famous example is the Hoyle state which is an excited state of ^{12}C, see Refs. 91–93. Antisymmetrization of the total wave function (Pauli blocking) is responsible that in contrast to the low-density Hoyle state (large rms radius $\approx$ 4.29 fm) where α-like clusters are well established, in the dense ground state ^{12}C (rms radius ≈ 2.65 fm) the uncorrelated product state (Slater determinant) is a good approximation. Clustering is also visible in low-lying breathing excitations of nuclei as well as in nuclear reactions. Another signature of quartetting my be the Wigner contribution[94] to the binding energy of near $Z = N$ nuclei.

A long-standing issue of correlations in nuclei is the α decay of heavy nuclei where a preformation of the α particle is assumed, see Refs. 95, 96 and further references given there. α-like correlations appear in the low-density tails at the surface of heavy nuclei which are α emitters. For example, in ^{212}Po a quartet $\{n_\uparrow, n_\downarrow, p_\uparrow, p_\downarrow\}$ moves on top of the double-magic ^{208}Pb core nucleus. Outside a critical radius $r_{\rm cr} = 7.44$ fm the nucleon density of the core nucleus is small, $n_B < 0.03$ fm^{-3}, so that an α-like bound state can be formed, whereas inside the core nucleus the quartet is described by uncorrelated single-nucleon states. Improving the local density approach, a rigorous description of the quartet state moving on top of a core nucleus has to be worked out in future.

3.6. *Astrophysics*

An important application for the nuclear matter EoS is the structure and the evolution of compact objects. Simulations of supernova explosions are performed recently to compare with observed signals. The thermodynamic parameters of the scenario of core collapse, Figure 2, are found in the warm dense matter regime. Cluster formation is relevant to reproduce a consistent EoS.[20]

In particular, the physics of core collapse supernovae enters the parameter region where cluster formation with $A \leq 4$ in the subsaturation region occurs.[14] The presence of clusters modifies the thermodynamic properties and affects, for instance, the neutrino transport.[21,26,27,97] Whereas previous approaches[18,19] considered only α particle formation, recently also other light elements have been taken into account, within a quantum statistical model[20] or using the excluded volume concept.[25]

A detailed knowledge about the supernova explosion including the neutrino transport is necessary to answer different questions such as the emission of matter by explosions or the cooling rates of pre-neutron stars, which is influenced by cluster formation and the occurrence of quantum phases. The formation of heavy elements is an essential, yet unsolved problem in astrophysics. It needs a hot, neutron-rich environment. At present it remains unclear to determine the site, for instance supernova explosions or neutron star mergers, where the r-process occurs.[98,99]

Another topic is the structure of neutron stars. Parameter values $\{T, n_B, Y_p\}$ appear in the crust where cluster formation occurs. Heavy nuclei are immersed in an environment consisting mainly of neutrons and electrons. At increasing density, pasta phases are formed. The crust/core transition is presently discussed,[34] in particular the existence of a first-order phase transition. Inside the core, clusters (nuclei) are dissolved because of Pauli blocking.

4. Results and discussion

Based on a quasiparticle concept, the present work aims at deriving the EoS for warm dense matter in the subsaturation region, incorporating the known low-density virial expansions as well as mean-field theories near saturation density. Different ingredients have been used:

(i) A cluster-virial expansion describes not only the different bound state contribution to the EoS, like the NSE, but takes into account also the contribution of the continuum. For $A = 2$, according to the Beth-Uhlenbeck formula the contribution of the continuum is given by the two-nucleon scattering phase shifts. Introducing single-nucleon quasiparticle energies, double counting of the mean-field terms has to be avoided.

(ii) Medium modified bound state energies and scattering phase shifts are used. They result from the solution of a few-nucleon wave equation which contains mean-field single-nucleon energy shifts as well as Pauli blocking terms. Both should be calculated taking into account self-consistently correlations and bound state formation in the medium. Analytical expressions are found in Ref. 3.

(iii) In homogeneous (stellar) matter, screening of the Coulomb interaction contributes to the medium modification of quasiparticle energies, in particular for large clusters, and pasta-like structures in the phase transition region.

Whereas for charged particle systems the Coulomb interaction is exactly known and a systematic quantum statistical treatment is well elaborated, see, e.g., Ref. 2, a fundamental expression for the N-N interaction is not at our disposal. Nevertheless, the quasiparticle properties are well defined from correlation functions which can, in principle, be measured. A semi-empirical approach has been used to calculate the quasiparticle properties after introducing an effective N-N interaction adjusted to known properties of nuclear systems.

In the zero-density limit, we can avoid the solution of the A-nucleon wave equation (4) using the empirical energies $E^0_{A\nu P}$[65] to evaluate the EoS (9) resulting in the NSE. The second virial coefficient is determined by the measured scattering phase shifts.

Similarly, empirical data for properties near the saturation density are used to parametrize the single-nucleon quasiparticle energy shifts $E_\tau(\mathbf{p}; T, n_B, Y_p)$ by a Skyrme force or RMF expressions. The same is also possible for the quasiparticle energy shifts $E_{A\nu P}(T, n_B, Y_p)$ of the light clusters d, t, h, α where empirical values for the rms radius or more details about the wave function can be used.[71] Evidence for clustering at low densities and medium modification of the binding energy are obtained from HIC[12,74] and nuclear structure investigations.[91–93]

Less investigations have been performed for the light metals $5 \leq A \leq 11$. Because their binding energies are weak (^{8}Be is unbound) and strongly influenced by the medium, see,[73] their abundances are strongly reduced if comparing with the NSE, see the results of HIC experiments[100] as well as calculations for stellar matter[101,102] and for cosmic rays.[103,104]

Comparing with the light elements, the internal structure of heavy elements $A \geq 12$ (including excited states) is not drastically influenced by medium effects. The interaction with the surrounding nucleons is determined by Pauli blocking which may be reflected, e.g., by the concept of excluded volume. In addition to the strong interaction also the Coulomb interaction has to be treated, see Sec. 2.6. In particular, it is of relevance for large clusters and for pasta structures in the region of phase instability. Future investigations are necessary the include heavy elements as well as pasta structure formation, especially in the region which is characterized by the thermodynamic instability of symmetric nuclear matter. Another challenging issue for a general EoS is the account of quantum condensates (pairing, quartetting) at low temperatures.

5. Outlook

For the interpretation of HIC results, the thermodynamic relations such as the EoS describing infinite systems in thermodynamic equilibrium are not directly connected with the measured cluster yields and their energy spectra because the nuclear system is inhomogeneous in space and, because of strong non-equilibrium, inhomogeneous in time. An adequate description should consider kinetic equations for the distribution functions (Wigner functions) of all clusters which have as equilibrium solutions not the ideal Fermi gas (as the Boltzmann equation) but an appropriate approximation of the EoS, see Table 2. Thus, the EoS containing quasiparticle clusters (medium-modified nuclei) may be considered as a prerequisite to formulate a transport code for nonequilibrium evolution. This is described by the extended von Neumann equation for the statistical operator $\varrho(t) = \lim_{\varepsilon \to 0} \varrho_\varepsilon(t)$,[105]

$$\frac{\partial}{\partial t}\varrho_\varepsilon(t) + \frac{i}{\hbar}\left[H, \varrho_\varepsilon(t)\right] = -\varepsilon\left(\varrho_\varepsilon(t) - \varrho_{\rm rel}(t)\right). \tag{37}$$

The relevant statistical operator $\varrho_{\rm rel}(t)$ is obtained from the maximum of entropy reproducing the local, time dependent composition with parameter values $T(\mathbf{r},t), \mu_n(\mathbf{r},t), \mu_p(\mathbf{r},t)$, but contains in addition the cluster distribution functions $f_{A\nu}^{\rm Wigner}(\mathbf{p},\mathbf{r},t)$ as relevant observables.[106,107]

Even if we can define a freeze-out state (i.e., temperature and chemical potentials) which determines the main features of the composition, further reaction and decay processes will occur before the cluster yields, observed in the detectors, are established. For instance, excited and unbound (e.g., ^{8}Be) nuclei will decay. An open question is the fate of n-n and $n-p$ continuum correlations after expansion of the fireball. Partially they contribute to the single nucleon distribution, but $n-p$ continuum correlations may also contribute to the deuteron distribution.[3] Future work is necessary to devise a transport theory for HIC which is compatible with the thermodynamic properties and the EoS, described in this work, as equilibrium solution.[108–110]

In this context it is also of interest to find optimal parameter sets $\{T, n_B, Y_p\}$ for the reproduction of observed abundances of clusters. This has been done for HIC where the yields of light clusters have been used to infer the thermodynamic parameter values.[11,111] In contrast to a simple chemical equilibrium such as the Albergo thermometer or densitometer which is connected with the ideal mixture of different components (NSE), density effects are of relevance. The freeze-out parameters represent a state during the evolution where local thermodynamic equilibrium ceases to be realized approximately. The further evolution is characterized by the different cluster distribution functions. It is determined by collisions, reactions, and decay processes.

This description can also be applied to astrophysical abundances of elements. Only the gross properties of elemental distribution are described by a freeze-out approach. Details are related to further reactions during cooling and expansion, forming local (with respect to the N-Z plane) deviations. Based on the cluster distribution functions $f_{A\nu}^{\rm Wigner}(\mathbf{p}, \mathbf{r}, t)$ as relevant observables, a reaction network can describe this stage of evolution of the nuclear system.

As an example, we can compare solar element abundances $X_A = n_A/n_B$[112] with calculations of the abundances for hot dense matter. Reasonable agreement with the gross behavior of the solar abundances was obtained with parameter values temperature $T = 5$ MeV, nucleon density $n_B = 0.016$ fm^{-3} .[5,73,101,102] Other parameter values, for instance $T = 5$ MeV, $n_B = 0.0156$ fm^{-3} as well as $T = 5.5$ MeV, $n_B = 0.0168$ fm^{-3} have been associated with the chemical composition of different stars.[104] Cosmic ray abundances was parametrized with a higher temperature $T = 5.88$ MeV, $n_B = 0.018$ fm^{-3}.[103] The asymmetry variable $Y_p = 0.5$ was not optimized.

Medium modifications are of relevance for those parameter values. Note that the origin of elements is not completely solved until now. In particular, the site where heavy elements are produced, is not identified yet. In this connection it is of interest to ask for parameter sets $\{T, n_B, Y_p\}$ which optimally reproduce the observed abundances.

Acknowledgments

The author is thankful for discussions with D. Blaschke, M. Hempel, J. Natowitz, W.U. Schroeder, A. Sedrakian, P. Schuck, S. Typel, and H.H. Wolter.

References

1. W. Ebeling, W.D. Kraeft, and D. Kremp, *Ergebnisse der Plasmaphysik und Gaselektronik*, Vol. V (Berlin: Akademie-Verlag 1976).
2. W.D. Kraeft, D. Kremp, W. Ebeling, and G. Röpke, *Quantum Statistics of Charge Particle Systems* (Berlin: Akademie-Verlag 1986).
3. G. Röpke, Phys. Rev. C **92**, 054001 (2015).
4. G. Röpke, Phys. Part. Nucl. **46**, 772 (2015).
5. G. Röpke, Wiss. Z. Univ. Rostock **33**, 33 (1984).

6. E.M. Burbidge, C.R. Burbidge, W.A. Fowler, and F. Hoyle, Rev. Mod. Phys. **29**, 547 (1957).
7. J. Knoll, L. Münchow, G. Röpke, and H. Schulz, Phys. Letters, B **112**, 13 (1982).
8. S. Nagamiya *et al.*, Phys. Rev. C, **24**, 971 (1981).
9. S. Das Gupta and A.Z. Mekjian, Phys. Rep. **72**, 131 (1981).
10. A.Z. Mekjian, Phys. Rev. C, **17**, 1051 (1978).
11. J. Natowitz *et al.*, Phys. Rev. Lett. **104**, 202501 (2010).
12. L. Qin *et al.*, Phys. Rev. Lett. **108**, 172701 (2012).
13. G.E. Brown, H.A. Bethe, and G. Baym, Nucl. Phys. A **375**, 481 (1982).
14. T. Fischer *et al.*, Astrophys. J. Suppl. **194**, 39 (2011).
15. T. Fischer *et al.*, Eur. Phys. J. A **50**, 46 (2014).
16. A.S. Botvina and I.N. Mishustin, Nucl. Phys. A **843**, 98 (2010).
17. D.H.E. Gross, Phys. Rep. **279**, 119 (1997).
18. J.M. Lattimer and F.D. Swesty, Nucl. Phys. **A 535**, 331 (1991).
19. H. Shen, H. Toki, K. Oyamatsu, and K. Sumiyoshi, Progr. Theor. Phys. **100**, 1013 (1998); Nucl. Phys. **A637** 435 (1998).
20. K. Sumiyoshi and G. Röpke, Phys. Rev. C **77**, 055804 (2008).
21. A. Arcones *et al.*, Phys. Rev. C **78**, 015806 (2008).
22. S. Heckel, P.P. Schneider, and A. Sedrakian, Phys. Rev. C **80**, 015805 (2009).
23. S. Typel *et al.*, Phys. Rev. C **81**, 015803 (2010).
24. Ad.R. Raduta and F. Gulminelli, Phys. Rev. C **82**, 065801 (2010).
25. M. Hempel and J. Schaffner-Bielich, Nucl. Phys. A **837**, 210 (2010).
26. S. Furusawa *et al.*, Astrophys. J. **738**, 178 (2011); **772**, 95 (2013).
27. E. O' Connor *et al.*, Phys. Rev. C **75**, 055803 (2007).
28. G. Shen, C.J. Horowitz, and S. Teige, Phys. Rev. C **83**, 035802 (2011).
29. M. Ferreira and C. Providencia Phys. Rev. C **85**, 055811 (2012).
30. S.S. Avancini *et al.* Phys. Rev. C **85**, 035806 (2012).
31. S. Talahmeh and H.R. Jaqaman, J. Phys. G: Nucl. Part. Phys. **40**, 015103 (2013).
32. A. Abdul-Rahman *et al.*, J. Phys. G: Nucl. Part. Phys. **42**, 055111 (2015).
33. N. Buyukcizmeci *et al.*, Nucl. Phys. **A 907**, 13 (2013).
34. H. Pais, S. Chiacchiera, and C. Providencia, Phys. Rev. C **91**, 055801 (2015).
35. M. Hempel *et al.*, Phys. Rev. C **88**, 014906 (2013).
36. P. Papakonstantinou, J. Margueron, F. Gulminelli, and Ad. R. Raduta, Phys. Rev. C **88**, 045805 (2013).
37. F. Aymard, F. Gulminelli, and J. Margueron, Phys. Rev. C **89**, 065807 (2014).
38. M. Hempel, arXiv:1410.6337 [nucl-th]; Phys. Rev. C **91**, 055807 (2015).
39. N. Buyukcizmeci, A.S. Botvina, and I.N. Mishustin, Astrophys. J. **789**, 33 (2014).
40. S. Typel, M. Oertel, and T. Klähn, Phys. of Particles and Nuclei **46**, 633 (2015).
41. M.D. Voskresenskaya and S. Typel, Nucl. Phys. A **887**, 42 (2012).
42. G. Röpke, L. Münchow, and H. Schulz, Nucl. Phys. A **379**, 536 (1982).

43. G. Röpke, L. Münchow, and H. Schulz, Phys. Letters, B **110**, 21 (1982).
44. G. Röpke, M. Schmidt, L. Münchow, and H. Schulz, Nucl. Phys. A **399**, 587 (1983).
45. A.L. Fetter, J.D. Walecka, *Quantum Theory of Many-Particle Systems* (McGraw-Hill, New York, 1971).
46. A.A. Abrikosov, L.P. Gorkov, and D.I. Dzyaloshinski, *Methods of Quantum Field Theory in Statistical Physics* (Dover, New York, 1975).
47. G. Röpke, M. Schmidt, and H. Schulz, Nucl. Phys. A **424**, 594 (1984).
48. G. Röpke, N.-U. Bastian, D. Blaschke, T. Klähn, S. Typel, and H.H. Wolter, Nucl. Phys. A **897**, 70 (2013).
49. M. Schmidt *et al.*, Ann. Phys. **202**, 57 (1990).
50. Meng Jin, M. Urban, and P. Schuck, Phys. Rev. C, **82**, 024911 (2010).
51. T. Alm, G. Röpke, and M. Schmidt, Z. Phys. A (Hadrons and Nuclei) **337**, 355 (1990).
52. H. Stein, A. Schnell, T. Alm, and G. Röpke, Z. Phys. A **351**, 295 (1995).
53. M. Baldo, U. Lombardo, and P. Schuck, Phys. Rev. C **52**, 975 (1995).
54. L.P. Kadanoff and G. Baym, *Quantum Statistical Mechanics* (New York: Benjamin 1962).
55. D.Q. Lamb, J.M. Lattimer, C.J. Pethick, and D.G. Ravenhall, Phys. Rev. Letters **41**, 1623 (1978); Nucl. Phys. A **360**, 459 (1981).
56. M. Barranco and J.R. Buchler, Phys. Rev. C, **22**, 1729 (1980); **24**, 827 (1981); Ap. J. (Letters), **245**, L109 (1981).
57. B. Friedman and V.R. Pandharipande, Nucl. Phys. A **361**, 502 (1981).
58. B.K. Jennings, S. Das Gupta, and N. Mobed, Phys. Rev. C, **25**, 278 (1982).
59. G. Röpke, Phys. Letters, B **121**, 223 (1983).
60. M.F. El Eid and W. Hillebrandt, Astr. Ap. Suppl. **42**, 215 (1980).
61. T.J. Mazurek, J.M. Lattimer, and G. E. Brown, Ap. J. **229**, 713 (1979).
62. L. Münchow *et al.*, J. Phys, G: Nucl. Phys. **8**, L135 (1982).
63. H. Schulz, L. Münchow, G. Röpke, and M. Schmidt, Phys. Letters, B **119**, 12 (1982).
64. H. Schulz, G. Röpke, and M. Schmidt, Z. Phys. A — Atoms and Nuclei **310**, 139 (1983).
65. G. Audi, A. Wapstra, and C. Thibault, Nucl. Phys. **729**, 337 (2003).
66. C. Fuchs and H.H. Wolter, Eur. Phys. J. A **30**, 5 (2006).
67. D. Vautherin and D.M. Brink, Phys. Rev. C, **5**, 626 (1972).
68. M. Dutra *et al.*, Phys. Rev. C **85**, 035201 (2012).
69. S. Typel, Phys. Rev. C **71**, 064301 (2005).
70. S. Typel and H.H. Wolter, Nucl. Phys. A **656**, 331 (1999).
71. G. Röpke, Nucl. Phys. A **867**, 66 (2011).
72. G. Röpke, Phys. Rev. C **79**, 014002 (2009).
73. G. Röpke, J. Phys.: Conf. Series **436**, 012070 (2013).
74. M. Hempel *et al.*, Phys. Rev. C **91**, 045805 (2015).
75. V.V. Burov, Yu.N. Eldyshev, V.K. Lukyanov, and Yu.S. Pol, Dubna-preprint E4-8029, Joint Institute for Nuclear Research, Dubna 1974.

76. A. Bohr and B. Mottelson, *Nuclear Structure* (New York, Amsterdam: Benjamin 1963).
77. G. Röpke, Wiss. Z. Univ. Rostock **31**, 61 (1982).
78. C.J. Horowitz and A. Schwenk, Nucl. Phys. **A 776**, 55 (2006).
79. K. Hagel *et al.*, Phys. Rev. Lett. **108**, 062702 (2012).
80. W.A. Küpper, G. Wegmann, and E.R. Hilf, Ann. Phys. **88**, 454 (1974).
81. J. Tõke and W.U. Schröder, Phys. Rev. C **79**, 064622 (2009).
82. J. Tõke and W.U. Schröder, Phys. Lett. B **734**, 364 (2014).
83. J.R. Buchler and R.I. Epstein, Ap. J. (Letters) **235**, L91 (1980).
84. B.A. Li *et al.*, Phys. Rep. **464**, 113 (2008).
85. K. Hagel, J.B. Natowitz, and G. Röpke, Eur. Phys. J. A **50**, 39 (2014).
86. R. Wada *et al.*, Phys. Rev. C **85**, 064618 (2012).
87. S. Kowalski *et al.*, Phys. Rev. C **75**, 014601 (2007).
88. G. Watanabe *et al.*, Phys. Rev. Lett. **103**, 035802 (2006).
89. G. Röpke, A. Schnell, P. Schuck, and P. Nozières, Phys. Rev. Lett. **80**, 3177 (1998).
90. T. Sogo, R. Lazauskas, G. Röpke, and P. Schuck, Phys. Rev. C **79**, 051301 (2009).
91. Y. Funaki, H. Horiuchi, and A. Tohsaki, Prog. Part. Nucl. Phys. **82**, 78 (2015).
92. A. Tohsaki, H. Horiuchi, P. Schuck, G. Röpke, Phys. Rev. Lett. **87**, 192501 (2001).
93. Y. Funaki *et al.*, Phys. Rev. C **77**, 064312 (2008).
94. G. Röpke, A. Schnell, P. Schuck, and U. Lombardo, Phys. Rev. C **61**, 024306 (2000).
95. G. Röpke *et al.*, Phys. Rev. C **90**, 034304 (2014).
96. Chang Xu *et al.*, Phys. Rev. C **93**, 011306 (2016).
97. S. Nasu *et al.*, Astrophys. J. **801**, 78 (2015).
98. A. Arcones *et al.*, Astr. Astrophys. **522**, A25 (2010).
99. O. Korobkin *et al.*, Monthly Not. Roy. Astr. Soc. **426**, 1940 (2012).
100. G. Röpke *et al.*, Phys. Rev. C **31**, 1556 (1985).
101. G. Röpke, Phys. Lett. B **185**, 281 (1987).
102. G. Röpke, Wiss. Z. Univ. Rostock **32**, 30 (1983).
103. G. Röpke, D. Blaschke, and H. Schulz, Astrophys. Space Sci. **95**, 417 (1983).
104. D. Blaschke and G. Röpke, Wiss. Z. Univ. Rostock **32**, 34 (1983).
105. D.N. Zubarev, V.G. Morozov, and G. Röpke, *Statistical Mechanics of Nonequilibrium Processes*, vol. 1,2, (Akademie-Verlag Berlin, 1996).
106. D.N. Zubarev, V.G. Morozov, I.P. Omelyan, and M.V. Tokarchuk, Theoret. Math. Phys. **96**, 997 (1993).
107. G. Röpke and H. Schulz, Nucl. Phys. A **477**, 472 (1988).
108. L.P. Csernai and J.I. Kapusta, Phys. Rep. **131**, 223 (1986).
109. P. Danielewicz and Q. Pan, Phys. Rev. C **46**, 2002 (1992).
110. C. Kuhrts *et al.*, Phys. Rev. C **63**, 034605 (2001).
111. G. Röpke *et al.*, Phys. Rev. C **88**, 024609 (2013).
112. A.G.W. Cameron, Space Sci. Rev. **15**, 121 (1973).

CHAPTER 3

NUCLEAR CLUSTERING IN FERMIONIC MOLECULAR DYNAMICS

Hans Feldmeier*,†,‡ and Thomas Neff*,§

* *GSI Helmholtzzentrum für Schwerionenforschung GmbH, Planckstraße 1, 64291 Darmstadt, Germany*
† *Frankfurt Institute for Advanced Studies, Max-von-Laue Straße, 60348 Frankfurt, Germany*
‡ *h.feldmeier@gsi.de*
§ *t.neff@gsi.de*

Clustering plays an important role in the structure of nuclei, especially for light nuclei in the p-shell. In nuclear cluster models these degrees of freedom are introduced explicitly. In the Resonating Group Method or in the Generator Coordinate Method the clusters are built from individual nucleons interacting via an effective nucleon-nucleon interaction; the total wave function is antisymmetrized. Fermionic Molecular Dynamics (FMD) goes beyond pure cluster models. It is a microscopic many-body approach using a Gaussian wave packet basis that includes the harmonic oscillator shell model and Brink-type cluster model wave functions as special cases. Clustering is not imposed but appears dynamically in the calculations. The importance of clustering for the understanding of bound states, resonances and scattering states is illustrated with examples discussing the charge radii of the Neon isotopes, the ^{3}He(α,γ)^{7}Be capture reaction and the cluster states in the ^{12}C continuum.

1. General considerations

1.1. *Clusters, thresholds and the resonating group method*

Let us first look at the textbook example for a nuclear cluster, the atomic nucleus of ^{4}He or α-particle. It is well bound with a binding energy of $E_1^{0^+} = -28.30$ MeV. As the ^{4}He nucleus is a rather compact system with a charge radius of about 1.67 fm all nucleons are so close that for each of the six possible pairs the nucleons are within the range of their nuclear interaction. When all pairs occupy the same $l = 0$ state, i.e., the coordinate

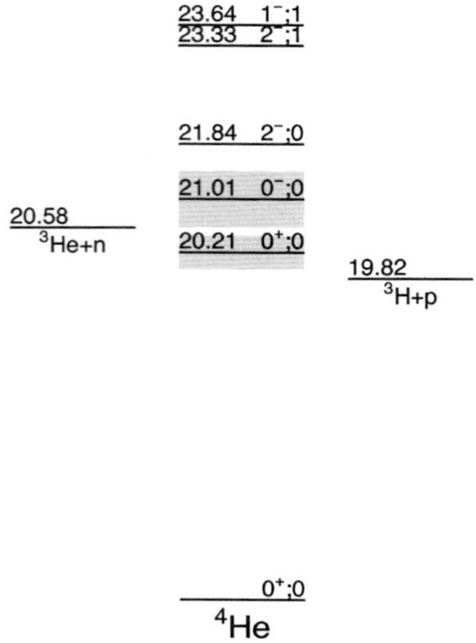

Fig. 1. Energy spectrum of ^{4}He with thresholds. Labels indicate J^π; T, energies in MeV (adapted from TUNL). For the resonances at 20.11 and 21.01 MeV the widths are indicated explicitly by shaded areas. The resonances above are rather broad and their widths overlap. The position of the ground state is not to scale.

space part is completely symmetric under particle permutations, the Pauli principle allows three pairs of nucleons with $S = 1$, $T = 0$ and three pairs with $S = 0$, $T = 1$. On the other hand these two channels exhibit the strongest attraction in the nucleon-nucleon force. Therefore it is not surprising that the ground state is tightly bound and when calculated with a realistic interaction it consists to about 85% of this simple configuration, which in shell model language corresponds to filling the lowest $0s$-shell with four nucleons.

The other important aspect that makes the α-particle an ideal cluster is that there is no further bound state in ^{4}He. The first excited states appear above 20 MeV and are already resonances, $E_2^{0^+} = 20.21$ MeV, width $\Gamma = 0.5$ MeV and $E_1^{0^-} = 21.01$ MeV, $\Gamma = 0.84$ MeV (see Fig. 1). This implies that the α-particle is not easily excited and difficult to polarize. Therefore one expects that the α-cluster structure survives as a four-body correlation also in heavier nuclei.

A special role is reserved for α-nuclei that have an equal even number of protons and neutrons and can be thought of being composed of $A/4$ α-clusters.[1] The idea of finding pronounced α-cluster states at the corresponding thresholds is nicely summarized in the Ikeda diagram.[2]

Generally one encounters noticeable cluster structures at energies in the vicinity of thresholds for breakup of a nucleus into various clusters, $C \to A + B$. Let us first consider the case where the energy E is just above the lowest threshold ($E > E(A) + E(B) - E(C)$) where the eigenstates of the A-body system C describe elastic scattering $A + B \to A + B$.

At large distance $|\vec{r}_{AB}| = |\vec{r}_A - \vec{r}_B|$ between A and B it is quite natural to describe the scattering system with RGM states

$$|\Phi_C\rangle = \int d^3 r_{AB} \, |\Phi_{AB}; \vec{r}_{AB} \rangle \varphi_{AB}(\vec{r}_{AB}), \tag{1}$$

where $\varphi_{AB}(\vec{r}_{AB})$ represents the wave function characterizing the relative motion between the center-of-mass (c.m.) positions of clusters A and B. The wave function of the many-body RGM basis state $\big| \Phi_{AB}; \vec{r}_{AB} \big\rangle$ labeled by the relative distance $\vec{r}_{AB} = \vec{r}_A - \vec{r}_B$ between the c.m. of cluster A and B

$$\big\langle \xi_A, \xi_B, \vec{r} \,\big|\, \Phi_{AB}; \vec{r}_{AB} \big\rangle = \widehat{\mathcal{A}} \left\{ \Phi_A(\xi_A) \Phi_B(\xi_B) \, \delta^3(\vec{r} - \vec{r}_{AB}) \right\} \tag{2}$$

is an antisymmetrized product of a localized wave function for the relative motion times the intrinsic wave functions $\Phi_A(\xi_A)$ and $\Phi_B(\xi_B)$. The antisymmetrizer $\widehat{\mathcal{A}}$ projects on a state that is antisymmetric under all particle permutations, $(\widehat{\mathcal{A}}^2 = \widehat{\mathcal{A}}, \widehat{\mathcal{A}}^\dagger = \widehat{\mathcal{A}})$. $\Phi_A(\xi_A)$ and $\Phi_B(\xi_B)$ denote respectively intrinsic eigenstates of A and B in terms of their sets of intrinsic variables

$$\xi_A = \{\vec{\xi}_i; i = 1, \ldots, A_A - 1\} \cup \{\sigma_i, \tau_i; i = 1, \ldots, A_A\}$$

and

$$\xi_B = \{\vec{\xi}_j; j = A_A + 1, \ldots, A_C - 1\} \cup \{\sigma_j, \tau_j; j = A_A + 1, \ldots, A_C\}.$$

$A_C = A_B + A_B$ denote the mass numbers. There are altogether $A_C - 1$ independent coordinates, $A_A - 1$ in cluster A and $A_B - 1$ in cluster B plus the relative distance $\vec{r}$. Due to translational invariance there is no dependence on the total center-of-mass coordinate. Therefore in the following it will not be mentioned anymore, except when needed. Furthermore angular momentum coupling is not explicitly denoted here, but of course used in calculations.

As the antisymmetric RGM wave function $\big\langle \xi_A, \xi_B, \vec{r} \,\big|\, \Phi_{AB}; \vec{r}_{AB} \big\rangle$ is not a single Slater determinant it is not an easy task to perform the antisymmetrization for larger particle numbers. For example transposing a nucleon from A with one from B affects in general all intrinsic positions $\vec{\xi}_i$ in set ξ_A and all $\vec{\xi}_j$ in set ξ_B as they are defined in a translationally invariant way with respect to the distance $\vec{r}_A - \vec{r}_B$ of the clusters A and B.

The acronym RGM stands for Resonating Group Method[3,4] introduced by J. A. Wheeler in 1937 where he proposed an antisymmetric many-body state consisting of groups of nucleons (clusters) that have a fixed internal structure but can move with respect to each other.

For energies above a threshold and at distances outside the range of the nuclear interaction between nucleons belonging to A or B the relative wave function $\varphi_{\mathrm{AB}}(\vec{r}_{\mathrm{AB}})$ has to match continuously the ingoing and outgoing Coulomb scattering states with the appropriate phase shift. Or in case of a narrow resonance described by a Gamow state it has to match an outgoing Coulomb wave with a complex relative momentum k, which encodes the width Γ.

Being totally antisymmetric under particle permutations, the many-body basis state $\left| \Phi_{\mathrm{AB}}; \vec{r}_{\mathrm{AB}} \right\rangle$ is a legitimate state at all $\vec{r}_{\mathrm{AB}}$, but for large overlaps of A and B at small $\vec{r}_{\mathrm{AB}}$, where the Pauli principle strikes, the antisymmetric part left over after the projection with $\widehat{\mathcal{A}}$ may be rather small. For a resonance, where the clusters A and B merge and stay together for a prolonged time, one expects in the interior the energy eigenstate to resemble more compact configurations typical for the compound nucleus C, while for intermediate distances one anticipates that deformed or polarized clusters A and B mix in. Finally the clusters separate at large distances as RGM states being in their ground states (or excited states if energy conservation allows).

In ^{4}He one encounters such a situation for

$$E(^3\mathrm{H}) + E(\mathrm{p}) - E(^4\mathrm{He}) \;<\; E \;<\; E(^3\mathrm{He}) + E(\mathrm{n}) - E(^4\mathrm{He})$$

(see Fig. 1). The 4-body system forms a resonance at $E(0_2^+) = 20.21$ MeV, width $\Gamma = 0.5$ MeV, very close to the threshold. Fig. 1 also reveals that there is a second threshold at $E(^3\mathrm{He}) + E(\mathrm{n}) - E(^4\mathrm{He}) = 20.58$ MeV just above $E(0_2^+) = 20.21$ MeV. This is quite an interesting situation as the channel $^3\mathrm{He} + \mathrm{n}$ is not open yet, but very close in energy. Because the two thresholds are so close we expect mixing between the two cluster configurations such that at larger distances the RGM wave function looks like a coupled channel one, namely

$$\left| \Phi_{^4\mathrm{He}} \right\rangle \overset{\text{large distance}}{\Longrightarrow} \int d^3r \left| \Phi_{^3\mathrm{H}} \Phi_{\mathrm{p}}; \vec{r} \right\rangle \varphi_{^3\mathrm{H},\mathrm{p}}(\vec{r}) + \int d^3r \left| \Phi_{^3\mathrm{He}} \Phi_{\mathrm{n}}; \vec{r} \right\rangle \varphi_{^3\mathrm{He},\mathrm{n}}(\vec{r}). \tag{3}$$

For simplicity all spin and angular momentum couplings are not explicitly denoted. At large distances $\vec{r}$ the wave function has at the resonance energy the structure of a ^{3}H + p scattering state with phase shifted Coulomb wave functions $\varphi_{^3\mathrm{H,p}}(\vec{r})$. Getting closer the second component ^{3}He + n contributes with an exponentially decaying relative wave function $\varphi_{^3\mathrm{He,n}}(\vec{r}) \propto \exp(-k_{^3\mathrm{He,n}} r)$, where the relative momentum $k_{^3\mathrm{He,n}} = \sqrt{-2\mu\Delta}$ is given by the reduced mass μ and the energy difference $\Delta = E - E(^3\mathrm{He}) - E(\mathrm{n})$ to the threshold in this channel. For the second resonance with quantum numbers 0_1^-, where E is above both thresholds, both channel wave functions have to be matched to phase shifted scattering solutions for negative parity.

To summarize: If the many-body system has an eigenenergy just below a breakup threshold the nuclear system already "feels" the continuum cluster structure in the outer part of the nucleus and the nucleons condense into the respective clusters. As we will show later in actual calculations these cluster structures can prevail to rather small distances. If the eigenenergy slips above the threshold one gets a resonance with an asymptotic wave function describing the two clusters in relative motion. The inner region of the A-body system may, but need not, resemble the cluster structure. One should also keep in mind that the antisymmetrizer $\widehat{\mathcal{A}}$ blurs the picture for strongly overlapping clusters.

1.2. *Role of antisymmetrization*

We discuss now the effect of the antisymmetrizer and then introduce the Generator Coordinate Method (GCM) for setting up a Slater determinant basis in A-body space that is numerically less painful than the RGM basis. We shall illustrate this with the example of ^{8}Be, where there is only one threshold in the vicinity, so that coupled channels are not needed. From Fig. 2 one sees that the ground state of ^{8}Be is only 91.8 keV above the ^{4}He + ^{4}He threshold ($\alpha + \alpha$) and has the tiny width of $\Gamma = 5.6$ eV. The next threshold ^{7}Li + p at 17.26 MeV will not influence the structure of the low lying resonances.

The structure of ^{8}Be is a textbook example of clusterization. Having 4 neutrons and 4 protons the nuclear interaction chooses to first condense four nucleons into 2 strongly bound α-clusters, respectively, thus gaining two times -28.30 MeV, and then arranging the relative motion of two α-clusters such that the sum of kinetic energy, Coulomb repulsion and nuclear attraction is minimized. This implies that for a given inter-cluster potential the spatial localization of $\varphi_{\alpha\alpha}(\vec{r})$ should be washed out as much as possible to keep the kinetic energy low. The localization in orientation $\hat{r}$ is completely removed by putting the relative motion into an $L = 0$ state.

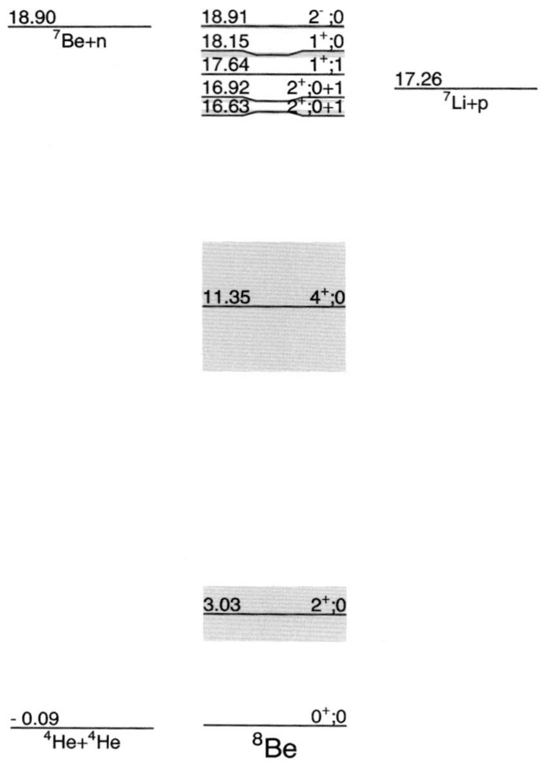

Fig. 2. Energy spectrum of ^{8}Be with thresholds. Labels indicate J^π, T, energies in MeV, widths are indicated by shaded areas (adapted from TUNL).

The RGM wave function denoting explicitly the couplings takes the form

$$\big\langle \xi_{\rm A}, \xi_{\rm B}, \vec{r} \,\big|\, \Phi_{\rm AB}^{(IL)J^\pi M} \big\rangle = \widehat{\mathcal{A}} \left\{ \left[\left[\Phi_{\rm A}^{I_{\rm A}}(\xi_{\rm A})\, \Phi_{\rm B}^{I_{\rm B}}(\xi_{\rm B}) \right]^I Y^L(\hat{r}) \right]^{J^\pi M} \varphi_{\rm AB}^L(r) \right\}, \tag{4}$$

where $r = |\vec{r}|$ and $\hat{r}$ denotes the direction of $\vec{r}$. In the case of ^{8}Be the wave functions $\Phi_{\rm A}^{I_{\rm A}}(\xi_{\rm A})$ and $\Phi_{\rm B}^{I_{\rm B}}(\xi_{\rm B})$ are α-cluster states. Their intrinsic spins $I_{\rm A} = I_{\rm B} = 0$ can only be coupled to the channel spin $I = 0$ which in turn is coupled with the angular momentum L of the relative motion to the total spin J. In this simple case $J = L$ and $\pi = (-1)^L$, where only even L are allowed because A and B are identical bosons.

There are also the typical shell model states where 2 protons and 2 neutrons are in the s-shell and the remaining nucleons in the p-shell. Such shell model states are seen as excited states above 16 MeV (cf. Fig. 2) with a small decay width because their structure is quite different so that they cannot easily decay into α-particles. Thus, first condensing four nucleons into α-clusters and then arranging the two α-clusters in an optimal way brings an energy advantage of 16 MeV over putting all 8 nucleons in a common mean field.

It should be noted that there is a large overlap between shell model and cluster model wave functions for the ground state band members. The cluster model wave functions at small distances are equivalent to shell model wave functions because in the limit $r \to 0$ antisymmetrization projects onto harmonic oscillator states (see Sec. 1.3 and Fig. 4). Therefore shell model configurations with $S = 0$ and $T = 0$ mix with the spatially extended cluster configurations where the energy is lowered by delocalization in the relative motion of the α-clusters. At variance with that, the states above 16 MeV that have a spin-flip nature with $S = 1$ or $T = 1$ quantum numbers are shell model configurations that do not mix with the cluster configurations.

In the 3-α nucleus ^{12}C the mean field or shell model configuration prevails, partly due to the contribution of the spin-orbit force, in the ground state but some reminiscence of cluster structure remains, as is also the case in ^{20}Ne and ^{24}Mg. In these nuclei α-cluster state appear as excited states near the energy of n α particles (see the discussion on ^{12}C in Sec. 5).

There is an important issue in the interpretation of $\varphi^0_{\alpha\alpha}(r)$. When the two clusters are well separated the antisymmetrizer $\widehat{\mathcal{A}}$ acts as a unit operator but when for smaller r the clusters are overlapping $\widehat{\mathcal{A}}$ begins to project away the non-antisymmetric part of the A-body wave function resulting first in a reduced amplitude and then for $r \to 0$ to oscillations.

Figure 3 shows the radial part of the relative wave function $\varphi_{\alpha\alpha}(\vec{r}) = \varphi^0_{\alpha\alpha}(r)\, Y^0_0(\hat{r})$ for the ground state of ^{8}Be in terms of a ^{4}He + ^{4}He cluster wave function (Eq. (4)) for two examples. For the intrinsic state of ^{4}He the most simple ansatz is taken with all nucleons in $0s$ relative motion for an oscillator parameter reproducing the experimental r.m.s. radius.

In the case of two α-clusters the channel spin is $I = 0$ because both clusters have spin $I_{\mathrm{A}} = I_{\mathrm{B}} = 0$ and hence the relative orbital angular momentum $L = J$ so that the angular momentum projected norm kernel for the RGM basis states $\left|\Phi_{\alpha\alpha}; \vec{r}\right\rangle$ (cf. Eq. (1)) projected on $L = J$ takes

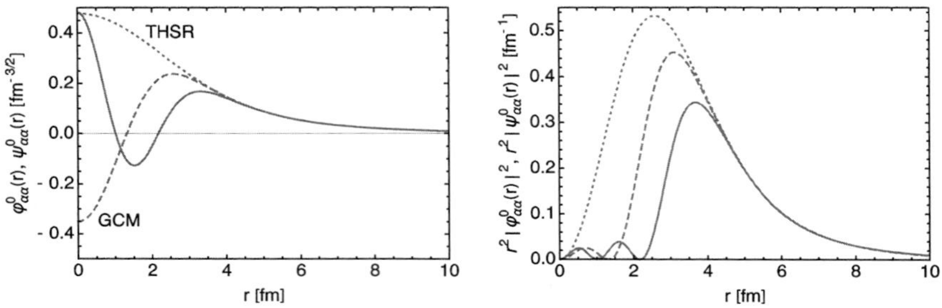

Fig. 3. (Left) GCM (dashed) and THSR (dotted) relative wave functions $\varphi^0_{\alpha\alpha}(r)$ before antisymmetrization and $\psi^0_{\alpha\alpha}(r)$ after antisymmetrization (coinciding full lines). (Right) Corresponding densities multiplied with r^2.

the form

$$n_J(r,r') = \int \mathrm{d}^2\Omega_r\, \mathrm{d}^2\Omega'_r \sum_M Y_M^{L=J}(\Omega_r) \left\langle \Phi_{\alpha\alpha};\vec{r} \,\middle|\, \Phi_{\alpha\alpha};\vec{r}' \right\rangle Y_M^{L=J}(\Omega'_r)^\star . \tag{5}$$

The wave function

$$\psi_{\alpha\alpha}^{L=J}(r) = \int_0^\infty r'^2 \mathrm{d}r'\, n_J^{1/2}(r,r')\, \varphi_{\alpha\alpha}^{L=J}(r') \tag{6}$$

takes the reduction due the Pauli principle into account and its absolute squared value represents the probability to find the α-clusters at distance r.

The RGM wave functions before, $\varphi^0_{\alpha\alpha}(r)$, and after antisymmetrization, $\psi^0_{\alpha\alpha}(r)$, shown in Fig. 3 solve the RGM equations with a nuclear Hamiltonian, that is they diagonalize the Hamiltonian in the many-body space spanned by the RGM basis states. One sees a vibrational zero-point motion of the α-clusters with a maximum probability around $r = 3.5$ fm and the almost exponential tail of the Whittaker function tunneling into the Coulomb barrier and reaching out to large distances.

The effects of antisymmetrization are seen when comparing $\varphi^0_{\alpha\alpha}(r)$ and $\psi^0_{\alpha\alpha}(r)$. They are quite different at small r and coincide at large r. The corresponding probabilities displayed on the right hand side of Fig. 3 show that at $r < 3.5$ fm a strong suppression sets in and below $r \approx 2.5$ fm the probability drops practically to zero. The antisymmetrized $\psi^0_{\alpha\alpha}(r)$ develops nodes reflecting the existence of Pauli-forbidden states.

At this point one should emphasize that $\varphi^0_{\alpha\alpha}(r)$ is not well defined for small r, which however does not matter because physical observables are not affected by this. This is illustrated by the fact that two quite different relative wave functions $\varphi^0_{\alpha\alpha}(r)$, one given by the THSR[5] ansatz which has no nodes and one which is obtained using GCM basis states, as described in the next section, with one node are mapped by the antisymmetrization onto to the very same $\psi^0_{\alpha\alpha}(r)$ (full line in Fig. 3). Therefore any reference to properties of the choice of $\varphi^0_{\alpha\alpha}(r)$ in the region where the clusters have strong overlap has no relevance.

For large r beyond $r \approx 5$ fm antisymmetrization is no longer essential and the norm kernel becomes local $n_J(r, r') \to N_J \delta(r - r')/(rr')$.

1.3. *Slater determinants and generator coordinate method*

The antisymmetrization which complicates the evaluation of Eq. (4) can be much more easily achieved if one uses Slater determinants to set up the many-body basis. These are often denoted as Generator Coordinate Method (GCM) states which are labeled and distinguished by generator coordinates. For example they can be generated by minimizing the energy of a Slater determinant in the Hartree-Fock method under a constraint that plays the role of the generate coordinate. Examples are various moments like mass quadrupole, octupole or charge dipole.

Another very simple example are Brink-type[6] cluster states that are the analogue to the RGM basis states discussed above. Here one does not use intrinsic states for the clusters but Slater determinants of localized single-particle states.

$$\left| \Psi_{\rm AB}; \vec{R}_{\rm A}, \vec{R}_{\rm B} \right\rangle = \widehat{\mathcal{A}} \Big\{ \left| \Psi_{\rm A}; \vec{R}_{\rm A} \right\rangle \otimes \left| \Psi_{\rm B}; \vec{R}_{\rm B} \right\rangle \Big\} \tag{7}$$

$\left| \Psi_{\rm A}; \vec{R}_{\rm A} \right\rangle$ and $\left| \Psi_{\rm B}; \vec{R}_{\rm B} \right\rangle$ are Slater determinants representing clusters A and B whose mean positions are shifted in space to $\vec{R}_{\rm A}$ and $\vec{R}_{\rm B}$, respectively. $\vec{R}_{\rm A}$ and $\vec{R}_{\rm B}$ are in this case the generator coordinates.

In general the wave function $\left\langle \xi_{\rm A}, \xi_{\rm B}, \vec{r}, \vec{r}_{cm} \,\middle|\, \Psi_{\rm AB}; \vec{R}_{\rm A}, \vec{R}_{\rm B} \right\rangle$ does not separate into intrinsic times relative motion as was the case for the RGM basis. Also the total c.m. motion can in general not be factorized out. This is a disadvantage when treating scattering problems where boundary conditions have to be imposed on the relative motion, but for bound states where the clusters have usually strong overlap and antisymmetrization matters both RGM and GCM states span very similar many-body Hilbert spaces. The total c.m. motion inherent in Slater determinants can be removed by means of projection. Narrow resonances may be treated in good approximation as bound states, for which a representation in terms of RGM states at large distances are not needed.

For ^{8}Be the most simple ansatz for the Brink state is a product of two α-clusters where the spatial part of the single-particle states are harmonic oscillator $0s$-states, $\exp\{-(\vec{r}_i - \vec{R}_{\mathrm{A,B}})^2/(2a)\}$, displaced by the respective mean positions, $\vec{R}_{\mathrm{A}}$ and $\vec{R}_{\mathrm{B}}$. For that simple case the many-body wave function can be written as

$$\begin{aligned}\big\langle \xi_{\mathrm{A}}, \xi_{\mathrm{B}}, \vec{r}, \vec{r}_{cm} \,\big|\, \Psi_{^8\mathrm{Be}}; \vec{R}_{\mathrm{A}}, \vec{R}_{\mathrm{B}} \big\rangle \\ = \widehat{\mathcal{A}}\Big\{ \Phi_\alpha(\xi_{\mathrm{A}})\, \Phi_\alpha(\xi_{\mathrm{B}})\, \varphi_{rel}(\vec{r}; \vec{R}^{rel}_{\mathrm{AB}}) \Big\}\, \varphi_{cm}(\vec{r}_{cm}; \vec{R}^{cm}_{\mathrm{AB}}) \end{aligned} \tag{8}$$

with relative and center-of-mass wave functions

$$\varphi_{rel}(\vec{r}; \vec{R}^{rel}_{\mathrm{AB}}) = \exp\left\{ -\frac{(\vec{r} - \vec{R}^{rel}_{\mathrm{AB}})^2}{2a/\mu_{\mathrm{AB}}} \right\}, \quad \vec{R}^{rel}_{\mathrm{AB}} = \vec{R}_{\mathrm{A}} - \vec{R}_{\mathrm{B}} \tag{9}$$

$$\varphi_{cm}(\vec{r}_{cm}; \vec{R}^{cm}_{\mathrm{AB}}) = \exp\left\{ -\frac{(\vec{r}_{cm} - \vec{R}^{cm}_{\mathrm{AB}})^2}{2a/(A_{\mathrm{A}} + A_{\mathrm{B}})} \right\}, \quad \vec{R}^{cm}_{\mathrm{AB}} = \frac{A_{\mathrm{A}}\vec{R}_{\mathrm{A}} + A_{\mathrm{B}}\vec{R}_{\mathrm{B}}}{A_{\mathrm{A}} + A_{\mathrm{B}}} \tag{10}$$

and the intrinsic wave functions

$$\Phi_\alpha(\xi_{\mathrm{A}}) = \widehat{\mathcal{A}} \prod_{i=1}^{4} \exp\left\{ -\frac{\vec{\xi}_i^{\,2}}{2a} \right\} \chi(\sigma_i, \tau_i) \quad \text{with } \vec{\xi}_i = \vec{r}_i - \frac{1}{A_{\mathrm{A}}} \sum_{l=1}^{A_{\mathrm{A}}} \vec{r}_l \tag{11}$$

$$\Phi_\alpha(\xi_{\mathrm{B}}) = \widehat{\mathcal{A}} \prod_{j=5}^{8} \exp\left\{ -\frac{\vec{\xi}_j^{\,2}}{2a} \right\} \chi(\sigma_j, \tau_j) \quad \text{with } \vec{\xi}_j = \vec{r}_j - \frac{1}{A_{\mathrm{B}}} \sum_{l=A_{\mathrm{A}}+1}^{A_{\mathrm{A}}+A_{\mathrm{B}}} \vec{r}_l, \tag{12}$$

which depend on the intrinsic coordinates $\vec{\xi}_i$ that are measured from the respective c.m. coordinates. The spin, isospin orientations are denoted by $\chi(\sigma_i, \tau_i)$.

Comparing to the RGM basis wave function in Eq. (2) one sees that the relative motion of the clusters is not sharply localized anymore but the δ-function is smeared out to the Gaussian $\varphi_{rel}(\vec{r}; \vec{R}^{rel}_{\mathrm{AB}})$ with mean position $\vec{R}^{rel}_{\mathrm{AB}}$ and a width given by the single-particle width parameter a divided by the reduced mass number μ_{AB} (= 2 in the example). One should be aware that the relative distance

$$\vec{r} = \frac{1}{A_{\mathrm{A}}} \sum_{i=1}^{A_{\mathrm{A}}} \vec{r}_i - \frac{1}{A_{\mathrm{B}}} \sum_{j=A_{\mathrm{A}}+1}^{A_{\mathrm{A}}+A_{\mathrm{B}}} \vec{r}_j$$

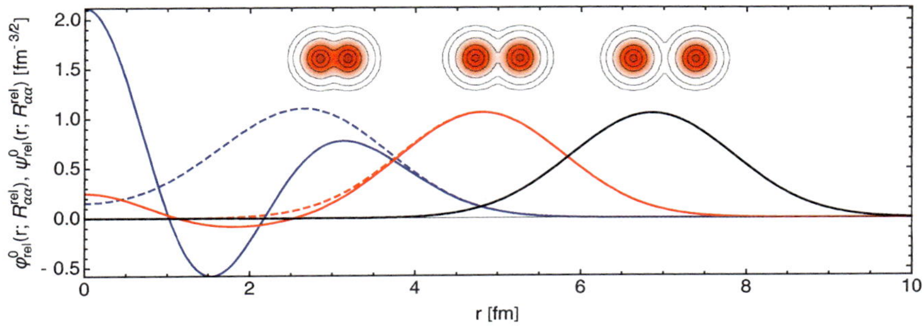

Fig. 4. Top: intrinsic one-body densities of Brink $\alpha - \alpha$ cluster states for mean distances $R^{rel}_{\alpha\alpha} = 3, 5, 7$ fm (blue, red, black). Bottom: corresponding relative wave functions $\varphi^0_{rel}(r; R^{rel}_{\alpha\alpha})$ before (dashed lines) and $\psi^0_{rel}(r; R^{rel}_{\alpha\alpha})$ after antisymmetrization (full lines) as function of distance r.

is not an independent variable anymore but depends on all single-particle positions. The second difference is that there occurs a total center-of-mass wave function, which depends on $\vec{r}_{cm} = \frac{1}{A_A + A_B} \sum_{i=1}^{A_A+A_B} \vec{r}_i$ and thus on all single-particle positions. This was not present in the RGM case. Because of translational invariance results do not depend on the overall center-of-mass motion and hence it was not considered at all in RGM.

In the general case the above factorization into intrinsic, relative, and c.m. motion does not hold. Slater determinants entangle in a spurious way intrinsic and c.m. motion, therefore in GCM this problem has to be addressed by means of projection on total c.m. momentum zero (see Sec. 2.2).

Figure 4 displays in the upper part the intrinsic one-body density distribution that is obtained when aligning $\vec{R}^{rel}_{AB}$ along the x-axis for $R^{rel}_{\alpha\alpha} \equiv |\vec{R}^{rel}_{AB}| = 3, 5, 7$ fm. In the lower part the corresponding relative wave functions $\varphi^0_{rel}(r; R^{rel}_{\alpha\alpha})$ and the antisymmetrized $\psi^0_{rel}(r; R^{rel}_{\alpha\alpha})$ are plotted as a function of r. They are obtained by projection of $\varphi_{rel}(\vec{r}; \vec{R}^{rel}_{\alpha\alpha})$ and $\psi_{rel}(\vec{r}; \vec{R}^{rel}_{\alpha\alpha})$ on $L = 0$. The relative wave packet at $R^{rel}_{\alpha\alpha} = 7$ fm is not influenced by antisymmetrization at all and the one at $R^{rel}_{\alpha\alpha} = 5$ fm very little. But the one for $R^{rel}_{\alpha\alpha} = 3$ fm, where the one-body density shows appreciable overlap of the α-clusters, is completely changed into an oscillating wave resembling a $2s$ harmonic oscillator state with two nodes. This way the Pauli principle is restored. It is interesting to note the similarity at small r to the antisymmetrized RGM state shown in Fig. 3. Apparently in the interior of the nucleus the antisymmetrization projects different wave functions onto the same state. The other information contained in this example is that for low energy a linear combination of GCM basis states taken at discrete values of $\vec{R}^{rel}_{\rm AB}$ will suffice to generate a smooth relative wave function that is completely equivalent to the RGM wave function.

To summarize: RGM states are easily matched to boundary conditions but increasingly difficult to antisymmetrize with growing particle number. GCM states are easy to antisymmetrize but are more difficult to match to scattering solutions. The GCM basis and the RGM basis are equivalent. For bound states and narrow resonances, treated in bound state approximation, the GCM basis is preferable because it is numerically easier, more flexible and can include also non-cluster states.

2. Fermionic molecular dynamics

Cluster models can only be understood as an approximation to a full description of the nucleus. The wave functions of the clusters are typically restricted to simple harmonic oscillator shell model functions. The description by purely cluster degrees of freedom is especially questionable at short distances where the clusters strongly overlap and where one expects a polarization of the clusters. Even in the case of ^{8}Be the α-cluster picture has to be modified. At short distances shell model configurations with a preferred occupation of $p_{3/2}$ over $p_{1/2}$ orbits due to the spin-orbit force should be important. In a pure cluster model such polarization and admixture effects are neglected and at best may be absorbed in the effective interaction. As these effects will strongly depend on the nucleus and even on the state under consideration it is not surprising that the effective interactions used in cluster models are not universal and are typically tuned for each nucleus. These issues become especially challenging for nuclei which are not α-nuclei.

Fermionic Molecular Dynamics (FMD)[7–11] is a many-body approach that can be understood as an extension of traditional cluster models. The effective degrees of freedom are here not restricted to the relative motion of a given set of clusters but the individual nucleons are considered as the degrees of freedom. The single-particle states are described by Gaussian wave-packets localized in phase space. In FMD clusters should emerge automatically as effective degrees of freedom if the system prefers a breakup into clusters. In this way FMD allows for a consistent description of clustering in all nuclei. An important ingredient is also the use of a realistic effective interaction that is not tuned to particular systems.

2.1. *Intrinsic states*

FMD is a fully microscopic many-body approach that uses Slater determinants as intrinsic many-body basis states

$$\left| Q \right\rangle = \widehat{\mathcal{A}}\{\left| q_1 \right\rangle \otimes \cdots \otimes \left| q_A \right\rangle\}. \tag{13}$$

The many-body state $\left| Q \right\rangle$ is given by A single-particle states $\left| q_k \right\rangle$

$$\left| q_k \right\rangle = \sum_{i=1}^{N} \left| a_{ki} \vec{b}_{ki} \right\rangle \otimes \left| \chi_{ki}^{\uparrow}, \chi_{ki}^{\downarrow} \right\rangle \otimes \left| \xi_k \right\rangle c_{ki}, \tag{14}$$

where N is typically one or two. The spatial part of the single-particle wave function is given by Gaussian wave packets

$$\langle \vec{x} \,|\, a, \vec{b} \rangle = \exp\left\{ -\frac{(\vec{x} - \vec{b})^2}{2a} \right\} \tag{15}$$

and a spin-part given by a spinor $\left| \chi^{\uparrow}, \chi^{\downarrow} \right\rangle$. A single-particle state describes either a proton or a neutron, given by $\left| \xi \right\rangle$. The parameters of the wave packet are the width a and the complex vector $\vec{b}$ that encodes the mean position and the mean momentum of the wave packet.

An important point for understanding the versatility of the FMD basis is that it contains both harmonic oscillator shell model and Brink-type cluster wave functions as special cases. Harmonic oscillator single-particle states can be obtained by linear combinations of Gaussians in the limit of zero displacement. These linear combinations are automatically included in the Slater determinant, as the Slater determinant is not changed under linear transformations of the Gaussian single-particle states. Brink-type cluster wave functions are generated by localizing clusters of nucleons in phase space.

Gaussian wave packets have also other favorable properties. Translations, boosts and rotations of Gaussian wave packets are again Gaussian wave packets (with transformed parameters). These properties of the Gaussian wave packets are inherited by the Slater determinants and it is straightforward to implement these transformations. Furthermore matrix elements of operators that are polynomial in coordinate or momentum space or have a Gaussian radial dependence can be evaluated analytically. A technical disadvantage is the fact that the Gaussian single-particle states are non-orthogonal which requires the calculation of overlap matrices.

2.2. *Restoration of symmetries*

The Slater determinants $\left| Q \right\rangle$ in general do not posses the symmetries of the Hamiltonian with respect to translations, rotations and reflections. These symmetries can be restored however by projection on parity, angular momentum, and total linear momentum:

$$\left| Q; J^\pi MK; \vec{P}=0 \right\rangle = \widehat{P}^\pi \widehat{P}^J_{MK} \widehat{P}^{\vec{P}=0} \left| Q \right\rangle, \tag{16}$$

with the projection operators for parity

$$\widehat{P}^\pi = \frac{1}{2}(1+\pi\widehat{\Pi}), \tag{17}$$

three dimensional angular momentum projection

$$\widehat{P}^J_{MK} = \frac{2J+1}{8\pi^2} \int d\Omega \; {D^J_{MK}}^\star(\Omega) \; \widehat{R}(\Omega) \tag{18}$$

with the rotation operator $\widehat{R}(\Omega)$ and the Wigner D-functions, and the projection on total linear momentum

$$\widehat{P}^{\vec{P}} = \frac{1}{(2\pi)^3} \int d^3X \; \exp\{-i(\widehat{\vec{P}} - \vec{P}) \cdot \vec{X}\}. \tag{19}$$

As the Hamiltonian commutes with the symmetry operations it is sufficient to project once and calculate projected matrix elements for intrinsic basis states $\left| Q^{(a)} \right\rangle$ and $\left| Q^{(b)} \right\rangle$ for the Hamiltonian

$$\mathsf{H}^{J^\pi}_{aK,bK'} = \left\langle Q^{(a)} \right| (\widehat{H} - \widehat{T}_{\mathrm{cm}}) \widehat{P}^\pi \widehat{P}^J_{KK'} \widehat{P}^{\vec{P}=0} \left| Q^{(b)} \right\rangle \tag{20}$$

and the overlap

$$\mathsf{N}^{J^\pi}_{aK,bK'} = \left\langle Q^{(a)} \right| \widehat{P}^\pi \widehat{P}^J_{KK'} \widehat{P}^{\vec{P}=0} \left| Q^{(b)} \right\rangle. \tag{21}$$

Multiconfiguration mixing eigenstates of the Hamiltonian are finally obtained by solving the generalized eigenvalue problem in a set of basis states $\{\left| Q^{(a)} \right\rangle\}$

$$\sum_{bK'} \mathsf{H}^{J^\pi}_{aK,bK'} \Psi^{J^\pi\alpha}_{bK'} = E^{J^\pi\alpha} \sum_{bK'} \mathsf{N}^{J^\pi}_{aK,bK'} \Psi^{J^\pi\alpha}_{bK'} \,. \tag{22}$$

2.3. *Determination of basis states*

The non-orthogonal and continuous nature of the wave packet basis does not lead to a unique choice of basis states. The basic idea in FMD is to generate a small number of optimized basis states by a variational procedure. This can be done on different levels of sophistication and effort. The simplest and numerically cheapest approach is a variation on the level of a single Slater determinant as in a mean-field calculation. Here one minimizes the energy of the intrinsic many-body state $\left| Q \right\rangle$ with respect to the parameters of the single-particle states

$$\min_{\{q_\nu\}} \frac{\left\langle Q \right| \widehat{H} - \widehat{T}_{\mathrm{cm}} \left| Q \right\rangle}{\left\langle Q \middle| Q \right\rangle}. \tag{23}$$

Symmetries can then be restored by projection, defining the *projection after variation* (PAV) procedure. This is numerically cheap as the projection only has to be performed for the final basis state and in general works reasonably well for the description of deformed states that form a rotational band. However even in such cases correlation energies are underestimated.

An improved description is obtained by *variation after projection* (VAP). Here the energy of the projected state

$$\min_{\{q_\nu, C_K\}} \frac{\sum_{K,K'} C_K{}^{\star} \left\langle Q \right| (\widehat{H} - \widehat{T}_{\mathrm{cm}}) \widehat{P}^{\pi} \widehat{P}^{J}_{KK'} \widehat{P}^{\vec{P}=0} \left| Q \right\rangle C_{K'}}{\sum_{K,K'} C_K{}^{\star} \left\langle Q \right| \widehat{P}^{\pi} \widehat{P}^{J}_{KK'} \widehat{P}^{\vec{P}=0} \left| Q \right\rangle C_{K'}} \tag{24}$$

is minimized with respect to the single-particle parameters q_ν and the K-mixing parameters C_K. The VAP procedure is essential for states that have a different intrinsic structure than the mean-field state. This can be related to clustering but also to single-particle excitations. The VAP procedure will be performed for all quantum numbers of interest which provides a set of intrinsic basis states $\left\{ \left| Q^{(a)} \right\rangle \right\}$. The diagonalization of the Hamiltonian in the basis states obtained by projecting this set on the desired symmetries will further improve the description of the individual eigenstates due to the possible admixture of other configurations with the same quantum numbers.

A further improvement can be obtained by adding basis states obtained by variation under constraints. Constraints are related to collective degrees of freedom that are treated as generator coordinates, like the radius or quadrupole deformation. Take the case of a nucleus that has a prolate and an oblate minimum. VAP will provide the configuration with the lowest energy. The other minimum can however be found by minimizing the energy under constraints on the quadrupole deformation. Constraints are also very useful to obtain basis states for loosely bound systems of cluster or halo nature. The VAP minimum corresponds typically to a rather compact configuration. To include extended configurations necessary for the description of the tail of the wave function the variation can be performed under constraints on the radius.

2.4. *Realistic effective interaction*

An important ingredient in the FMD approach is the use of a realistic effective interaction. Realistic interactions, as used in *ab initio* approaches, reproduce the nucleon nucleon scattering data up to the pion production threshold and the deuteron properties. However this does not allow a unique determination of the interaction. Whereas the long-range properties of the interaction are determined by pion dynamics the short-range behavior of the interaction is not fully constrained by the scattering data. In the Argonne v_{18} interaction[12] the short-range behavior is modeled phenomenologically as a local interaction, whereas in interactions derived in chiral EFT[13,14] the short-range behavior is described by non-locally regularized contact terms. Despite these differences all realistic interactions show a strong short-range repulsion and a strong tensor force that make the application of realistic interactions in FMD or other many-body approaches difficult or impossible.

As in many *ab initio* approaches a unitary transformation that decouples the low- and high-momentum components is used to derive an effective low-momentum interaction. In FMD we use the Unitary Correlation Operator Method (UCOM) that explicitly introduces central and tensor correlations. The UCOM approach has the advantage that it provides an explicit operator representation of the transformed interaction.[15–17] The unitary transformation is performed in two-body approximation. Therefore two-body properties like nucleon nucleon phase shifts are conserved, but differences will appear in many-body system due to omitted three- and higher body terms in the transformed interaction. For light nuclei the two-body UCOM interaction works remarkably well as there appears to be a partial cancellation between the omitted three-body terms and genuine three-body forces that are not included. However the UCOM interaction shows deficiencies with respect to spin-orbit properties and for the saturation of heavier nuclei.

3. Clustering in bound states

A nice example for the importance of clustering is given by the Neon isotopes. Our investigation[18] was triggered by experimental measurements of the charge radii for $^{17-22}$Ne by the COLLAPS collaboration. Of particular interest here was the structure of ^{17}Ne which has a very small two-proton separation energy of 0.93 MeV. In three-body calculations assuming an ^{15}O core a large s^2 component and therefore halo-like structures were predicted, whereas shell model calculations based on Coulomb displacement energies and magnetic moments predicted only a small s^2 component.

In the FMD calculations we performed variation after parity projection calculations with an additional constraint on the radius as generator coordinate. For ^{17}Ne we found two minima that correspond essentially to an ^{15}O core with two protons in a mixed s^2/d^2 configuration. These configurations strongly mix and in the multiconfiguration mixing calculation we find that the s^2 admixture in the ^{17}Ne wave function is about 42%. Due to the small two-proton separation energy and the missing centrifugal barrier in the s-orbit the proton distribution is rather extended which explains the large charge radius of ^{17}Ne. The left part of Fig. 5 shows the extended tail of the proton distributions compared to the compact neutron distribution. The intrinsic density of the dominant FMD configuration further illustrates the correlations of the two valence protons. They like to form a pair on one side of the ^{15}O core. This asymmetric shape can only be obtained by a linear combination of s^2 and d^2 configurations. Going from ^{17}Ne to ^{18}Ne we find a similar picture with an ^{16}O core and two protons in s^2 and d^2

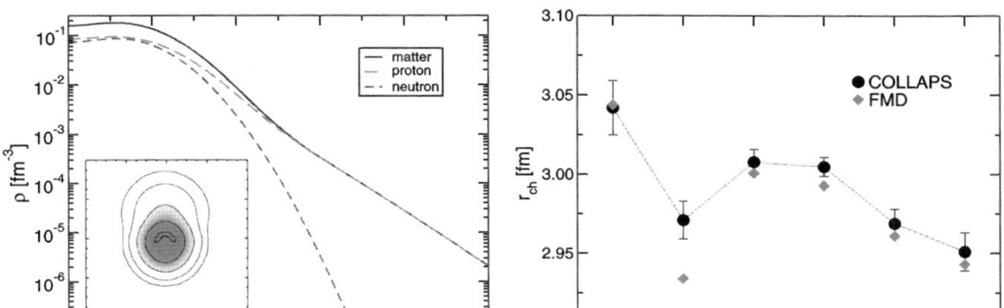

Fig. 5. (Left) ^{17}Ne proton and neutron point density distributions. The inset shows the intrinsic density of the dominant FMD configuration. (Right) Charge radii for $^{17-22}$Ne calculated in FMD compared to experimental results from COLLAPS.

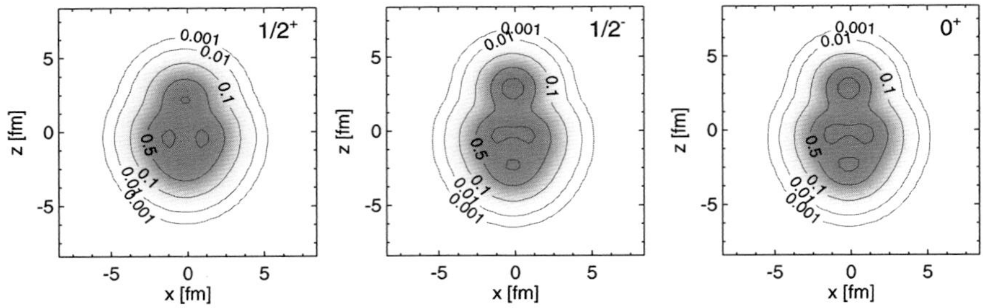

Fig. 6. Intrinsic densities of dominant configurations for the $1/2^+$ ground state of ^{19}Ne (left), the $1/2^-$ state in ^{19}Ne (middle) and the 0^+ ground state of ^{20}Ne.

configurations. However the admixture of the s^2 component is only 15% explaining the much smaller charge radius of ^{18}Ne.

Interestingly the charge radii of ^{19}Ne and ^{20}Ne are again much larger. Another surprising observation is that experimentally the $1/2^-$ state in ^{19}Ne is almost degenerate with the natural parity $1/2^+$ state. These observations, together with the rather low thresholds, already hint at the importance of ^{4}He and ^{3}He cluster configurations. In the FMD calculations we therefore not only included configurations optimized for positive parity but also configurations optimized for negative parity as shown in Fig. 6. Indeed we find in the multiconfiguration mixing calculations that in the ^{19}Ne ground state extended configurations with both ^{4}He+^{15}O and ^{3}He+^{16}O cluster configurations contribute and are responsible for the large charge radius. In ^{20}Ne the ^{4}He+^{16}O threshold is only 4.5 MeV above the ground state and we find a significant admixture of extended cluster configurations. In the heavier Neon isotopes the admixture of ^{4}He cluster configurations still exists but is much smaller than in ^{19}Ne and ^{20}Ne and correspondingly the charge radii are smaller.

Similar studies have been performed for the Lithium[19] and Beryllium isotopes.[20–22] Also here the interplay between single-particle, especially the formation of halos, and cluster structure is reflected in the evolution of the charge radii.

4. Clusters and nuclear reactions

Obviously clusters play an essential role in the description of nuclear reactions as the initial (and possibly final) states are given by asymptotically free clusters. Within the FMD approach the first application was the

^{3}He(α,γ)^{7}Be radiative capture reaction.[23] To calculate the reaction cross section one needs the initial scattering states, the final bound states and the electromagnetic transition matrix elements. To perform the calculation the model space is divided into two regions. In the external region bound and scattering states are described by ^{3}He and ^{4}He clusters in their FMD ground states. In the interaction region additional FMD configurations obtained by variation after parity and angular momentum projection on $1/2^+$, $3/2^+$, $5/2^+$, $1/2^-$, $3/2^-$, $5/2^-$ and $7/2^-$ are included. The radius is used in addition as a generator coordinate. These FMD configurations go beyond the simple cluster model picture and are essential for obtaining polarized cluster configurations.

The bound and scattering eigenstates are determined by diagonalization of the Hamiltonian in the full space consisting of the internal and external region with the corresponding boundary conditions. In the external region only the Coulomb interaction between the clusters has to be considered and the system can be described as point-like clusters with the relative motion given by Coulomb wave functions depending on scattering energy and phase shift or by Whittaker functions depending on the bound state energies. The matching between the microscopic world of A-nucleon wave functions and the world of point-like clusters is done using the microscopic R-matrix method of the Brussels group.[24] Technically this requires a rewriting of the FMD Brink-type wave functions as RGM wave functions as discussed in Sec. 1.3. As the width parameters in the ^{4}He and ^{3}He cluster wave functions are different it is essential to perform the projection on total linear momentum. Only then the total wave function factorizes into the intrinsic wave functions of the clusters, the wave function describing their relative motion and the total (plane wave) motion of the center of mass.

The properties of the bound states agree well with experimental data, but only if the polarized configurations are included in the model space. With the frozen cluster configurations alone, ^{7}Be is bound only by 200 keV, in the full model space the $3/2^-$ state is bound by 1.49 MeV and the $1/2^-$ state by 1.31 MeV with respect to the cluster threshold. The calculated charge radius of 2.67 fm is also in good agreement with the experimental value of 2.647(17) fm. For the phase shifts of the S- and D-waves we also find good agreement with experimental data. Again it is important to include the polarized configurations.

The calculated cross section for ^{3}He(α,γ)^{7}Be, shown in the left part of Fig. 7 in form of the astrophysical S-factor agrees remarkably well with

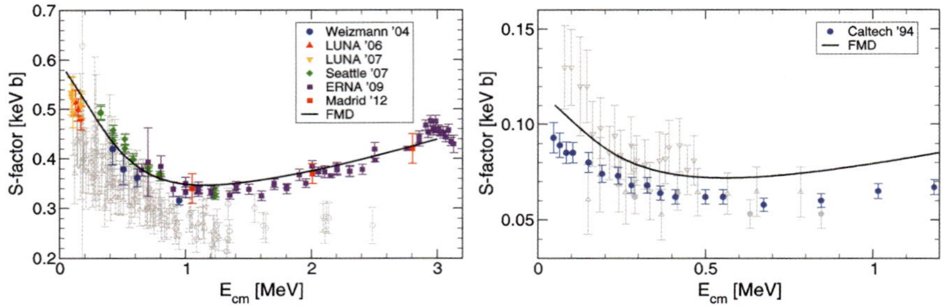

Fig. 7. Astrophysical S-factors for the ^{3}He(α,γ)^{7}Be and ^{3}H(α, γ)^{7}Li capture reactions. Light gray data points indicate older experimental data.

recent experimental data, both with respect to the energy dependence and the absolute normalization. In case of the mirror reaction ^{3}H(α,γ)^{7}Li, shown in the right part of Fig. 7 the calculated energy dependence agrees perfectly with the data but the normalization is too large by about 15%.

5. Clustering in resonance states

The structure of ^{12}C above the three-α threshold poses a challenge for nuclear theory. The ground state band can be well described in the no-core shell model using a harmonic oscillator single-particle basis. However, many of the states in the continuum have a well developed cluster structure, and these states are completely missing in the no-core shell model.[25] Microscopic α-cluster models have been able to reproduce many properties of these continuum states.[26] On the other hand, the cluster model is an idealization. Experimentally Gamow-Teller as well as $M1$ and $E1$ transitions into continuum states can be observed. Within a cluster model such transitions are forbidden, indicating that for a full picture both shell and cluster structure have to be included in a theoretical description.

Earlier FMD calculations[10,27,28] as well as AMD calculations[29] investigated both the ground state band and the cluster states but treated states in the continuum in a bound-state approximation. This might be justified for the very narrow Hoyle state at 7.65 MeV but is certainly very unreliable for the broad resonances observed higher up in energy.

To address these questions we extended our approach with a proper treatment of the continuum in a similar way as explained in the last section. We first performed a study within the microscopic α-cluster model with full antisymmetrization and a phenomenological two-body interaction[30] similar to earlier calculations by Descouvemont and Baye[31] and Arai.[32] In this cluster model the internal region the Hilbert space is built from three-α configurations on a triangular grid without any restrictions. In the external region ^{8}Be+^{4}He configurations are added. In principle we have to deal with a real three-body continuum. However, in ^{8}Be the Coulomb repulsion of the two α particles is compensated by the nuclear attraction which lowers the energy surface of the ^{8}Be+^{4}He configurations compared to free α particles significantly. It turns out that it is important to include not only the ^{8}Be ground state but also the excited 2^+ state. The inclusion of additional ^{8}Be states changes the final results however only slightly.

The microscopic R-matrix method[24] is used to match the microscopic wave functions in the internal region to the asymptotic behavior described by point-like ^{8}Be- and α-clusters. For bound states the asymptotics is given by Whittaker functions, for resonances we match to purely outgoing Coulomb wave functions. The energies of these Gamow states are complex, with the real part giving the resonance position and the imaginary part the resonance width. We can also obtain scattering states with real energies by matching to linear combinations of incoming and outgoing Coulomb wave functions which provides the full scattering matrix. For the lower lying resonances shown in Fig. 8 the resonances are well separated and resonance

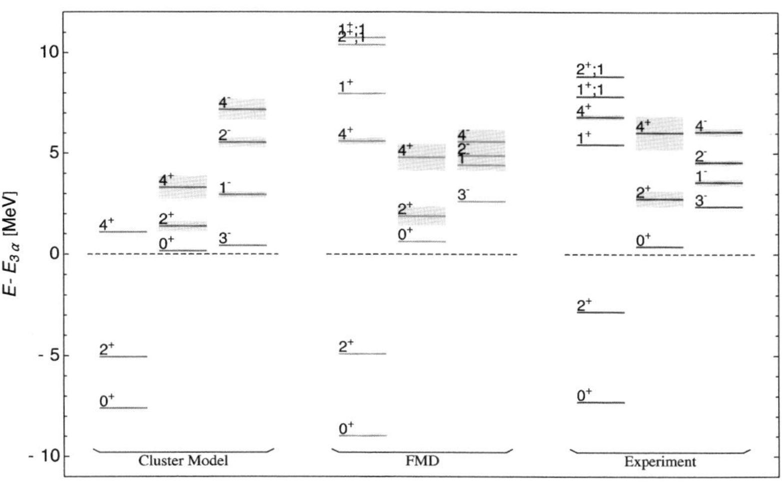

Fig. 8. ^{12}C energy spectra obtained with a microscopic cluster model (left) and FMD (middle) compared with experiment (right). Energies are given with respect to the $3 - \alpha$ threshold. Resonance widths are indicated by shaded areas.

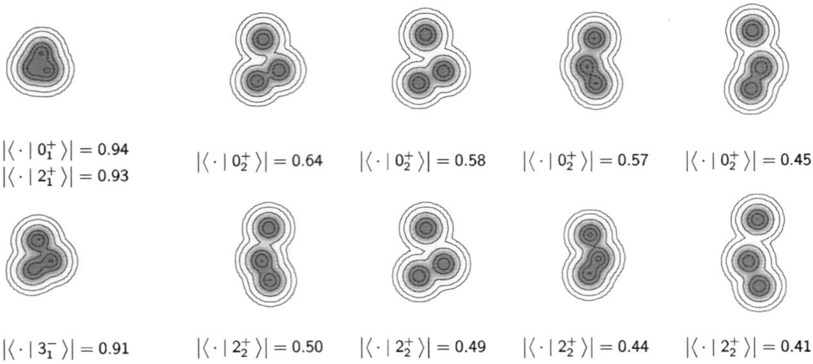

Fig. 9. (Left) Intrinsic densities of dominant FMD basis states and the amplitudes of these basis states in the 0_1^+, 2_1^+ and 3_1^- eigenstates. (Right) Intrinsic densities of FMD basis states that contribute with large amplitudes to the strongly clustered 0_2^+ and 2_2^+ states. Note that the basis states are not orthogonal.

parameters obtained from the Gamow states and from the phase shifts are consistent. If one goes higher in energy the situation becomes more complicated and it is mostly impossible to isolate individual resonances. It is however possible to calculate transition strengths as a function of energy.

Recently we also added the continuum to the full FMD calculation. Here basis states in the internal region are obtained by using variation after projection on angular momentum and parity. For each spin we first vary the parameters of the many-body state to obtain the lowest possible energy. A second basis state is then obtained by minimizing the energy of the second state with respect to its parameters keeping the first state fixed. We further increase the model space by using the radii of the intrinsic states as generator coordinates. In Fig. 9, the intrinsic densities for some basis states that have a large overlap with the eigenstates are shown. A triangular structure with strongly overlapping α-clusters (that also has a large overlap with a p-shell shell model wave function) can already be seen in the ground state band members 0_1^+ and 2_1^+. An underlying α-structure is also visible in the 3_1^- state but with a larger spatial extension. For the 0_2^+ Hoyle state and the 2_2^+ state the spatial extension is even larger and the ^{8}Be+α structure is reflected in the large overlap with several configurations with more or less open triangles.

The ^{8}Be clusters that are needed for the description in the external region are obtained by diagonalization in a basis of FMD many-body states and of α-α configurations, treating them as pseudostates. We include two 0^+ states, two 2^+ states and a 4^+ state for ^{8}Be.

In Fig. 8, we compare the spectra containing bound states and resonances obtained with the microscopic cluster model and with FMD. The cluster model can not describe spin-flip states like the 1^+ states or the 2^+ $(T = 1)$ state. The FMD calculations show in general a good agreement with experimental observations. For example we obtain the 4^+ from the ground state band, the 4^+ state of the Hoyle state band and the 4^- at roughly the same energy, in good agreement with experiment. Future studies will focus on transitions into the continuum. Apart from the investigations of the monopole strength,[27] recent experiments studied transitions into the second 2^+ state and the 1^- state.[33] Also Gamow-Teller transitions from ^{12}B and ^{12}N and electromagnetic transitions from the 2^+ $(T = 1)$ state into the ^{12}C continuum have been measured carefully.[34]

References

1. E. Wigner, *Phys. Rev.* **51**, 106 (1937).
2. K. Ikeda, N. Takigawa, and H. Horiuchi, *Prog. Theor. Phys. Suppl.* **E68**, 464 (1968).
3. J. A. Wheeler, *Phys. Rev.* **52**, 1107 (1937).
4. Y. C. Tang, M. LeMere and D. R. Thompson, *Phys. Rep.* **47**, 167 (1978).
5. Y. Funaki, H. Horiuchi, A. Tohsaki, P. Schuck and G. Röpke, *Prog. Theor. Phys.* **108**, 297 (2002).
6. D. M. Brink, In *Proceedings of the International School of Physics 'Enrico Fermi' course XXXVI*, p. 247 (1965).
7. H. Feldmeier, *Nucl. Phys. A* **515**, 147 (1990).
8. H. Feldmeier and J. Schnack, *Rev. Mod. Phys.* **72**, 655 (2000).
9. T. Neff and H. Feldmeier, *Nucl. Phys. A* **738**, 357 (2004).
10. R. Roth, T. Neff, H. Hergert and H. Feldmeier, *Nucl. Phys. A* **745**, 3 (2004).
11. T. Neff and H. Feldmeier, *Eur. Phys. J Special Topics* **156**, 69 (2008).
12. R. B. Wiringa, V. G. J. Stoks and R. Schiavilla, *Phys. Rev. C* **51**, 38 (1995).
13. D. R. Entem and R. Machleidt, *Phys. Rev. C* **68**, 041001 (2003).
14. E. Epelbaum, W. Glöckle and U.-G. Meißner, *Nucl. Phys. A* **747**, 362 (2005).
15. T. Neff and H. Feldmeier, *Nucl. Phys. A* **713**, 311 (2003).
16. R. Roth, T. Neff and H. Feldmeier, *Prog. Part. Nucl. Phys.* **65**, 50 (2010).
17. D. Weber, H. Feldmeier, H. Hergert and T. Neff, *Phys. Rev. C* **89**, 034002 (2014).
18. W. Geithner, T. Neff, G. Audi, K. Blaum, P. Delahaye, H. Feldmeier, S. George, C. Guénaut, F. Herfurth, A. Herlert, S. Kappertz, M. Keim, A. Kellerbauer, H.-J. Kluge, M. Kowalska, P. Lievens, D. Lunney, K. Marinova, R. Neugart, L. Schweikhard, S. Wilbert and C. Yazidjian, *Phys. Rev. Lett.* **101**, 252502 (2008).

19. W. Nörtershäuser, T. Neff, R. Sánchez and I. Sick, *Phys. Rev. C* **84**, 024307 (2011).
20. M. Žáková, Z. Andjelkovic, M. L. Bissell, K. Blaum, G. W. F. Drake, C. Geppert, M. Kowalska, J. Krämer, A. Krieger, M. Lochmann, T. Neff, R. Neugart, W. Nörtershäuser, R. Sánchez, F. Schmidt-Kaler, D. Tiedemann, Z.-C. Yan, D. T. Yordanov and C. Zimmermann, *J. Phys. G* **37**, 055107 (2010).
21. A. Krieger, K. Blaum, M. L. Bissell, N. Frömmgen, C. Geppert, M. Hammen, K. Kreim, M. Kowalska, J. Krämer, T. Neff, R. Neugart, G. Neyens, W. Nörtershäuser, C. Novotny, R. Sánchez and D. T. Yordanov, *Phys. Rev. Lett.* **108**, 142501 (2012).
22. S. Terashima, I. Tanihata, R. Kanungo, A. Estradé, W. Horiuchi, F. Ameil, J. Atkinson, Y. Ayyad, D. Cortina-Gil, I. Dillmann, A. Evdokimov, F. Farinon, H. Geissel, G. Guastalla, R. Janik, M. Kimura, R. Knoebel, J. Kurcewicz, Y. A. Litvinov, M. Marta, M. Mostazo, I. Mukha, T. Neff, C. Nociforo, H. J. Ong, S. Pietri, A. Prochazka, C. Scheidenberger, B. Sitar, Y. Suzuki, M. Takechi, J. Tanaka, J. Vargas, J. S. Winfield and H. Weick, *Prog. Theor. Exp. Phys.* **2014**, 101D02 (2014).
23. T. Neff, *Phys. Rev. Lett.* **106**, 042502 (2011).
24. P. Descouvemont and D. Baye, *Rep. Prog. Phys.* **73**, 036301 (2010).
25. P. Maris, J. P. Vary, A. Calci, J. Langhammer, S. Binder and R. Roth, *Phys. Rev. C* **90**, 014314 (2014).
26. M. Kamimura, *Nucl. Phys. A* **351**, 456 (1981).
27. M. Chernykh, H. Feldmeier, T. Neff, P. von Neumann-Cosel and A. Richter, *Phys. Rev. Lett.* **98**, 032501 (2007).
28. M. Chernykh, H. Feldmeier, T. Neff, P. von Neumann-Cosel and A. Richter, *Phys. Rev. Lett.* **105**, 022501 (2010).
29. Y. Kanada-En'yo, *Prog. Theor. Phys.* **117**, 655 (2007).
30. T. Neff and H. Feldmeier, *J. Phys. Conf. Ser.* **569**, 012062 (2014).
31. P. Descouvemont and D. Baye, *Phys. Rev. C* **36**, 54 (1987).
32. K. Arai, *Phys. Rev. C* **74**, 064311 (2006).
33. W. R. Zimmerman, M. W. Ahmed, B. Bromberger, S. C. Stave, A. Breskin, V. Dangendorf, T. Delbar, M. Gai, S. S. Henshaw, J. M. Mueller, C. Sun, K. Tittelmeier, H. R. Weller and Y. K. Wu, *Phys. Rev. Lett.* **110**, 152502 (2013).
34. S. Hyldegaard, M. Alcorta, B. Bastin, M. J. G. Borge, R. Boutami, S. Brandenburg, J. Büscher, P. Dendooven, C. A. Diget, P. Van Duppen, T. Eronen, S. P. Fox, L. M. Fraile, B. R. Fulton, H. O. U. Fynbo, J. Huikari, M. Huyse, H. B. Jeppesen, A. S. Jokinen, B. Jonson, K. Jungmann, A. Kankainen, O. S. Kirsebom, M. Madurga, I. Moore, A. Nieminen, T. Nilsson, G. Nyman, G. J. G. Onderwater, H. Penttilä, K. Peräjärvi, R. Raabe, K. Riisager, S. Rinta-Antila, A. Rogachevskiy, A. Saastamoinen, M. Sohani, O. Tengblad, E. Traykov, Y. Wang, K. Wilhelmsen, H. W. Wilschut and J. Äystö, *Phys. Rev. C* **81**, 024303 (2010).

CHAPTER 4

COMPARISON OF EQUATION OF STATE MODELS WITH DIFFERENT CLUSTER DISSOLUTION MECHANISMS

Helena Pais

CFisUC, Department of Physics, University of Coimbra,
P-3004-516 Coimbra, Portugal,
pais.lena@uc.pt

Stefan Typel

Institut f. Kernphysik, Technische Universität Darmstadt,
Schlossgartenstr. 9, D-64289 Darmstadt, Germany
GSI Helmholtzzentrum für Schwerionenforschung GmbH,
Planckstraße 1, D-64291 Darmstadt, Germany,
s.typel@gsi.de

The appearance of nuclear clusters in stellar matter at densities below nuclear saturation is an important feature in the modeling of the equation of state for astrophysical applications. There are different theoretical concepts to describe the dissolution of nuclei with increasing density and temperature. In this contribution, the predictions of two approaches are compared: the medium dependent change of the nuclear masses in a generalized relativistic density functional approach and the excluded-volume mechanism in a statistical model. Both approaches use the same description for the interaction between the nucleons. The composition of neutron star matter, in particular the occurrence of light and heavy nuclei, and its thermodynamic properties are studied.

1. Introduction

In astrophysical simulations of core-collapse supernovae (CCSN)[1–4] and the description of compact star properties,[5–10] the equation of state (EoS) is an essential ingredient. It provides the information on thermodynamic properties of strongly interacting matter with baryons and leptons as basic degrees of freedom.[11–20] The description of such matter at densities below the nuclear saturation density $n_{\rm sat} \approx 0.16$ fm^{-3} and temperatures $T \lesssim 15\,{\rm MeV}$ represents a particular challenge for theoretical models, where the short-range strong interaction and the long-range electromagnetic interaction have to be considered explicitly. The competition of these forces with the entropy leads to the formation and dissolution of inhomogeneous structures on mesoscopic length scales with different geometries. This is in contrast to the liquid-gas phase transition in pure nuclear matter where the Coulomb interaction and the charge of particles are neglected, see, e.g., Refs. 21–32 for details. In stellar matter the phenomenon of frustration is observed since the available space of thermodynamic variables is reduced due to the specific condition of charge neutrality.

In dilute matter at temperatures above 2 MeV and with densities up to approximately 0.001 $n_{\rm sat}$, light nuclei (deuterons, tritons, helions, α-particles) are the most relevant nuclear species that are formed as many-nucleon correlations.[27,33–38] With increasing densities and at lower temperatures, also heavier nuclei with increasing mass numbers appear and the chemical composition of matter changes. Finally, when the saturation density $n_{\rm sat}$ is approached, the occurrence of so-called "pasta phases" with several different geometries is expected,[39–47] before the system converts to uniform matter composed of nucleons, electrons and muons.

The complex structure of stellar matter is particularly important in astrophysical applications. The formation of clusters and pasta phases in CCSN matter affects the neutrino opacity.[41,42,48–50] This quantity plays a central role in the energy transport and deposition. Thus it can affect the development of a shock wave during the supernova collapse[40,51] and the cooling of the proto neutron star (PNS).[52,53] The crust cooling will also affect its thickness, and, consequently, the moment of inertia, which will have direct influence on the interpretation of pulsar glitches.[54] The composition of matter also determines the structure of the crust of cold catalyzed neutron stars where the additional condition of β-equilibrium fixes the isospin asymmetry of the nuclear subsystem.[55–61]

Statistical models with nucleons, nuclei and charged leptons as degrees of freedom are a popular approach to describe stellar matter at subsaturation densities.[16,28,62–71] The interaction between the constituents is often incorporated on the basis of nonrelativistic Skyrme or relativistic density functionals, employing a mean-field picture. In most cases, only the strong interaction between the nucleons is considered in this way. In order to model the transition from clustered matter to uniform matter at high densities and temperatures, the dissolution of nuclei has to be implemented in the theoretical approach. Various prescriptions are employed to suppress the occurrence of clusters.

A widely used heuristic approach is the excluded-volume mechanism using a geometrical picture.[16,28,72,73] A finite volume, usually proportional to the mass number A, is attributed to each nucleon or nucleus, leaving a reduced volume for the motion of the remaining particles. As a consequence, the system lowers the fraction of larger-size constituents in favor of smaller-size particles with increasing density. Hence, nuclei are converted to nucleons. The effect of the finite particle volume on the system properties can be interpreted as the action of an effective repulsive, hard-core interaction. The excluded-volume approach is implemented, e.g., in an EoS model that considers an ensemble of nucleons, nuclei and electrons in statistical equilibrium.[16] The interaction between the nucleons is described in a relativistic mean-field (RMF) model. For this model, EOS tables for a variety of RMF parametrizations[16,20,28,32] are available online.[74] They include the full distribution of light and heavy nuclei and cover a broad range in density, temperature, and isospin asymmetry.

In a more microscopic description, nuclei will dissolve with increasing density, mainly as a result of the Pauli principle.[33,72,75–79] At high densities, it becomes difficult to form clusters as many-nucleon correlations, since most of the momentum space is already occupied by background nucleons. The essential consequence is a reduction of the cluster binding energy with increasing density. This effect becomes smaller, however, at higher temperatures due to a more dilute population of states in momentum space. Such medium-dependent mass shifts can be extracted from microscopic calculations of few-body correlations in dense matter. They are used in a parametrized form in density functionals for stellar matter. A reduction of the binding energy causes a lowering of the particle abundance simply because of statistical reasons. The mass-shift method is realized in the generalized relativistic density functional (gRDF) approach,[33,80,81] which is an extension of a conventional RMF model for nuclear matter with density dependent (DD) couplings.[82]

In this work, we compare the formation and dissolution of light and heavy clusters in dilute matter within these two models: the statistical model with excluded-volume mechanism,[16] denoted as HS model in the following, and the gRDF approach with cluster mass shifts.[33,80,81] For this comparison, it is important to note that the same parametrization (DD2) of the RMF interaction is employed in both models such that differences in the predictions can be attributed to the cluster description. Besides the density dependence of the cluster suppression, we will investigate the predictions of the two models as a function of the temperature. Here, different procedures to include thermal excitations of the nuclei, which depend on the employed level densities, become relevant.

The paper is organized as follows. In Section 2, the theoretical formulation of the excluded-volume mechanism and its particular realization in the HS model and the gRDF approach are discussed. Predictions of the HS and gRDF models for neutron star matter, i.e. stellar matter in β-equilibrium, are presented in Section 3. The main emphasis is on the comparison of the chemical composition, in particular the occurrence of light and heavy nuclei. Also thermodynamic properties are briefly compared. Conclusions are given in Section 4.

2. Theoretical formalism

All thermodynamic properties of stellar matter can be determined once a suitable thermodynamic potential is specified. In most cases, a Helmholtz free energy density $f(T, \{n_i\})$ or a grand canonical potential density $\omega(T, \{\mu_i\})$ is given. They depend on the temperature T and the number densities n_i or chemical potentials μ_i of all constituents i. In both cases, the excluded-volume mechanism and the medium-dependence of mass shifts can be incorporated in the formalism. However, one has to make sure that the theory is thermodynamically consistent, i.e. the usual definitions of thermodynamic quantities and relations for derivatives hold after the modification of the original thermodynamic potential.

A general formulation of the excluded-volume mechanism and mass shifts is presented in the following Subsection 2.1, assuming relativistic kinematics of the particles and general particle statistics. This model incorporates effects from the interaction between particles in the framework of a RMF approach with density dependent couplings. In Subsection 2.3, the relation of this formulation to the HS model is established and differences are explored. The formalism of the gRDF approach is presented in Subsection 2.2.

2.1. *RMF model with excluded-volume mechanism and mass/energy shifts*

The excluded-volume mechanism is a simple means to model a repulsive interaction between particles. A very general theoretical formulation was presented in Ref. 73 that is adapted to the present application. Every particle is assumed to have a finite volume that reduces the total volume V of the system to the available volume V_i for the motion of a particle i. Hence, this available volume is

$$V_i = V\Phi_i \tag{1}$$

with the available volume fraction Φ_i. In the following it is supposed that the functions Φ_i only depend on the number densities $n_j = N_j/V$ of all particles j in the system. The geometric picture of rigid spheres in the excluded-volume mechanism corresponds to the choice

$$\Phi_i = 1 - \sum_j v_{ij} n_j \tag{2}$$

with coefficients v_{ij} that are connected to the radii R_i of the particles as

$$v_{ij} = \frac{2\pi}{3}\left(R_i + R_j\right)^3 . \tag{3}$$

This formulation is symmetric in the indices i and j as required by the consistency with the virial equation of state at low densities for particles with different radii.[73] Other functional dependencies than (2) are permissible in general. However, the rigid sphere picture does not apply any more in this case. The particle number densities $n_i = n_i^{(v)}$ are the vector densities in the relativistic description. They have to be distinguished from the scalar densities $n_i^{(s)}$. Both will be defined below.

The total grand canonical potential density of the system can be written as

$$\omega(T, \{\mu_i\}) = \sum_i \omega_i + \omega_{\text{meson}} - \omega^{(r)} \tag{4}$$

with three distinct terms. The single quasi-particle contribution

$$\omega_i = \Phi_i \omega_i^{(0)} \tag{5}$$

contains the available volume fraction Φ_i and the standard expression

$$\omega_i^{(0)}(T, m_i^*, \mu_i^*) = -T\frac{g_i}{\sigma_i}\int \frac{d^3k}{(2\pi)^3} \ln\left\{1 + \sigma_i \exp\left[-\frac{E_i(k, m_i^*) - \mu_i^*}{T}\right]\right\} \tag{6}$$

with the particle degeneracy factor g_i, the effective mass

$$m_i^* = m_i - S_i, \tag{7}$$

and the effective chemical potential

$$\mu_i^* = \mu_i - V_i. \tag{8}$$

The appearance of the scalar potential S_i and the vector potential V_i are typical for the RMF approach. The quantity σ_i encodes the particle statistics. The case $\sigma_i = +1$ corresponds to Fermi-Dirac particles and $\sigma_i = -1$ to Bose-Einstein particles. In the limit $\sigma_i \to 0$, the result for Maxwell-Boltzmann statistics

$$\omega_i^{(0)}(T, m_i^*, \mu_i^*) = -T g_i \int \frac{d^3k}{(2\pi)^3} \exp\left[-\frac{E_i(k, m_i^*) - \mu_i^*}{T}\right] \tag{9}$$

is recovered. Effects of the finite particle volumes are taken into account by the density dependent prefactor Φ_i in (5). For alternative formulations of the excluded-volume mechanism see Ref. 73.

The relativistic relation

$$E_i(k, m_i^*) = \sqrt{k^2 + (m_i^*)^2} \tag{10}$$

connects the momentum k with the energy E_i of each particle, which is considered as a quasi-particle since scalar and vector potentials S_i and V_i appear in the definition of the effective mass (7) and the effective chemical potential (8). These potentials are given by

$$S_i = C_\sigma g_{i\sigma} n_\sigma + C_\delta g_{i\delta} n_\delta - \Delta m_i \tag{11}$$

and

$$V_i = C_\omega g_{i\omega} n_\omega + C_\rho g_{i\rho} n_\rho + \Delta E_i + D_i V_{\text{meson}}^{(r)} + V_i^{(r)} + W_i^{(r)} \tag{12}$$

with contributions due to the coupling of the particles to meson fields, density dependent mass shifts Δm_i, energy shifts ΔE_i and rearrangement terms $V_{\text{meson}}^{(r)}$, $V_i^{(r)}$, and $W_i^{(r)}$. The factors

$$C_j = \frac{\Gamma_j^2}{m_j^2} \tag{13}$$

are found from the meson-nucleon couplings Γ_j and masses m_j of the considered mesons $j = \omega, \sigma, \rho, \delta$ as usual in RMF models. The couplings Γ_j themselves depend on a number density

$$\tilde{n} = \sum_i D_i n_i^{(v)} \tag{14}$$

with factors D_i to be specified below. The source densities in equations (11) and (12) are given by

$$n_j = \sum_i g_{ij} n_i^{(v)} \tag{15}$$

for the Lorentz vector mesons $j = \omega, \rho$ and

$$n_j = \sum_i g_{ij} n_i^{(s)} \tag{16}$$

for the scalar mesons $j = \sigma, \delta$ with appropriate factors g_{ij} and vector and scalar densities $n_i^{(v)}$ and $n_i^{(s)}$, respectively, that are defined below.

The meson rearrangement potential

$$V_{\text{meson}}^{(r)} = \frac{1}{2}\left(C'_\omega n_\omega^2 + C'_\rho n_\rho^2 - C'_\sigma n_\sigma^2 - C'_\delta n_\delta^2\right), \tag{17}$$

in Eq. (12) with

$$C'_j = \frac{dC_j}{d\tilde{n}} \tag{18}$$

and the excluded-volume rearrangement term

$$V_i^{(r)} = \sum_j \omega_j \frac{\partial \ln \Phi_j}{\partial n_i^{(v)}} \tag{19}$$

also appear in the contribution

$$\omega^{(r)} = \omega_{\text{meson}}^{(r)} + \omega_\Phi^{(r)} + \omega_\Delta^{(r)} \tag{20}$$

to the total grand canonical density (4) with

$$\omega_{\text{meson}}^{(r)} = V_{\text{meson}}^{(r)} \tilde{n} \tag{21}$$

and

$$\omega_\Phi^{(r)} = \sum_i n_i^{(v)} V_i^{(r)}. \tag{22}$$

The rearrangement term

$$W_i^{(r)} = \sum_j \left(n_j^{(s)} \frac{\partial \Delta m_j}{\partial n_i^{(v)}} + n_j^{(v)} \frac{\partial \Delta E_j}{\partial n_i^{(v)}} \right) \tag{23}$$

in the vector potential (12), which is caused by the density dependence of the mass shifts and the energy shifts, is contained in the contribution

$$\omega_\Delta^{(r)} = \sum_i n_i^{(v)} W_i^{(r)} \tag{24}$$

to (20). Finally, the meson term in (4) is given by

$$\omega_{\text{meson}} = -\frac{1}{2}\left(C_\omega n_\omega^2 + C_\rho n_\rho^2 - C_\sigma n_\sigma^2 - C_\delta n_\delta^2\right). \tag{25}$$

The appearance of the rearrangement contributions guarantees the thermodynamic consistency of the model. In particular, the particle number densities

$$n_i = -\left.\frac{\partial \omega}{\partial \mu_i}\right|_{T,\{\mu_{j\neq i}\}} = n_i^{(v)} = g_i \Phi_i \int \frac{d^3k}{(2\pi)^3}\, f_i \tag{26}$$

and the scalar densities

$$n_i^{(s)} = \left.\frac{\partial \omega}{\partial m_i}\right|_{T,\{\mu_j\}} = g_i \Phi_i \int \frac{d^3k}{(2\pi)^3}\, f_i \frac{m_i^*}{\sqrt{k^2 + (m_i^*)^2}} \tag{27}$$

with the distribution function

$$f_i(T,k,m_i^*,\mu_i^*) = \left\{\exp\left[\frac{E_i(k,m_i^*) - \mu_i^*}{T}\right] + \sigma_i\right\}^{-1} \tag{28}$$

assume the usual form for quasi-particles, however, including the available volume fractions Φ_i as a prefactor. Because the functions Φ_i do not depend on the scalar densities in the models considered in the present paper, there are no rearrangement contributions in the scalar potential (11). For the case of a dependence on the scalar densities, see Ref. 73.

If the relativistic energy (10) is replaced by the nonrelativistic approximation

$$E_i(k,m_i^*) = m_i^* + \frac{k^2}{2m_i^*}, \tag{29}$$

the number density (26) and the single quasi-particle grand canonical potential density in Maxwell-Boltzmann statistics (9) can be given analytically with the ideal gas results

$$n_i = n_i^{(v)} = \frac{g_i \Phi_i}{(\lambda_i^*)^3} \exp\left(\frac{\mu_i^* - m_i^*}{T}\right) \tag{30}$$

and

$$\omega_i(T,m_i^*,\mu_i^*) = -T n_i^{(v)} \tag{31}$$

that contain the effective thermal wavelength

$$\lambda_i^* = \lambda(m_i^*,T) = \sqrt{\frac{2\pi}{m_i^* T}} \tag{32}$$

depending on the temperature and the effective mass.

For massless particles, $\omega_i^{(0)}$ can also be calculated analytically, e.g., as

$$\omega_\gamma^{(0)} = -g_\gamma \frac{\pi^2}{90} T^4 \tag{33}$$

for photons with degeneracy factor $g_\gamma = 2$.

Irrespective of the particle statistics and the dispersion relation for $E_i(k, m_i^*)$, the grand canonical potential density (4) can be written as

$$\omega(T, \{\mu_i\}) = \sum_i \omega_i + \omega_{\text{meson}} - \omega_{\text{meson}}^{(r)} - \omega_\Phi^{(r)} - \omega_\Delta^{(r)} \tag{34}$$

taking the decomposition (20) into account. The first and the next-to-last term can be combined as

$$\sum_i \omega_i - \omega_\Phi^{(r)} = \sum_i \left[\omega_i - n_i^{(v)} V_i^{(r)}\right] = \sum_j \omega_j \sum_i \left[\delta_{ij} - n_i^{(v)} \frac{\partial \ln \Phi_j}{\partial n_i^{(v)}}\right]. \tag{35}$$

With the choice (2) for the available volume fractions Φ_i, one finds in particular

$$\begin{aligned} \sum_i \omega_i - \omega_\Phi^{(r)} &= \sum_j \Phi_j \omega_j^{(0)} \sum_i \left[\delta_{ij} + n_i^{(v)} \frac{v_{ji}}{\Phi_j}\right] \\ &= \sum_j \omega_j^{(0)} \left[\Phi_j + \sum_i n_i^{(v)} v_{ji}\right] = \sum_j \omega_j^{(0)} \end{aligned} \tag{36}$$

and thus

$$\omega(T, \{\mu_i\}) = \sum_i \omega_i^{(0)} + \omega_{\text{meson}} - \omega_{\text{meson}}^{(r)} - \omega_\Delta^{(r)} \tag{37}$$

with the original single quasi-particle contributions (6) without a prefactor and no explicit rearrangement term from the excluded-volume mechanism. Nevertheless, the dependence on the available volume fractions Φ_i enters through the rearrangement term $V_i^{(r)}$ in the vector potentials V_i.

The entropy density of the system is obtained by the standard thermodynamic derivative

$$\begin{aligned} s &= - \left.\frac{\partial \omega}{\partial T}\right|_{\{\mu_i\}} \\ &= - \sum_i g_i \Phi_i \int \frac{d^3k}{(2\pi)^3} \left[f_i \ln f_i + \frac{1 - \sigma_i f_i}{\sigma_i} \ln\left(1 - \sigma_i f_i\right)\right] - \sum_i \Phi_i \frac{\omega_i^{(0)}}{g_i} \frac{\partial g_i}{\partial T} \end{aligned} \tag{38}$$

with the usual contribution and a term that takes an explicit temperature dependence of the degeneracy factors g_i into account, e.g., for nuclei, see below.

2.2. gRDF model

In this extension of a RMF model with density dependent couplings, nucleons, nuclei, electrons, muons and photons are considered as the relevant particle species. In addition, two-nucleon correlations in the continuum are included as effective quasi-particles like nuclei with temperature dependent resonance energies or masses. Nucleons, electrons, and muons are treated as fermions including their antiparticles. Photons are added as massless bosons as usual. The correct Bose-Einstein or Fermi-Dirac statistics are taken into account for light nuclei (^{2}H = 'd', ^{3}H = 't', ^{3}He = 'h', and ^{4}He = 'α') and two-nucleon correlations in the $^3S_1$(np) and $^1S_0$(np, nn, and pp) channels. Heavy nuclei with mass number $A > 4$ are described with nonrelativistic kinematics assuming Maxwell-Boltzmann statistics. Experimental rest masses m_i are used as far as available. For nuclei they are taken from the 2012 Atomic Mass Evaluation[83] if available, or they are calculated in the DZ10 model.[84] All nuclei within the neutron and proton driplines are included in the calculations. The driplines were determined from the neutron and proton separation energies after removing the Coulomb contribution of a homogeneously charge sphere of radius $r_C = 1.25 \cdot A^{1/3}$ fm to the total binding energies.

Since chemical equilibrium is assumed for the full ensemble, the chemical potential of every particle is given by

$$\mu_i = B_i\mu_b + Q_i\mu_q + L_i^{(e)}\mu_{l^{(e)}} + L_i^{(\mu)}\mu_{l^{(\mu)}} \tag{39}$$

with baryon numbers B_i, charge numbers Q_i, electron lepton numbers $L_i^{(e)}$, and muon lepton numbers $L_i^{(\mu)}$ and corresponding independent chemical potentials of the conserved currents. The quantum numbers are specified in Table 1 for all elementary particles of the model and nuclei $i = (A, Z)$ with mass numbers A, charge numbers Z, and neutron numbers $N = A - Z$. The chemical potential of an antiparticle is the negative of that of a particle. The lepton chemical potentials are taken to be equal, i.e.,

$$\mu_l = \mu_{l^{(e)}} = \mu_{l^{(\mu)}}\,, \tag{40}$$

in equation (39).

Table 1. Quantum numbers of elementary particles and nuclei considered in the EoS models.

quantum number	n	$\bar{n}$	p	$\bar{p}$	(A,Z)	e^-	e^+	μ^-	μ^+	γ
B_i	+1	−1	+1	−1	A	0	0	0	0	0
Q_i	0	0	+1	−1	Z	−1	+1	−1	+1	0
$L_i^{(e)}$	0	0	0	0	0	+1	−1	0	0	0
$L_i^{(\mu)}$	0	0	0	0	0	0	0	+1	−1	0

All nucleons, whether they are free or bound in nuclei, couple to the meson fields with coupling factors g_{ij} in equations (15) and (16). Explicitly they are given by

$$g_{i\omega} = \begin{cases} B_i & \text{if} \quad i = n, p, \bar{n}, \bar{p} \\ Af(A) & \text{if} \quad i = (A,Z) \end{cases} \tag{41}$$

$$g_{i\sigma} = \begin{cases} |B_i| & \text{if} \quad i = n, p, \bar{n}, \bar{p} \\ Af(A) & \text{if} \quad i = (A,Z) \end{cases} \tag{42}$$

$$g_{i\rho} = \begin{cases} B_i & \text{if} \quad i = n, \bar{n} \\ -B_i & \text{if} \quad i = p, \bar{p} \\ (A-2Z)f(A) & \text{if} \quad i = (A,Z) \end{cases} \tag{43}$$

$$g_{i\delta} = 0 \tag{44}$$

with a scaling function $f(A)$. It is defined as

$$f(A) = \begin{cases} 1 & \text{if} \quad A \leq 4 \\ 1 - \left[1 - \left(\frac{4}{A}\right)^{1/3}\right]^3 & \text{if} \quad A > 4 \end{cases} \tag{45}$$

in order to scale the coupling of heavy nuclei according to their surface size. The factors D_i in the density (14), which determine the density dependent coupling strengths Γ_j in equation (13), are identical to the baryon numbers B_i, hence $\tilde{n}$ is just the total baryon density

$$n_b = \sum_i B_i n_i^{(v)} \tag{46}$$

of the system. The explicit form of the density dependent couplings in the DD2 parametrization is described in Ref. 33.

All particles are assumed to be pointlike, i.e., the available volume fractions Φ_i are all constant equal to one and hence there are no rearrangement contribution $V_i^{(r)}$ and $\omega_\Phi^{(r)}$ to the vector potentials (12) and the grand canonical potential density (34), respectively. Instead of the excluded volume mechanism, density and temperature dependent mass shifts Δm_i are introduced in the scalar potential (11) in order to generate the dissolution of nuclei. In contrast, explicit energy shifts ΔE_i are not introduced in the vector potential (12). The total mass shift includes two contributions

$$\Delta m_i = \Delta m_i^{(\mathrm{Coul})} + \Delta m_i^{(\mathrm{strong})} \tag{47}$$

in the gRDF model. The Coulomb contribution is due to the screening of the Coulomb field by the electrons and muons. It is given by

$$\Delta m_i^{(\mathrm{Coul})}(\{n_i^{(v)}\}) = -\frac{3}{5}\frac{Q_i^2\alpha}{R_i}\left(\frac{3}{2}x_i - \frac{1}{2}x_i^3\right) \tag{48}$$

in Wigner-Seitz (WS) approximation for a nucleus $i = (A, Z)$ with radius R_i that is estimated as

$$R_i = R_0 A_i^{1/3} \tag{49}$$

using the radius parameter $R_0 = 1.25$ fm. The quantity $\alpha = e^2/(\hbar c)$ in (48) denotes the fine structure constant. The ratio

$$x_i = \frac{R_i}{R_{i,q}} \tag{50}$$

in equation (48) contains the WS cell radius

$$R_{i,q} = \left(-\frac{3Q_i}{4\pi n_q^{(l)}}\right)^{1/3} \tag{51}$$

with the total leptonic charge density

$$n_q^{(l)} = \sum_{i=e^-,e^+,\mu^-,\mu^+} Q_i n_i = -n_{e^-} + n_{e^+} - n_{\mu^-} + n_{\mu^+} \tag{52}$$

that is negative. Because the system is charge neutral, the negative of $n_q^{(l)}$ is identical to the total hadronic charge density

$$n_q^{(h)} = \sum_{i=n,\bar{n},p,\bar{p},(A,Z)} Q_i n_i^{(v)} = Y_q n_b \tag{53}$$

with the charge fraction Y_q. The strong shift in equation (47) is given by

$$\Delta m_i^{(\mathrm{strong})}(T, \{n_i^{(v)}\}) = f_i(n_i^{(\mathrm{eff})}, n_i^{(\mathrm{diss})}) B_i^{(0)} \tag{54}$$

with the binding energy $B_i^{(0)}$ of the nucleus in the vacuum and a shift function f_i, which depends on the effective density

$$n_i^{(\text{eff})} = \frac{2}{A_i}\left[Z_i Y_q + (N_i)(1 - Y_q)\right] n_b \tag{55}$$

and the dissolution density $n_i^{(\text{diss})}$. The functional form of the shift function is chosen as

$$f_i = \begin{cases} x & \text{if} \quad x \leq 1 \\ x + \frac{(x-1)^3(y-1)}{3(y-x)} & \text{if} \quad x > 1 \text{ and } x < y \end{cases} \tag{56}$$

with the two parameters

$$x = \frac{n_i^{(\text{eff})}}{n_i^{(\text{diss})}} \tag{57}$$

and

$$y = \frac{n_{\text{sat}}}{n_i^{(\text{diss})}}. \tag{58}$$

For light nuclei, the dissolution density is

$$n_i^{(\text{diss})} = \frac{B_i^{(0)}}{\delta B_i(T)} \tag{59}$$

where the quantity $\delta B_i(T)$ is defined in Ref. 33. For nucleon-nucleon scattering states, the shift function of the deuteron is used. For heavy nuclei, the parametrisation

$$x = \frac{n_i^{(\text{eff})}}{n_{\text{sat}}} y \tag{60}$$

with

$$y = 3 + \frac{28}{A} \tag{61}$$

is used such that

$$n_i^{(\text{diss})} = \frac{n_{\text{sat}}}{y}. \tag{62}$$

For $A = 4$, the approximate dissolution density of α-particles at zero temperature is reproduced. With increasing mass number A, the dissolution density $n_i^{(\text{diss})}$ approaches one third of the saturation density. The shift function (56) replaces the corresponding function in equation (72) of Ref. 33. In the present form, the binding energy of a nucleus decreases linearly with the density until the binding threshold is reached. At higher densities, f_i grows more rapidly since the state becomes a resonance. The strong increase also avoids the reäppearance of nuclei, in particular light clusters, at very high densities.

The degeneracy factors g_i of the elementary particles and light nuclei (d, t, h, α) are given by

$$g_i = 2J_i + 1 \tag{63}$$

with the spin J_i of the particle (in the ground state). The contribution of excited states is considered for heavy nuclei with $A > 4$ by introducing temperature dependent degeneracies

$$g_i(T) = \left[2J_i + 1 + \int_0^{E_{\max}} d\varepsilon\, \varrho_i(\varepsilon) \exp\left(-\frac{\varepsilon + \Delta_i}{T}\right)\right] \gamma(T) \tag{64}$$

with two factors. The first term contains the contribution of the ground state and of excited states with a density of states $\varrho_i(\varepsilon)$. The second factor $\gamma(T)$ is introduced to model the dissolution of heavy nuclei at high temperatures. Experimental values for the ground state spin J_i of nuclei are used as far as known. Otherwise $J_i = 0$, $J_i = 1/2$ or $J_i = 1$ are assumed for even-even, odd-even/even-odd or odd-odd nuclei, respectively. The density of states is given by a modified Fermi gas formula[85]

$$\varrho_i(\varepsilon) = \frac{\sqrt{\pi}}{24} \frac{a_i}{\sqrt{a_i^{(n)} a_i^{(p)}}} \frac{\exp\left(\beta_i \varepsilon + \frac{a_i}{\beta_i}\right)}{\left(\beta_i \varepsilon^3\right)^{1/2}} \frac{1 - \exp\left(-\frac{a_i}{\beta_i}\right)}{\left[1 - \frac{1}{2}\beta_i \varepsilon \exp\left(-\frac{a_i}{\beta_i}\right)\right]^{1/2}} \tag{65}$$

with the level density parameters

$$a_i^{(j)} = \frac{\pi^2}{3} g_i^{(j)} \tag{66}$$

for neutrons and protons ($j = n, p$) in a nucleus i and their sum $a_i = a_i^{(n)} + a_i^{(p)}$. The factor

$$g_i^{(j)} = \frac{m_j k_i^{(j)}}{2\pi^2} \tag{67}$$

depends on the Fermi momenta

$$k_i^{(j)} = \left[\frac{3\pi^2 N_i^{(j)}}{V_i}\right]^{1/3} \tag{68}$$

with the number of nucleons ($N_i^{(n)} = N$, $N_i^{(p)} = Z$) in the nucleus $i = (A, Z)$ and its effective volume

$$V_i = \frac{4\pi}{3} R_i^3 \tag{69}$$

with radius $R_i = 1.4 \cdot A^{1/3}$ fm. For given a_i and ϵ, the ratio a_i/β_i and thus β_i are determined by solving the equation

$$\left(\frac{a_i}{\beta_i}\right)^2 = a_i\varepsilon \left[1 - \exp\left(-\frac{a_i}{\beta_i}\right)\right] . \tag{70}$$

Then all quantities in equation (65) are known and the density of states can be calculated. In contrast to usual Fermi gas models, $\varrho_i(\epsilon)$ does not diverge for $\varepsilon \to 0$ but assumes a finite value. The maximum energy $E_{\max}$ in (65) is given by the minimum of the neutron and proton separation energies of the nucleus and $\Delta_i = \delta_i/A^{1/3}$ MeV is a simple pairing correction with $\delta_i = 0, 1, 2$ for even-even, odd-even/even-odd and odd-odd nuclei, respectively. The dissolution factor in equation (64) is written as

$$\gamma(T) = \begin{cases} \exp\left[-\left(\frac{T}{T_{\text{fl}}-T}\right)^2\right] & \text{if} \quad T < T_{\text{fl}} \\ 0 & \text{if} \quad T \geq T_{\text{fl}} \end{cases} \tag{71}$$

with the flashing temperature $T_{\text{fl}} = 11.26430$ MeV of the DD2-RMF model. It is defined by the condition that the pressure $p(T, n_b)$ attains a minimum of $p = 0$ as a function of the baryon density n_b for constant temperature in symmetric nuclear matter. T_{fl} is slightly lower than the critical temperature $T_{\text{cr}} = 13.72384$ MeV of the liquid-gas phase transition at which $\partial p/\partial n_b|_{T_{\text{cr}}} = 0$. For $T < T_{\text{cr}}$ uniform nuclear matter is globally unstable and for $T < T_{\text{fl}}$ locally unstable to density fluctuations.

Because the system is presumed to be charge neutral, the leptonic charge density (52) has to compensate the hadronic charge density (53) such that $n_q^{(l)} + n_q^{(h)} = 0$. This condition reduces the dimension of the thermodynamic space of variables and determines the leptonic chemical potential μ_l. In order to find the thermodynamic properties of the ensemble for given baryon density n_b and hadronic charge fraction Y_q, the remaining independent chemical potentials μ_b and μ_q are varied such that n_b and Y_q are reproduced when the coupled set of equations for the densities and potentials are solved self-consistently. Predictions of the gRDF model for a number of thermodynamic quantities and for the chemical composition of matter are available in tabular form according to the CompOSE[74] format.

The table covers temperatures from 0.1 MeV to 100 MeV, baryon densities from 10^{-10} fm^{-3} to 1 fm^{-3}, and hadronic charge fractions from 0.01 to 0.60.

2.3. *HS model*

The statistical model of Hempel and Schaffner-Bielich,[16] denoted HS model in the following, is formulated using a canonical ensemble of nucleons, nuclei, electrons and photons by specifying the free energy density

$$f(T,\{n_i\}) = \omega(T,\{\mu_i\}) + \sum_i \mu_i n_i. \tag{72}$$

It can be obtained from the grand canonical potential density (4) by applying a Legendre transformation to exchange the chemical potentials μ_i with the particle number densities n_i as natural variables. There are different variants of the HS model,[16,20,28,32] depending on the choice of the parametrization of the nuclear interaction and the table of nuclei. Here we consider the version of Ref. 32 with the DD2 parametrization.

The set of particle degrees of freedom is similar to that of the gRDF model, however, muons and two-nucleon scattering correlations are not taken into account in the HS model. An extensive table of nuclei is considered with masses taken from the 2003 Atomic Mass Evaluation[86] if available or otherwise from the microscopic-macroscopic FRDM model.[87] Effects of the nuclear interaction are included in a RMF model with density-dependent couplings in the DD2 parametrization[33] identical to the gRDF approach of the previous section. However, in the HS model, nucleons inside nuclei do not couple to the mesons, i.e., the coupling factors in equations (15) and (16) are defined as

$$g_{i\omega} = \begin{cases} B_i & \text{if} \quad i = n, p, \bar{n}, \bar{p} \\ 0 & \text{if} \quad i = (A, Z) \end{cases} \tag{73}$$

$$g_{i\sigma} = \begin{cases} |B_i| & \text{if} \quad i = n, p, \bar{n}, \bar{p} \\ 0 & \text{if} \quad i = (A, Z) \end{cases} \tag{74}$$

$$g_{i\rho} = \begin{cases} B_i & \text{if} \quad i = n, \bar{n} \\ -B_i & \text{if} \quad i = p, \bar{p} \\ 0 & \text{if} \quad i = (A, Z) \end{cases} \tag{75}$$

$$g_{i\delta} = 0. \tag{76}$$

Correspondingly, only nucleons are considered in the sum (14) with $D_i = B_i$ that appear as argument $\tilde{n}$ in the meson-nucleon couplings Γ_j.

For the degeneracy factors g_i of the nuclei, the form

$$g_i(T) = 2J_i + 1 + \int_0^{E_{\max}} d\varepsilon\, \varrho_i(\varepsilon) \exp\left(-\frac{\varepsilon}{T}\right) \tag{77}$$

is employed with the empirical density of states

$$\varrho_i(\varepsilon) = \frac{c_1}{A^{5/3}} \exp\left(\sqrt{2a(A)\varepsilon}\right) \tag{78}$$

containing the mass dependent level density parameter

$$a(A) = \frac{A}{8}\left(1 - \frac{c_2}{A^{1/3}}\right) \text{MeV}^{-1} \tag{79}$$

and constants $c_1 = 0.2$ MeV^{-1} and $c_2 = 0.8$. The maximum energy $E_{\max}$ for the integration in (77) is given by the binding energy of the specific nucleus such that only bound excited states are taken into account. The ground state spins in (77) are taken as $J_i = 0$ and $J_i = 1$ for even A and odd A nuclei, respectively.

The HS model uses the concept of finite sizes for nucleons and nuclei with two realizations of the excluded-volume mechanism in order to incorporate the dissolution of nuclei with increasing density of the system. All other particles (electrons, photons) are treated as pointlike. For nuclei $i = (A, Z)$ the available volume fractions are given by the standard form (2) with volume parameters $v_{ij} = B_j/n_{\text{sat}}$ (j = nucleon or nucleus) proportional to the baryon number B_j. This leads to identical forms

$$\Phi_{(A,Z)} = \kappa = \begin{cases} 1 - \frac{n_b}{n_{\text{sat}}} & \text{if} \quad n_b \leq n_{\text{sat}} \\ 0 & \text{if} \quad n_b > n_{\text{sat}} \end{cases} \tag{80}$$

with the baryon density (46) independent of the nucleus (A, Z). The available volume fraction $\Phi_{(A,Z)}$ vanishes for $n_b \geq n_{\text{sat}}$ and hence the occurence of nuclei at densities above saturation is prohibited. For nucleons $i = n, p, \bar{n}, \bar{p}$ the filling factor

$$\xi = 1 - \sum_{(A,Z)} A \frac{n_{(A,Z)}}{n_{\text{sat}}} \tag{81}$$

is introduced in the HS model where only nuclei contribute in the sum. Hence, nucleons do not block the volume that is available for themselves. They only feel the occurrence of nuclei. However, the factor ξ cannot be considered as an available volume fraction in the sense of Section 2.1 and thus we set $\Phi_i = 1$ for nucleons in the following.

As in the gRDF model, the interaction of the nuclei with the background electrons is taken into account in the HS model using the Wigner-Seitz approximation. This can be formulated as a mass shift, which is given by equation (48). However, in the HS model, the mass shift Δm_i is replaced by an explicit energy shift for nuclei

$$\Delta E_i(\{n_i^{(v)}\}) = -\frac{3}{5}\frac{Q_i^2\alpha}{R_i}\left(\frac{3}{2}x_i - \frac{1}{2}x_i^3\right) \tag{82}$$

of the same form as in equation (48). The radius of a nucleus is calculated from

$$R_{(A,Z)} = \left(\frac{3A}{4\pi n_{\text{sat}}}\right)^{1/3} . \tag{83}$$

Since the Coulomb energy shift (82) depends only on the electron and positron densities, there are only energy rearrangement contributions for leptons. Mass shifts of nuclei are not considered in the HS model.

Using the formalism of Section 2.1, the free energy density (72) of the HS model can be deduced. Since the available volume fractions Φ_i are either constant one or assume the standard form (2), the simplified form (37) of the grand canonical potential density can be used to obtain $f(T,\{n_i\})$. It is convenient to examine the contributions of different particles to the free energy density separately, i.e., we write

$$f(T,\{n_i\}) = \sum_{(A,Z)} f_{(A,Z)} + f_{\text{nuc}} + f_e + f_\gamma \tag{84}$$

with the contribution of nuclei, nucleons (including mesons), electrons, and photons. Since nuclei are treated as nonrelativistic particles with Maxwell-Boltzmann statistics, we can use (30) and solve it for the chemical potential

$$\mu_i = T \ln\left[\frac{n_i\,(\lambda_i^*)^3}{g_i\Phi_i}\right] + m_i^* + V_i. \tag{85}$$

We note that $V_i = \Delta E_i$ and $m_i^* = m_i$ for nuclei. The first term in equation (84) is given by

$$\sum_{(A,Z)} f_{(A,Z)} = \sum_{(A,Z)} \left(\omega_{(A,Z)}^{(0)} + \mu_{(A,Z)} n_{(A,Z)}\right)$$

$$= \sum_{(A,Z)} \left\{ T \ln \left[\frac{n_{(A,Z)} \lambda^3_{(A,Z)}}{g_{(A,Z)} \Phi_{(A,Z)}} \right] - T + m_{(A,Z)} + V_i \right\} n_{(A,Z)}$$

$$= \sum_{(A,Z)} f^{(0)}_{(A,Z)}(T, n_{(A,Z)}) - T \sum_{(A,Z)} n_{(A,Z)} \ln \kappa + f_{\text{Coul}} \quad (86)$$

with the free energy density

$$f_i^{(0)}(T, n_i) = \left[T \ln \left(\frac{n_i \lambda_i^3}{g_i} \right) - T + m_i \right] n_i \quad (87)$$

for an ideal gas of particles i with rest mass m_i and thermal wavelength $\lambda_i = \sqrt{2\pi/(m_i T)}$. The explicit excluded-volume term in (86) contains the available volume fraction $\kappa = \Phi_{(A,Z)}$. The Coulomb term

$$f_{\text{Coul}} = \sum_{(A,Z)} n_{(A,Z)} \Delta E^{(\text{Coul})}_{(A,Z)} \quad (88)$$

arises due to the energy shifts (82).

The nucleonic contribution to (84)

$$f_{\text{nuc}}(T, n_n, n_p, n_{\bar{n}}, n_{\bar{p}}) = \sum_{i=n,p,\bar{n},\bar{p}} \left(\omega_i^{(0)} + \mu_i n_i \right) + \omega_{\text{meson}} - \omega^{(r)}_{\text{meson}} \quad (89)$$

depends only on the temperature and the densities of the nucleons. This form is not used in the HS model directly but the rescaled contribution

$$\xi f_{\text{nuc}}(T, n'_n, n'_p, n'_{\bar{n}}, n'_{\bar{p}}) \quad (90)$$

with the scaled nucleon densities

$$n'_i = \frac{n_i}{\xi} \quad (91)$$

by introducing the filling factor (81). If nuclei exist in the system, the available volume for the nucleons is reduced, e.i., $\xi < 1$. Correspondingly, the effective density in the available volume has to be increased as compared to the densities in the total volume and the obtained results for the free energy density needs to be adjusted again to the total volume leading to the prefactor ξ in equation (90).

The screening of the Coulomb potential due to the electrons affects not only the energies of nuclei but also the thermodynamics of the electrons and positrons themselves. The scalar potential S_i vanishes but the vector potential is given by a rearrangment term

$$V_i = W_i^{(r)} = \sum_{(A,Z)} n_{(A,Z)} \frac{\partial \Delta E_{(A,Z)}}{\partial n_i}$$

$$= \mp \sum_{(A,Z)} n_{(A,Z)} \frac{3}{10} \frac{Z^2 \alpha}{R_{(A,Z)} n_e} \left[x_{(A,Z)} - x^3_{(A,Z)} \right] \tag{92}$$

with the net electron number density $n_e = n_{e^-} - n_{e^+}$. The negative (positive) sign applies to electrons (positrons). These particles get effective chemical potentials $\mu_i^* = \mu_i - V_i$ with the correct sign change of μ_i and V_i for the particle to antiparticle conversion. The electronic contribution to the total free energy density assumes the form

$$f_e = \sum_{i=e^-,e^+} \left(\omega_i^{(0)} + \mu_i n_i \right) - \omega_\Delta^{(r)} \tag{93}$$

with the explicit rearrangement term

$$\omega_\Delta^{(r)} = \sum_{i=e^-,e^+} n_i W_i^{(r)} = - \sum_{(A,Z)} n_{(A,Z)} \frac{3}{10} \frac{Z^2 \alpha}{R_{(A,Z)}} \left[x_{(A,Z)} - x^3_{(A,Z)} \right]. \tag{94}$$

3. Results for neutron star matter

Properties of stellar matter are available for the gRDF and HS model for a large range of densities, temperatures and isospin asymmetries. In the present study they were extracted from EoS tables in the CompOSE format using the FORTRAN code `compose.f90`. In order to compare the two models, only a section through the full space of these variables is considered in the following. We restrict ourselves to the conditions of fully catalyzed neutron star matter, i.e., charge-neutral matter in β equilibrium without neutrinos. Then, the hadronic charge fraction Y_q is determined and the baryon density n_b and the temperature T remain as independent variables. In Fig. 1 the variation of Y_q as a function of n_b and T is depicted for the gRDF and the HS model. Regions of different charge fraction values are color-coded in steps of 0.05. For baryon densities below approx. 10^{-7} fm^{-3} and temperatures between 0.5 MeV and 3 MeV, Y_q is above 0.6 and thus outside the range of the EoS tables. In general, there is a rather good agreement of the predictions between the two models. For constant temperature, there is a decrease of the charge fraction with increasing baryon density as

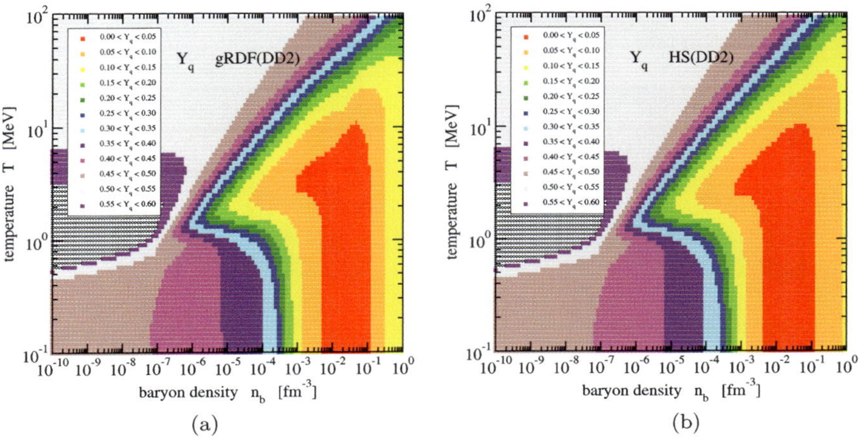

Fig. 1. Hadronic charge fraction Y_q of neutron star matter for the gRDF model (a) and the HS model (b). The area of datapoints outside the EoS table (Y_q larger than 0.6) is indicated by crosses.

long as n_b is lower than approx. $4 \cdot 10^{-2}$ fm^{-3}. For $T > 10$ MeV a further decrease of Y_q with n_b is observed, whereas the charge fraction increases at lower T for the highest baryon densities. Above the nuclear saturation density, a somewhat larger difference in Y_q between the models can be noticed because the gRDF model considers muons in contrast to the HS model. The lowest charge fractions with $Y_q < 0.05$ are found for $T < 10$ MeV in the baryon density range from roughly $5 \cdot 10^{-3}$ fm^{-3} to 10^{-1} fm^{-3}. At the lowest temperatures, the hadronic charge fraction evolves with n_b as expected for cold neutron stars indicating the gradual neutronization of matter with increasing depth.

3.1. *Chemical composition*

The change of the hadronic charge fraction is accompanied with a change of the chemical composition in neutron star matter. The gRDF and HS models predict to some extent differences for both light and heavy nuclei. We first consider the isospin zero light nuclei. In Fig. 2 the mass fractions

$$X_i = A_i \frac{n_i}{n_b} \tag{95}$$

of deuterons and α particles, respectively, are depicted in the full range of baryon densities and temperatures. The contour lines of the color-coded

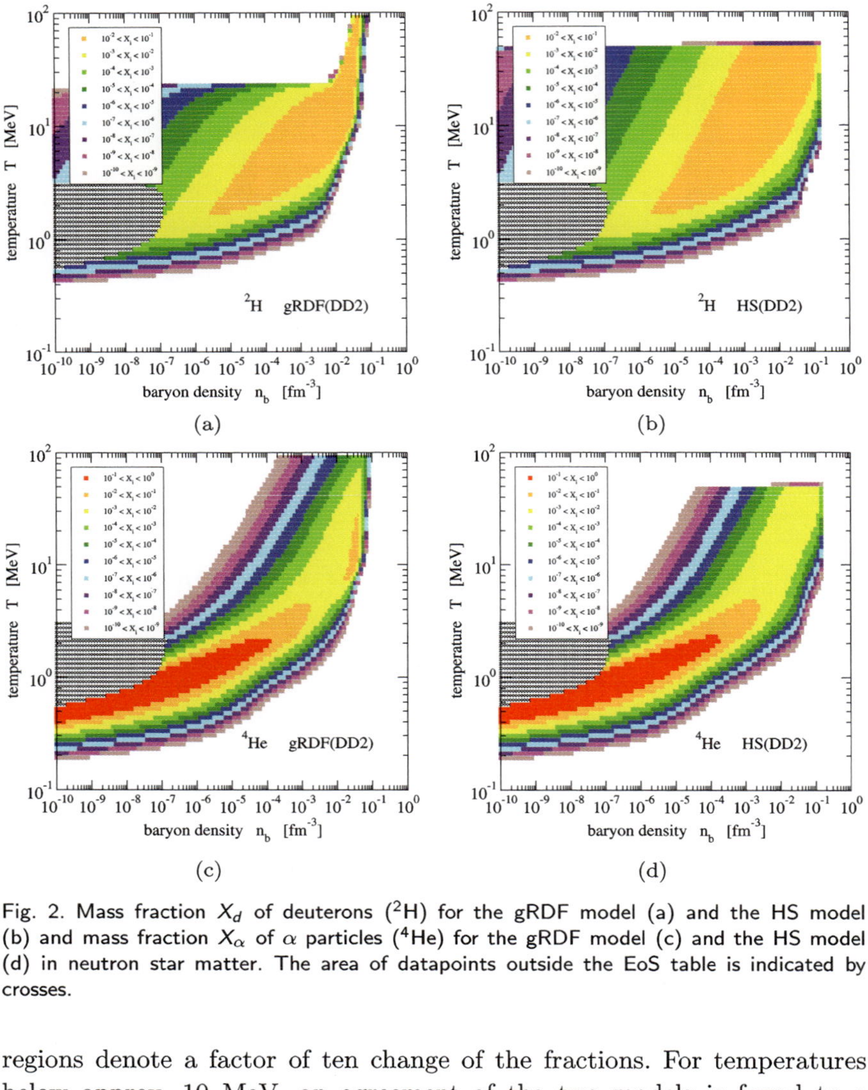

Fig. 2. Mass fraction X_d of deuterons (^{2}H) for the gRDF model (a) and the HS model (b) and mass fraction X_α of α particles (^{4}He) for the gRDF model (c) and the HS model (d) in neutron star matter. The area of datapoints outside the EoS table is indicated by crosses.

regions denote a factor of ten change of the fractions. For temperatures below approx. 10 MeV, an agreement of the two models is found to a large extent. In most cases, the α-particle fraction exceeds that of the deuteron, except for densities above 10^{-2} fm^{-3} and temperatures larger than approx. 4 MeV. The maximum fractions of deuterons and, in particular α particles, are located in a rather narrow band of increasing density and temperature. At temperatures $T > 20$ MeV, the largest differences between the models are observed. In the gRDF model, the net deuteron fraction decreases quickly with increasing temperature, for baryon densities $n_b < 10^{-2}$ fm^{-3}, because the ground state contribution is compensated by the contribution of the corresponding scattering channel resulting in a disappearance of deuterons at higher temperatures. Contributions from two-nucleon scattering states are not considered in the HS model. Hence, the deuteron survives to much higher temperatures. For α-particles, a compensating effect due to scattering states is neglected in both models because the ground state is much stronger bound as compared to the deuteron. In the HS model, the occurrence of clusters is suppressed at temperatures above 50 MeV by introducing an artificial cutoff. This is most apparent for the α-particle mass fractions in the bottom of Fig. 2.

The observations for the deuteron and the α particle apply in a similar way to the ^{3}H and ^{3}He nuclei, with isospin 1/2 and opposite neutron-to-proton composition. A comparison in Fig. 3 clearly shows that the more neutron-rich nucleus ^{3}H is more likely to be found in neutron star matter with low hadronic charge fraction Y_q, c.f. Fig. 1, than the proton-rich nucleus ^{3}He. In the gRDF model, both the triton and the helion are predicted to reappear at the highest temperatures for baryon densities below approx. 10^{-6} fm^{-3}. This can be explained by a simple thermodynamic consequence since for $T > 70$ MeV binding energies have a rather minor effect on the distribution functions. A similar feature is expected for the HS model, if the cutoff for considering clusters was not introduced at 50 MeV.

Larger differences in the predictions for light nuclei in the gRDF and HS model are visible for densities above 10^{-3} fm^{-3}. This becomes more apparent in Fig. 4, where the evolution of the mass fractions with density is depicted for several isotherms and a restricted density range, as compared to Figures 2 and 3. At densities lower than approx. 10^{-3} fm^{-3}, the predictions of the two models are almost identical. Here, mass-shifts or excluded-volume effects hardly affect the mass fractions of light nuclei. But at higher densities, distinct differences appear. The HS model with excluded-volume mechanism predicts a more gradual dissolution of the light clusters with density than the gRDF model with mass shifts.

At temperatures below approx. $0.5 - 1.0$ MeV, light clusters are hardly found in neutron star matter. The variation of the mass fractions with temperature for three fixed values of the baryon number density n_b are shown in Fig. 5. At low densities of 10^{-4} fm^{-3} or 10^{-3} fm^{-3}, there are only minor differences between the gRDF and HS model in the mass fractions. Only

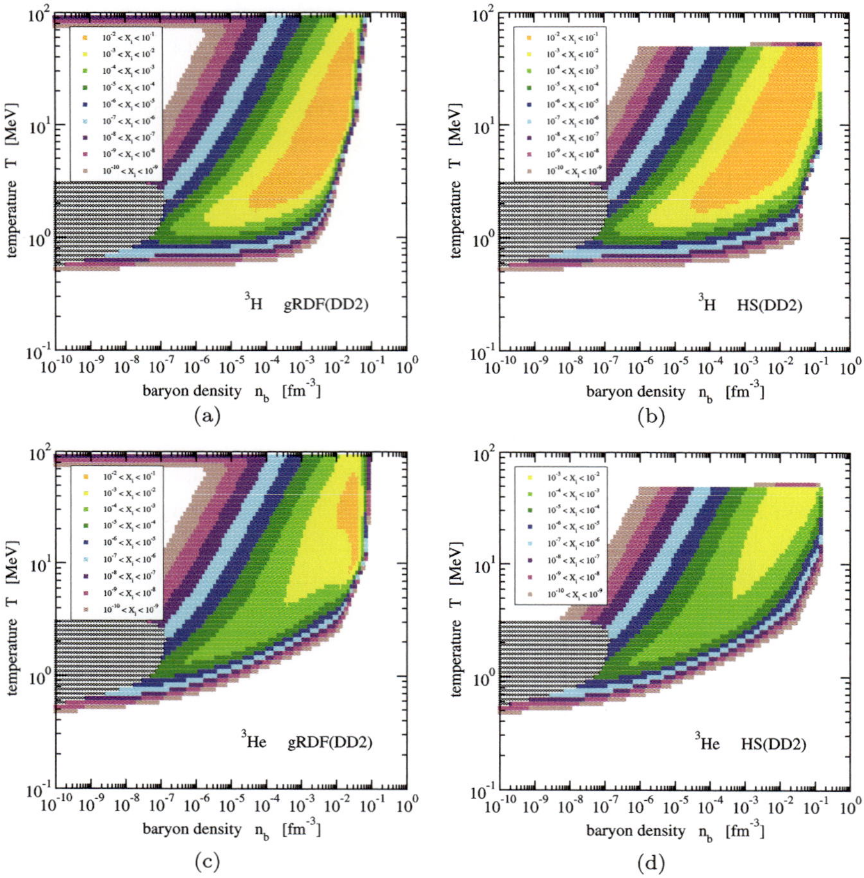

Fig. 3. Mass fraction X_t of tritons (^{3}H) for the gRDF model (a) and the HS model (b) and mass fraction X_h of helions (^{3}He) for the gRDF model (c) and the HS model (d) in neutron star matter. The area of datapoints outside the EoS table is indicated by crosses.

at 10^{-2} fm^{-3} the predictions of the models differ more strongly. There are larger discrepancies at lower temperatures because here differences in the cluster masses or excluded-volume effects are more relevant in the thermodynamic description. In case of the gRDF model, the net deuteron fraction rapidly decreases with temperature, when T becomes larger than approx. 20 MeV. This is caused by the continuum contribution in the np($^3S_1$) channel that compensates the deuteron ground state contribution.

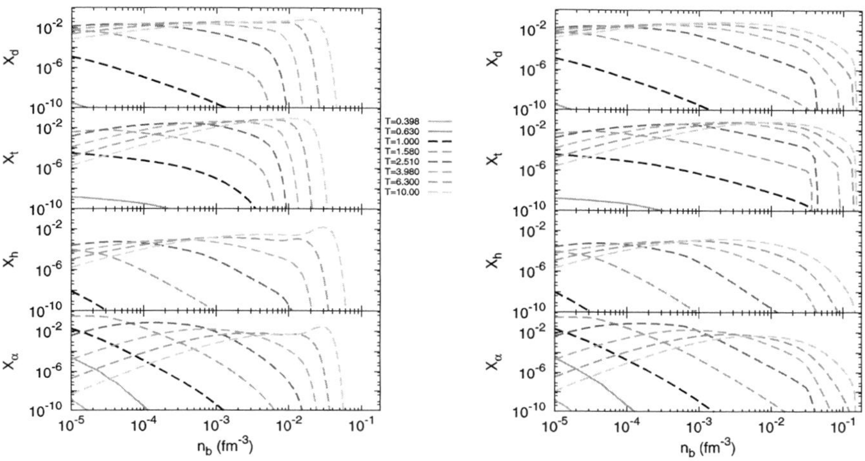

(a) (b)

Fig. 4. (Color online) Mass fractions of the light clusters, ^2H (d), ^3H (t), ^3He (h) and ^4He (α) for the gRDF model (a) and for the HS model (b) for several isotherms.

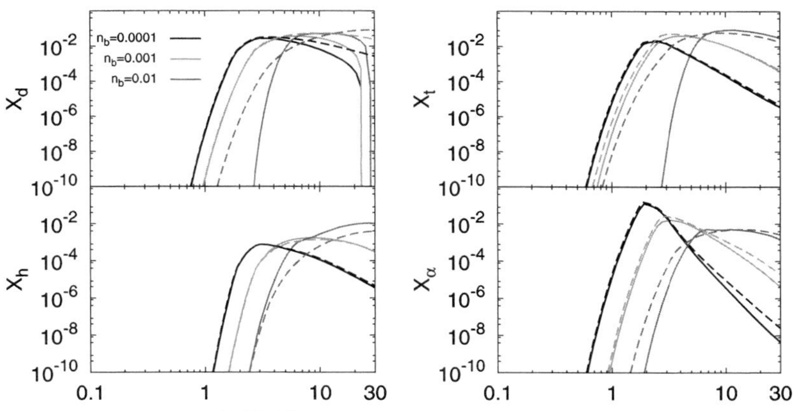

Fig. 5. (Color online) Mass fractions of the light clusters, ^2H (d), ^3H (t), ^3He (h) and ^4He (α), for the gRDF model (full lines) and the HS model (dashed lines) as a function of the temperature for different values of the baryon number density.

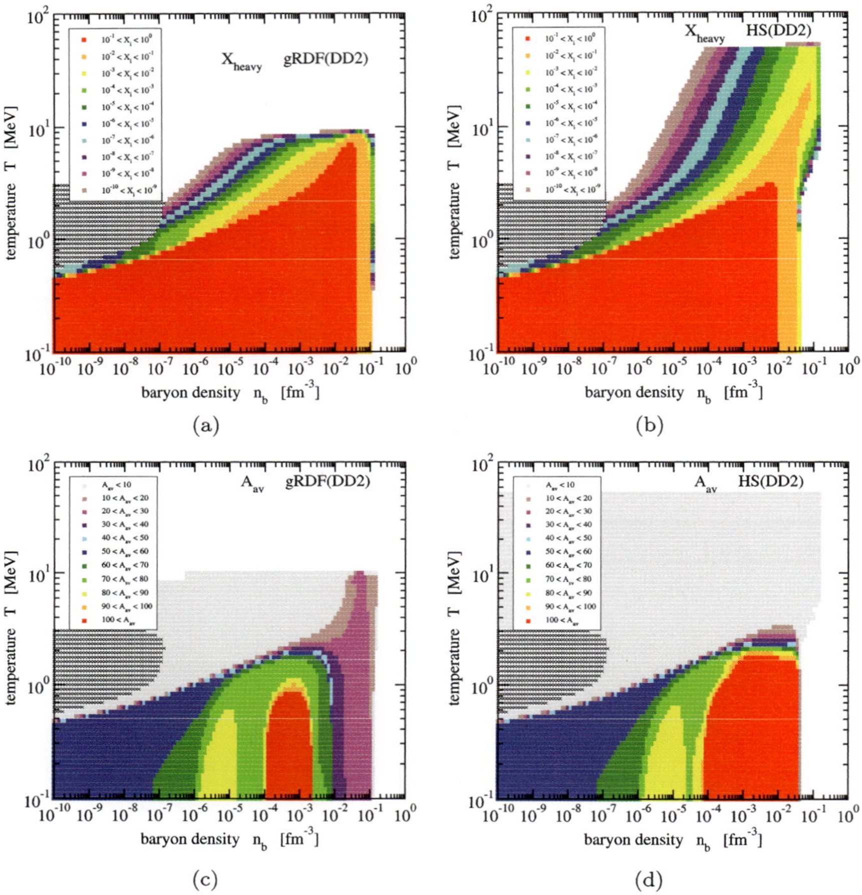

Fig. 6. Mass fraction X_{heavy} of heavy nuclei for the gRDF model (a) and the HS model (b) and average mass number A_{av} of heavy nuclei for the gRDF model (c) and the HS model (d) in neutron star matter. The area of datapoints outside the EoS table is indicated by crosses.

The chemical composition of neutron star matter at low temperatures is dominated by heavy nuclei with mass numbers $A > 4$. This is evident from panels (a) and (b) of Fig. 6, that depict the total mass fraction of heavy nuclei

$$X_{\text{heavy}} = \sum_{(A,Z),A>4} X_{(A,Z)} \tag{96}$$

with the mass fraction $X_{(A,Z)} = An_{(A,Z)}$ of individual nuclei as a function of baryon density and temperature. Below $T \approx 2$ MeV, the gRDF model and the HS model look very similar but the dissolution of heavy clusters occurs at lower densities in the HS model, in the range from 10^{-2} fm^{-3} to $5 \cdot 10^{-2}$ fm^{-3}, whereas heavy nuclei survive up to densities somewhat above 10^{-1} fm^{-3}. The disappearance of heavy nuclei with increasing temperature for constant baryon density shows some distinct differences, when the two models are compared. The $\gamma(T)$ factor in the degeneracy factor (64) of the gRDF model causes the heavy clusters to be removed from the system above approx. 10 MeV. In the HS model, they appear at even higher temperatures, since the excluded-volume mechanism as a geometric concept is only depending on the density but not on the temperature. The reduction of X_{heavy} in the HS model with increasing temperature is only a consequence of statistics. In panel (b) of Fig. 6 we notice again the artificial cluster cutoff at 50 MeV in this model.

In order to assess the importance of heavy nuclei in neutron star matter, not only the mass fraction X_{heavy}, but also the size of the nuclei has to be examined. In panels (c) and (d) of Fig. 6 the variation of the average mass number

$$A_{\text{av}} = \frac{1}{X_{\text{heavy}}} \sum_{(A,Z),A>4} AX_{(A,Z)} \tag{97}$$

is illustrated, where the contour lines indicate a change by ten units. For baryon densities below 10^{-5} fm^{-3}, both models are very similar, and the expected increase of the average mass number with density is observed for the lowest temperatures. Above 10^{-5} fm^{-3}, however, the models exhibit significant differences. In the gRDF model, the highest average mass numbers are found for temperatures below 1 MeV, in the density range from 10^{-4} fm^{-3} to $2 \cdot 10^{-3}$ fm^{-3}. At even higher densities, A_{av} continuously decreases to smaller values, before the heavy clusters dissolve, cf. panel (a) of Fig. 6. The situation is very different in the HS model. Here, the heaviest clusters survive almost until they disappear when the density increases. When the temperature increases above 3 MeV, the average mass number of heavy nuclei quickly shrinks, and the chemical composition is governed by light clusters.

The change of the average mass numbers A_{av} is accompanied by a similar variation of the average charge number Z_{av} and the average neutron number N_{av} of heavy nuclei. These quantities are depicted in Fig. 7 for the gRDF model and the HS model in the same style as A_{av} in Fig. 6. The

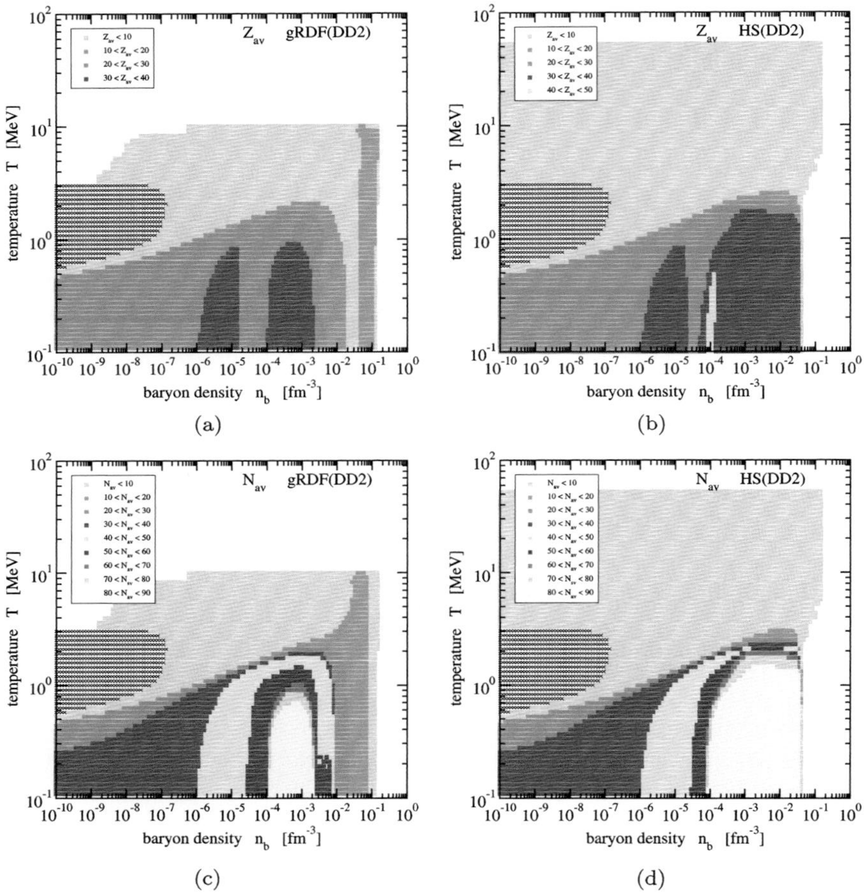

Fig. 7. (Color online) Average charge number $Z_{\rm av}$ of heavy nuclei for the gRDF model (a) and the HS model (b) and average neutron number $N_{\rm av}$ of heavy nuclei for the gRDF model (c) and the HS model (d) in neutron star matter. The area of datapoints outside the EoS table is indicated by crosses.

difference between the models are again clearly visible. Both $Z_{\rm av}$ and $N_{\rm av}$ decrease smoothly with increasing density above $2 \cdot 10^{-3}$ fm^{-3} for constant temperature below 2 MeV in the gRDF model, and the maximum values are found in the density range from 10^{-4} fm^{-3} to $2 \cdot 10^{-3}$ fm^{-3}. In the HS model, large average charge and neutron number appear also at higher densities. Because the matter is very neutron rich, cf. Fig. 1, $N_{\rm av}$ reaches larger values than $Z_{\rm av}$. At the lowest temperatures, the sequence of charge and neutron

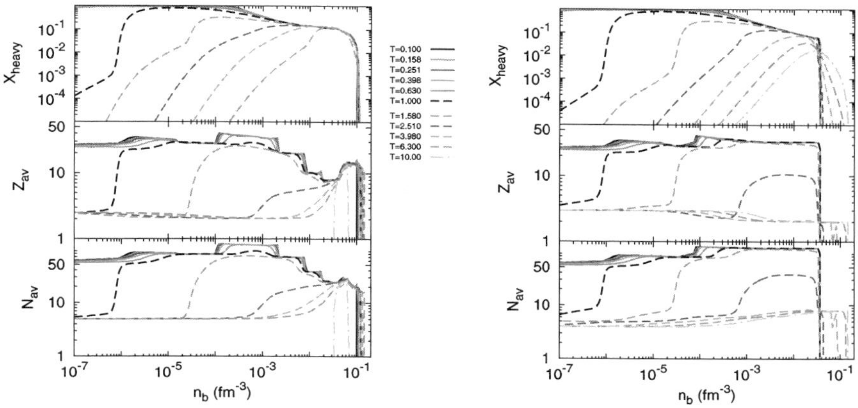

(a) (b)

Fig. 8. (Color online) Mass fraction X_{heavy}, average charge number Z_{av} and average neutron number N_{av} of heavy nuclei as a function of the baryon number density n_b for constant temperatures in the gRDF model (a) and in HS model (b).

numbers with increasing baryon density is similar to that expected in the crust of cold neutron stars.

The variation of X_{heavy}, Z_{av} and N_{av} with increasing density for several isotherms in geometric progression is shown in Fig. 8, for a restricted range of densities. Heavy nuclei dominate the chemical composition at temperatures below approx. 0.6 MeV and baryon densities below approx. $2 \cdot 10^{-4}$ fm^{-3}, the neutron drip density in cold neutron star matter. In the HS model, heavy nuclei disappear abruptly at densities of about one fifth of the saturation density at low temperatures, but survive for temperatures above 1 MeV also at higher densities. They dissolve at almost the same density of $n_{\text{sat}}/3$, irrespective of the temperature in the gRDF model. Shell effects in the average charge and neutron numbers are clearly visible at low temperatures since the distribution of heavy nuclei is dominated by a few species. The jumps at the magic numbers are washed out with increasing temperature. A comparison of the gRDF results with those of the HS model shows that the size of the heavy clusters gradually reduces with increasing density in the former model but stays almost constant up to the dissolution density in the latter model.

The fraction of heavy nuclei and their average mass number as a function of the temperature for fixed baryon densities is depicted in Fig. 9. The dissolution of heavy nuclei with increasing temperature is evident with a

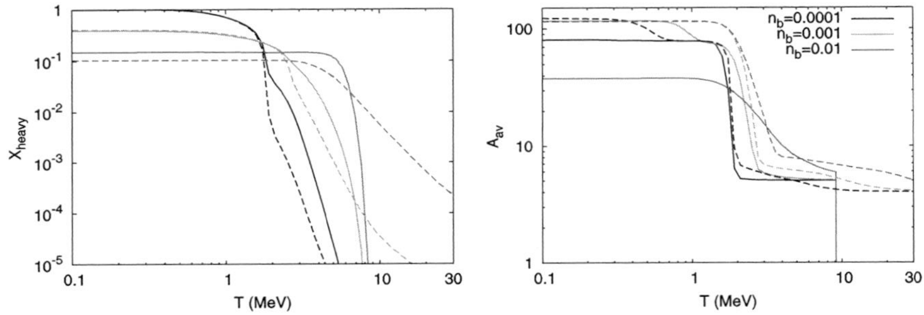

Fig. 9. (Color online) Mass fraction X_{heavy} and average mass number A_{av} of heavy nuclei as a function of the temperature T for constant baryon number densities n_b in the gRDF model (full lines) and in the HS model (dashed lines).

steeper decrease in the gRDF model, where heavy nuclei are suppressed, if T approaches 10 MeV. At the highest density of 10^{-2} fm^{-3}, differences in X_{heavy} between the models, even at low temperatures, are observed. The evolution of the average mass fraction with temperature also exhibits clear differences, depending on the chosen value of n_b in accordance with the lower panel of Fig. 6. The sudden change of A_{av} with the temperature at low T is again related to shell effects that are taken care of in both models due to the use of realistic mass tables.

3.2. *Thermodynamic properties*

Besides the chemical composition of neutron star matter, the predictions of the gRDF model and the HS model for thermodynamic quantities can be compared. As representative examples, the pressure p and the entropy per baryon $S/A = s/n_b$ are depicted in Fig. 10 as a function of baryon density and temperature. The pressure and entropy per baryon along lines that separate differently colored regions differ by a factor of ten. There is a systematic increase of the pressure with n_b and T reaching the largest values at baryon densities approaching 1 fm^{-3}. In contrast, the highest entropy per baryon is found at the highest temperatures close to 100 MeV, at the lowest baryon densities. In general, S/A decreases with decreasing temperature and increasing baryon density. There is an exception from this trend in the region close to the line of constant $S/A = 1$, where the chemical composition of the matter rapidly changes. Overall there is a surprising agreement of the pressure and entropy per baryon between the two models, despite the very different approaches to model the dissolution of

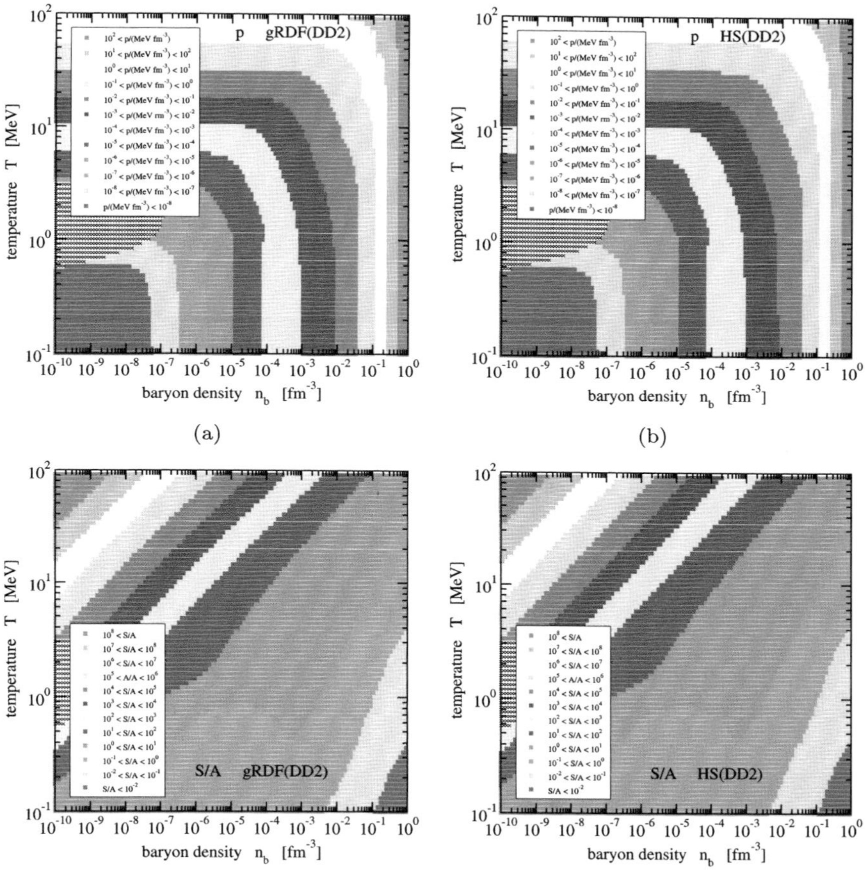

Fig. 10. (Color online) Pressure p for the gRDF model (a) and the HS model (b) and entropy per baryon S/A for the gRDF model (c) and the HS model (d) in neutron star matter. The area of datapoints outside the EoS table is indicated by crosses.

nuclei. A similar concordance of thermodynamic quantities, such as the free energy per baryon or chemical potentials is observed. Hence, we refrain from showing the corresponding figures.

4. Conclusions

In the present work, we presented and compared two different theoretical approaches to model the equation of state and in particular the formation and dissolution of nuclear clusters in stellar matter: a statistical model with excluded-volume effects (HS model) and a generalized relativistic density functional (gRDF model) with an in-medium change of the cluster masses. The theoretical formalism of the two approaches was presented in detail and the essential differences were delineated. Both models use a relativistic mean-field description for the nucleonic part with the same parametrisation of the density dependent meson-nucleon couplings, but they differ in the treatment of the cluster degrees of freedom. Data on the chemical composition and on thermodynamic properties are available in tabular form for both models in a wide range of baryon densities, temperatures and isospin asymmetries in the CompOSE format. The corresponding tables are taken from the CompOSE website.

We were interested mainly in the evolution of the light and heavy clusters with density and temperature. For that purpose, we studied neutron star matter, i.e. charge-neutral matter in β equilibrium, and observed that the gRDF and the HS models predict some differences for the mass fractions and average sizes of the clusters. Both models behave similarly for low temperatures and low densities where mass shifts or excluded-volume effects are negligible. When the baryon density approaches the nuclear saturation density, the dissolution of clusters is more gradual in the gRDF model than in the HS model. Heavy clusters disappear rather abruptly with increasing density in the HS model, whereas in the gRDF model, their size gradually reduces until they melt.

Above temperatures of 50 MeV, light and heavy clusters are artificially suppressed in the HS model. In the gRDF model, heavy clusters are removed from the system when the temperature exceeds approximately 10 MeV due to the specific temperature dependence of the degeneracy factors. In the HS model, they do appear at even higher temperatures, since the excluded-volume mechanism does not depend on the temperature, but only on the density.

Despite the differences in the cluster abundances and properties, both models show overall a very good agreement of the thermodynamic properties. Thus it can be expected that dynamical simulations of core-collapse supernovae or neutron star mergers are only marginally affected by the choice of the cluster description in the equation of state as long as the interaction model for the nucleonic part is identical. However, the chemical composition of stellar matter could affect processes such as the neutrino transport or nucleosynthesis in the above mentioned astrophysical scenarios. These consequences of the theoretical cluster description can be explored in future studies.

Acknowledgments

This work was supported by NewCompStar, COST Action MP1304, by the Helmholtz Association (HGF) through the Nuclear Astrophysics Virtual Institute (VH-VI-417) and by the DFG through Grant No. SFB1245 H.P. is supported by FCT under Project No. SFRH/BPD/95566/2013. She is very thankful to S.T. and the Theory group at GSI for the kind hospitality during her stay there.

References

1. A. Mezzacappa, Ascertaining the Core Collapse Supernova Mechanism: The State of the Art and the Road Ahead, *Ann. Rev. Nucl. Part. Sci.* **55**, 467–515 (2005). doi: 10.1146/annurev.nucl.55.090704.151608.
2. H.-T. Janka, K. Langanke, A. Marek, G. Martinez-Pinedo and B. Mueller, Theory of Core-Collapse Supernovae, *Phys. Rept.* **442**, 38–74 (2007). doi: 10.1016/j.physrep.2007.02.002.
3. H.-T. Janka, Explosion Mechanisms of Core-Collapse Supernovae, *Ann. Rev. Nucl. Part. Sci.* **62**, 407–451 (2012). doi: 10.1146/annurev-nucl-102711-094901.
4. A. Burrows, Colloquium: Perspectives on core-collapse supernova theory, *Rev. Mod. Phys.* **85**, 245 (2013). doi: 10.1103/RevModPhys.85.245.
5. N. K. Glendenning, *Compact stars: Nuclear physics, particle physics, and general relativity*. Springer, New York (1997).
6. F. Weber, Strange quark matter and compact stars, *Prog.Part.Nucl.Phys.* **54**, 193–288 (2005). doi: 10.1016/j.ppnp.2004.07.001.
7. P. Haensel, A. Y. Potekhin and D. G. Yakovlev, *Neutron stars 1: Equation of state and structure* (2007).
8. A. Y. Potekhin, The physics of neutron stars, *Phys. Usp.* **53**, 1235–1256 (2010). doi: 10.3367/UFNe.0180.201012c.1279. [Usp. Fiz. Nauk180,1279 (2010)].
9. J. M. Lattimer, The nuclear equation of state and neutron star masses, *Ann.Rev.Nucl.Part.Sci.* **62**, 485–515 (2012). doi: 10.1146/annurev-nucl-102711-095018.
10. K. Hebeler, J. M. Lattimer, C. J. Pethick and A. Schwenk, Equation of state and neutron star properties constrained by nuclear physics and observation, *Astrophys. J.* **773**, 11 (2013). doi: 10.1088/0004-637X/773/1/11.

11. J. M. Lattimer and F. D. Swesty, A Generalized equation of state for hot, dense matter, *Nucl. Phys.* **A535**, 331–376 (1991). doi: 10.1016/0375-9474(91)90452-C.
12. H. Shen, H. Toki, K. Oyamatsu and K. Sumiyoshi, Relativistic equation of state of nuclear matter for supernova explosion, *Prog. Theor. Phys.* **100**, 1013 (1998). doi: 10.1143/PTP.100.1013.
13. H. Shen, H. Toki, K. Oyamatsu and K. Sumiyoshi, Relativistic equation of state of nuclear matter for supernova and neutron star, *Nucl. Phys.* **A637**, 435–450 (1998). doi: 10.1016/S0375-9474(98)00236-X.
14. T. Klähn *et al.*, Constraints on the high-density nuclear equation of state from the phenomenology of compact stars and heavy-ion collisions, *Phys. Rev.* **C74**, 035802 (2006). doi: 10.1103/PhysRevC.74.035802.
15. J. M. Lattimer and M. Prakash, Neutron Star Observations: Prognosis for Equation of State Constraints, *Phys.Rept.* **442**, 109–165 (2007). doi: 10.1016/j.physrep.2007.02.003.
16. M. Hempel and J. Schaffner-Bielich, Statistical Model for a Complete Supernova Equation of State, *Nucl. Phys.* **A837**, 210–254 (2010). doi: 10.1016/j.nuclphysa.2010.02.010.
17. H. Shen, H. Toki, K. Oyamatsu and K. Sumiyoshi, Relativistic Equation of State for Core-Collapse Supernova Simulations, *Astrophys. J. Suppl.* **197**, 20 (2011). doi: 10.1088/0067-0049/197/2/20.
18. G. Shen, C. J. Horowitz and S. Teige, A New Equation of State for Astrophysical Simulations, *Phys. Rev.* **C83**, 035802 (2011). doi: 10.1103/PhysRevC.83.035802.
19. G. Shen, C. J. Horowitz and E. O'Connor, A Second Relativistic Mean Field and Virial Equation of State for Astrophysical Simulations, *Phys. Rev.* **C83**, 065808 (2011). doi: 10.1103/PhysRevC.83.065808.
20. A. W. Steiner, M. Hempel and T. Fischer, Core-collapse supernova equations of state based on neutron star observations, *Astrophys. J.* **774**, 17 (2013). doi: 10.1088/0004-637X/774/1/17.
21. M. Barranco and J. R. Buchler, Thermodynamic properties of hot nucleonic matter, *Phys.Rev.* **C22**, 1729–1737 (1980). doi: 10.1103/PhysRevC.22.1729.
22. H. Muller and B. D. Serot, Phase transitions in warm, asymmetric nuclear matter, *Phys.Rev.* **C52**, 2072–2091 (1995). doi: 10.1103/PhysRevC.52.2072.
23. K. Bugaev, M. I. Gorenstein, I. Mishustin and W. Greiner, Exactly soluble model for nuclear liquid gas phase transition, *Phys.Rev.* **C62**, 044320 (2000). doi: 10.1103/PhysRevC.62.044320.
24. F. Gulminelli, P. Chomaz, A. H. Raduta and A. R. Raduta, The Influence of Coulomb on the liquid gas phase transition and nuclear multifragmentation, *Phys. Rev. Lett.* **91**, 202701 (2003). doi: 10.1103/PhysRevLett.91.202701.
25. C. Ducoin, P. Chomaz and F. Gulminelli, Role of isospin in the nuclear liquid-gas phase transition, *Nucl.Phys.* **A771**, 68–92 (2006). doi: 10.1016/j.nuclphysa.2006.03.005.

26. C. Ducoin, P. Chomaz and F. Gulminelli, Isospin fractionation: Equilibrium versus spinodal decomposition, *Nucl. Phys.* **A781**, 407–423 (2007). doi: 10.1016/j.nuclphysa.2006.11.047.
27. C. Ducoin, P. Chomaz and F. Gulminelli, Isospin-dependent clusterization of Neutron-Star Matter, *Nucl.Phys.* **A789**, 403–425 (2007). doi: 10.1016/j.nuclphysa.2007.03.006.
28. M. Hempel, T. Fischer, J. Schaffner-Bielich and M. Liebendorfer, New Equations of State in Simulations of Core-Collapse Supernovae, *Astrophys. J.* **748**, 70 (2012). doi: 10.1088/0004-637X/748/1/70.
29. S. Furusawa, S. Yamada, K. Sumiyoshi and H. Suzuki, A new baryonic equation of state at sub-nuclear densities for core-collapse simulations, *Astrophys. J.* **738**, 178 (2011). doi: 10.1088/0004-637X/738/2/178.
30. M. Hempel, V. Dexheimer, S. Schramm and I. Iosilevskiy, Noncongruence of the nuclear liquid-gas and deconfinement phase transitions, *Phys. Rev.* **C88** (1), 014906 (2013). doi: 10.1103/PhysRevC.88.014906.
31. S. Furusawa, K. Sumiyoshi, S. Yamada and H. Suzuki, New equations of state based on the liquid drop model of heavy nuclei and quantum approach to light nuclei for core-collapse supernova simulations, *Astrophys. J.* **772**, 95 (2013). doi: 10.1088/0004-637X/772/2/95.
32. T. Fischer, M. Hempel, I. Sagert, Y. Suwa and J. Schaffner-Bielich, Symmetry energy impact in simulations of core-collapse supernovae, *Eur. Phys. J.* **A50**, 46 (2014). doi: 10.1140/epja/i2014-14046-5.
33. S. Typel, G. Röpke, T. Klähn, D. Blaschke and H. H. Wolter, Composition and thermodynamics of nuclear matter with light clusters, *Phys. Rev.* **C81**, 015803 (2010). doi: 10.1103/PhysRevC.81.015803.
34. C. J. Horowitz and A. Schwenk, Cluster formation and the virial equation of state of low-density nuclear matter, *Nucl. Phys.* **A776**, 55–79 (2006). doi: 10.1016/j.nuclphysa.2006.05.009.
35. S. Heckel, P. P. Schneider and A. Sedrakian, Light nuclei in supernova envelopes: a quasiparticle gas model, *Phys. Rev.* **C80**, 015805 (2009). doi: 10.1103/PhysRevC.80.015805.
36. M. Ferreira and C. Providencia, Description of light clusters in relativistic nuclear models, *Phys. Rev.* **C85**, 055811 (2012). doi: 10.1103/PhysRevC.85.055811.
37. S. S. Avancini, C. C. Barros, Jr., D. P. Menezes and C. Providencia, α particles and the pasta phase, *Phys. Rev.* **C82**, 025808 (2010). doi: 10.1103/PhysRevC.82.025808.
38. S. S. Avancini, C. C. Barros, Jr, L. Brito, S. Chiacchiera, D. P. Menezes and C. Providencia, Light clusters in nuclear matter and the 'pasta' phase, *Phys. Rev.* **C85**, 035806 (2012). doi: 10.1103/PhysRevC.85.035806.
39. D. G. Ravenhall, C. J. Pethick and J. R. Wilson, Structure of matter below nuclear saturation density, *Phys. Rev. Lett.* **50**, 2066–2069 (1983). doi: 10.1103/PhysRevLett.50.2066.
40. R. D. Williams and S. E. Koonin, Sub-saturation phases of nuclear matter, *Nucl. Phys. A.* **435**, 844 (1985).

41. C. J. Horowitz, M. A. Perez-Garcia, J. Carriere, D. K. Berry and J. Piekarewicz, Nonuniform neutron-rich matter and coherent neutrino scattering, *Phys. Rev.* **C70**, 065806 (2004). doi: 10.1103/PhysRevC.70.065806.
42. C. J. Horowitz, M. A. Perez-Garcia, D. K. Berry and J. Piekarewicz, Dynamical response of the nuclear 'pasta' in neutron star crusts, *Phys. Rev.* **C72**, 035801 (2005). doi: 10.1103/PhysRevC.72.035801.
43. T. Maruyama, T. Tatsumi, D. N. Voskresensky, T. Tanigawa and S. Chiba, Nuclear pasta structures and the charge screening effect, *Phys. Rev.* **C72**, 015802 (2005). doi: 10.1103/PhysRevC.72.015802.
44. G. Watanabe, T. Maruyama, K. Sato, K. Yasuoka and T. Ebisuzaki, Simulation of transitions between 'Pasta' phases in dense matter, *Phys. Rev. Lett.* **94**, 031101 (2005). doi: 10.1103/PhysRevLett.94.031101.
45. H. Sonoda, G. Watanabe, K. Sato, K. Yasuoka and T. Ebisuzaki, Phase diagram of nuclear "pasta" and its uncertainties in supernova cores, *Phys. Rev.* **C77**, 035806 (2008). doi: 10.1103/PhysRevC.77.035806, 10.1103/PhysRevC.81.049902. [Erratum: Phys. Rev.C81,049902(2010)].
46. H. Pais and J. R. Stone, Exploring the Nuclear Pasta Phase in Core-Collapse Supernova Matter, *Phys. Rev. Lett.* **109**, 151101 (2012). doi: 10.1103/PhysRevLett.109.151101.
47. F. Grill, H. Pais, C. Providncia, I. Vidaa and S. S. Avancini, Equation of state and thickness of the inner crust of neutron stars, *Phys. Rev.* **C90**(4), 045803 (2014). doi: 10.1103/PhysRevC.90.045803.
48. C. J. Horowitz, M. A. Perez-Garcia and J. Piekarewicz, Neutrino - pasta scattering: The Opacity of nonuniform neutron - rich matter, *Phys. Rev.* **C69**, 045804 (2004). doi: 10.1103/PhysRevC.69.045804.
49. H. Sonoda, G. Watanabe, K. Sato, T. Takiwaki, K. Yasuoka and T. Ebisuzaki, The impact of nuclear pasta on neutrino transport in collapsing cores, *Phys. Rev.* **C75**, 042801 (2007). doi: 10.1103/PhysRevC.75.042801.
50. M. D. Alloy and D. P. Menezes, Nuclear 'pasta phase' and its consequences on neutrino opacities, *Phys. Rev.* **C83**, 035803 (2011). doi: 10.1103/PhysRevC.83.035803.
51. S. Furusawa, H. Nagakura, K. Sumiyoshi and S. Yamada, The influence of inelastic neutrino reactions with light nuclei on the standing accretion shock instability in core-collapse supernovae, *Astrophys. J.* **774**, 78 (2013). doi: 10.1088/0004-637X/774/1/78.
52. P. Haensel, Composition of dense matter and neutrino cooling of neutron stars, *Acta Phys.Polon.* **B25**, 373–399 (1994).
53. D. Page and S. Reddy, Forecasting neutron star temperatures: predictability and variability, *Phys. Rev. Lett.* **111**(24), 241102 (2013). doi: 10.1103/PhysRevLett.111.241102.
54. B. Haskell and A. Melatos, Models of Pulsar Glitches, *Int. J. Mod. Phys.* **D24**(03), 1530008 (2015). doi: 10.1142/S0218271815300086.
55. E. Salpeter, Energy and Pressure of a Zero-Temperature Plasma, *Astrophys.J.* **134**, 669–682 (1961). doi: 10.1086/147194.

56. G. Baym, C. Pethick and P. Sutherland, The Ground state of matter at high densities: Equation of state and stellar models, *Astrophys.J.* **170**, 299–317 (1971). doi: 10.1086/151216.
57. P. Haensel and B. Pichon, Experimental nuclear masses and the ground state of cold dense matter, *Astron.Astrophys.* **283**, 313 (1994).
58. F. Douchin and P. Haensel, A unified equation of state of dense matter and neutron star structure, *Astron. Astrophys.* **380**, 151 (2001). doi: 10.1051/0004-6361:20011402.
59. S. B. Ruester, M. Hempel and J. Schaffner-Bielich, The outer crust of non-accreting cold neutron stars, *Phys. Rev.* **C73**, 035804 (2006). doi: 10.1103/PhysRevC.73.035804.
60. J. M. Pearson, S. Goriely and N. Chamel, Properties of the outer crust of neutron stars from Hartree-Fock-Bogoliubov mass models, *Phys. Rev.* **C83**, 065810 (2011). doi: 10.1103/PhysRevC.83.065810.
61. R. N. Wolf *et al.*, Plumbing Neutron Stars to New Depths with the Binding Energy of the Exotic Nuclide ^{82}Zn, *Phys. Rev. Lett.* **110**(4), 041101 (2013). doi: 10.1103/PhysRevLett.110.041101.
62. A. S. Botvina and I. N. Mishustin, Formation of hot heavy nuclei in supernova explosions, *Phys. Lett.* **B584**, 233–240 (2004). doi: 10.1016/j.physletb.2004.01.064.
63. A. S. Botvina and I. N. Mishustin, Statistical approach for supernova matter, *Nucl. Phys.* **A843**, 98–132 (2010). doi: 10.1016/j.nuclphysa.2010.05.052.
64. S. I. Blinnikov, I. V. Panov, M. A. Rudzsky and K. Sumiyoshi, The equation of state and composition of hot, dense matter in core-collapse supernovae, *Astron. Astrophys.* **535**, A37 (2011). doi: 10.1051/0004-6361/201117225.
65. A. R. Raduta and F. Gulminelli, Thermodynamics of clusterized matter, *Phys. Rev.* **C80**, 024606 (2009). doi: 10.1103/PhysRevC.80.024606.
66. A. R. Raduta and F. Gulminelli, Statistical description of complex nuclear phases in supernovae and proto-neutron stars, *Phys. Rev.* **C82**, 065801 (2010). doi: 10.1103/PhysRevC.82.065801.
67. F. Gulminelli and A. R. Raduta, Ensemble in-equivalence in supernova matter within a simple model, *Phys. Rev.* **C85**, 025803 (2012). doi: 10.1103/PhysRevC.85.025803.
68. N. Buyukcizmeci *et al.*, A comparative study of statistical models for nuclear equation of state of stellar matter, *Nucl. Phys.* **A907**, 13–54 (2013). doi: 10.1016/j.nuclphysa.2013.03.010.
69. N. Buyukcizmeci, A. S. Botvina and I. N. Mishustin, Tabulated equation of state for supernova matter including full nuclear ensemble, *Astrophys. J.* **789**, 33 (2014). doi: 10.1088/0004-637X/789/1/33.
70. A. R. Raduta, F. Aymard and F. Gulminelli, Clusterized nuclear matter in the (proto-)neutron star crust and the symmetry energy, *Eur. Phys. J.* **A50**, 24 (2014). doi: 10.1140/epja/i2014-14024-y.
71. V. V. Sagun, A. I. Ivanytskyi, K. A. Bugaev and I. N. Mishustin, The statistical multifragmentation model for liquid-gas phase transition

with a compressible nuclear liquid, *Nucl. Phys.* **A924**, 24–46 (2014). doi: 10.1016/j.nuclphysa.2013.12.012.
72. M. Hempel, J. Schaffner-Bielich, S. Typel and G. Röpke, Light Clusters in Nuclear Matter: Excluded Volume versus Quantum Many-Body Approaches, *Phys. Rev.* **C84**, 055804 (2011). doi: 10.1103/PhysRevC.84.055804.
73. S. Typel, Variations on the excluded-volume mechanism, *Eur. Phys. J.* **A52** (1), 16 (2016). doi: 10.1140/epja/i2016-16016-3.
74. T. Klähn, M. Oertel and S. Typel. CompOSE: Compstar Online Supernova Equations of State. http://compose.obspm.fr (2013–2014).
75. K. Sumiyoshi and G. Roepke, Appearance of light clusters in post-bounce evolution of core-collapse supernovae, *Phys. Rev.* **C77**, 055804 (2008). doi: 10.1103/PhysRevC.77.055804.
76. G. Röpke, Parametrization of light nuclei quasiparticle energy shifts and composition of warm and dense nuclear matter, *Nucl. Phys.* **A867**, 66–80 (2011). doi: 10.1016/j.nuclphysa.2011.07.010.
77. G. Röpke, N. U. Bastian, D. Blaschke, T. Klähn, S. Typel and H. H. Wolter, Cluster virial expansion for nuclear matter within a quasi-particle statistical approach, *Nucl. Phys.* **A897**, 70–92 (2013). doi: 10.1016/j.nuclphysa.2012.10.005.
78. G. Röpke, Quantum statistical calculation of cluster abundances in hot dense matter, *J. Phys. Conf. Ser.* **436**, 012070 (2014). doi: 10.1088/1742-6596/436/1/012070. [436,012070(2013)].
79. G. Röpke, Nuclear matter equation of state including two-, three-, and four-nucleon correlations, *Phys. Rev.* **C92**(5), 054001 (2015). doi: 10.1103/PhysRevC.92.054001.
80. M. D. Voskresenskaya and S. Typel, Constraining mean-field models of the nuclear matter equation of state at low densities, *Nucl. Phys.* **A887**, 42–76 (2012). doi: 10.1016/j.nuclphysa.2012.05.006.
81. S. Typel, Equation of State in a Generalized Relativistic Density Functional Approach. In *Compact Stars in the QCD Phase Diagram IV Prerow, Germany, September 26-30, 2014* (2015). URL `https://inspirehep.net/record/1358250/files/arXiv:1504.01571.pdf`.
82. S. Typel and H. H. Wolter, Relativistic mean field calculations with density dependent meson nucleon coupling, *Nucl. Phys.* **A656**, 331–364 (1999). doi: 10.1016/S0375-9474(99)00310-3.
83. M. Wang, G. Audi, A. Wapstra, F. Kondev, M. Mac-Cormick, X. Xu and B. Pfeiffer, The AME2012 atomic mass evaluation (II). Tables, graphs and references, *Chinese Physics C.* **36**, 1603 (Dec., 2012). doi: 10.1088/1674-1137/36/12/003.
84. J. Duflo and A. P. Zuker, Microscopic mass formulae, *Phys. Rev.* **C52**, R23 (1995). doi: 10.1103/PhysRevC.52.R23.
85. M. K. Grossjean and H. Feldmeier, Level density of a Fermi gas with pairing interactions, *Nucl. Phys.* **444**, 113–132 (1985).

86. G. Audi, A. Wapstra and C. Thibault, The AME2003 atomic mass evaluation: (II). Tables, graphs and references, *Nucl. Phys.* **729**, 337 (2003). doi: 10.1016/j.nuclphysa.2003.11.003.
87. P. Moller, J. R. Nix, W. D. Myers and W. J. Swiatecki, Nuclear ground state masses and deformations, *Atom. Data Nucl. Data Tabl.* **59**, 185–381 (1995). doi: 10.1006/adnd.1995.1002.

CHAPTER 5

CLUSTERING AND PASTA PHASES IN NUCLEAR DENSITY FUNCTIONAL THEORY

Bastian Schuetrumpf and Chunli Zhang

FRIB Laboratory, Michigan State University, East Lansing, Michigan 48824, USA

Witold Nazarewicz

Department of Physics and Astronomy and FRIB Laboratory, Michigan State University, East Lansing, Michigan 48824, USA
Faculty of Physics, University of Warsaw, 02-093 Warsaw, Poland

Nuclear density functional theory (DFT) is the tool of choice in describing properties of complex nuclei and intricate phases of bulk nucleonic matter. It is a microscopic approach based on an energy density functional representing the nuclear interaction. An attractive feature of nuclear DFT is that it can be applied to both finite nuclei and pasta phases appearing in the inner crust of neutron stars. While nuclear pasta clusters in a neutron star can be easily characterized through their density distributions, the level of clustering of nucleons in a nucleus can often be difficult to assess. To this end, we use the concept of nucleon localization. We demonstrate that the localization measure provides us with fingerprints of clusters in light and heavy nuclei, including fissioning systems. Furthermore we investigate the rod-like pasta phase using twist-averaged boundary conditions, which enable calculations in finite volumes accessible by state of the art DFT solvers.

1. Introduction

Nuclear clustering is a ubiquitous phenomenon,[1–5] but its comprehensive understanding still eludes us. On the one hand, the very occurrence of cluster states at low excitation energies around cluster-decay thresholds[6] is a consequence of an openness of the nuclear many-body system.[7] On the other hand, the properties of cluster states, once they are formed, can be well described within a mean-field picture,[8–13] and their characteristics can be traced back to the symmetries and geometry of the nuclear mean-field.[14,15] The quantitative theoretical description of clustering requires the use of advanced many-body, open-system framework employing realistic interactions, and there has been a significant progress in this area.[5,16] If one is aiming at a global characterization of cluster states throughout the nuclear landscape and for neutron stars, a good starting point is nuclear density functional theory (DFT)[17] based on an energy density functional (EDF) representing the effective in-medium nuclear interaction.

The degree of clustering in nuclei is difficult to assess quantitatively in DFT as the single particle wave functions are generally spread throughout the nucleus and the resulting nucleonic distributions are rather crude indicators of cluster structures. Therefore, in this study, we utilize a different measure, called spatial localization, which is a better signature of clustering and cluster shell structure. The localization, originally introduced in the context of many-electron systems,[18–21] has recently been applied to nuclear structure elucidations.[22]

The range of clustering phenomena in nuclear physics extends well beyond finite nuclei. In compact stellar objects such as neutron stars or supernova, nuclear matter is present on the macroscopic scale. Especially remarkable are the phases of bulk nucleonic matter just below the nuclear saturation density. Here nuclear matter can lump into clusters, which have shapes of rods, slabs, and other complex, often very intricate, geometries. Those structures are usually referred to as nuclear pasta.[23,24]

Nuclear pasta phases have been studied in semi-classical models,[25–33] as well as in quantum mechanical models, including DFT with and without time-dependence.[34–41] In this work, we study nuclear rods with twist-averaged boundary conditions (TABC),[42–44] which provide the most general solutions for states in a periodic potential. TABC minimize finite-volume effects drastically and enable precise calculations within the chosen model. TABC have been utilized in the context of the crust of neutron stars within DFT and Quantum Monte Carlo approaches.[45–48]

This paper is organized as follows. We outline the nuclear DFT framework used in Sec. 2. Section 3 defines the spatial localization of nucleons. In Sec. 4 we describe the DFT+TABC approach in the context of infinite systems. In Sec. 5 we apply the localization measure to finite nuclei, both light and heavy, and compare the DFT results with those obtained within the harmonic oscillator model. Section 6 shows the DFT+TABC predictions for the rod phase in the neutron star crust. We summarize our results and present the outlook for the future in Sec. 7.

2. Nuclear density functional theory

Nuclear DFT models, including self-consistent mean-field models based on the Hartree-Fock (HF) or Hartree-Fock-Bogoliubov (HFB) approximation, are designed to describe nuclei all over the nuclear chart. The HFB equations are derived from the variational principle and the resulting nuclear wave function is represented by a product state. In DFT, the interaction between the nucleons is approximated by an EDF, which can be written as an expansion in local one-body densities and currents, such as, e.g., the particle density, kinetic density, and spin-orbit density. The EDF parameters have to be optimized to selected experimental data. In this work we use Skyrme type[17] EDFs. For finite nuclei, we employ the optimized parametrizations SkM* [49] (^{164}Fm and ^{132}Sn) UNEDF1-HFB[50] (light nuclei, no pairing). For the nuclear rod calculations we compare the SLy6[51] and TOV-min[52] parametrizations, both of which were designed to perform well for neutron-rich matter.

For finite nuclei, we utilize the DFT solvers HFBTHO[53] and HFODD.[54] Both codes expand the self-consistent wave functions in a harmonic oscillator basis. While with HFODD all self-consistent mean-field symmetries can be broken, HFBTHO assumes axial geometry. For infinite systems, the harmonic oscillator representation of wave functions cannot be used. Here we use the symmetry-unconstrained DFT solver Sky3d[55] based on the fast Fourier transform method, in which the wave functions are represented on an equidistant Cartesian 3D grid.

3. Spatial localization

The localization measure has been originally introduced to characterize chemical bonds[18–21] and has also been demonstrated to be useful for characterizing clusters in light nuclei.[22] The localization can be defined through a probability $R_{q\sigma}(\boldsymbol{r}, \delta)$ of finding a pair of nucleons at point $\boldsymbol{r}$ having the same spin σ and isospin $q = n, p$ within a small radius δ. As discussed in,[18] this probability can be written as:

$$R_{q\sigma}(\boldsymbol{r}, \delta) \approx \frac{1}{3}\left(\tau_{q\sigma} - \frac{1}{4}\frac{|\boldsymbol{\nabla}\rho_{q\sigma}|^2}{\rho_{q\sigma}} - \frac{\boldsymbol{j}^2_{q\sigma}}{\rho_{q\sigma}}\right)\delta^2 + \mathcal{O}(\delta^3) \tag{1}$$

where the particle density $\rho_{q\sigma}$, kinetic density $\tau_{q\sigma}$, density gradient $\boldsymbol{\nabla}\rho_{q\sigma}$, and the current density $\boldsymbol{j}_{q\sigma}$ are defined as:

$$\rho_{q\sigma}(\boldsymbol{r}) = \sum_{\alpha\in q} v_\alpha^2 |\psi_\alpha(\boldsymbol{r}\sigma)|^2 \tag{2a}$$

$$\tau_{q\sigma}(\boldsymbol{r}) = \sum_{\alpha\in q} v_\alpha^2 |\boldsymbol{\nabla}\psi_\alpha(\boldsymbol{r}\sigma)|^2 \tag{2b}$$

$$\boldsymbol{\nabla}\rho_{q\sigma}(\boldsymbol{r}) = 2\sum_{\alpha\in q} v_\alpha^2 \mathrm{Re}[\psi_\alpha^*(\boldsymbol{r}\sigma)\boldsymbol{\nabla}\psi_\alpha(\boldsymbol{r}\sigma)] \tag{2c}$$

$$\boldsymbol{j}_{q\sigma}(\boldsymbol{r}) = \sum_{\alpha\in q} v_\alpha^2 \mathrm{Im}[\psi_\alpha^*(\boldsymbol{r}\sigma)\boldsymbol{\nabla}\psi_\alpha(\boldsymbol{r}\sigma)], \tag{2d}$$

with v_α^2 being the occupations of canonical orbits $\psi_\alpha(\boldsymbol{r}\sigma)$. Henceforth, we take the expression in parentheses of Eq. (1) as a localization measure. This expression is neither dimensionless nor normalized. A natural choice for normalization is the Thomas-Fermi kinetic density $\tau_{q\sigma}^{\mathrm{TF}} = \frac{3}{5}(6\pi^2)^{2/3}\rho_{q\sigma}^{5/3}$. With this normalization, one defines a reversed and normalized localization measure,

$$\mathcal{C}_{q\sigma}(\boldsymbol{r}) = \left[1 + \left(\frac{\tau_{q\sigma}\rho_{q\sigma} - \frac{1}{4}|\boldsymbol{\nabla}\rho_{q\sigma}|^2 - \boldsymbol{j}_{q\sigma}^2}{\rho_{q\sigma}\tau_{q\sigma}^{\mathrm{TF}}}\right)^2\right]^{-1}, \tag{3}$$

known for electronic systems as the electron localization function (ELF). In the following, we shall refer to $\mathcal{C}_{q\sigma}(\boldsymbol{r})$ as the nucleon localization function (NLF). A large value of NLF indicates that the probability of finding two nucleons with the same spin and isospin at the same location is very low. Thus the nucleon localization is large at that point. In particular the alpha particle has perfect localization for all combinations of spin and isospin.[22] The value of $\mathcal{C} = 1/2$ is characteristic of a nearly homogeneous Fermi gas as found in nuclear matter.

The above definition of the NLF works well in regions with non-zero nucleonic density. When the local densities become very small in the regions outside the range of the nuclear mean field, numerical artifacts become dominant, because the numerator and denominator in Eq. (3) are both close to zero (see, e.g., Fig. 1 of Ref. 22). Consequently, for finite nuclei, we multiply the NLF by a normalized particle density $\mathcal{C}(\boldsymbol{r}) \to \mathcal{C}(\boldsymbol{r})[\rho_{q\sigma}(\boldsymbol{r})/\max(\rho_{q\sigma}(\boldsymbol{r}))]$. Such a procedure is not necessary for the nuclear pasta case, because particle densities do not vanish in the regions between the nuclear clusters in pasta phases.

4. Twist-averaged boundary conditions

The code Sky3d is based on Fourier transforms for derivatives. Therefore the native boundary conditions are periodic boundary conditions (PBC). While PBC are well suited for densities and fields for most nuclear pasta calculations, like the nuclear rods discussed in Sec. 6, they represent a constraint on the HF single-particle states. According to the Floquet-Bloch theorem, the most general wave function in a periodic potential can be written as:

$$\psi_{\alpha\boldsymbol{q}}(\boldsymbol{r}) = u_{\alpha\boldsymbol{q}}(\boldsymbol{r})e^{\mathrm{i}\boldsymbol{q}\boldsymbol{r}}, \tag{4}$$

where $\psi_{\alpha\boldsymbol{q}}(\boldsymbol{r})$ is the single-particle wave function labeled by quantum numbers α, $\boldsymbol{q}$ is the wave vector that determines the boundary condition, and $u_{\alpha\boldsymbol{q}}(\boldsymbol{r})$ is a periodic function of $\boldsymbol{r}$. These wave functions obey the boundary conditions

$$\psi_{\alpha\boldsymbol{\theta}}(\boldsymbol{r} + \boldsymbol{T}_i) = e^{\mathrm{i}\theta_i}\psi_{\alpha\boldsymbol{\theta}}(\boldsymbol{r}), \tag{5}$$

where $\boldsymbol{\theta}$ is the set of Bloch angles (or twists) and $\boldsymbol{T}_i$ $(i = 1, 2, 3)$ stands for the three lattice vectors.

In the DFT+TABC approach, the expectation value of an observable $\hat{O}$ is obtained by averaging over the twists:

$$\langle\hat{O}\rangle = \int \frac{d^3\boldsymbol{\theta}}{\pi^3}\,\langle\Psi_{\boldsymbol{\theta}}|\hat{O}|\Psi_{\boldsymbol{\theta}}\rangle, \tag{6}$$

where $\Psi_{\boldsymbol{\theta}}$ is the HF/HFB state. The twists θ_i change between zero (PBC) and π (anti-PBC), as the time-reversal symmetry is assumed.[44] In this work, we use the DFT+TABC implementation described in Ref. 47.

5. Finite nuclei

5.1. *Harmonic oscillator model*

Because of the large binding energy of ^{4}He, α-cluster states are prominent in alpha-conjugate nuclei. For those light systems, the deformed harmonic oscillator (HO) model can serve as a rough guidance.[8,10,15,56] To this end, we first study the NLF using the wave functions of the axial HO with the deformation parameter $\eta = \omega_\perp/\omega_z$, where $\omega_\perp$ and ω_z are HO frequencies. In the examples shown, we focus on prolate-deformed nuclei ($\eta > 1$). We only show results for the $\sigma = 1$ spin component, because of the assumed time-reversal symmetry. Since the Coulomb force is neglected, the results for protons and neutrons are identical.

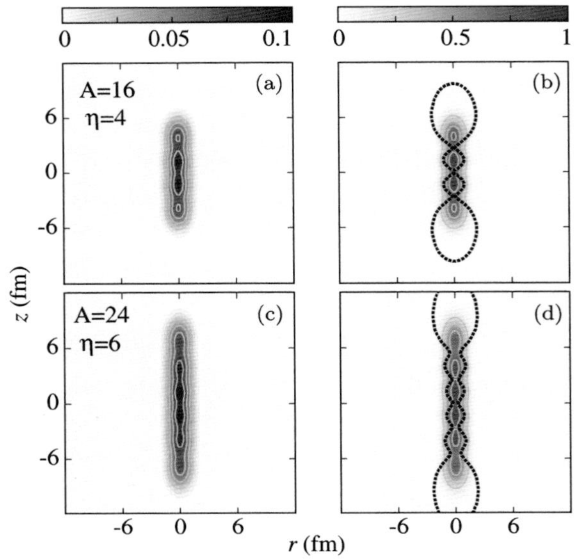

Fig. 1. Linear alpha-chains in the deformed HO model: 4-α chain in ^{16}O (top) and 6-α chain in ^{24}Mg (bottom). Left panels show the particle density of a fixed spin and isospin (in nucleons/fm^3). Right panels show the corresponding NLF (dimensionless), masked with the density form factor as described in Sec. 3. Black dotted lines are the contour lines of the original definition (3) corresponding to the value of $\mathcal{C} = 0.9$.

In Fig. 1, we show two typical examples of elongated configurations in ^{16}O and ^{24}Mg. In order to simulate α-chain configurations, we choose η according to the α particle content. Such a deformation produces the supershell structure associated with the η-fold SU(3) dynamical symmetry of the rational HO, which stabilizes the formation of α-cluster chains.[15] While a separation into α particles is difficult to see in the particle density plot, especially for ^{24}Mg, the NLF clearly reveals four maxima for ^{16}O and six maxima for ^{24}Mg, with localizations close to one. This means that the nucleons are very localized for each spin/isospin component, implying the presence of aligned α particles.

As a second example, in Fig. 2 we show a hyperdeformed ($\eta = 3$) configuration in ^{36}Ar. While the particle density hardly shows clustering, the localization shows large values, especially at the tips of the nucleus. The structure in between, corresponds to a deformed ^{28}Si. It also exhibits cluster structures at $z = 0$ and $z \approx \pm 2\,\mathrm{fm}$. The black dotted line represents the $\mathcal{C} = 0.9$ contour of the original localization measure.

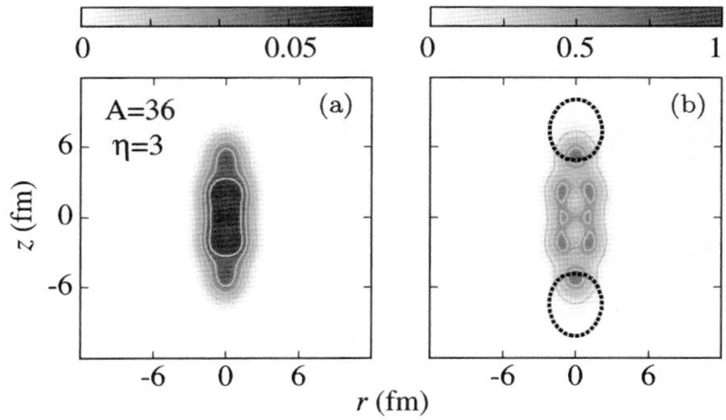

Fig. 2. Similar to Fig. 1 but for ^{36}Ar with $\eta = 3$.

5.2. *DFT description*

As a next step we want to compare the simple HO model predictions with the DFT results. To this end, in Fig. 3 we show the HF energy of ^{36}Ar as a function of the quadrupole deformation β_2. The three local minima are predicted at $\beta_2 \approx 0.1$, 0.5, and 0.8. The corresponding neutron densities and NLFs are also displayed in Fig. 3. (Clustering in ^{36}Ar has also been studied in the DFT calculations of Ref. 13). The weakly-deformed ground state at $\beta_2 \approx 0.1$ does not show any structure in the density. Its NLF exhibits a maximum in the center and an enhancement at the tips. This distribution constitutes a unique fingerprint of the shell structure of ^{36}Ar that is clearly missing in the density plot. The configuration B is less deformed than that calculated with the HO in Fig. 2. However, its NLF is similar. In particular, the localization enhancement at the tips reveals the presence of alpha clustering. The central structure shows two rings of the enhanced surface localization. Unlike shape B, shape C has a more uniform NLF, with the localization peaked around the nuclear surface where the contribution from one specific single-particle orbit is likely to dominate.

We now proceed to heavier nuclei. As an illustrative example, we show in Fig. 4 the distributions computed for the doubly-magic system ^{132}Sn. The particle densities do not exhibit any pronounced shell structure, except perhaps for a small depression of ρ_p in the interior. The NLFs, however, show clear patterns of concentric rings of enhanced localization that are distinct for protons and neutrons. While the proton localization has two radial maxima and a central depression, the neutron localization has two radial maxima and a central maximum. Therefore, the additional closed shell in

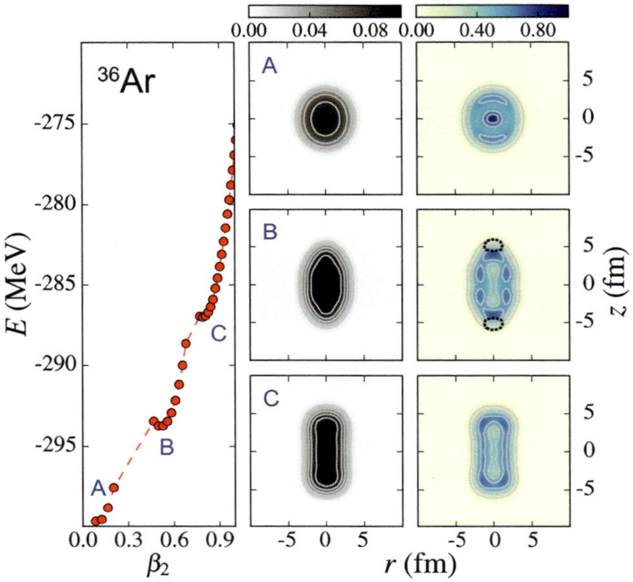

Fig. 3. Left: Total HF energy of ^{36}Ar calculated with the functional UNEDF1-HFB versus the quadrupole deformation β_2. The neutron densities corresponding to the three local minima, marked A, B, and C are shown in the middle panel (in nucleons/fm^3) and the associated neutron localization is shown in the right panel. Black dotted lines are the contour lines of the original definition (3) corresponding to the value of $\mathcal{C} = 0.9$.

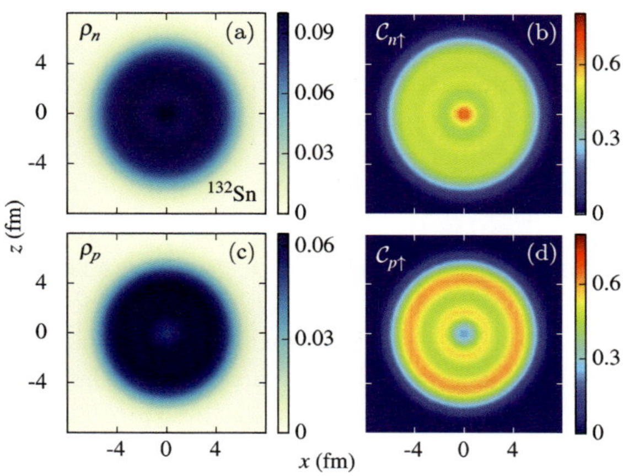

Fig. 4. HFB calculations for ^{132}Sn with SkM*: (a) neutron density, (b) neutron localization, (c) proton density, and (d) proton localization. The particle densities are in nucleons/fm^3.

the neutrons produces an extra region of high localization. Of course, unlike in the atomic systems,[18] the total number of shells can not be directly read from the number of peaks in the NLF, because the radial distributions of wave functions belonging to different nucleonic shells vary fairly smoothly. Nevertheless, as one can assess from Fig. 4, each magic number leaves a strong imprint on the localization.

An application where shell-structure imprints on the NLF are important as they can reveal valuable structural information is nuclear fission. The idea is to recognize the NLF imprint of the fission fragment as it is formed along the fission path. To illustrate this concept, we choose the symmetric fission path of ^{264}Fm, a subject of several recent DFT studies.[57–59] The results of our HFB SkM* calculations are shown in Fig. 5. As the constraining quadrupole moment Q_{20} gets larger, the particle densities

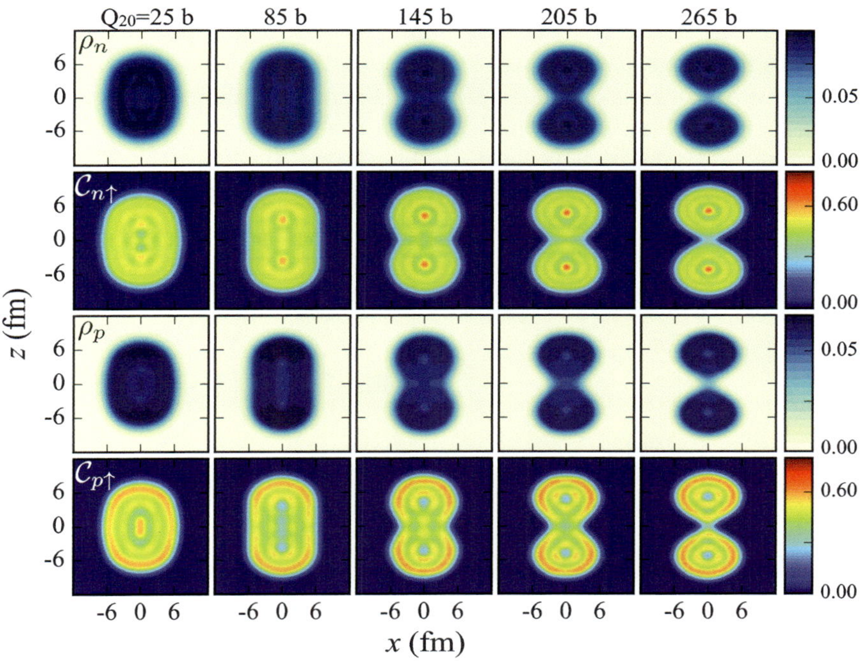

Fig. 5. HFB calculations for ^{264}Fm with SkM* for five deformed configurations along the symmetric fission pathway. Shown are: neutron density ρ_n (top); neutron localization $\mathcal{C}_{n\uparrow}$ (second row); proton density ρ_p (third row); and proton localization $\mathcal{C}_{p\uparrow}$ (bottom). Quadrupole moments Q_{20} characterizing HFB solutions are indicated. The particle densities are in nucleons/fm^3.

become increasingly elongated. A neck develops at $Q_{20} \approx 145\,\text{b}$, and the scission point is reached at $Q_{20} \approx 265\,\text{b}$ above which ^{264}Fm splits into two ^{132}Sn fragments. By comparing the results of Fig. 4 for ^{132}Sn one can see the gradual development of the ^{132}Sn clusters within the fissioning nucleus. This example shows in a dramatic way that the NLF can serve as an excellent fingerprint of developing cluster structures in heavy nuclei.

6. Nuclear rods

In the neutron star crust, or at higher temperatures and higher proton fractions in supernova explosions, exotic pasta phases of nucleonic matter develop. In this study, we use the DFT+TABC method to analyze the nuclear rod (spaghetti) phase as a representative example. We compare two different packings of rods at a constant mean density $\bar{\rho} = 0.0358\,\text{fm}^{-3}$: square and hexagonal packing. Since our computations are performed in rectangular boxes, the unit cell contains one rod for square packing but two rods for hexagonal packing. The computational cells with one periodic repetition of the rods in the x- and y-directions are shown in Fig. 6. The box length L_z in the z-direction, in which the density of the rods is translationally invariant, is sufficiently large so that the finite-volume effects are negligible with TABC. (For a detailed discussion of the performance of

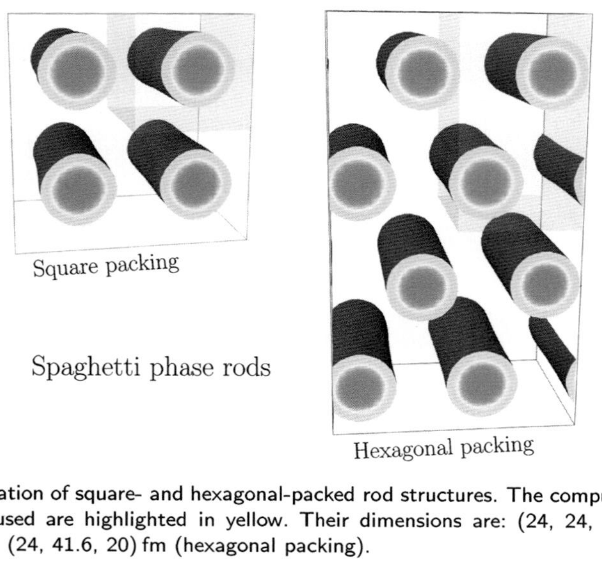

Fig. 6. Illustration of square- and hexagonal-packed rod structures. The computational boxes (L_x, L_y, L_z) used are highlighted in yellow. Their dimensions are: (24, 24, 20) fm (square packing) and (24, 41.6, 20) fm (hexagonal packing).

TABC for pasta matter, please see Ref. 47.) The calculations are performed with four different twists in the z-direction. In the calculations with no neutron background gas present, a one-dimensional averaging over the twists in the z-direction is sufficient. Otherwise a three-dimensional averaging must be used.

Since the geometries are different for square and hexagonal packing, and different numbers of rods are present in a computational box, to compare the results of these two variants of calculations we introduce a one-dimensional rod density

$$\rho_{\rm rod} = \frac{A}{\#\text{ of rods} \cdot L_z}, \tag{7}$$

where A is the number of nucleons in the computational box. In addition to packing geometry, we also compare predictions of different EDFs: SLy6 and TOV-min. The parametrization SLy6[51] was optimized to the equation of state of neutron matter and data on magic nuclei; here, constraints were applied to the saturation point of symmetric matter, its compressibility, and its symmetry energy. The EDF TOV-min[52] was fitted to binding energies, diffraction radii, charge radii and other observables in finite nuclei. Additionally the constraints were imposed on the mass-radius relation for neutron stars to conform with the latest observations; hence, TOV-min is expected to perform well for neutron-rich systems.

Figure 7 shows the nuclear matter equation of state for TOV-min and SLy6 at the two proton fractions $X_P = 1/3$ and $X_P = 1/6$, which we use

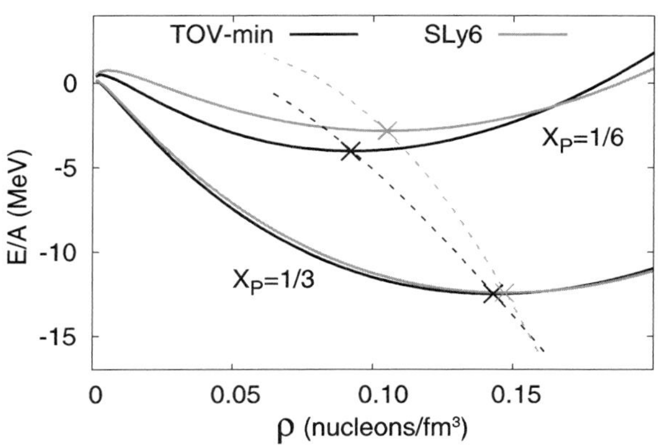

Fig. 7. Binding energy per nucleon for TOV-min and SLy6 versus the mean density of uniform nuclear matter. The saturation point is marked. The trajectory of saturation points is shown as a dashed line for $X_P \in [0.1, 0.5]$.

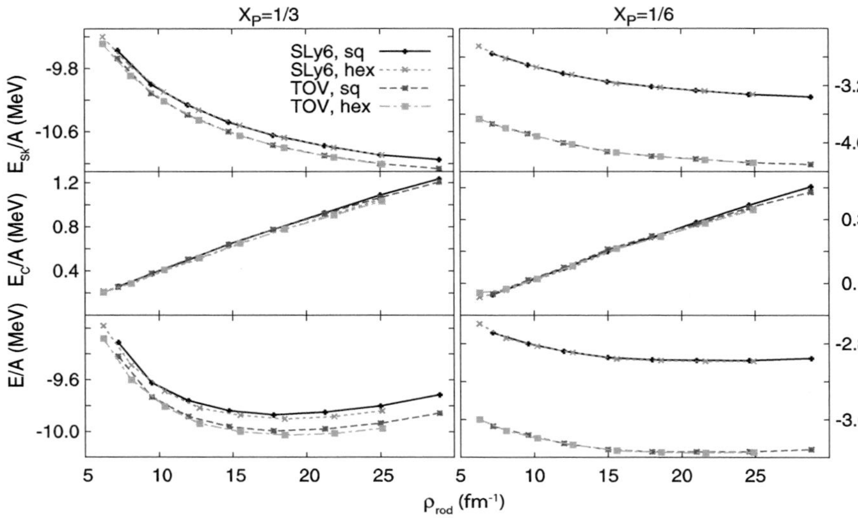

Fig. 8. Total, Coulomb, and Skyrme energies for squared and hexagonal packed rods in dependence on the 1-D rod density ρ_{rod} as defined in Eq. 7. Left panel shows results for $X_P = 1/3$ right panel shows $X_P = 1/6$.

for our rod calculations. While the binding energy is larger for TOV-min at lower proton fractions, both forces are fairly similar at higher values of X_P.

In Fig. 8, the total, Coulomb, and Skyrme energies of SLy6 and TOV-min are shown as a function of ρ_{rod} for the two different values of X_P, and different packing geometries. For $X_P = 1/3$ both functionals yield similar results, with TOV-min producing a slightly stronger binding than SLy6. For the Skyrme EDF, different packings of the rods give very close results, since the rods are not connected. The Skyrme energy is decreasing with larger rods, because the volume term becomes dominant over the surface term. However, the long-ranged Coulomb force is affected by the packing geometry, and the difference is about 5% of the total Coulomb energy. The total energy has a minimum near $\rho_{\text{rod}} = 18\,\text{fm}^{-1}$. At the lower proton fraction of $X_P = 1/6$ the Skyrme energies strongly depend on the EDF used. The rods calculated with TOV-min are about 1 MeV lower in energy than those with SLy6; this is due to the difference in the density dependence of the symmetry energies between these EDFs. At this proton fraction, a neutron background forms, as studied in detail in[40] and is shown in Fig. 9. The background density is approximately 0.01 nucleons/fm^3. Although the rods are embedded in the neutron gas for both geometries, the energies

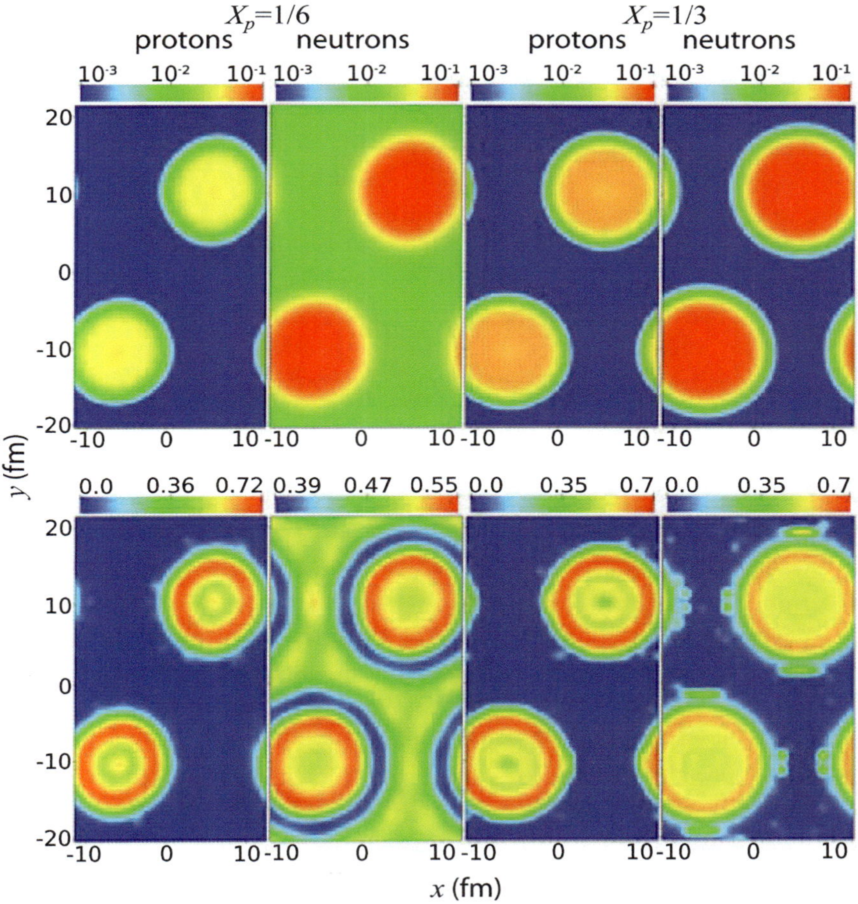

Fig. 9. Proton and neutron densities (top) and localizations (bottom) at the hexagonal geometry for $X_P = 1/6$ (left) and $X_P = 1/3$ (right) calculated with DFT+TABC using TOV-min.

associated with the nuclear forces, E_{Sk}, are very close. This suggests that the neutron background is mostly decoupled from the bound rod structures. As the total contribution of the Coulomb term is reduced, the impact of different packing geometries is even less important. The minimum of the total energy shifts to the larger rod density, $\rho_{\text{rod}} \approx 20\,\text{fm}^{-1}$.

Particle densities computed with TOV-min for hexagonal packing are shown in Fig. 9 (top). The results for SLy6, and for square packing, are very similar; hence, they are not shown. A small deformation of the proton densities at $X_P = 1/6$ can be seen. Most likely, it is a residual finite-volume effect, which has not been fully removed by TABC. For $X_P = 1/3$ the twist-averaged shapes are almost perfectly circular.

The corresponding NLFs are shown in Fig. 9 (bottom). Some irregularities resulting from residual finite-volume and grid-discretization effects are seen, as the localization is a more sensitive measure than the local density. The localizations exhibit pronounced radial oscillations resulting in ring-like structures, which are fingerprints of shell effects associated with the rod phase. It is interesting to note that the neutrons at $X_P = 1/6$ show a markedly different localization. Since the neutron gas is present in this case, the localization tends to be around the Fermi-gas value of $\mathcal{C} = 0.5$. At the inner boundaries of the rods, the localization is enhanced to a value of about 0.55, but right outside the rod, the localization drops to about 0.39.

7. Conclusions

In this work we presented some recent developments pertaining to the DFT description of nucleon clustering in both finite nuclei and pasta phases. We demonstrated that the nucleon localization is a superb indicator of clustering in light nuclei, in heavy fissioning systems, and in crustal structures of neutron stars. In particular, the pattern of NLFs can serve as a fingerprint of the single-particle shell structure associated with cluster configurations.

To learn more about pasta structures, we employed the DFT+TABC approach to compare hexagonal and square packing geometries of the rod phase. We conclude that hexagonal packing is slightly favored by the Coulomb force. The localization of rods depends little on details when no neutron gas is present. In the presence of neutron background, the NLF values become close to the Fermi-gas limit of 0.5, and the structure is washed out, with the main NLF variations happening at the rod's boundaries.

Future studies will include α clustering in medium-mass nuclei as well as clustering in heavy nuclei, in particular in fission. Clustering effects and NLFs will also be studied for various nuclear pasta phases.

Acknowledgements

Useful discussions with P.-G. Reinhard and A.S. Umar and the help of E. Olsen are gratefully acknowledged. This material is based upon work supported by the U.S. Department of Energy, Office of Science under Award Numbers DOE-DE-NA0002847 (the Stewardship Science Academic Alliances program), DE-SC0013365 (Michigan State University), and DE-SC0008511 (NUCLEI SciDAC-3 collaboration). This work used computational resources of the Institute for Cyber-Enabled Research at Michigan State University and the National Institute for Computational Sciences (NICS).

References

1. C. Beck, ed., *Clusters in Nuclei, vol 1.* vol. 818, *Lecture Notes in Physics*, Springer, Berlin, Heidelberg (2010).
2. C. Beck, ed., *Clusters in Nuclei, vol 2.* vol. 848, *Lecture Notes in Physics*, Springer, Berlin, Heidelberg (2012).
3. C. Beck, ed., *Clusters in Nuclei, vol 3.* vol. 875, *Lecture Notes in Physics*, Springer, Berlin, Heidelberg (2014).
4. W. von Oertzen, M. Freer and Y. Kanada-Enyo, Nuclear clusters and nuclear molecules, *Phys. Rep.* **432**, 43 (2006). doi: http://dx.doi.org/10.1016/j.physrep.2006.07.001.
5. R. G. Lovas, Z. Dombrádi, G. G. Kiss, A. T. Kruppa and G. Lévai, eds., *10th International Conference on Clustering Aspects of Nuclear Structure and Dynamics, J. Phys. Conf. Ser.*, vol. 436 (2013).
6. K. Ikeda, N. Takigawa and H. Horiuchi, The systematic structure-change into the molecule-like structures in the self-conjugate 4n nuclei, *Prog. Theor. Phys. Suppl.* **E68**, 464 (1968). doi: 10.1143/PTPS.E68.464.
7. J. Okołowicz, M. Płoszajczak and W. Nazarewicz, On the origin of nuclear clustering, *Prog. Theor. Phys. Suppl.* **196**, 230 (2012). doi: 10.1143/PTPS.196.230.
8. G. Leander and S. Larsson, Potential-energy surfaces for the doubly even N = Z nuclei, *Nucl. Phys. A.* **239**, 93 (1975). ISSN 0375-9474. doi: http://dx.doi.org/10.1016/0375-9474(75)91136-7.
9. H. Flocard, P. H. Heenen, S. J. Krieger and M. S. Weiss, Configuration space, cranked Hartree-Fock calculations for the nuclei 16O, 24Mg and 32S, *Prog. Theor. Phys.* **72**, 1000 (1984). doi: 10.1143/PTP.72.1000.
10. M. Freer, R. Betts and A. Wuosmaa, Relationship between the deformed harmonic oscillator and clustering in light nuclei, *Nucl. Phys. A* **587**, 36 (1995). doi: http://dx.doi.org/10.1016/0375-9474(94)00820-D.
11. T. Ichikawa, J. A. Maruhn, N. Itagaki and S. Ohkubo, Linear chain structure of four-α clusters in ^{16}O, *Phys. Rev. Lett.* **107**, 112501 (2011). doi: 10.1103/PhysRevLett.107.112501.
12. J.-P. Ebran, E. Khan, T. Nikšić and D. Vretenar, How atomic nuclei cluster, *Nature*, **487**, 341 (2012).

13. J.-P. Ebran, E. Khan, T. Nikšić and D. Vretenar, Density functional theory studies of cluster states in nuclei, *Phys. Rev. C* **90**, 054329 (2014). doi: 10.1103/PhysRevC.90.054329.
14. K. T. Hecht, Relation between cluster and shell-model wave functions, *Phys. Rev. C* **16**, 2401 (1977). doi: 10.1103/PhysRevC.16.2401.
15. W. Nazarewicz and J. Dobaczewski, Dynamical symmetries, multiclustering and octupole susceptibility in superdeformed and hyperdeformed nuclei, *Phys. Rev. Lett.* **68**, 154 (1992). doi: 10.1103/PhysRevLett.68.154.
16. S. Elhatisari, D. Lee, G. Rupak, E. Epelbaum, H. Krebs, T. A. Lähde, T. Luu and U.-G. Meißner, Ab initio alpha–alpha scattering, *Nature*, **528**, 111 (2015). doi: 10.1038/nature16067.
17. M. Bender, P.-H. Heenen and P.-G. Reinhard, Self-consistent mean-field models for nuclear structure, *Rev. Mod. Phys.* **75**, 121–180 (2003). doi: 10.1103/RevModPhys.75.121.
18. A. D. Becke and K. E. Edgecombe, A simple measure of electron localization in atomic and molecular systems, *J Chem. Phys.* **92**, 5397 (1990).
19. A. Savin, R. Nesper, S. Wengert and T. F. Fässler, ELF: The electron localization function, *Angew. Chem. Int. Ed. Engl.* **36**, 1808 (1997). doi: 10.1002/anie.199718081.
20. A. Scemama, P. Chaquin and M. Caffarel, Electron pair localization function: A practical tool to visualize electron localization in molecules from quantum monte carlo data, *J. Chem. Phys.* **121**, 1725 (2004). doi: http://dx.doi.org/10.1063/1.1765098.
21. T. Burnus, M. A. L. Marques and E. K. U. Gross, Time-dependent electron localization function, *Phys. Rev. A* **71**, 010501 (2005). doi: 10.1103/PhysRevA. 71.010501.
22. P.-G. Reinhard, J. A. Maruhn, A. S. Umar and V. E. Oberacker, Localization in light nuclei, *Phys. Rev. C* **83**, 034312 (2011). doi: 10.1103/PhysRevC.83.034312.
23. D. G. Ravenhall, C. J. Pethick and J. R. Wilson, Structure of matter below nuclear saturation density, *Phys. Rev. Lett.* **50**, 2066–2069 (1983). doi: 10.1103/PhysRevLett.50.2066.
24. M. Hashimoto, H. Seki and M. Yamada, Shape of nuclei in the crust of neutron star, *Prog. Theor. Phys.* **71**(2), 320–326 (1984). doi: 10.1143/PTP.71.320.
25. C. O. Dorso, P. A. Giménez Molinelli and J. A. López, Topological characterization of neutron star crusts, *Phys. Rev. C* **86**, 055805 (2012). doi: 10.1103/PhysRevC.86.055805.
26. R. Williams and S. Koonin, Sub-saturation phases of nuclear matter, *Nucl. Phys.* **435**(3?4), 844–858 (1985). ISSN 0375-9474. doi: 10.1016/0375-9474(85)90191-5.
27. M. Okamoto, T. Maruyama, K. Yabana and T. Tatsumi, Nuclear "pasta" structures in low-density nuclear matter and properties of the neutron-star crust, *Phys. Rev. C* **88**, 025801 (2013).

28. H. Pais, S. Chiacchiera and C. Providência, Light clusters, pasta phases, and phase transitions in core-collapse supernova matter, *Phys. Rev. C* **91**, 055801 (2015). doi: 10.1103/PhysRevC.91.055801.
29. A. S. Schneider, C. J. Horowitz, J. Hughto and D. K. Berry, Nuclear "pasta" formation, *Phys. Rev. C* **88**, 065807 (2013). doi: 10.1103/PhysRevC.88.065807.
30. A. S. Schneider, D. K. Berry, C. M. Briggs, M. E. Caplan and C. J. Horowitz, Nuclear "waffles", *Phys. Rev. C* **90**, 055805 (2014).
31. T. Maruyama *et al.*, Quantum molecular dynamics approach to the nuclear matter below the saturation density, *Phys. Rev. C* **57**, 655–665 (Feb, 1998). doi: 10.1103/PhysRevC.57.655.
32. H. Sonoda, G. Watanabe, K. Sato, K. Yasuoka and T. Ebisuzaki, Phase diagram of nuclear "pasta" and its uncertainties in supernova cores, *Phys. Rev. C* **77**, 035806 (Mar, 2008). doi: 10.1103/PhysRevC.77.035806.
33. G. Watanabe, H. Sonoda, T. Maruyama, K. Sato, K. Yasuoka and T. Ebisuzaki, Formation of nuclear "pasta" in supernovae, *Phys. Rev. Lett.* **103**, 121101 (2009).
34. P. Bonche and D. Vautherin, A mean-field calculation of the equation of state of supernova matter, *Nucl. Phys. A* **372**(3), 496–526 (1981). ISSN 0375-9474. doi: 10.1016/0375-9474(81)90049-X.
35. P. Magierski and P.-H. Heenen, Structure of the inner crust of neutron stars: Crystal lattice or disordered phase?, *Phys. Rev. C* **65**, 045804 (2002). doi: 10.1103/PhysRevC.65.045804.
36. P. Gögelein and H. Müther, Nuclear matter in the crust of neutron stars, *Phys. Rev. C* **76**, 024312 (2007). doi: 10.1103/PhysRevC.76.024312.
37. W. G. Newton and J. R. Stone, Modeling nuclear "pasta" and the transition to uniform nuclear matter with the 3d Skyrme-Hartree-Fock method at finite temperature: Core-collapse supernovae, *Phys. Rev. C* **79**, 055801 (2009).
38. H. Pais and J. R. Stone, Exploring the nuclear pasta phase in core-collapse supernova matter, *Phys. Rev. Lett.* **109**, 151101 (2012). doi: 10.1103/PhysRevLett. 109.151101.
39. B. Schuetrumpf, M. A. Klatt, K. Iida, J. A. Maruhn, K. Mecke and P.-G. Reinhard, Time-dependent Hartree-Fock approach to nuclear "pasta" at finite temperature, *Phys. Rev. C* **87**, 055805 (2013). doi: 10.1103/PhysRevC.87.055805.
40. B. Schuetrumpf, K. Iida, J. A. Maruhn and P.-G. Reinhard, Nuclear "pasta matter" for different proton fractions, *Phys. Rev. C* **90**, 055802 (2014). doi: 10.1103/PhysRevC.90.055802.
41. B. Schuetrumpf, M. A. Klatt, K. Iida, G. E. Schröder-Turk, J. A. Maruhn, K. Mecke and P.-G. Reinhard, Appearance of the single gyroid network phase in "nuclear pasta" matter, *Phys. Rev. C* **91**, 025801 (2015). doi: 10.1103/PhysRevC.91.025801.
42. C. Gros, The boundary condition integration technique: results for the hubbard model in 1d and 2d, *Z. Phys. B* **86**(3), 359–365 (1992).

43. C. Gros, Control of the finite-size corrections in exact diagonalization studies, *Phys. Rev. B* **53**, 6865–6868 (1996). doi: 10.1103/PhysRevB.53.6865.
44. C. Lin, F. H. Zong and D. M. Ceperley, Twist-averaged boundary conditions in continuum quantum monte carlo algorithms, *Phys. Rev. E* **64**, 016702 (2001). doi: 10.1103/PhysRevE.64.016702.
45. B. Carter, N. Chamel and P. Haensel, Entrainment coefficient and effective mass for conduction neutrons in neutron star crust: simple microscopic models, *Nuclear Physics A* **748**(3–4), 675–697 (2005). ISSN 0375-9474. doi: http://dx.doi.org/10.1016/j.nuclphysa.2004.11.006.
46. N. Chamel, S. Naimi, E. Khan and J. Margueron, Validity of the Wigner-Seitz approximation in neutron star crust, *Phys. Rev. C* **75**, 055806 (2007). doi: 10.1103/PhysRevC.75.055806.
47. B. Schuetrumpf and W. Nazarewicz, Twist-averaged boundary conditions for nuclear pasta Hartree-Fock calculations, *Phys. Rev. C* **92**, 045806 (2015). doi: 10.1103/PhysRevC.92.045806.
48. F. Gulminelli, T. Furuta, O. Juillet and C. Leclercq, Boundary conditions for star matter and other periodic fermionic systems, *Phys. Rev. C* **84**, 065806 (2011). doi: 10.1103/PhysRevC.84.065806.
49. J. Bartel, P Quentin, M. Brack, C. Guet, and H.-B. Håkansson, Towards a better parametrisation of Skyrme-like effective forces: A critical study of the SkM force, *Nucl. Phys. A.* **386**, (1), 79 – 100 (1982). ISSN 0375-9474
50. N. Schunck, J. D. McDonnell, J. Sarich, S. M. Wild and D. Higdon, Error analysis in nuclear density functional theory, *J. Phys. G* **42**(3), 034024 (2015).
51. E. Chabanat *et al.*, A skyrme parametrization from subnuclear to neutron star densities part ii. nuclei far from stabilities, *Nucl. Phys. A* **635**(1–2), 231–256 (1998). ISSN 0375-9474.
52. J. Erler, C. J. Horowitz, W. Nazarewicz, M. Rafalski and P.-G. Reinhard, Energy density functional for nuclei and neutron stars, *Phys. Rev. C* **87**, 044320 (2013). doi: 10.1103/PhysRevC.87.044320.
53. M. Stoitsov, N. Schunck, M. Kortelainen, N. Michel, H. Nam, E. Olsen, J. Sarich and S. Wild, Axially deformed solution of the Skyrme-Hartree–Fock–Bogoliubov equations using the transformed harmonic oscillator basis (II) hfbtho v2.00d: A new version of the program, *Computer Physics Communications.* **184**(6), 1592 – 1604 (2013). ISSN 0010-4655. doi: http://dx.doi.org/10.1016/j.cpc.2013.01.013.
54. N. Schunck, J. Dobaczewski, J. McDonnell, W. Satuła, J. Sheikh, A. Staszczak, M. Stoitsov and P. Toivanen, Solution of the Skyrme–Hartree–Fock–Bogolyubov equations in the cartesian deformed harmonic-oscillator basis.: (VII) HFODD (V2.49t): A new version of the program, *Computer Physics Communications*, **183**(1), 166–192 (2012). ISSN 0010-4655. doi: http://dx.doi.org/10.1016/j.cpc.2011.08.013.
55. J. Maruhn, P.-G. Reinhard, P. Stevenson and A. Umar, The TDHF code sky3d, *Comput. Phys. Commun.* **185**(7), 2195–2216 (2014). ISSN 0010-4655. doi: http://dx.doi.org/10.1016/j.cpc.2014.04.008.

56. W. Rae, Clustering phenomena and shell effects in nuclear structure and reactions, *Int. J. Mod. Phys. A* **03**, 1343 (1988). doi: 10.1142/S0217751X880 00576.
57. A. Staszczak, A. Baran, J. Dobaczewski and W. Nazarewicz, Microscopic description of complex nuclear decay: Multimodal fission, *Phys. Rev. C* **80**, 014309 (2009). doi: 10.1103/PhysRevC.80.014309.
58. J. Sadhukhan, J. Dobaczewski, W. Nazarewicz, J. A. Sheikh and A. Baran, Pairing-induced speedup of nuclear spontaneous fission, *Phys. Rev. C* **90**, 061304 (2014). doi: 10.1103/PhysRevC.90.061304.
59. C. Simenel and A. S. Umar, Formation and dynamics of fission fragments, *Phys. Rev. C* **89**, 031601 (2014). doi: 10.1103/PhysRevC.89.031601.

CHAPTER 6

FRAGMENTATION OF NEUTRON STAR MATTER

Pablo N. Alcain and Claudio O. Dorso

Departamento de Fsica, FCEyN, UBA and IFIBA,
Conicet, Pabellón 1, Ciudad Universitaria,
1428 Buenos Aires, Argentina

1. Introduction

Neutron stars appear in the universe as the remains of the death of a massive star (but not too massive). Stars die when the internal thermonuclear reactions in the star are no longer able to balance the gravitational compression. At this stage, stars undergo gravitational collapse, followed by a shock wave that ejects most of the mass of the star leaving behind a dense core. During this collapse a big part of electrons and protons turn into neutrons through electron capture emitting neutrinos. Because of this, the residual mass tends to have an excess of neutrons over protons, hence the name neutron star. It should be emphasized that neutron stars are electrically balanced objects: their net charge is zero.

The mass of a neutron star ranges between 1 and 3 solar masses, its radius is of the order of a 10^{-5} solar radius or about 10 km, which gives it an average density of 10^{15} g/cm^3 or about 20 times that of normal nuclei. Energy considerations indicate that the cores of neutron stars are topped by a *crust* of about 1 km thick, where the neutrons produced by β-decay form neutron rich nuclear matter, immersed in a sea of electrons. Its structure can be divided in two parts, according to current models :[1,2] the *crust*, about 1.5 km thick and with a density of up to half the normal nuclear density ρ_0; and the *core*, where the structure is still unknown and remains highly speculative.[3]

The density of such crust ranges from normal nuclear density ($\approx 3\times10^{14}$ g/cm^3 or $\approx \rho_0 = 0.15$ fm^{-3}) at a depth $d \approx 1$ km to the neutron drip density ($\approx 4 \times 10^{11}$ g/cm^3) at $d \approx 0.5$ km, to, finally, an even lighter mix of neutron rich nuclei also embedded in a sea of electrons with densities decreasing down to practically zero in the neutron star envelope. Ravenhall *et al.* in Ref. 4 and Hashimoto *et al.* in Ref. 5 proposed that the neutron star crust is composed by the structures known as *nuclear pasta.*

Several models have been developed to study nuclear pasta, and they have shown that these structures arise due to the interplay between nuclear and Coulomb forces in an infinite medium. Nevertheless, the dependence of the observables on different thermodynamic quantities has not been studied in depth. The original works of Ravenhall *et al.*[4] and Hashimoto *et al.*[5] used a compressible liquid drop model, and have shown that the now known as *pasta phases* — *lasagna*, *spaghetti* and *gnocchi* — are solutions to the ground state of neutron star matter. From then on, different approaches have been taken, that we roughly classify in two categories: mean field or microscopic.

Mean field works include the Liquid Drop Model, by Lattimer *et al.*,[1] Thomas-Fermi, by Williams and Koonin,[6] among others .[7–12] Microscopic models include Quantum Molecular Dynamics, used by Maruyama *et al.*[13,14] and by Watanabe *et al.*,[15] Simple Semiclassical Potential, by Horowitz *et al.*[16] and Classical Molecular Dynamics, used in our previous works.[17] Of the many models used to study neutron stars, the advantages of classical or semiclassical models are the accessibility to position and momentum of all particles at all times and the fact that no specific shape is hardcoded in the model, as happens with most mean field models. This allows the study of the structure of the nuclear medium from a particle-wise point of view. Many models exist with this goal, like quantum molecular dynamics,[13] simple-semiclassical potential[16] and classical molecular dynamics.[18] In these models the Pauli repulsion between nucleons of equal isospin is either hard-coded in the interaction or as a separate term.[19]

Semiclassical models with molecular dynamics have been used to study the *nuclear pasta* regime, with mainly two parametrizations of the interaction: Simple Semiclassical Potential,[16] Quantum Molecular Dynamics[13] and Classical Molecular Dynamics.[17,20]

Multifragmentation in nuclear systems has been studied before,[21,22] but mostly with nuclear matter (without Coulomb interaction). In a recent work by Caplan *et al.*,[23] expanding neutron star matter has been studied as possible explanations for nucleosynthesis in neutron star mergers.

When two neutron stars collide, a neutron star merger happens. The compression of neutron star matter as a possible source for r-process nuclei was first discussed in Ref. 24. According to hydrodynamic models,[25] these have typical expansions coefficients of $\eta = 10^{-21}\,\mathrm{c/fm} < \eta < 4 \cdot 10^{-20}\,\mathrm{c/fm}$. Inspired by the neutron star merger, we perform a study on the fragmentation of expanding neutron star matter. In Section 2 we explore different models and, in Subsection 2.2 we define the model we use. Section 4 explains how we simulate the expansion of the system. To analyze fragments, we describe in Section 5 the cluster recognition algorithm, with emphasis on the identification of infinite fragments.

2. Nucleon Dynamics

To study the nuclear structure of stellar crusts is necessary understand the behavior of nucleons, i.e. protons and neutrons, at densities and temperatures as those encountered in the outer layers of stars. This knowledge comes from decades of study of nucleon dynamics through the use of computational models that have evolved from simple to complex. In this section we review this evolution, presenting a synopsis of existing models and culminating with the model used in this work, namely classical molecular dynamics (CMD).

2.1. *Evolution of models*

Several approaches have been used to simulate the behavior of nuclear matter, especially in reactions. The initial statistical studies of the 1980's, which lacked fragment-fragment interactions and after-breakup fragment dynamics,[26] were followed in the early 1990s by more robust studies that incorporated these effects. All these approaches, however, were based on idealized geometries and missed important shape fluctuation and reaction-kinematic effects. On the transport-theory side, models were developed in the late 1990s based on classical, semiclassical, and quantum-models; because of their use in the study of stellar nuclear systems nowadays, a side-by-side comparison of these transport models is in order.

Starting with the semiclassical, a class of models known generically as BUU is based on the Vlasov-Nordheim equation[27] also known as the Boltzmann-Uehling-Uhlenbeck.[28] These models numerically track the time evolution of the Wigner function, $f(\mathbf{r}, \mathbf{p})$, under a mean potential $U(\mathbf{r})$ to obtain a description of the probability of finding a particle at a point in phase space. The semiclassical interpretation constitutes BUU's main advantage; the disadvantages come from the limitation of using only a mean field which does not lead to cluster formation. To produce fragments, fluctuations must be added by hand, and more add-ons are necessary for secondary decays and other more realistic features.

On the quantum side, the molecular dynamics models, known as QMD, generically speaking, solve the equations of motion of nucleon wave packets moving within mean fields (derived from Skyrme potential energy density functional). The method allows the imposition of a Pauli-like blocking mechanism, use of isospin-dependent nucleon-nucleon cross sections, momentum dependence interactions and other variations to satisfy the operator's tastes. The main advantage is that QMD is capable of producing fragments, but at the cost of a poor description of cluster properties and the need of sequential decay codes external to QMD to cut the fragments and de-excite them ;[29] normally, clusters are constructed by a coalescence model based on distances and relative momenta of pairs of nucleons. Variations of these models applied to stellar crusts can be found elsewhere in this volume.

Of particular importance to this article is the classical molecular dynamics (CMD) model, brainchild of the Urbana group [18] and designed to reproduce the predictions of the Vlasov-Nordheim equation while providing a more complete description of heavy ion reactions. As it will be described in more detail in the next section, the model is based on the Pandharipande potential which provides the "nuclear" interaction through a combination of Yukawa potentials selected to correspond to infinite nuclear matter with proper equilibrium density, energy per particle, and compressibility.

Problems common to both BUU and QMD are the failure to produce appropriate number of clusters, and the use of hidden adjustable parameters, such as the width of wave packets, number of test particles, modifications of mean fields, effective masses and cross sections. These problems are not present in the classical molecular dynamics model, which without any adjustable parameters, hidden or not, is able to describe the dynamics of the reaction in space from beginning to end and with proper energy, space, and time units. It intrinsically includes all particle correlations at all levels: 2-body, 3-body, etc. and can describe nuclear systems ranging from highly correlated cold nuclei (such as two approaching heavy ions), to hot and dense nuclear matter (nuclei fused into an excited blob), to phase transitions (fragment and light particle production), to hydrodynamics flow (after-breakup expansion) and secondary decays (nucleon and light particle emission).

The only apparent disadvantage of the CMD is the lack of quantum effects, such as the Pauli blocking, which at medium excitation energies stops the method from describing nuclear structure correctly. Fortunately, in collisions, the large energy deposition opens widely the phase space available for nucleons and renders Pauli blocking practically obsolete,[30] while in stellar environments, at extremely low energies and with frozen-like structures, momentum-transferring collisions cease to be an important factor in deciding the stable configuration of the nuclear matter. Independent of that, the role of quantum effects in two body collisions is guaranteed to be included by the effectiveness of the potential in reproducing the proper cross sections, furthermore, an alternate fix is the use of momentum dependent potentials (as introduced by Dorso and Randrup[19]) when needed.

Although no theories can yet claim to be a perfect description of nuclear matter, all approaches have their advantages and drawbacks and, if anything can be said about them, is that they appear to be complementary to each other.

2.2. *Classical molecular dynamics*

In this work, we study fragmentation of Neutron Star Matter under pasta-like conditions with the classical molecular dynamics model CMD. It has been used in several heavy-ion reaction studies to: help understand experimental data;[31] identify phase-transition signals and other critical phenomena ;[30,32–35] and explore the caloric curve[36] and isoscaling.[37,38] CMD uses two two-body potentials to describe the interaction of nucleons, which are a combination of Yukawa potentials:

$$
\begin{aligned}
V_{np}^{\mathrm{CMD}}(r) &= v_r \exp(-\mu_r r)/r - v_a \exp(-\mu_a r)/r \\
V_{nn}^{\mathrm{CMD}}(r) &= v_0 \exp(-\mu_0 r)/r
\end{aligned}
$$

where V_{np} is the potential between a neutron and a proton, and V_{nn} is the repulsive interaction between either nn or pp. The cutoff radius is $r_c = 5.4\,\mathrm{fm}$ and for $r > r_c$ both potentials are set to zero. The Yukawa parameters μ_r, μ_a and μ_0 were determined to yield an equilibrium density of $\rho_0 = 0.16\,\mathrm{fm}^{-3}$, a binding energy $E(\rho_0) = 16\,\mathrm{MeV/nucleon}$ and a compressibility of 250 MeV.

To simulate an infinite medium, we used this potential with $N = 11000$ particles under periodic boundary conditions, with different proton fraction (i.e. with $x = Z/A = 0.2 < x < 0.4$) in cubical boxes with sizes adjusted to have densities $\rho = 0.05\,\mathrm{fm}^{-3}$ and $\rho = 0.08\,\mathrm{fm}^{-3}$. These simulations have been done with LAMMPS,[39] using its GPU package.[40]

2.2.1. *Ground state nuclei*

Although the $T = 0$ state of this classical nuclear matter at normal densities is a simple cubic solid, nuclear systems can be mimicked by adding enough kinetic energy to the nucleons. To study nuclei, for instance, liquid-like spherical drops with the right number of protons and neutrons are constructed confined in a steep spherical potential and then brought to the "ground" state by cooling them slowly from a rather high temperature until they reach a self-contained state. Removing the confining potential, the system is further cooled down until a reasonable binding energy is attained. The remaining kinetic energy of the nucleons helps to resemble the Fermi motion. Figure 1 shows the binding energies of "ground-state nuclei" obtained with the mass formula and with CMD; see Ref. 38 for details.

2.2.2. *Collisions*

With respect to collisions, these potentials are known to reproduce nucleon-nucleon cross sections from low to intermediate energies,[18] and it has been used extensively in studying heavy ion collisions (see Refs. 31 and 35). For such reactions, two "nuclei" are boosted against each other at a desired

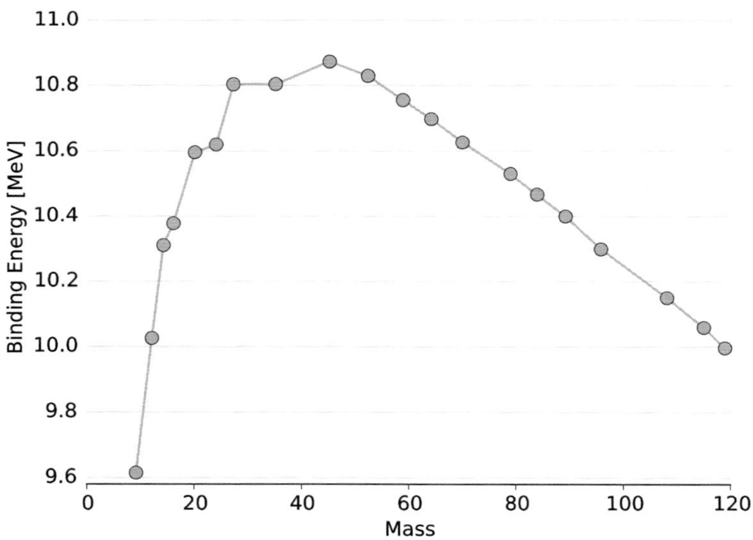

Fig. 1. Binding energies of ground-state nuclei obtained with CMD model, extracted from Ref. 38.

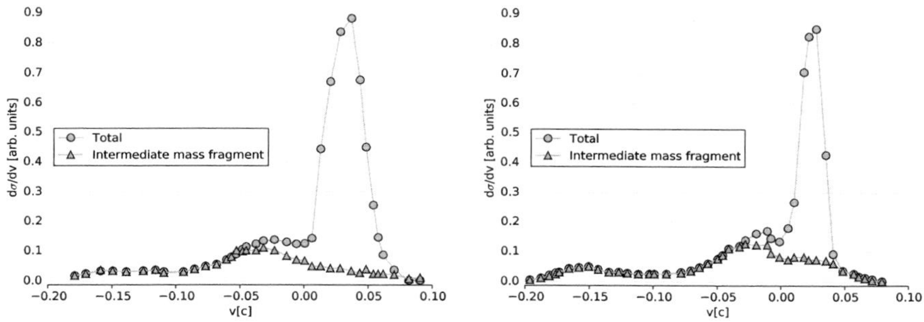

(a) η=0 .0001fm/c (b) $\eta = 0.0001$ fm/c

Fig. 2. (a) Experimental and (b) simulated parallel velocity distributions for $^{58}Ni + C$ collisions.

energy. From collision to collision, the projectile and target are rotated with respect to each other at random values of the Euler angles. The evolution of the system is followed using a velocity-Verlet algorithm with energy conservation better than 0.01%. At any point in time, the nucleon information, i.e., position and momenta, can be turned into fragment information by identifying the clusters and free particles; several such cluster recognition algorithms have been developed by our collaborator, C.O. Dorso, and they are well described in the literature.[41,42]

The method yields mass multiplicities, momenta, excitation energies, secondary decay yields, etc. comparable to experimental data.[31,43] Figure 10, for instance, shows experimental and simulated parallel velocity distributions for particles obtained from mid-peripheral and peripheral $^{58}Ni + C$ collisions performed at the Coupled Tandem and Super-Conducting Cyclotron accelerators of AECL at Chalk River.[31]

2.2.3. *Thermostatic properties of nuclear matter*

To study thermal properties of static nuclear matter, these drops or infinite systems, nucleons are positioned at random, but with a selected density, in a container and "heated". After equilibration, the system can then be used to extract macroscopic variables. Repeating these simulations for a wide range of density and temperature values, information about the energy per nucleon, $\epsilon(\rho, T)$, can be obtained and used to construct analytical fits in the spirit of those pioneered by Bertsch, Siemens, and Kapusta ;[44–46] these fits in turn can be used to derive other thermodynamic variables, such as pressure, etc.

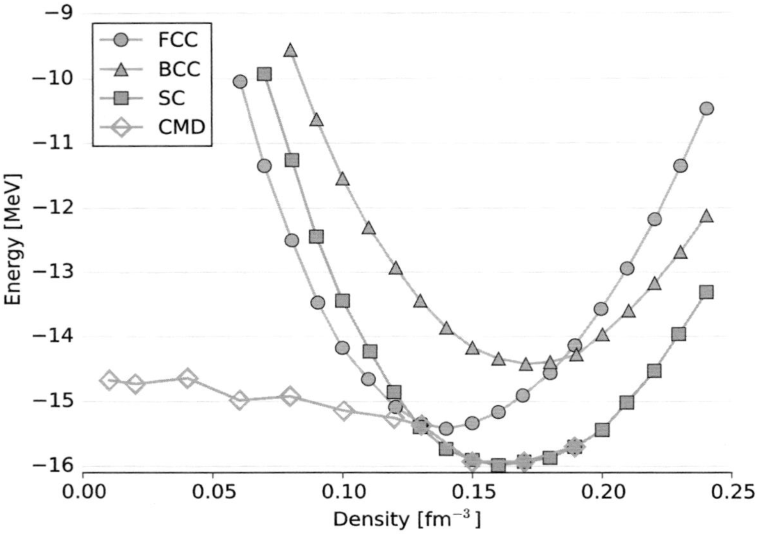

Fig. 3. Nuclear matter energy per particle of cold matter calculated with CMD. Full symbols are for homogeneous systems, while open circles denote the emergence of the pseudo-pasta.

Figure 3 shows the results of the method as applied by Giménez-Molinelli *et al.*[47] for stiff cold infinite nuclear matter. The open symbols correspond to CMD calculations at low temperatures, showing a departure from the imposed homogeneous solutions obtained with the full symbols for different crystal symmetries. This shows the emergence of pseudo pastas in nuclear matter.

For finite systems, Fig. 4 demonstrates the feasibility of using CMD to study thermal properties such as the caloric curve (i.e. the temperature–excitation energy relationship) for a system of 80 nucleons equilibrated at four different densities; see Ref. 38 for complete details.

2.3. *Coulomb interaction in the model*

Since a neutralizing electron gas embeds the nucleons in the neutron star crust, the Coulomb forces among protons are screened. The model we used to model this screening effect is the Thomas-Fermi approximation, used with various nuclear models .[13,17,48] According to this approximation,

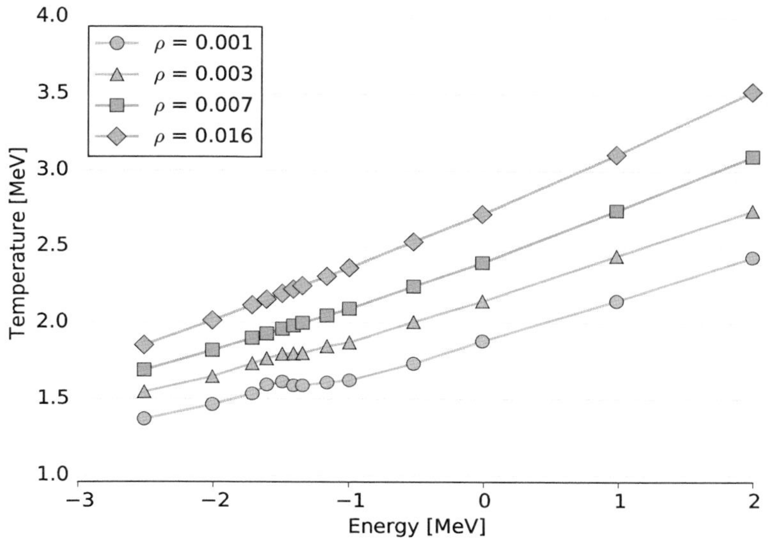

Fig. 4. Caloric curve of a system equilibrated at four different densities calculated with CMD.

protons interact via a Yukawa-like potential, with a screening length λ:

$$V_{TF}(r) = q^2 \frac{e^{-r/\lambda}}{r}.$$

Theoretical estimates for the screening length λ are $\lambda \sim 100\,\text{fm}$,[49] but we set the screening length to $\lambda = 20\,\text{fm}$. This choice was based on previous studies,[50] where we have shown that this value is enough to adequately reproduce the expected length scale of density fluctuations for this model, while larger screening lengths would be a computational difficulty. We analyze the opacity to neutrinos of the structures for different proton fractions and densities.

3. Neutron Star Matter at low densities and temperature

When we consider the system with a screened Coulomb interaction as described in 2.3, a very interesting phenomena takes place, described with detail in Ref. 50. At sub-saturation densities, the system displays an inhomogeneous structure known as nuclear pasta, characterized by the emergence of multiple structure per simulation cell, that can be roughly classified

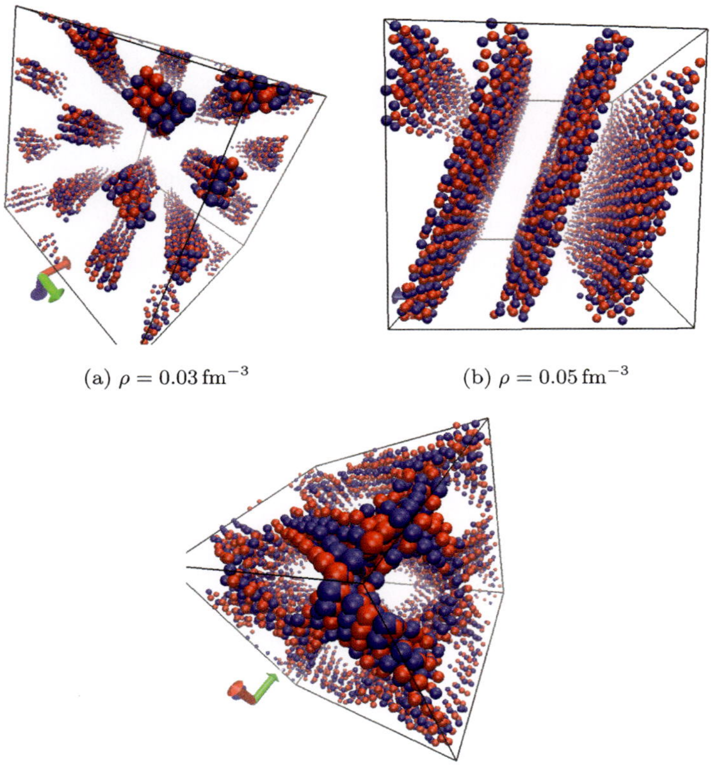

(a) $\rho = 0.03\,\mathrm{fm}^{-3}$ (b) $\rho = 0.05\,\mathrm{fm}^{-3}$

(c) $\rho = 0.08\,\mathrm{fm}^{-3}$

Fig. 5. (a) Spaghetti, (b) lasagna and (c) tunnels obtained with the CMD method, for the symmetric case $x = 0.5$ and low temperature $T = 0.5\,\mathrm{MeV}$ extracted from Ref. 50

as *gnocchi*, *spaghetti*, *lasagna* and tunnels. As an example, in Figure 5 we show the configurations of some of these pastas.

4. Expansion

To simulate an expanding system we scale linearly with time the length of the box in every dimension,

$$L(t) = L_0(1 + \eta\, t)$$

This, however, is not enough to expand the system collectively. We also need the particles inside the cell to expand like the box. In order to accomplish this, based on Ref. 51, we add to each particle a velocity v_{exp} dependent on the position in the box:

$$\mathbf{v_{exp}} = \eta\, \mathbf{r}$$

We can see from this expression that the particles in the edge of the box will have an expanding velocity equal to that of the box.

Another effect to consider of this expansion is that when a particle crosses a boundary its velocity has to change according to the velocity of the expanding box. For example, if the particle crosses the left-hand boundary of the periodic box, the velocity of the image particle $v_i^\dagger$ on the right-hand must be modified $v_i^\dagger = v_i + L_0\, \eta$.

5. Cluster recognition

In typical configurations we have not only the structure known as nuclear pasta, but also a nucleon gas that surrounds the nuclear pasta. In order to properly characterize the pasta phases, we must know which particles belong to the pasta phases and which belong to this gas. To do so, we have to find the clusters that are formed along the simulation.

One of the algorithms to identify cluster formation is Minimum Spanning Tree (MST). In MST algorithm, two particles belong to the same cluster $\{C_n^{\mathrm{MST}}\}$ if the relative distance of the particles is less than a cutoff distance r_{cut}:

$$i \in C_n^{\mathrm{MST}} \Leftrightarrow \exists j \in C_n \mid r_{ij} < r_{cut}$$

This cluster definition works correctly for systems with no kinetic energy, and it is based in the attractive tail of the nuclear interaction. However, if the particles have a non-zero momentum, we can have a situation of two particles that are closer than the cutoff radius, but with a large relative kinetic energy.

To deal with situations of non-zero temperatures, we need to take into account the relative momentum among particles. One of the most sophisticated methods to accomplish this is the Early Cluster Recognition Algorithm (ECRA).[52] In this algorithm, the particles are partitioned in different disjoint clusters C_n^{ECRA}, with the total energy in each cluster:

$$\epsilon_n = \sum_{i \in C_n} K_i^{CM} + \sum_{i,j \in C_n} V_{ij}$$

where K_i^{CM} is the kinetic energy relative to the center of mass of the cluster. The set of clusters $\{C_n\}$ then is the one that minimizes the sum of all the cluster energies $E_{\text{partition}} = \sum_n \epsilon_n$.

ECRA algorithm can be easily used for small systems,[53] but being a combinatorial optimization, it cannot be used in large systems. While finding ECRA clusters is very expensive computationally, using simply MST clusters can give extremely biased results towards large clusters. We have decided to go for a middle ground choice, the Minimum Spanning Tree Energy (MSTE) algorithm.[17] This algorithm is a modification of MST, taking into account the kinetic energy. According to MSTE, two particles belong to the same cluster $\{C_n^{\text{MSTE}}\}$ if they are energy bound:

$$i \in C_n^{\text{MSTE}} \Leftrightarrow \exists j \in C_n : V_{ij} + K_{ij} \leq 0$$

While this algorithm doesn't yield the same theoretically sound results from ECRA, it still avoids the largest pitfall of nave MST implementations for the temperatures used in this work.

5.1. *Infinite Clusters*

We developed an algorithm for the recognition of infinite clusters across the boundaries. We explain here in detail the implementation for MST clusters in 2D, being the MSTE and 3D extension straightforward. In Figure 6 we see a schematical representation of 2D clusters recognized in a periodic cell, labeled from 1 to 6 (note that these clusters don't connect yet through the periodic walls).

In order to find the connections of these clusters through the boundaries, we draw a labeled graph of the clusters, where we connect clusters depending on whether they connect or not through a wall and label such connection with the wall label. For example, we begin with cluster 1. It connects with cluster 2 going out through the E wall, therefore we add a $1 \to 2$ connection labeled as E. Symmetrically, we add a $2 \to 1$ connection labeled as W. Now we go for the pair 1-3. It connects going out through the S wall, so we add $1 \to 3$ labeled as S and $3 \to 1$ labeled as N. Cluster 1 does not connect with 4, 5, or 6, therefore those are the only connections we have. Once we've done that, we get the graph of Figure 7.

We now wonder whether these subgraphs represent an infinite cluster or not not. In order to have an infinite clusters, we need to have a loop (the opposite is not true: having a loop is not enough to have an infinite cluster, as we can see in subragph 5–6), so we first identify loops and mark them

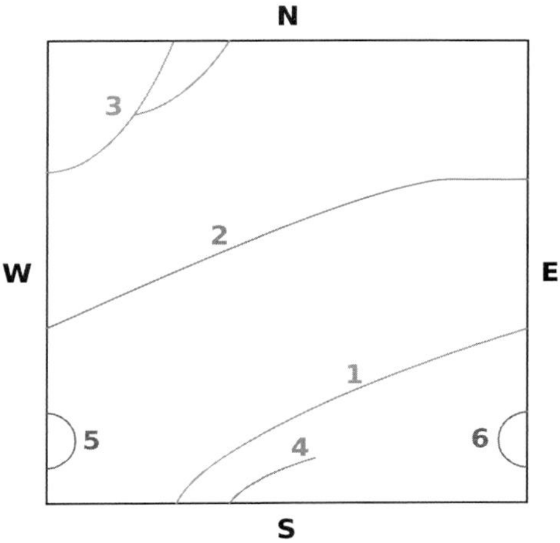

Fig. 6. (Color online) Schematical representation of 2D clusters, recognized only in the cell and not through the periodic walls, labeled as N, S, W, E. The clusters inside the cell are labeled from 1 to 6.

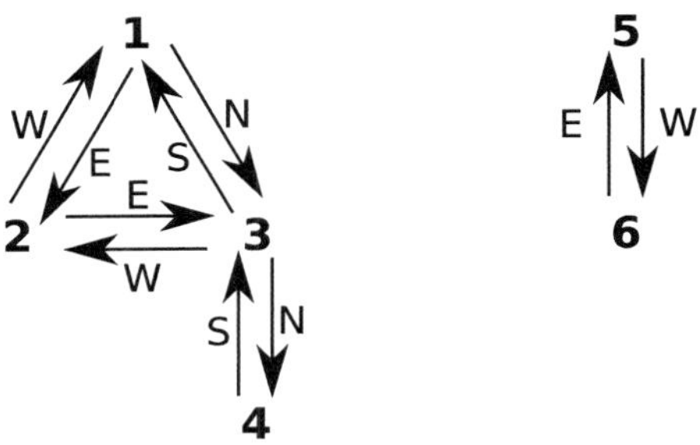

Fig. 7. Graph of the clusters with connections labeled by the wall of the boundary they connect through. The graph can be divided in 2 subgraphs that don't connect: 1–2–3–4 and 5–6. Each of these subgraphs is as cluster when periodic boundary conditions are considered.

as candidates for infinite clusters. Every connection adds to a loop (since the graph connections are back and forth), but we know from inspecting the Figure 7 that the cluster 1–2–3 is infinite. Finding out what makes, in the graph, the cluster 1–2–3 infinite is key to identify infinite clusters. And the key feature of cluster 1–2–3 is that its loop 1–2–3–1 can be transversed through the walls E–E–S, while loops like 5–6 can be transversed only through E–W. Now, in order for the cluster to be infinite, we need it to extend infinitely in (at least) one direction. So once we have the list of walls of the loop, we create a magnitude I associated to each loop that is created as follows: beginning with $I = 0$, we add a value M_i if there is (at least one) i wall. The values are: $M_E = 1$, $M_W = -1$, $M_N = 2$, $M_S = -2$. If I is nonzero, then the loop is infinite. For example, for the loop E–E–S, we have E and S walls, so $I = M_E + M_S = 3$ and the loop is infinite. For the loop E–W, $I = M_E + M_W = 0$, and the loop is finite.

6. Results

In Figure 8 we show the initial and final states for the expansion of $N = 11000$ particle primordial cell (see caption for details) for the case of low velocity expansion. It can be immediately seen that though the initial configuration shows a compact particle distribution, the final configuration consists of *gnocchi* structures (almost spherical fragments) with a mass of about 80 particles.

In Figure 9 we show the fraction of particles in the primordial cell that form part of an infinite cluster (*Infinite Fragment Fraction*, IFF). It can be easily seen that in the early stages of the evolution, due to the fact that the temperature is low, most of the system in the primordial cell belongs to the infinite clusters. But as the system evolves according to the expansion

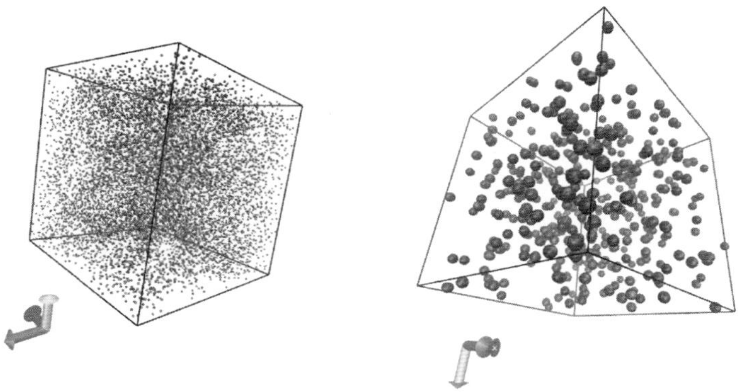

(a) $\eta = 0.0001\,\mathrm{fm/c}$ (b) $\eta = 0.0005\,\mathrm{fm/c}$

Fig. 8. (Color online) Snapshots of a system in the initial configuration (a) and the final expanded system (b). In the expanded system we see clearly formed *gnocchi* clusters.

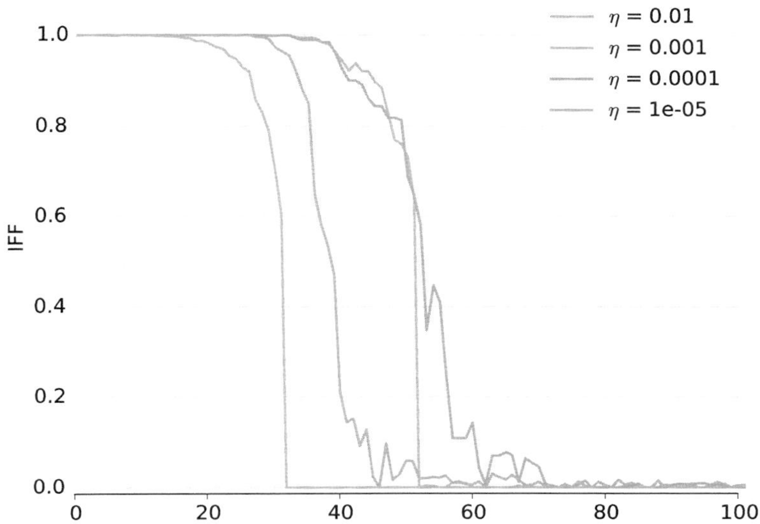

Fig. 9. (Color online) Infinite Fragment Fraction (see text) as a function of the length of the edge of the primordial cell. For every expansion rate displayed the IFF goes to zero in the asymptotic regime.

rate as explained above, the IFF goes down and goes to zero rather quickly, meaning that there is no more infinite fragment in the system. See caption for details.

Figure 10 shows the asymptotic fragment mass distribution for 4 different expansion rates. In this case, the MSTE algorithm has been applied over the primordial cell, taking into account periodic boundary conditions knowing that there is no infinite fragment as shown in Fig. 9. It can be seen that as the homogeneous expansion velocity increases, the fragment mass distribution displays the familiar transition from U-shaped to exponential decay. Somewhere in between, a power law distribution is to be expected. In particular, Fig. 10 shows that with an expansion of $\eta = 0.009\,\text{fm/c}$ we are close to a power law distribution. It is interesting to note that at variance with percolation or Lennard-Jones systems, due to the presence of the Coulomb long range repulsion term, it is not possible to see an infinite cluster in the asymptotic regime. Moreover, together with the rather large rate of expansion, we do not expect to find big clusters.

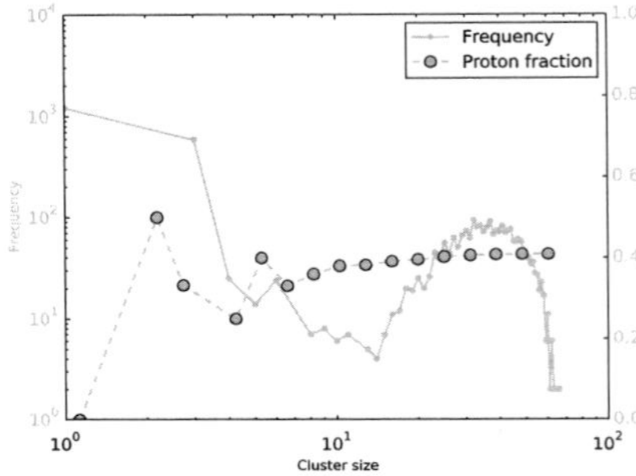

(a) $\eta = 0.0001\,\mathrm{fm/c}$

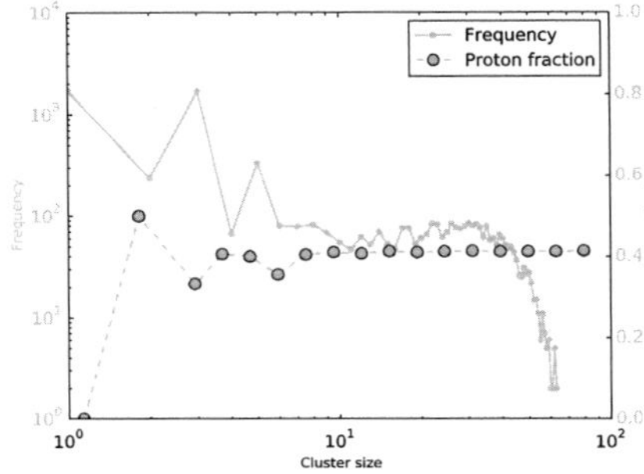

(b) $\eta = 0.0005\,\mathrm{fm/c}$

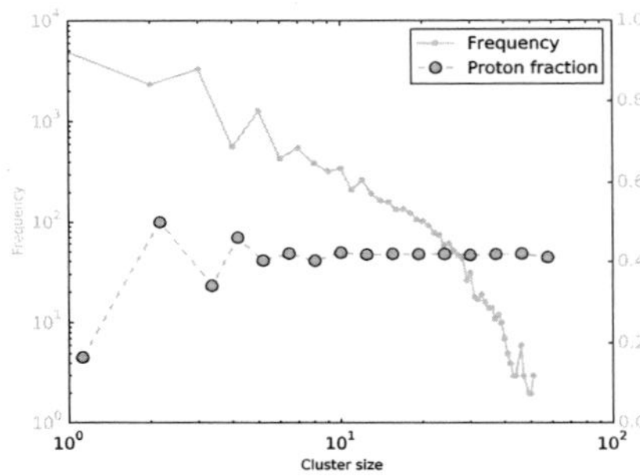

(c) $\eta = 0.009\,\mathrm{fm/c}$

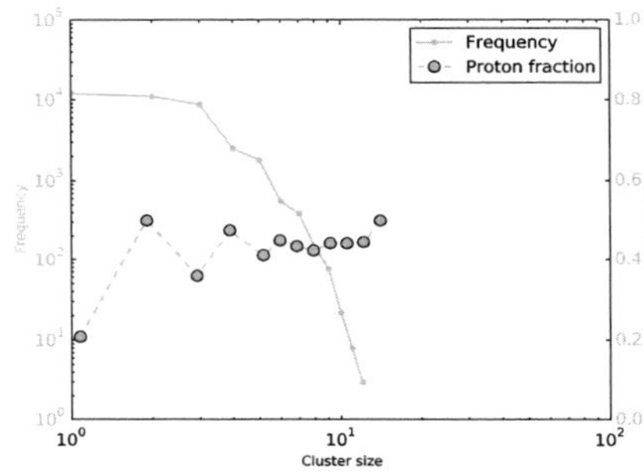

(d) $\eta = 0.03\,\mathrm{fm/c}$

Fig. 10. (Color online) Fragment mass distribution. Labels grow with larger expansion rates. System consisting of 11000 particles with $x = 0.4$ and initial density $\rho_0 = 0.08\,\mathrm{fm}^{-3}$.

7. Discussion and Concluding Remarks

In this work we have performed numerical experiments of homogeneously expanding systems with not very small amount of particles in the primordial cell. In order to analyze the fragment structure of such a system as a function of time, we have developed a graph-based tool for the identification of infinite fragments for any definition of percolation-like (i.e., additive) clusters. Once this formalism is applied to the above mentioned simulations, we have been able to identify the region in which a power-law distribution of masses is expected. The fragment mass distribution shapes range from U-like to exponential decay.

We are currently performing simulations with a larger number of particles to properly characterize the critical behavior of the system.

References

1. D. Page, J. M. Lattimer, M. Prakash and A. W. Steiner, Minimal Cooling of Neutron Stars: A New Paradigm, *ApJS.* **155**(2), 623 (Dec., 2004). ISSN 0067-0049. doi: 10.1086/424844. URL http://iopscience.iop.org/0067-0049/155/2/623.
2. U. Geppert, M. Küker, and D. Page, Temperature distribution in magnetized neutron star crusts, *Astronomy & Astrophysics.* **426**(1), 11 (2004). doi: 10.1051/0004-6361:20040455.
3. S. Woosley and T. Janka, The physics of core-collapse supernovae, *Nature Physics.* **1**(3), 147–154 (Dec., 2005). ISSN 1745-2473. doi: 10.1038/nphys172. URL http://www.nature.com/nphys/journal/v1/n3/abs/nphys172.html.
4. D. G. Ravenhall, C. J. Pethick, and J. R. Wilson, Structure of Matter below Nuclear Saturation Density, *Phys. Rev. Lett.* **50**(26), 2066–2069 (June, 1983). doi: 10.1103/PhysRevLett.50.2066. URL http://link.aps.org/doi/10.1103/PhysRevLett.50.2066.
5. M.-a. Hashimoto, H. Seki and M. Yamada, Shape of Nuclei in the Crust of Neutron Star, *Prog. Theor. Phys.* **71**(2), 320–326 (Feb., 1984). ISSN 0033-068X, 1347-4081. doi: 10.1143/PTP.71.320. URL http://ptp.oxfordjournals.org/content/71/2/320.
6. R. D. Williams and S. E. Koonin, Sub-saturation phases of nuclear matter, *Nuclear Physics A.* **435**(3–4), 844–858 (Mar., 1985). ISSN 0375-9474. doi: 10.1016/0375-9474(85)90191-5. URL http://www.sciencedirect.com/science/article/pii/0375947485901915.
7. K. Oyamatsu, Nuclear shapes in the inner crust of a neutron star, *Nuclear Physics A.* **561**(3), 431–452 (Aug., 1993). ISSN 0375-9474. doi: 10.1016/0375-9474(93)90020-X. URL http://www.sciencedirect.com/science/article/pii/037594749390020X.
8. C. P. Lorenz, D. G. Ravenhall, and C. J. Pethick, Neutron star crusts, *Phys. Rev. Lett.* **70**(4), 379–382 (Jan., 1993). doi: 10.1103/PhysRevLett.70.379. URL http://link.aps.org/doi/10.1103/PhysRevLett.70.379.
9. K. S. Cheng, C. C. Yao and Z. G. Dai, Properties of nuclei in the inner crusts of neutron stars in the relativistic mean-field theory, *Phys. Rev. C.* **55**(4), 2092–2100 (Apr., 1997). doi: 10.1103/PhysRevC.55.2092. URL http://link.aps.org/doi/10.1103/PhysRevC.55.2092.
10. G. Watanabe, K. Iida, and K. Sato, Thermodynamic properties of nuclear "pasta" in neutron star crusts, *Nuclear Physics A.* **676** (1–4), 455–473 (Aug., 2000). ISSN 0375-9474. doi: 10.1016/S0375-9474(00)00197-4. URL http://www.sciencedirect.com/science/article/pii/S0375947400001974.
11. G. Watanabe and K. Iida, Electron screening in the liquid-gas mixed phases of nuclear matter, *Phys. Rev. C.* **68**(4), 045801 (Oct., 2003). doi: 10.1103/PhysRevC.68.045801. URL http://link.aps.org/doi/10.1103/ PhysRevC.68.045801.

12. K. Nakazato, K. Oyamatsu and S. Yamada, Gyroid Phase in Nuclear Pasta, *Phys. Rev. Lett.* **103**(13), 132501 (Sept., 2009). doi: 10.1103/PhysRevLett.103.132501. URL http://link.aps.org/doi/10.1103/PhysRevLett.103.132501.
13. T. Maruyama, K. Niita, K. Oyamatsu, T. Maruyama, S. Chiba and A. Iwamoto, Quantum molecular dynamics approach to the nuclear matter below the saturation density, *Phys. Rev. C.* **57**(2), 655–665 (Feb., 1998). doi: 10.1103/PhysRevC.57.655. URL http://link.aps.org/doi/10.1103/PhysRevC.57.655.
14. T. Kido, T. Maruyama, K. Niita and S. Chiba, MD simulation study for nuclear matter, *Nuclear Physics A.* **663–664**, 877c–880c (Jan., 2000). ISSN 0375-9474. doi: 10.1016/S0375-9474(99)00736-8. URL http://www.sciencedirect.com/science/article/pii/S0375947499007368.
15. G. Watanabe, K. Sato, K. Yasuoka and T. Ebisuzaki, Structure of cold nuclear matter at subnuclear densities by quantum molecular dynamics, *Phys. Rev. C.* **68** (3), 035806 (Sept., 2003). doi: 10.1103/PhysRevC.68.035806. URL http://link.aps.org/doi/10.1103/PhysRevC.68.035806.
16. C. Horowitz, M. Prez-Garca, J. Carriere, D. Berry and J. Piekarewicz, Nonuniform neutron-rich matter and coherent neutrino scattering, *Phys. Rev. C.* **70** (6), 065806 (Dec., 2004). doi: 10.1103/PhysRevC.70.065806. URL http://link.aps.org/doi/10.1103/PhysRevC.70.065806.
17. C. O. Dorso, P. A. Gimnez Molinelli, and J. A. Lpez, Topological characterization of neutron star crusts, *Phys. Rev. C.* **86**(5), 055805 (Nov., 2012). doi: 10.1103/PhysRevC.86.055805. URL http://link.aps.org/doi/10. 1103/PhysRevC.86.055805.
18. R. J. Lenk, T. J. Schlagel and V. R. Pandharipande, Accuracy of the Vlasov-Nordheim approximation in the classical limit, *Phys. Rev. C.* **42**(1), 372–385 (July, 1990). doi: 10.1103/PhysRevC.42.372. URL http://link.aps.org/doi/10.1103/PhysRevC.42.372.
19. C. Dorso and J. Randrup, Classical simulation of nuclear systems, *Physics Letters B.* **215**(4), 611–616 (Dec., 1988). ISSN 0370-2693. doi: 10.1016/0370-2693(88)90030-5. URL http://www.sciencedirect.com/science/article/pii/0370269388900305.
20. P. N. Alcain, P. A. Gimnez Molinelli and C. O. Dorso, Beyond nuclear "pasta": Phase transitions and neutrino opacity of new "pasta" phases, *Phys. Rev. C.* **90**(6), 065803 (Dec., 2014). doi: 10.1103/PhysRevC.90.065803. URL http://link.aps.org/doi/10.1103/PhysRevC.90.065803.
21. A. Bonasera, M. Bruno, C. O. Dorso and F. Mastinu, Critical phenomena in nuclear fragmentation (2000). URL http://inis.iaea.org/Search/search.aspx?orig_q=RN:31053937.
22. S. Chikazumi, T. Maruyama, S. Chiba, K. Niita and A. Iwamoto, Quantum molecular dynamics simulation of expanding nuclear matter and nuclear multifragmentation, *Phys. Rev. C.* **63**(2), 024602 (Jan., 2001). doi: 10.1103/PhysRevC.63.024602. URL http://link.aps.org/doi/10. 1103/PhysRevC.63.024602.

23. M. E. Caplan, A. S. Schneider, C. J. Horowitz, and D. K. Berry, Pasta nucleosynthesis: Molecular dynamics simulations of nuclear statistical equilibrium, *Phys. Rev. C.* **91**(6), 065802 (June, 2015). doi: 10.1103/PhysRevC.91.065802. URL http://link.aps.org/doi/10.1103/PhysRevC.91.065802.
24. J. M. Lattimer and D. N. Schramm, Black-hole-neutron-star collisions, *The Astrophysical Journal Letters.* **192**, L145–L147 (Sept., 1974). ISSN 0004-637X. doi: 10.1086/181612. URL http://adsabs.harvard.edu/abs/1974ApJ...192L.145L.
25. S. Goriely, A. Bauswein, and H.-T. Janka, r-process Nucleosynthesis in Dynamically Ejected Matter of Neutron Star Mergers, *ApJ.* **738**(2), L32 (2011). ISSN 2041-8205. doi: 10.1088/2041-8205/738/2/L32. URL http://stacks.iop. org/2041-8205/738/i=2/a=L32.
26. H. W. Barz, J. P. Bondorf, D. Idier, and I. N. Mishustin, Cluster production and correlations in nuclear molecular dynamics, *Physics Letters B.* **382**(4), 343–348 (Aug., 1996). ISSN 0370-2693. doi: 10.1016/0370-2693(96)00692-2. URL http://www.sciencedirect.com/science/article/pii/0370269396006922.
27. L. W. Nordheim, On the kinetic method in the new statistics and its application in the electron theory of conductivity, *Proceedings of the Royal Society of London. Series A.* **119**(783), 689–698 (1928). URL http://rspa.royalsocietypublishing.org/content/119/783/689.full.pdf.
28. E. A. Uehling and G. E. Uhlenbeck, Transport Phenomena in Einstein-Bose and Fermi-Dirac Gases. I, *Phys. Rev.* **43**(7), 552–561 (Apr., 1933). doi: 10.1103/PhysRev.43.552. URL http://link.aps.org/doi/10.1103/PhysRev.43.552.
29. A. Polanski, S. Petrochenkov and V. Uzhinsky, Development of a quantum molecular dynamic (QMD) model to describe fission and fragment production, *Radiat Prot Dosimetry.* **116**(1–4), 582–584 (Dec., 2005). ISSN 0144-8420, 1742-3406. doi: 10.1093/rpd/nci205. URL http://rpd.oxfordjournals.org/content/116/1-4/582.
30. J. A. Lpez and C. Dorso, *Lectures Notes on Phase Transformations in Nuclear Matter.* WORLD SCIENTIFIC (Aug., 2000). ISBN 978-981-02-4007-3 978-981-281-340-4. URL http://www.worldscientific.com/worldscibooks/10.1142/4169.
31. A. Chernomoretz, L. Gingras, Y. Larochelle, L. Beaulieu, R. Roy, C. St-Pierre, and C. O. Dorso, Quasiclassical model of intermediate velocity particle production in asymmetric heavy ion reactions, *Phys. Rev. C.* **65**(5), 054613 (May, 2002). doi: 10.1103/PhysRevC.65.054613. URL http://link.aps.org/doi/10.1103/PhysRevC.65.054613.
32. A. Barranon, C. O. Dorso and J. A. Lopez, Searching for criticality in nuclear fragmentation, *Revista mexicana de fsica.* **47**, 93–97 (2001). ISSN 0035-001X. URL http://cat.inist.fr/?aModele=afficheN&cpsidt=13424850.
33. C. O. Dorso and J. A. Lpez, Selection of critical events in nuclear fragmentation, *Phys. Rev. C.* **64**(2), 027602 (July, 2001). doi: 10.1103/PhysRevC.64.027602. URL http://link.aps.org/doi/10.1103/PhysRevC.64.027602.

34. A. Barranón, R. Cárdenas, C. O. Dorso, and J. A. López, The critical exponent of nuclear fragmentation, *APH N.S., Heavy Ion Physics.* **17**(1), 59–73 (Jan., 2003). ISSN 1219-7580, 1588-2675. doi: 10.1556/APH.17.2003.1.8. URL http://link.springer.com/article/10.1556/APH.17.2003.1.8.
35. A. Barrañón, C. O. Dorso and J. A. López, Time dependence of isotopic temperatures, *Nuclear Physics A.* **791**(1–2), 222–231 (July, 2007). ISSN 0375-9474. doi: 10.1016/j.nuclphysa.2007.04.008. URL http://www.sciencedirect.com/science/article/pii/S0375947407004769.
36. A. Barrañón, J. Roa, and J. López, Entropy in the nuclear caloric curve, *Phys. Rev. C.* **69** (1), 014601 (Jan., 2004). doi: 10.1103/PhysRevC.69.014601. URL http://link.aps.org/doi/10.1103/PhysRevC.69.014601.
37. C. O. Dorso, C. R. Escudero, M. Ison, and J. A. López, Dynamical aspects of isoscaling, *Phys. Rev. C.* **73**(4), 044601 (Apr., 2006). doi: 10.1103/PhysRevC.73.044601. URL http://link.aps.org/doi/10.1103/PhysRevC. 73.044601.
38. C. A. Dorso, P. A. G. Molinelli and J. A. López, Isoscaling and the nuclear EoS, *J. Phys. G: Nucl. Part. Phys.* **38**(11), 115101 (Nov., 2011). ISSN 0954-3899. doi: 10.1088/0954-3899/38/11/115101. URL http://iopscience.iop.org/0954-3899/38/11/115101.
39. S. Plimpton, Fast Parallel Algorithms for Short-Range Molecular Dynamics, *Journal of Computational Physics.* **117** (1), 1–19 (Mar., 1995). ISSN 0021-9991. doi: 10.1006/jcph.1995.1039. URL http://www.sciencedirect.com/ science/article/pii/S002199918571039X.
40. W. M. Brown, A. Kohlmeyer, S. J. Plimpton and A. N. Tharrington, Implementing molecular dynamics on hybrid high performance computers – Particle–particle particle-mesh, *Computer Physics Communications.* **183**(3), 449–459 (Mar., 2012). ISSN 0010-4655. doi: 10.1016/j.cpc.2011.10.012. URL http://www.sciencedirect.com/science/article/pii/S0010465511003444.
41. C. O. Dorso and J. Aichelin, When and how are fragments formed in heavy ion collisions?, *Physics Letters B.* **345**(3), 197–202 (Feb., 1995). ISSN 0370-2693. doi: 10.1016/0370-2693(94)01632-M. URL http://www.sciencedirect.com/science/article/pii/037026939401632M.
42. A. Strachan and C. O. Dorso, Time scales in fragmentation, *Phys. Rev. C.* **55**(2), 775–787 (Feb., 1997). doi: 10.1103/PhysRevC.55.775. URL http://link.aps.org/doi/10.1103/PhysRevC.55.775.
43. M. Belkacem, P. F. Mastinu, V. Latora, A. Bonasera, M. D'Agostino, M. Bruno, J. D. Dinius, M. L. Fiandri, F. Gramegna, D. O. Handzy, W. C. Hsi, M. Huang, G. V. Margagliotti, P. M. Milazzo, C. P. Montoya, G. F. Peaslee, R. Rui, C. Schwarz, G. Vannini and C. Williams, Searching for the nuclear liquid-gas phase transition in Au+Au collisions at 35 MeV/nucleon, *Phys. Rev. C.* **54** (5), 2435–2444 (Nov., 1996). doi: 10.1103/PhysRevC.54.2435. URL http://link.aps.org/doi/10.1103/PhysRevC.54.2435.
44. G. Bertsch and P. J. Siemens, Nuclear fragmentation, *Physics Letters B.* **126** (1), 9–12 (June, 1983). ISSN 0370-2693. doi: 10.1016/0370-2693(83)90004-7. URL http://www.sciencedirect.com/science/article/pii/0370269383900047.

45. J. Kapusta, Deuteron and entropy production and the nuclear liquid-gas phase transition, *Phys. Rev. C.* **29**(5), 1735–1743 (May, 1984). doi: 10.1103/PhysRevC.29.1735. URL http://link.aps.org/doi/10.1103/ PhysRevC.29.1735.
46. J. A. Lopez and P. J. Siemens, Nuclear fragmentation, *Nuclear Physics A.* **431**(4), 728–744 (Dec., 1984). ISSN 0375-9474. doi: 10.1016/0375-9474(84)90277-X. URL http://www.sciencedirect.com/science/article/pii/037594748490277X.
47. P. A. Giménez Molinelli, J. I. Nichols, J. A. López and C. O. Dorso, Simulations of cold nuclear matter at sub-saturation densities, *Nuclear Physics A.* **923**, 31–50 (Mar., 2014). ISSN 0375-9474. doi:10.1016/j.nuclphysa.2014.01.003. URL http://www.sciencedirect.com/science/article/pii/S0375947414000049.
48. C. J. Horowitz, M. A. Pérez-García, and J. Piekarewicz, Neutrino-"pasta" scattering: The opacity of nonuniform neutron-rich matter, *Phys. Rev. C.* **69**(4), 045804 (Apr., 2004). doi: 10.1103/PhysRevC.69.045804. URL http://link.aps.org/doi/10.1103/PhysRevC.69.045804.
49. A. L. Fetter and J. D. Walecka, *Quantum Theory of Many-particle Systems.* Courier Dover Publications (2003). ISBN 978-0-486-42827-7.
50. P. N. Alcain, P. A. Giménez Molinelli, J. I. Nichols and C. O. Dorso, Effect of Coulomb screening length on nuclear "pasta" simulations, *Phys. Rev. C.* **89**(5), 055801 (May, 2014). doi: 10.1103/PhysRevC.89.055801. URL http://link.aps.org/doi/10.1103/PhysRevC.89.055801.
51. C. O. Dorso and A. Strachan, Onset of fragment formation in periodic expanding systems, *Phys. Rev. B.* **54**(1), 236–243 (July, 1996). doi: 10.1103/PhysRevB.54.236. URL http://link.aps.org/doi/10.1103/PhysRevB.54.236.
52. C. Dorso and J. Randrup, Early recognition of clusters in molecular dynamics, *Physics Letters B.* **301** (4), 328–333 (Mar., 1993). ISSN 0370-2693. doi: 10.1016/0370-2693(93)91158-J. URL http://www.sciencedirect.com/science/article/pii/037026939391158J.
53. C. O. Dorso and P. E. Balonga, Fluctuation dynamics of fragmenting spherical nuclei, *Phys. Rev. C.* **50** (2), 991–995 (Aug., 1994). doi: 10.1103/PhysRevC.50.991. URL http://link.aps.org/doi/10.1103/PhysRevC.50.991.

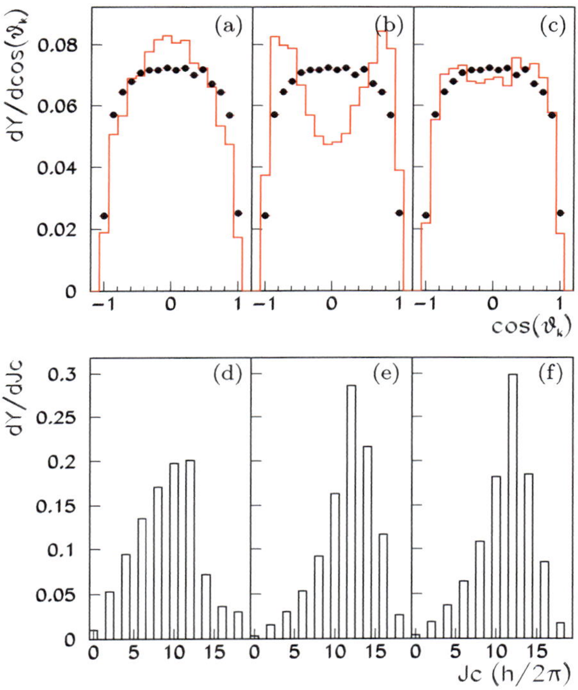

Fig. 14. Analysis of dissipative $^{15}O+2\alpha$ complete events for the $^{12}C+^{12}C$ reaction. Upper part: the experimental (dots) $cos(\theta_k)$ distribution is compared to the calculated ones (red lines) according to different emission sequences (see text). Lower part: Compound angular momentum distribution in the unfiltered HFℓ calculations. Panel (a) and (d): n is the first emitted particle. Panel (b) and (e): n is the second emitted particle. Panel (c) and (f): n is the third emitted particle. All distributions are normalized to unitary area. Figure taken from Ref. 21

sequences. The corresponding angular momentum distributions of the initial ^{24}Mg compound are also displayed in bottom panels of the same figure.

When the neutron is the first particle emitted from ^{24}Mg (panel a) and the two α's correspond to the last two steps of the de-excitation chain, the distribution is single-peaked. Indeed the evaporation sequence corresponds to the lowest average angular momentum for the initial compound (panel d), and moreover the angular momentum available for the first α emission is reduced by the first step of the de-excitation. Conversely, if the two α's are the first and last emitted particles, with the neutron being emitted in the intermediate step (panel b), the distribution shows two clear peaks. This is consistent with two sequential emissions from a high J source and a low J

CHAPTER 7

CLUSTERING IN STABLE AND EXOTIC LIGHT NUCLEI

C. Beck

Institut Pluridiciplinaire Hubert Curien,
Départment de Recherches Subatomiques,
UMR7178, CNRS-IN2P3 et Université de Strasbourg,
IPHC Bat.20/2, 23, rue du Loess B.P. 28
F-67037 Strasbourg, France
christian.beck@iphc.in2p3.fr

Since the pioneering discovery of molecular resonances in the ^{12}C+^{12}C reaction more than half a century ago a great deal of research work has been undertaken in alpha clustering. Our knowledge on physics of nuclear molecules has increased considerably and nuclear clustering remains one of the most fruitful domains of nuclear physics, facing some of the greatest challenges and opportunities in the years ahead. The occurrence of "exotic" shapes and Bose-Einstein alpha condensates in light $N = Z$ alpha-conjugate nuclei is investigated. Various approaches of the superdeformed and hyperdeformed bands associated with quasimolecular resonant structures are presented. Evolution of clustering from stability to the drip-lines is examined: clustering aspects are, in particular, discussed for light exotic nuclei with large neutron excess such as neutron-rich Oxygen isotopes with their complete spectroscopy.

1. Introduction

In the last decades, one of the greatest challenges in nuclear science is the understanding of the clustered structure of nuclei from both the experimental and theoretical perspectives.[1–7] Our knowledge on physics of nuclear molecules has increased considerably and nuclear clustering remains one of the most fruitful domains of nuclear physics, facing some of the greatest challenges and opportunities in the years ahead. Besides the well known series of Cluster Conferences,[8–10] a series of workshops on the state of the art in nuclear cluster physics was started. The first one was held in Strasbourg in 2008,[11] the second one in Brussels[12] in 2010 and the last one in

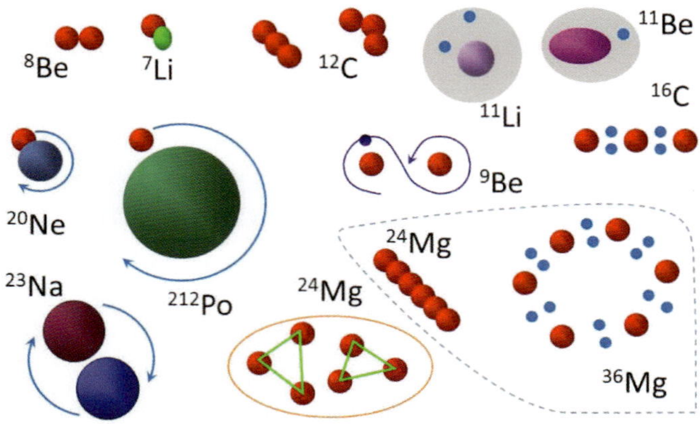

Fig. 1. Different types of clustering behaviour identified in nuclei, from small clusters outside a closed shell, to complete condensation into α particles, to halo nucleons outside of a normal core, have been discussed the last two or three decades.[14,15] This figure was adapted from Refs. 13 and 14 courtesy from W. Catford.

Yokohama[13] in 2014. Figure 1 (taken from the cover of Ref. 13) summarizes the different types of clustering discussed during the last two or three decades.[14,15] Most of these structures were investigated in an experimental context by using either some new approaches or developments of older methods.[16] Starting in the 1960s the search for resonant structures in the excitation functions for various combinations of light α-cluster ($N = Z$) nuclei in the energy regime from the Coulomb barrier up to regions with excitation energies of $E_x = 20 - 50$ MeV remains a subject of contemporary debate.[1] These resonances have been interpreted in terms of nuclear molecules.[1]

The question of how quasimolecular resonances may reflect continuous transitions from scattering states in the ion-ion potential to true cluster states in the compound systems was still unresolved in the 1990s.[1] In many cases, these resonant structures have been associated with strongly-deformed shapes and with α-clustering phenomena,[18–20] predicted from the Nilsson-Strutinsky approach, the cranked α-cluster model,[18,19] or other mean-field calculations.[20,21] In light α-like nuclei clustering is observed as a general phenomenon at high excitation energy close to the α-decay thresholds.[18,19,22] This exotic behavior has been perfectly illustrated by the famous "Ikeda-diagram" for $N{=}Z$ nuclei in 1968,[23] which has been recently modified and extended by von Oertzen[24,25] for neutron-rich nuclei, as shown in the left panel of Fig. 2. Despite the early inception of cluster studies, it is

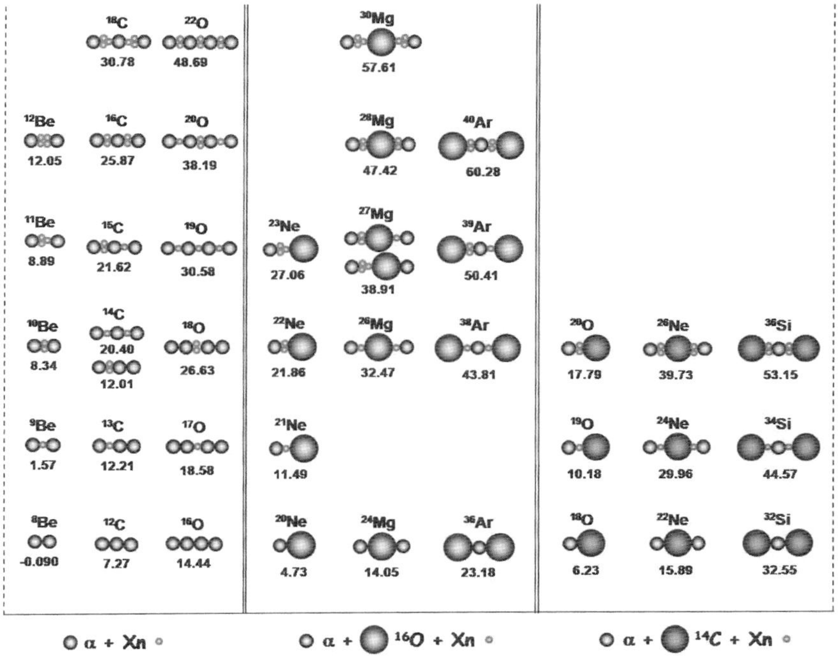

Fig. 2. Schematic illustration of the structures of molecular shape isomers in light neutron-rich isotopes of nuclei consisting of α-particles, ^{16}O- and ^{14}C-clusters plus some covalently bound neutrons (Xn means X neutrons).[26] The so called "Extended Ikeda-Diagram" with α-particles (left panel) and ^{16}O-cores (middle panel) can be generalized to ^{14}C-cluster cores (right panel. The lowest line of each configuration corresponds to parts of the original Ikeda diagram. However, because of its deformation, the ^{12}C nucleus is not included, as it was earlier. The numbers represent the threshold energy dissociating the ground state into the respective cluster configuration. Threshold energies are given in MeV. This figure has been adapted courtesy from W. von Oertzen.[26]

only recently that radioactive ion beams experiments, with great helps from advanced theoretical works, enabled new generation of studies, in which data with variable excess neutron numbers or decay thresholds are compared to predictions with least or no assumptions of cluster cores. Some of the predicted but elusive phenomena, such as molecular orbitals or linear chain structures, are now gradually coming to light.

Clustering is a general feature[26] not only observed in light neutron-rich nuclei,[27–30] but also in halo nuclei[31,32] such as ^{11}Li[33] or ^{14}Be,[34] for instance. The problem of cluster formation has also been treated extensively for very heavy systems by R.G. Gupta,[21] by D. Poenaru, V. Zagrebaev and W. Greiner[35,36] and by C. Simenel[37] where giant molecules and collinear ternary fission may co-exist.[38] Finally, signatures of alpha clustering have also been predicted and/or discovered in light nuclei surviving from intermediate-energy[39] to ultrarelativistic-energy[40,41] nuclear collisions. The topic of clustering in nuclei benefits of intense theoretical activity where new experimental information has come to light very recently. Several status reports were already given in conferences and their written contributions can be found in their proceedings.[15,42,43]

In this short review article, only few selected topics will be presented, I will limit myself first to the light ^{12}C, ^{16}O, ^{20}Ne and ^{24}Mg α-like nuclei in Section 2, then to alpha clustering, nuclear molecules and large deformations for heavier light nuclei in Section 3. The search for electromagnetic transitions and alpha condensates in heavier α-like nuclei will be discussed in Sections 4 and 5, respectively, and, finally, clustering effects in light neutron-rich nuclei (oxygen isotopes) will be presented in Section 6 before the summary, conclusions and outlook of Section 7 are briefly proposed.

2. Renewed interest in the spectroscopy of ^{12}C, ^{16}O, ^{20}Ne and ^{24}Mg α-like nuclei

The ground state of ^{8}Be is the most simple and convincing example of α-clustering in light nuclei as suggested by several theoretical models and appears naturally in *ab initio* calculations.[44,45] The picture of the ^{8}Be nucleus prediced by the No Core Shell model[44] as being a dumbbell-shaped configuration of two alpha particles closely resembles the superdeformed (SD) shapes known to arise in heavier nuclei in the actinide mass region. This dumbbell-like structure gives rise to a rotational band, from which the moment of inertia is found to be commensurate with an axial deformation of 2:1. The possible of large deformation of light α-conjugate nuclei with SD, hyperdeformed (HD) and linear-chain configurations will be discussed in following Sections.

2.1. ^{12}*C nucleus "Hoyle" state*

The renewed interest in ^{12}C was mainly focused to a better understanding of the nature of the so called "Hoyle" state,[46–48] the excited 0^+ state at 7.654 MeV that can be described in terms of a bosonic condensate, a cluster state and/or a α-particle gas.[49–51] The resonant "Hoyle" state[46] (without it carbon would not exist as proved in 1957 by an experimental group at

Caltech[47]) is regarded as the prototypical alpha-cluster state whose existence is of great importance for the nucleosynthesis of ^{12}C within stars. The structure of this state has been thoroughly investigated with theoretically modelled with both *ab initio*[44,45,52–57] and cluster models.[58–60] Much experimental progress has been achieved recently as far as the spectroscopy of ^{12}C near and above the α-decay threshold is concerned.[61–66] More particularly, the 2_2^+ "Hoyle" rotational excitation in ^{12}C has been observed by several experimental groups.[61,63]

The most convincing experimental result comes from measurements of the ^{12}C(γ,α)^{8}Be reaction performed at the HIGS facility.[63] The measured angular distributions of the alpha particles are consistent with an L=2 pattern including a dominant 2^+ component. This 2_2^+ state that appears at around 10 MeV is considered to be the 2^+ excitation of the "Hoyle" state (in agreement with the previous experimental investigation of Itoh *et al.*[61]) according to the α cluster[67] and α condensation models.[49] On the other hand, the experiment ^{12}C(α,α)^{12}C* carried out at the Birmingham cyclotron,[65] UK, populates a new state compatible with an equilateral triangle configuration of three α particles. Still, the structure of the "Hoyle" state remains controversial as experimental results of its direct decay into three α particles are found to be in disagreement.[68–74]

2.2. ^{16}*O nucleus*

In the study of Bose-Einstein Condensation (BEC), that will be presented in more detail in Section 5, the α-particle states in light $N = Z$ nuclei,[49–51] are of great importance. At present, the search for an experimental signature of BEC in ^{16}O is of highest priority. Furthermore, *ab initio* calculations[75] predict that nucleons are arranged in a tetrahedral configuration of alpha clusters. A state with the structure of the "Hoyle" state[46] in ^{12}C coupled to an α particle is predicted in ^{16}O at about 15.1 MeV (the 0_6^+ state), the energy of which is $\approx$ 700 keV above the 4α-particle breakup threshold.[76] However, any state in ^{16}O equivalent to the "Hoyle" state[46] in ^{12}C is most certainly going to decay exclusively by particle emission with very small γ-decay branches, thus, very efficient particle-γ coincidence techniques will have to be used in the near future to search for them. BEC states are expected to decay by alpha emission to the "Hoyle" state and could be found among the resonances in α-particle inelastic scattering on ^{12}C decaying to that state. In 1967 Chevallier *et al.*[77] could excite these states in the α-particle transfer channel leading to the ^{8}Be–^{8}Be final state and proposed that a structure corresponding to a rigidly rotating linear arrangement of four alpha particles may exist in ^{16}O. At this time this experimental observation was considered as the equivalent of the three alpha chain states postulated by Morinaga for the ^{12}C nucleus.[78] However, very recently, a more sophisticated experimental setup was used at Notre Dame:[79] although the excitation function is generally in good agreement with the previous results[77] a phase shift analysis of the angular distributions does not provide evidence to support the reported hypothesis of a 4α-chain state configuration.

2.3. ^{20}*Ne and* ^{24}*Mg nuclei*

Experimental investigations are still underway to understand the nuclear structure of high spin states of both ^{16}O and ^{20}Ne nuclei for instance at Notre Dame and/or iThemba Labs[80] facilities. Another possibility might be to perform Coulomb excitation measurements with intense ^{16}O and ^{20}Ne beams at intermediate energies. The nucleus viewed as a collection of α-particles has been discussed all over the mass table since a long time and it was only recently shown that clear deviations from statistical models in the decay of excited ^{24}Mg exist.[39,81–83] As far as the theory is concerned, a diversity of cluster/symmetry models (cluster models, *ab initio* calculations, BEC etc.) which make concrete predictions in terms of alpha clustering in light nuclei is available on the market. What is presently missing is a clearly defined procedure for relating these abstract predictions to the observed level schemes and clear criteria for what constitutes discriminating evidence for a particuliar model. A more detailed discussion of BEC will be proposed in one of the forthcoming section (Section 5).

The search for exotic chain-like structures in light α-conjugate nuclei remains an exciting prospect. Experiments reporting tentative evidence of α-chains in ^{12}C.[78] ^{16}O,[77,79,80] ^{20}Ne[80] and ^{24}Mg[84] have been largely unsubstantiated, and the view is that such structure have not yet been definitively observed experimentally.

3. Alpha clustering, nuclear molecules and large deformations

The real link between superdeformation/hyperdemormation (SD/HD), nuclear molecules and alpha clustering[20,85,86] is of particular interest, since nuclear shapes with major-to-minor axis ratios of 2:1 have the typical ellipsoidal elongation for light nuclei i.e. with quadrupole deformation parameter $\beta_2 \approx 0.6$. Furthermore, the structure of possible octupole-unstable 3:1 nuclear shapes - hyperdeformation with $\beta_2 \approx 1.0$ - has also been discussed for actinide nuclei[86] in terms of clustering phenomena. Typical examples for possible relationship between quasimolecular bands and extremely deformed (SD/HD) shapes have been widely discussed in the literature for $A = 20 - 60\alpha$-conjugate $N = Z$ nuclei, such as ^{28}Si,[87–92] ^{32}S,[20,88,93–95] ^{36}Ar,[86,96–101] ^{40}Ca,[102–106] ^{44}Ti,[20,107,108] ^{48}Cr[109,110] and ^{56}Ni.[111–114]

Excitation functions have been measured over a wide range energies for many reactions. Norrby discussed a study of ^{32}S via the ^{28}Si+α reaction[94] that revealed 30 new level assignements spanning 132 resonances in total.

In fact, highly deformed shapes and SD rotational bands have been discovered in several light α-conjugate nuclei, such as ^{36}Ar and ^{40}Ca by using γ-ray spectroscopy techniques.[96,97,102] In particular, the extremely deformed rotational bands in ^{36}Ar (shown as crosses in Fig. 3) might be comparable in shape to the quasimolecular bands observed in both ^{12}C+^{24}Mg (shown as open triangles) and ^{16}O+^{20}Ne (shown as full rectangles) reactions. These resonances belong to a rotational band, with a

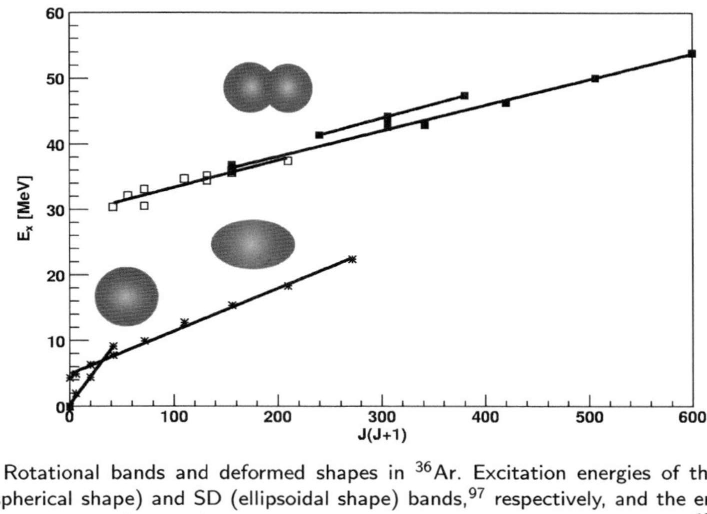

Fig. 3. Rotational bands and deformed shapes in ^{36}Ar. Excitation energies of the ground state (spherical shape) and SD (ellipsoidal shape) bands,[97] respectively, and the energies of HD (dinuclear shape) band from the quasimolecular resonances observed in the ^{12}C+^{24}Mg (open rectangles)[98,115–117] and ^{16}O+^{20}Ne (full rectangles)[118,119] reactions are plotted as a function of J(J+1). This figure has been adapted from Refs. 96,98,101.

moment of inertia close to that of a HD band provided by both the cranked α-cluster model[19] and the Nilsson-Strutinsky calculations. The fact that similar quasi-molecular states observed in the two reactions fall on the same rotational band gives further support to our interpretation of the ^{36}Ar composite system resonances. An identical conclusion was reached for the ^{40}Ca composite system where SD bands have been discovered.[96] Therefore, similar investigations are underway for heavier α-like composite systems such as ^{44}Ti,[20] ^{48}Cr[109] and ^{56}Ni.[111,114]

Ternary clusterizations in light α-like composite systems are also predicted theoretically, but were not found experimentally in ^{36}Ar so far.[96] On the other hand, ternary fission of ^{56}Ni — related to its HD shapes — was identified from out-of-plane angular correlations measured in the ^{32}S+^{24}Mg reaction with the Binary Reaction Spectrometer (BRS) at the VIVITRON Tandem facility of the IPHC, Strasbourg.[120] This finding[120] is not limited to light N=Z compound nuclei, true ternary fission[35,38,121] can also occur for very heavy[38,121] and superheavy[122] nuclei.

4. Electromagnetic transitions as a probe of quasimolecular states and clustering in light nuclei

Clustering in light nuclei is traditionally explored through reaction studies, but observation of electromagnetic transitions can be of high value in establishing, for example, that highly-excited states with candidate cluster structure do indeed form rotational sequences.

4.1. *^{16}O nucleus*

There is a renewed interest in the spectroscopy of the ^{16}O nucleus at high excitation energy.[96,99] Exclusive data were collected on ^{16}O in the inverse kinematics reaction ^{24}Mg+^{12}C studied at $E_{lab}(^{24}\mathrm{Mg})$ = 130 MeV with the BRS in coincidence with the EUROBALL IV installed at the VIVITRON facility.[96,99] From the α-transfer reactions (both direct transfer and deep-inelastic orbiting collisions[123]), new information has been deduced on branching ratios of the decay of the 3^+ state of ^{16}O at 11.085 MeV $\pm$ 3 keV. The high-energy level scheme of ^{16}O shown in Refs. 96,99 indicated that this state does not α-decay because of its non-natural parity (in contrast to the two neighbouring 4^+ states at 10.36 MeV and 11.10 MeV), but it γ decays to the 2^+ state at 6.92 MeV (54.6 $\pm$ 2 %) and to the 3^- state at 6.13 MeV (45.4%). By considering all the four possible transition types of the decay of the 3^+ state (i.e. E1 and M2 for the $3^+ \rightarrow 3^-$ transition and, M1 and E2 for the $3^+ \rightarrow 2^+$ transition), our calculations yield the conclusion that $\Gamma_{3^+} < 0.23$ eV, a value fifty times lower than known previously, which is an important result for the well studied ^{16}O nucleus.[96,99] Clustering effects in the light neutron-rich oxygen isotopes $^{17-20}$O will also be discussed in Section 5.

Alpha clustering plays an important role in the description of the ground state and excited states of light nuclei in the *p* shell. For heavier nuclei, in the *sd*-shell, cluster configurations may be based on heavier substructures like ^{12}C, ^{14}C and ^{16}O as shown by the "Extended Ikeda-diagram" proposed in Fig. 2. This was already well discussed to appear in ^{24}Mg(^{12}C-^{12}C) and ^{28}Si(^{12}C-^{16}O) both theoretically and experimentally.

4.2. ^{28}*Si nucleus*

The case of the mid-*sd*-shell nucleus ^{28}Si is of particular interest as it shows the coexistence of deformed and cluster states at rather low energies.[89,90] Its ground state is oblate, with a partial α-^{24}Mg structure, two prolate normal deformed bands are found, one built on the 0_2^+ state at 4.98 MeV and on the 0_3^+ state at 6.69 MeV. The SD band candidate with a pronounced α-^{24}Mg structure is suggested.[89] In this band, the 2^+ (9.8 MeV), 4^+ and 6^+ members are well identified as can be clearly observed in Fig. 4.

In the following we will briefly discuss a resonant cluster band which is predicted to start close to the Coulomb barrier of the ^{12}C+^{16}O collision, i.e. around 25 MeV excitation energy in ^{28}Si. We have studied the ^{12}C(^{16}O,γ)^{28}Si radiative capture reaction at five resonant energies around the Coulomb barrier by using the zero degree DRAGON spectrometer installed at Triumf, Vancouver.[124,125] Details about the setup, that has been optimized for the ^{12}C(^{12}C,γ)^{24}Mg radiative capture reaction in our of previous DRAGON experiments, can be found in Ref. 126. The ^{12}C(^{16}O,γ)^{28}Si data clearly show[124,125] the direct feeding of the prolate 4_3^+ state at 9.16 MeV and the octupole deformed 3^- at 6.88 MeV. This state is the band head of an octupole band which mainly decays to the ^{28}Si oblate ground state with a strong E$_3$ transition. Our results are very similar to what has been measured for the ^{12}C+^{12}C radiative capture reaction above the Coulomb barrier in the first DRAGON experiment[126] where the enhanced feeding of the ^{24}Mg prolate band has been measured for a 4^+-2^+ resonance at $E_{c.m.} = 8.0$ MeV near the Coulomb barrier.

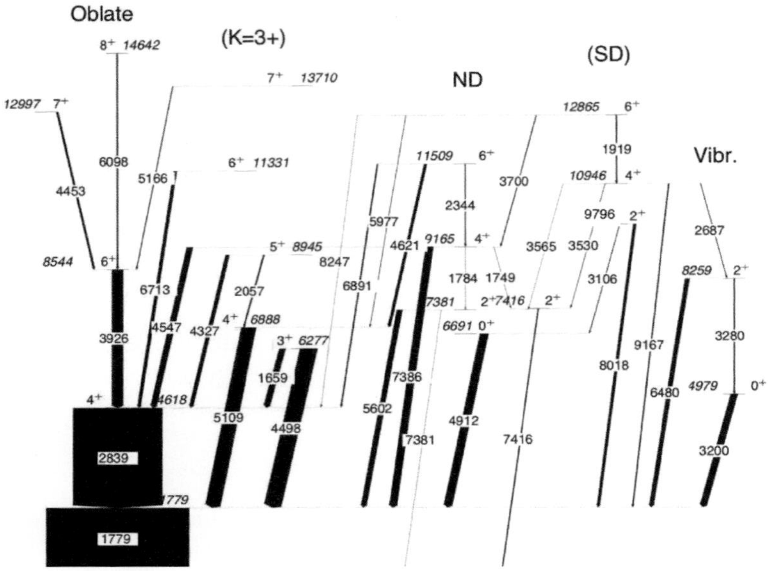

Fig. 4. Subset of positive-parity levels in ^{28}Si derived from the analysis of a Gammasphere experiment. Excited states and transitions energies are labeled with their energy in keV, while the width of the arrows corresponds to the relatve intensity of the observed transitions. The different structures are labeled according to previous assignements as oblate, prolate (ND), vibrational and with different K values. This figure has been adapted from Ref. 89 courtesy from D.G. Jenkins.

At the lowest energy of ^{12}C+^{16}O radiative capture reaction, an enhanced feeding from the resonance $J^{\pi} = 2^{+}$ and 1^{+} T=1 states around 11 MeV is observed in ^{28}Si. Again this is consistent with ^{12}C+^{12}O radiative capture reaction data where $J^{\pi} = 2^{+}$ has been assigned to the entrance resonance and an enhanced decay has been measured via intermediate 1^{+} T=1 states around 11 MeV in ^{24}Mg. A definitive scenario for the decay of the resonances at these low bombarding energies in both systems will come from the measurement of the γ decay spectra with a γ-spectrometer with better resolution than BGO but still rather good efficiency such as $LaBr_3$ (lanthanum) crystals (see also the forthcoming subsection).

4.3. ^{12}C+^{12}C *resonances*

A further area where electromagnetic transitions would be of great interest in support of cluster models is in the case of the quasi-molecular resonances observed in the $^{12}C+^{12}C$ reaction.[1] The width of these resonances were $\approx$ 100 keV, indicating the formation of a ^{24}Mg intermediate system with a lifetime significantly longer than the nuclear crossing time. These resonances were subsequently interpreted as $^{12}C+^{12}C$ cluster states.

There has been only one valient attempt to directly observe transitions in this reaction by Haas *et al.*[127] focussing on transitions between 10^+ and 8^+ resonant states at a bombarding energy $E(^{12}C) = 32$ MeV chosen to populate a known and isolated 10^+ resonance. Position Sensitive Detectors (PSD) were mounted on either side of the beam axis at the center of the Chateau de Cristal array of 74 barium fluoride detectors. Triple γ-^{12}C-^{12}C coincidences were recorded and it was possible to observe a few events in the expected energy window corresponding to the $10^+ \rightarrow 8^+$ transitions as shown by Fig. 5. However, the data were not sufficiently clean to rule

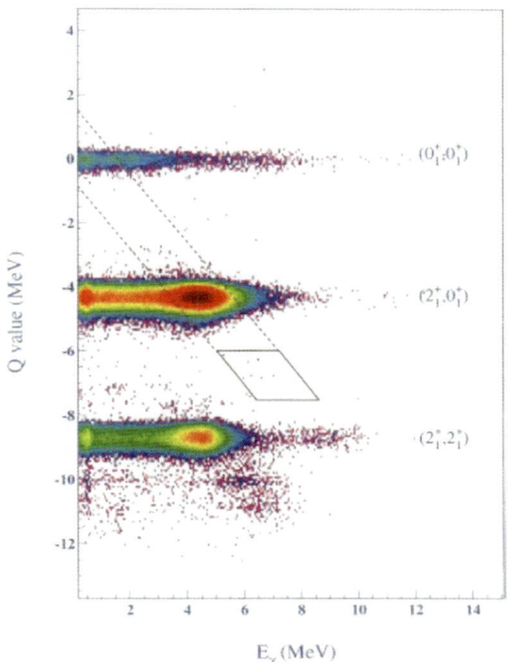

Fig. 5. Reaction Q-value of the $^{12}C(^{12}C,^{12}C)^{12}C)$ reaction at $E_{lab} = 32.9$ MeV versus γ-ray energies. The spectrum has been obtained with fragment-fragment-γ coincidence condition and a γ multiplicity M = 1 as explained in the text (see also Ref. 127 for more details). This figure has been adapted from Ref. 127.

out these events as due to the experimental background. The measurement reported only an upper limit (for the radiative partial width of 1.2 ± 10^{-5}) given the extreme challenges of eliminating all background displayed in Fig. 5.

It will be very interesting to revisit this earlier experiment taking advantage of new experimental techniques and developments in detector technology for the detection of gamma rays and/or fragments. For example, novel scintillator materials like lanthanum bromide offer superior resolution for the gamma ray of interest while improved silicon detector performance and solid angle coverage could lead to significant improvements both in sensitivity and in statistics.

5. Condensation of α clusters in light nuclei

In principle the nucleus is a quasi-homogeneus collection of protons and neutrons, which adopts a spherical configuration i.e. a spherical droplet of nuclear matter. For light nuclei the nucleons are capable to arrange themselves into clusters of a bosonic character. The very stable α-particle is the most favorable light nucleus for quarteting — α clustering — to occur in dense nuclear matter. These cluster structures have indeed a crucial role in the synthesis of elements in stars. The so called "Hoyle" state,[46,48] the main portal through which ^{12}C is created in nucleosynthesis with a pronounced three-α-cluster structure, is the best exemple of α clustering in light nuclei. In α clustering a geometric picture can be proposed in the framework of point group symmetries.[41] For instance, in ^{8}Be the two α clusters are separated by as much as $\approx$ 2fm, ^{12}C exhibits a triangle arrangement of the three α particles $\approx$ 3fm apart, ^{16}O forms a tetrahedron, etc. Evidence for tetrahedral symmetries in ^{16}O was given by the algebraic cluster model.[128] Such kind of symmetries are rather well illustrated by the schematic picture of the ^{20}Ne nucleus proposed in Fig. 6. More realistic is the density plot for ^{20}Ne nucleus calculated as an arrangement of two α particles with a ^{12}C core which is displayed in Fig. 7 to illustrate the enhancement of the symmetries of the α clustering.

In the study of the Bose-Einstein Condensation (BEC) the α-particle states were first described for ^{12}C and ^{16}O[49,129] and later on generalized to heavier light $N{=}Z$ nuclei.[50,51,130,131] The structure of the "Hoyle" state and the properties of its assumed rotational band have been studied very carefully from measurements of the ^{12}C(γ,3α) reaction performed at the HIGS

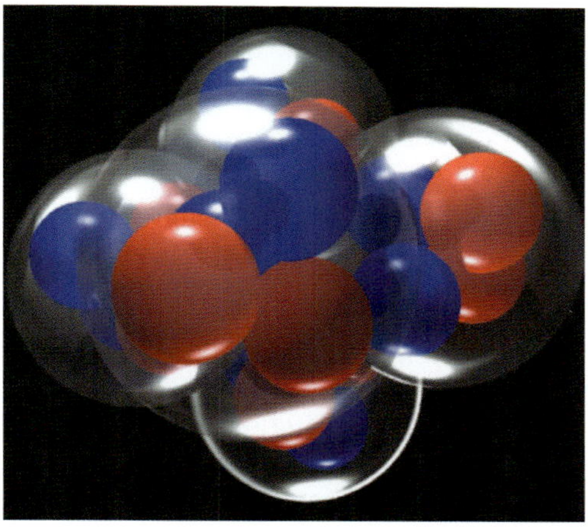

Fig. 6. Schematic illustration of the clustering arrangement of five alpha particles in the nucleus ^{20}Ne.

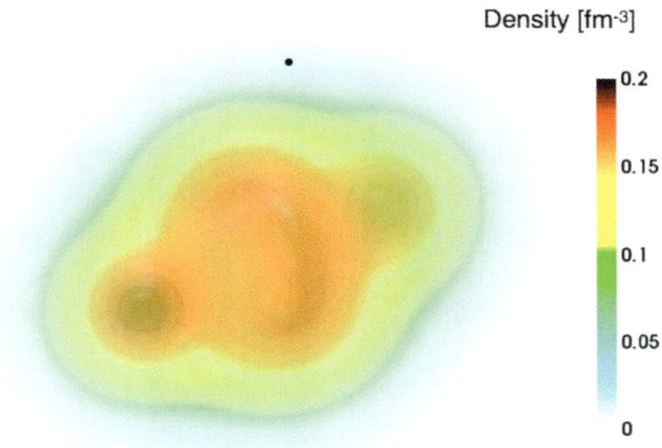

Fig. 7. Self-consistent ground-state densities of the nucleus ^{20}Ne as calculated with EDF (see text for details). Densities (in units of fm^{-3}) are plotted in the intrinsic frame of reference that coincides with the principal axes of the nucleus. This figure has been adapted from Ref. 130 courtesy from E. Kahn and J.-P. Ebran.

facility, TUNL.[63] At present, the search for an experimental signature of BEC in ^{16}O is of highest priority. A state with the structure of the "Hoyle" state in ^{12}C coupled to an α particle is predicted in ^{16}O at about 15.1 MeV (the 0_6^+ state), the energy of which is $\approx$ 700 keV above the 4α-particle breakup threshold:[76,132,133] in other words, this 0_6^+ state might be a good candidate for the dilute 4α gas state. However, any state in ^{16}O equivalent to the "Hoyle" state in ^{12}C is most certainly going to decay by particle emission with very small, probably un-measurable, γ-decay branches, thus, very efficient particle-detection techniques will have to be used in the near future to search for them.

BEC states are expected to decay by α emission to the "Hoyle" state and could be found among the resonances in α-particle inelastic scattering on ^{12}C decaying to that state or could be observed in an α-particle transfer channel leading to the ^{8}Be–^{8}Be final state. The attempts to excite these states by α inelastic scattering[61] was confirmed recently.[134] Another possibility, that has not been yet explored, might be to perform Coulomb excitation measurements with intense ^{16}O beams at intermediate energies.

Clustering of ^{20}Ne has also been described within the energy density functional theory[130] (EDF) as illustrated by Fig. 7 that displays axially and reflection symmetric self-consistent equilibrium nucleon density distributions. We note the well known quasimolecular α-^{12}C-α structure although clustering effects are less pronounced than the ones (schematically displayed in Fig. 5) predicted by Nilsson-Strutinsky calculations and even by mean-field calculations (including Hartree-Fock and/or Hartree-Fock-Bogoliubov calculations).[19–21,131]

The most recent work of Girod and Schuck[131] validates several possible scenarios for the influence of clustering effects as a function of the neutron richness that will trigger more experimental works. We describe in the following (i.e. in Section 6) recent experimental investigations on the Oxygen isotopes chain.

6. Clustering in light neutron-rich nuclei

As discussed previously, clustering is a general phenomenon observed also in nuclei with extra neutrons as it is presented in an "Extended Ikeda-diagram"[23] proposed by von Oertzen[24] (see the left panel of Fig. 2). With additional neutrons, specific molecular structures appear with binding effects based on covalent molecular neutron orbitals. In these diagrams α-clusters and ^{16}O-clusters (as shown by the middle panel of the diagram of Fig. 2) are the main ingredients. Actually, the ^{14}C nucleus may play similar role in clusterization as the ^{16}O one since it has similar properties as a cluster: (i) it has closed neutron p-shells, (ii) first excited states are well above $E^* = 6$ MeV, and (iii) it has high binding energies for α-particles.

The possibility of extending molecular structures from dimers (beryllium isotopes)[135] to trimers[136] has been investigated in detail for carbon isotopes.[137,138,141] Here the neutrons would be exchanged between the three centers (alpha particles). It is possible that the three α-particle configuration can align themselves in a linar fashion, or alternative collapse into a triangle arragment — in either case the neutrons being localised across the three centers. At present experimental evidence for such structures have been found in ^{13}C[137] and ^{14}C.[141] Possibly the best case for the linear arrangement — from a theoretical perspective[139,140] — is ^{16}C.[138]

A general picture of clustering and molecular configurations in light nuclei can also be drawn from a detailed investigation of the light oxygen isotopes with A $\geq$ 17. Here we will only present recent results on the even-even oxygen isotopes: ^{18}O[142] and ^{20}O.[143] But very striking cluster states have also been found in odd-even oxygen isotopes such as: ^{17}O[144] and ^{19}O.[145]

Figures 8 and 9 give an overview of all bands in ^{18}O and ^{20}O, respectively, as plots of excitation energies as a function of J(J + 1) together with

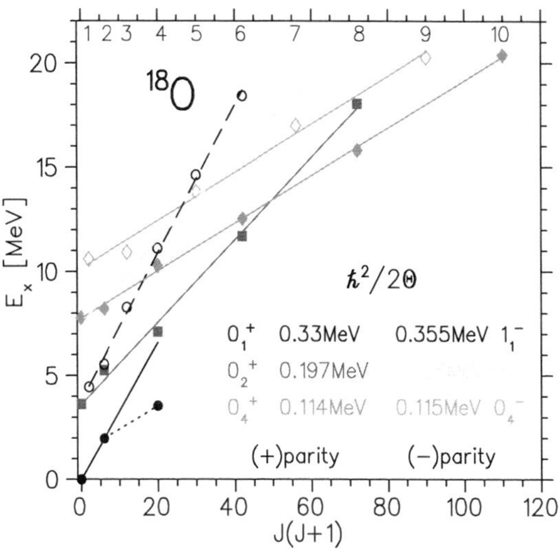

Fig. 8. Overview of six rotational band structures observed in ^{18}O. Excitation energy systematics for the members of the rotational bands forming inversion doublets with K=0 are plotted as a function of J(J+1). The curves are drawn to guide the eye for the slopes. The indicated slope parameters contain information on the moments of inertia. Square symbols correspond to cluster bands, whereas diamonds symbols correspond to molecular bands. This figure has been adapted from Ref. 142 courtesy from W. von Oertzen.

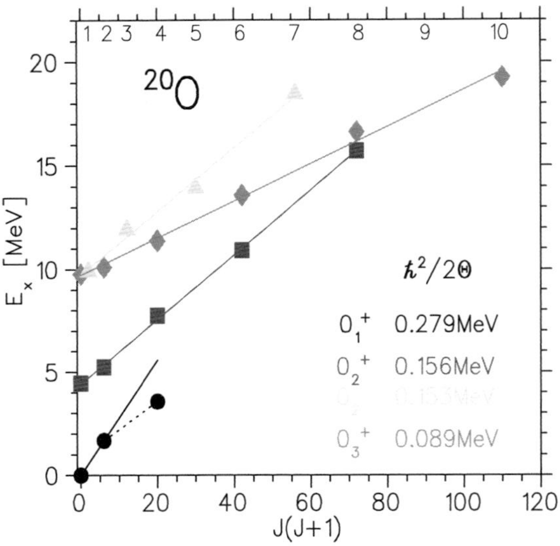

Fig. 9. Overview of four rotational band structures observed in ^{20}O. Excitation energy systematics for the members of the rotational bands forming inversion doublets with K=0 are plotted as a function of J(J + 1). The curves are drawn to guide the eye for the slopes. The indicated slope parameters contain information on the moments of inertia. Square and triangle symbols correspond to cluster bands, whereas diamonds symbols correspond to molecular bands. This figure has been adapted from Ref. 143 courtesy from W. von Oertzen.

their respective moments of inertia. In the assignment of the bands both the dependence of excitation energies on J(J + 1) and the dependence of measured cross sections on 2J + 1[143] were considered. Slope parameters obtained in a linear fit to the excitation energies[143] indicate the moment of inertia of the rotational bands given in the respective figures. The intrinsic structure of the cluster bands is reflection asymmetric, the parity projection gives an energy splitting between the partner bands. The assignments of the experimental molecular bands of ^{18}O are supported by both the Generator-Coordinate-Method[146] and the Antisymmetrized Molecular Dynamics (AMD) calculations.[147]

We can compare the bands of ^{20}O[143] shown in Fig. 9 with the ones of ^{18}O in Fig. 8. The first doublet ($K=0_2^{\pm}$) has a slightly larger moment of inertia (smaller slope parameter) in ^{20}O, which is consistent with its interpretation as ^{14}C–^{6}He or ^{16}C–^{4}He molecular structures (they start well below the thresholds of 16.8 MeV and 12.32 MeV, respectively). The second band, for which the negative parity partner is yet to be determined, has a slope parameter slightly smaller than in ^{18}O. This is consistent with the study of the bands in ^{20}O by Furutachi *et al.*,[147] which clearly establishes parity inversion doublets predicted by AMD calculations for the ^{14}C–^{6}He cluster and ^{14}C-2n-α molecular structures. The corresponding moments of inertia given in Fig. 5 are strongly suggesting large deformations for the cluster structures. We may conclude that the reduction of the moments of inertia of the lowest bands of ^{20}O is consistent with the assumption that the strongly bound ^{14}C nucleus having equivalent properties to ^{16}O, has a similar role as ^{16}O in relevant, less neutron rich nuclei. Therefore, the Ikeda-diagram[23] and the "extended Ikeda-diagram" consisting of ^{16}O cluster cores with covalently bound neutrons[24] must be further extended to include also the ^{14}C cluster cores as illustrated in Fig. 2.

7. Summary, conclusions and outlook

The link of alpha clustering, quasimolecular resonances, orbiting phenomena and extreme deformations (SD, HD, ...) has been discussed in this review article. Several examples emphasize the general connexion between molecular structure and deformation effects within the shell model, or rather the Nilsson model. However, we have also presented the BEC picture of light (and medium-light) α-like nuclei that appears to be an alternate way of understanding most of properties of nuclear clusters. New results regarding cluster and molecular states in neutron-rich oxygen isotopes in agreement with AMD predictions are finally summarized. Consequently, the "Extended Ikeda-diagram" has been further modified for light neutron-rich nuclei by inclusion of the ^{14}C cluster, similarly to the ^{16}O one. Of particular interest is the quest for the 4α states of ^{16}O near the ^{8}Be+^{8}Be and ^{12}C+α decay thresholds, which correspond to the so-called "Hoyle" state. The search for extremely elongated configurations (HD) in rapidly rotating medium-mass nuclei, which has been pursued by γ-ray spectroscopy measurements, will have to be performed in conjunction with charged-particle techniques in the near future since such states are most certainly going to decay by particle emission (see Refs. 16 and 120). Marked progress has been made in many traditional and novels subjects of nuclear cluster physics. The developments in these subjects show the importance of clustering among the basic modes of motion of nuclear many-body systems. All these open questions will require precise coincidence measurements[16] coupled with state-of-the-art theory.

Dedication and acknowledments

This review article is dedicated to the memory of my friends Alex Szanto de Toledo and Valery Zagrebaev who unexpectelly passed away in early 2015. I am very pleased to first acknowledge Walter Greiner (who celebrated his 80th birthday on October 29th, 2015) for his continuous support of the cluster physics.[1,11,35,36,148] I would like also to thank Christian Caron (Springer) for initiating in 2008 the series of the three volumes of *Lecture Notes in Physics* entitled "Clusters in Nuclei" and edited between 2010 and 2014.[3–5] All the authors of the 19 chapters of these volumes are warmly thanked for their fruitfull collaboration during the course of the project which is still in progress.

References

1. W. Greiner, Y. J. Park and W. Scheid, *Nuclear Molecules*, (ed.), World Scientific (1995).
2. M. Brenner, *Cluster Structure of Atomic Nuclei*, (ed.), Research Signpost (2010).
3. C. Beck, *Clusters in Nuclei, Vol. 1*, (ed.), Springer Verlag Berlin Heidelberg (2010).
4. C. Beck, *Clusters in Nuclei, Vol. 2*, (ed.), Springer Verlag Berlin Heidelberg (2012).
5. C. Beck, *Clusters in Nuclei, Vol. 3*, (ed.), Springer Verlag Berlin Heidelberg (2014).
6. D. S. Delion, *Theory of Particle and Cluster Emission*, (ed.), Springer Verlag Berlin Heidelberg (2010); *Lecture Notes in Physics* **819**, 1 (2010).
7. Y. Funaki, H. Horiuchi and A. Tohsaki, *Prog. Part. Nucl. Phys.* **82**, 78 (2015).
8. Proceedings of the 8th International Conference on Clustering Aspects of Nuclear Structure and Dynamics (CLUSTERS '03) 24-29 November 2003, Nara, Japan, (ed.), K.Ikeda, I. Tanihata, and H. Horiuchi; *Nucl. Phys. A* **738** (2004).
9. Proceedings of the 9th International Conference on Clustering Aspects of Nuclear Structure and Dynamics (CLUSTERS '07) 3-7 September 2007, Stratford upon Avon, UK, (ed.), M. Freer; *J. Phys.: Conference Series* **111** (2008).
10. Proceedings of the 10th International Conference on Clustering Aspects of Nuclear Structure and Dynamics (CLUSTERS '12) 24-28 September 2012, Debrecen, Hungary, (ed.), R. Lovas, Z. Dombradi, G. Levai, A. Kruppa, and G. Levai; *J. Phys.: Conference Series* **436** (2013).
11. Proceedings of the Workshop on the State of the Art in Nuclear Cluster Physics (SOTANCP 2008) 13-16 May 2008, Strasbourg, France, (ed.), C.

Beck, M. Dufour, and P. Schuck; in Special issue of *International Journal of Modern Physics E* **17** (2008).
12. Proceedings of the International Workshop on State of the Art in Nuclear Cluster Physics (SOTANCP 2010) 25-28 May 2010, Strasbourg, France, (ed.), P. Descouvemont, M. Dufour, and J.-M. Sparenberg; in Special issue of *International Journal of Modern Physics E* **20** (2010).
13. Proceedings of the International Workshop on State of the Art in Nuclear Cluster Physics (SOTANCP 2014) 26-30 May 2014, Yokohama, Japan, (ed.), T. Yamada and Y. Kanada-En'yo; *J. Phys.: Conference Series* **569** (2014).
14. W. Catford, arXiv:**1302.3849** (2013).
15. C. Beck, *J. Phys.: Conference Series* **569**, 012002 (2015); C. Beck, arXiv:**1408.0684** (2014).
16. P. Papka and C. Beck, *Clusters in Nuclei*, Vol.2, p.299, (ed.), C. Beck, *Lecture Notes in Physics* **848**, 299 (2012).
17. K. A. Erb and D. A. Bromley, *Treatise on Heavy Ion Science*, Vol. 3, p. 201, (ed.), Plenum, New York (1985).
18. M. Freer, *C. R. Physique* **4**, 475 (2003).
19. M. Freer, *Rep. Prog. Phys.* **70**, 2149 (2007).
20. H. Horiuchi, *Clusters in Nuclei*, Vol. 1, p.57, (ed.), C. Beck, *Lecture Notes in Physics* **818**, 57 (2010).
21. R. K. Gupta, *Clusters in Nuclei*, Vol.1, p.232, (ed.), C. Beck, *Lecture Notes in Physics* **818**, 232 (2010).
22. W. von Oertzen, M. Freer, and Y. Kanada-En'yo, *Phys. Rep.* **432**, 43 (2007).
23. H. Horiuchi and K. Ikeda, *Prog. Theor. Phys.* **40**, 277 (1968).
24. W. von Oerzten, *Eur. Phys. J. A* **11**, 403 (2001).
25. W. von Oerzten and H.-G. Bohlen, *C. R. Physique* **4**, 465 (2003).
26. W. von Oerzten and M. Milin, *Clusters in Nuclei*, Vol. 3, (ed.), C. Beck, p.147, *Lecture Notes in Physics* **875**, 147 (2014).
27. N. Orr and F. M. Marques, *C. R. Physique* **4**, 451 (2003); N. Orr and F. M. Marques, arXiv:**0303005**.
28. Y. Kanada-En'yo, M. Kimura, and H. Horiuchi, *C. R. Physique* **4**, 497 (2003).
29. Y. Kanada-En'yo and M. Kimura, *Clusters in Nuclei*, Vol.1, (ed.), C. Beck, p.129 *Lecture Notes in Physics* **818**, 129 (2010).
30. M. Ito and K. Ikeda, *Rep. Prog. Phys.* **77**, 096301 (2014).
31. I. Tanihata and R. Kanungo, *C. R. Physique* **4**, 437 (2003).
32. F. Nunes, *C. R. Physique* **4**, 489 (2003).
33. K. Ikeda *et al.*, *Clusters in Nuclei*, Vol. 1, p.165, (ed.), C. Beck, *Lecture Notes in Physics* **818**, 165 (2010); K. Ikeda *et al.*, arXiv:**1007.2474**.
34. N. Nakamura and Y. Kondo, *Clusters in Nuclei*, Vol. 2, p.67, (ed.), C. Beck, *Lecture Notes in Physics* **848**, 67 (2012).
35. V. Zagrebaev and W. Greiner, *Clusters in Nuclei*, Vol.1, p. 267, (ed.), C. Beck, *Lecture Notes in Physics* **818**, 267 (2010).

36. D. N. Poenaru and W. Greiner, *Clusters in Nuclei*, Vol.1, p.1, (ed.), C. Beck, *Lecture Notes in Physics* **818**, 1 (2010).
37. C. Simenel, *Clusters in Nuclei*, Vol.3, p. 95 (ed.), C. Beck , *Lecture Notes in Physics* **875**, 95 (2014); C. Simenel, arXiv:**1211.2387**.
38. D. Kamanin and Y. Pyatkov, *Clusters in Nuclei*, Vol. 3, p. 183, (ed.), C. Beck, *Lecture Notes in Physics* **875**, 183 (2014).
39. B. Borderie *et al.*, *Phys. Lett. B* **755**, 475 (2016); B. Borderie *et al.*, arXiv:**1603.01114**.
40. P. I. Zarubin, *Clusters in Nuclei*, Vol. 3, p. 183, (ed.), C. Beck, *Lecture Notes in Physics* **875**, 183 (2014); P. L. Zarubin, arXiv:**1309.4881**.
41. W. Broniowski and E. R. Arriola, Phys. Rev. Lett. **112**, 112501 (2014); W. Broniowski and E. R. Arriola, arXiv:**1312.0289**; P. Bozek *et al.*, *Phys. Rev. C* **90**, 064902 (2014); P. Bozek *et al.*, arXiv:**1410.7434**.
42. C. Beck, Proceedings of the First International African Symposium on Exotic Nuclei (IASEN '13) 2-6 December 2013, Cape Town, South Africa, in "Exotic Nuclei", (ed.), E. Cherepanov, Y. Penionzkhevich, R. Bark and J. Cornell (World Scientific) p.165 (20125); C. Beck, arXiv:**1402.3758**.
43. C. Beck, Proceedings of the First International Symposium on "New Horizons in Fundamental Physics — From Neutrons Nuclei via Superheavy Elements and Supercritical Fields to Neutron Stars and Cosmic Rays, Makutsi Safari Farm, South Africa, November 23–29, 2015; C. Beck, arXiv:**1603.02574**.
44. P. Navratil *et al.*, *Phys. Rev. Lett.* **84**, 5728 (2000).
45. R. B. Wiringa *et al.*, *Phys. Rev. C* **62**, 014001 (2000).
46. F. Hoyle *Astrophys. J. Suppl. Ser.* **1**, 121 (1954).
47. C. W. Cook *et al.*, *Phys. Rev.* **107**, 508 (1957).
48. M. Freer and H.O.U. Fynbo, *Prog. Part. Nucl. Phys.* **71**, 1 (2014).
49. A. Tohsaki *et al.*, *Phys. Rev. Lett.* **87**, 192501 (2001); A. Tohsaki *et al.*, arXiv:**nucl-th/0110014**.
50. W. von Oertzen, *Clusters in Nuclei*, Vol.1, p.102, (ed.), C. Beck, *Lecture Notes in Physics* **818**, 102 (2010); W. von Oertzen, arXiv:**1004.4247**.
51. T. Yamada *et al.*, *Clusters in Nuclei*, Vol.2, p.229, (ed.), Beck C, *Lecture Notes in Physics* **848**, 229 (2012); T. Yamada *et al.*, arXiv:**1103.3940**.
52. M. Chernykh *et al.*, *Phys. Rev. Lett.* **98**, 032501 (2007).
53. M. Chernykh *et al.*, *Phys. Rev. Lett.* **105**, 022501 (2010); M. Chernykh *et al.*, arXiv:**1004.3877**.
54. E. Epelbaum *et al.*, *Phys. Rev. Lett.* **106**, 192501 (2011); E. Epelbaum *et al.*, arXiv:**1101.2547**.
55. E. Epelbaum *et al.*, *Phys. Rev. Lett.* **109**, 252501 (2012); E. Epelbaum *et al.*, arXiv:**1208.1328**.
56. E. Epelbaum *et al.*, *Phys. Rev. Lett.* **110**, 112502 (2013).
57. A. C. Dreyfuss *et al.*, *Phys. Lett. B* **727**, 511 (2013); A. C. Dreyfuss *et al.*, arXiv:**1212.2255**.
58. Y. Fukuoka *et al.*, *Phys. Rev. C* **88**, 014321 (2013); Y. Fukuoka *et al.*, arXiv:**1304.5927**.

59. S. Ishikawa, *Phys. Rev. C* **90**, 061604 (2014); S. Ishikawa, arXiv:**1412.4176**.
60. Y. Funaki, *Phys. Rev. C* **92**, 021302 (2015); Y. Funaki, arXiv:**1408.5855**.
61. M. Itoh *et al.*, *Nucl. Phys. A* **738**, 268 (2004); *Phys. Rev. C* **84**, 054308 (2011).
62. M. Freer *et al.*, *Phys. Rev. C* **83**, 034314 (2011).
63. W. R. Zimmerman *et al.*, *Phys. Rev. Lett.* **110**. 152502 (2014); W. R. Zimmerman *et al.*, arXiv:**1303.1038**.
64. Tz. Kokalova *et al.*, *Phys. Rev. C* **87**, 057307 (2013).
65. D. J. Marin-Lambarri *et al.*, *Phys. Rev. Lett.* **113**, 012502 (2014); D. J. Marin-Lambarri *et al.*, arXiv:**1405.7445**.
66. D. Jenkins and S. Courtin, *J. Phys. G: Nucl. Part. Phys.* **42**, 034010 (2015).
67. E. Uegaki *et al.*, *Prog. Theor. Phys.* **57**, 1262 (1977).
68. M. Freer *et al.*, *Phys. Rev. C* **49**, R1751 (1994).
69. Ad. R. Raduta *et al.*, *Phys. Lett. B* **705** 65 (2011); Ad. R. Raduta *et al.*, arXiv:**1110.1617**.
70. J. Manfredi *et al.*, *Phys. Rev. C* **85**, 037603 (2012).
71. O. S. Kirsebom *et al.*, *Phys. Rev. Lett.* **108**, 202501 (2012).
72. T.K. Rana *et al.*, *Phys. Rev. C* **88**, 021601(R) (2013); T.K. Rana *et al.*, arXiv:**1203.3336**.
73. M. Itoh *et al.*, *Phys. Rev. Lett.* **113**,102501 (2014).
74. L. Morelli *et al.*, *J. Phys. G: Nucl. Part. Phys.* **43**, 045110 (2016).
75. E. Epelbaum *et al.*, *Phys. rev. Lett.* **112**, 102501 (2014); E. Epelbaum *et al.*, arXiv:**1312.7703**.
76. Y. Funaki *et al.*, *Phys. Rev. Lett.* **101**, 082502 (2008); Y. Funaki *et al.*, arXiv:**0802.3246**.
77. P. Chevallier *et al.*, *Phys. Rev.* **160**, 827 (1967).
78. H. Morinaga, *Phys. Rev.* **101**, 254 (1956).
79. N. Curtis *et al.*, *Phys. Rev. C* **88**, 064309 (2013).
80. P. Papka and Tz. Kokalova Tz (private communications); Tz. Kokalova *et al.*, *Phys. Rev. C***87**, 057309 (2013); J. A. Swartz *et al.*, *Phys. Rev. C* **91**, 034317 (2015).
81. G. Baiocco *et al.*, *Phys. Rev. C* **87**, 054614 (2013); G. Baiocco *et al.*, arXiv:**1301.5990**.
82. L. Morelli *et al.*, *J. Phys. G: Nucl. Part. Phys.* **41**, 075107 (2014); L. Morelli *et al.*, arXiv:**1404.3601**.
83. L. Morelli *et al.*, *J. Phys. G: Nucl. Part. Phys.* **41**, 075108 (2014); L. Morelli *et al.*, arXiv:**1404.3603**.
84. A. H. Wuosmaa *et al.*, *Phys. Rev. Lett.* **68**, 1295 (1992).
85. C. Beck, *Nucl. Phys. A* **738**, 24 (2004); C. Beck, arXiv:**nucl-ex/0401004** (2004); C. Beck, *Int. J. Mod. Phys.* **E13**, 9 (2004); C. Beck, arXiv:**nucl-ex/0401005** (2004); C. Beck *et al.*, *Nucl. Phys. A* **734**, 453 (2004).
86. J. Cseh *et al.*, *Phys. Rev. C* **80**, 034320 (2009).
87. Y. Taniguchi *et al.*, *Phys. Rev. C* **80**, 044316 (2009); Y. Taniguchi *et al.*, arXiv:**0907.0517**.

88. T. Ichikawa *et al.*, *Phys. Rev. C* **83**, 054319 (2011); T. Ichikawa *et al.*, arXiv:**1012.4311**.
89. D. G. Jenkins *et al.*, *Phys. Rev. C* **86**, 064308 (2012).
90. D. G. Jenkins, *J. Phys. G: Nucl. Part. Phys.* **43**, 023003 (2016).
91. D. G. Jenkins, *Clusters in Nuclei*, Vol.3, p.25, (ed.), C. Beck, *Lecture Notes in Physics* **875**, 25 (2014).
92. J. Darai, J. Cseh, and D. G. Jenkins, *Phys. Rev. C* **86**, 064309 (2012).
93. M. Kimura and H. Horiuchi, *Phys. Rev. C* **69** 051304 (2004); M. Kimura and H. Horiuchi, arXiv:**nucl-th/0311061**.
94. T. Lönnroth, *et al.*, *Eur. Phys. J. A* **46** 5, (2010).
95. D. Pandit *et al.*, *Phys. Rev. C* **81**, 061302 (2010).
96. C. Beck *et al.*, *AIP Conf. Proc.* **1098**, 207 (2008); C. Beck *et al.*, arXiv:**0811.0992**.
97. C. E. Svensson *et al. Phys. Rev. Lett.* **85**, 2693 (2000).
98. W. Sciani *et al.*, *Phys. Rev. C* **80**, 034319 (2009).
99. C. Beck *et al.*, *Phys. Rev. C* **80**, 034604 (2009); C. Beck *et al.*, arXiv:**0905.2901**.
100. C. Beck *et al.*, *Acta Phys. Pol. B* **42**, 747 (2011); C. Beck *et al.*, arXiv:**0111.3423**.
101. C. Beck, *J. Phys. Conf.* **436**, 012013 (2013); C. Beck *et al.*, arXiv:**1303.0960**.
102. E. Ideguchi *et al.*, *Phys. Rev. Lett.* **87**, 222501 (2001).
103. M. Rousseau *et al.*, *Phys. Rev. C* **66**, 034612 (2002); M. Rousseau *et al.*, arXiv:**nucl-ex/0206019**.
104. Y. Taniguchi *et al.*, *Phys. Rev. C* **76**, 044317 (2007).
105. M. Norrby *et al.*, (unpublished); Norrby M, private communication.
106. E. A. Benjamim *et al.*, *J. Phys. Conf.* **436**, 012015 (2013).
107. C. D. O'Leary *et al.*, *Phys. Rev. C* **61**, 064314 (2010).
108. M. Fukada *et al.*, *Phys. Rev. C* **80**, 5064613 (2009).
109. M. D. Salsac *et al.*, *Nucl. Phys.* **A 801**, 1 (2008).
110. A. Di Nitto *et al.*, *Phys. Rev. C* **93**, (2016).
111. R. Nouicer *et al.*, *Phys. Rev. C* **60**, 041303(R) (1999); R. Nouicer *et al.*, arXiv:**nucl-ex/9909009**; C. Beck *et al.*, *Phys. Rev. C* **63**, 014607 (2001).
112. D. Rudolph *et al.*, *Phys. Rev. Lett.* **82**, 3763 (1999).
113. C. Beck *et al.*, *Phys. Rev. C* **63**, 014607 (2001).
114. C. Bhattacharya *et al.*, *Phys. Rev. C* **65**, 014611 (2002); C. Bhattacharya *et al.*, arXiv:**nucl-ex/0108030** (2001).
115. M. Cindro *et al.*, *Phys. Lett.* **84B**, 55 (1979).
116. M. C. Mermaz *et al.*, *Nucl. Phys.* **A 429**, 351 (1984).
117. D. Pocanic and N. Cindro, *Nucl. Phys.* **A 433**, 531 (1985).
118. J. Schimizu *et al.*, *Phys. Lett.* **112B**, 323 (1982).
119. M. Gai *et al.*, *Phys. Rev. C* **30**, 925 (1984).
120. W. von Oertzen *et al.*, *Eur. Phys. J. A* **36**, 279 (2008).
121. Y. Pyatkov *et al.*, *Eur. Phys. J. A* **45**, 29 (2010); *Eur. Phys. J. A* **48**, 94 (2012).

122. V. I. Zagrebaev *et al.*, *Phys. Rev. C*, **81**, 044608 (2010).
123. S. J. Sanders, A. Szanto de Toledo, and C. Beck, *Phys. Rep.* **311**, 487 (1999); S. J. Sanders, A. Szanto de Toledo and C. Beck, arXiv:**nucl-ex/9904009**.
124. D. Lebhertz *et al.*, *Phys. Rev. C*, **85**, 034333 (2012).
125. A. Goasduff *et al.*, *Phys. Rev. C*, **89**, 014305 (2014).
126. D. G. Jenkins *et al.*, *Phys. Rev. C*, **76**, 044310 (2007).
127. F. Haas *et al.*, *Nuovo Cim. A*, **110**, 898 (1997).
128. R. Bijker and F. Iachello, *Phys. Rev. Lett.* **112**, 152501 (2014).
129. T. Suhara *et al.*, *Phys. Rev. Lett.* **112**, 062501 (2001).
130. J.-P. Ebran, E. Khan, T. Niksic, and D. Vretenar, *Nature* **487**, 341 (2012); J.-P. Ebran, E. Khan, T. Niksic, and D. Vretenar, arXiv:**1203.1244**; *Phys. Rev. C*, **87**, 044307 (2013).
131. M. Girod and P. Schuck, *Phys. Rev. Lett.* **111**, 132503 (2013); M. Girod and P. Schuck, arXiv:**1309.6104**.
132. M. Dufour *et al.*, *J. Phys. Conf.* **436**, 012005 (2013).
133. Y. Kanada-En'yo *Phys. Rev. C*, **89**, 024302 (2014).
134. M. Freer *et al.*, *Phys. Rev. C*, **86**, 034320 (2012).
135. W. von Oertzen, *Z. Phys. A*, **354**, 37 (1996); W. von Oertzen, *Z. Phys. A*, **357**, 355 (1997).
136. W. von Oertzen, *Nuovo Cim. A*, **110**, 895 (1997).
137. M. Milin *et al.*, *Eur. Phys. J. A* **14**, 295 (2002).
138. H. G. Bohlen *et al.*, *Phys. Rev. C* **68**, 054606 (2003).
139. N. Itagaki *et al.*, *Phys. Rev. C* **64**, 014301 (2001).
140. J. A. Maruhn *et al.*, *Nucl. Phys. A* **833**, 1 (2010).
141. W. von Oertzen *et al.*, *Eur. Phys. J. A* **21**, 193 (2004).
142. W. von Oertzen *et al.*, *Eur. Phys. J. A* **43**, 17 (2010).
143. W. von Oertzen *et al.*, *AIP Conf. Proc.* **1165**, 19 (2009).
144. M. Milin *et al.*, *Eur. Phys. J. A* **41**, 335 (2009).
145. H. G. Bohlen *et al.*, *Eur. Phys. J. A* **47**, 44 (2011).
146. P. Descouvemont and D. Baye, *Phys. Rev. C* **31**, 2274 (1985).
147. M. Furutachi *et al.*, *Prog. Theor. Phys.* **119**, 403 (2008).
148. W. Greiner, *Int. J. Mod. Phys. E* **17**, 2379 (2008).

CHAPTER 8

CLUSTERS IN ASTROPHYSICS

Michael Wiescher

Joint Institute for Nuclear Astrophysics and Department of Physics,
University of Notre Dame 225 Nieuwland Science Hall
Notre Dame, IN 46556, USA
wiescher.1@nd.edu

Tan Ahn

Joint Institute for Nuclear Astrophysics and Department of Physics,
University of Notre Dame 225 Nieuwland Science Hall
Notre Dame, IN 46556, USA
tan.ahn@nd.edu

The role of nuclear cluster configurations in nucleosynthesis processes is a topic of increasing interest for the nuclear astrophysics and nuclear structure community. While the discussion has traditionally focused on prime examples such as the Hoyle state and its role for the triple-alpha process, nuclear cluster structure has a much broader impact, both in quiescent as well as explosive stellar burning scenarios. This paper will present a summary of the role nuclear clusters play in hydrogen, helium as well as in heavy-ion burning stellar environments and will address possible experimental methods and techniques for improving the presently rather limited amount of specific nuclear cluster data that are necessary for an improved understanding of these often complex nucleosynthesis processes.

1. Introduction

The cluster structure of nuclei has been discussed for many decades as a possible phenomenon in nuclear structure physics. The idea of the cluster structure in nuclei through pair binding of single nucleons has already been proposed by Hafstad and Teller in 1938.[1] The occurrence of such cluster configurations would have enormous implications on scientific questions in nuclear astrophysics.

Nuclear astrophysics is concerned with the microscopic aspects of astrophysical phenomena in stars and stellar explosions. This includes the fundamental questions about the origin of of the elements in our universe, the source of energy in quiescent and explosive stellar burning, and the timescale of the different burning phases in such stellar events. These questions are typically associated with the reaction rates of nuclear processes in stellar environments. This includes both the rates for nuclear reaction and fusion processes as well as the rates for nuclear decay processes. Stellar life is characterized by a sequence of stellar burning processes that are determined by temperature and density conditions as well as by the available nuclear fuel in the burning zone.

The different burning phases during stellar evolution are determined by the energy production in the core of a star that generates the internal pressure which maintains the stability of the star against gravitational contraction or collapse. Nuclear energy is produced by nuclear reactions and decay processes that take place in the temperature and density conditions of stellar environments. Since the reactions that produce the energy are charged particle reactions, the reaction cross sections are subject to the Coulomb barrier. Therefore, the first stellar burning phase that is triggered in the interior of a gravitationally contracting pre-stellar cloud is hydrogen burning. Hydrogen burning is characterized by different reaction sequences such as the pp-chains[2] or the CNO cycles[3] by which hydrogen is converted to helium with the release of 25 MeV per fusion process. Most of these reactions are radiative capture processes. The reaction cross sections are determined by both resonant reaction and non-resonant direct capture contributions. The strength of the resonances scales with the single particle structure of the corresponding unbound states in the compound nucleus. Similar to that, the cross section of a direct capture transition to a bound state in the final nucleus scales directly with the single particle strength of the final state. Therefore both the reaction rates in the pp-chains as well as in the CNO cycles are determined by the single particle structure of the associated nuclei.

When the hydrogen fuel in the stellar core is exhausted the hydrogen burning region expands forming a shell around the hydrogen depleted core. The core itself contracts until density and temperature conditions are reached that allow helium induced reactions such as the triple-alpha process to occur that converts three ^{4}He nuclei to one ^{12}C nucleus. The efficiency of this reaction relies on the so-called Hoyle-state, the prime example for the relevance of a cluster state in nuclear astrophysics.[4] Besides this fusion process that dominates the energy production in helium burning stars, other α induced reactions take place such as $^{12}C(\alpha,\gamma)^{16}O$ that defines the carbon oxygen ratio in our universe,[5] and $^{22}Ne(\alpha,n)^{25}Mg$ that is next to $^{13}C(\alpha,n)^{16}O$ the dominant neutron source for the slow neutron capture or s-process. This is particularly important since the neutrons released from these sources are responsible for the production of about 50% of the heavy elements between iron and lead.[6] All these reactions are similarly characterized by α cluster configurations that determine the strength of resonances and direct reaction components alike. While hydrogen burning builds on the single particle configuration of light nuclei, helium burning is governed by the α cluster structure in light nuclei.

With the exhaustion of helium as nuclear fuel in the stellar core, helium burning continues in its outer shell, but the core contracts under its gravitational weight until conditions of sufficiently high temperature and density are reached that allow for fusion of single carbon nuclei to occur. This is the regime of carbon burning that will only be reached by stars of sufficient mass above $\approx 2\ M_{\odot}$. Carbon burning is dominated by the fusion of two ^{12}C nuclei. The fusion cross section is characterized by resonance-like structures that have been interpreted as molecular $^{12}C+^{12}C$ configurations.[7] Until recently the reaction rates for the different fusion channels have been estimated by averaging over the resonances using a potential model approach for calculating the reaction strength, e.g. Yakovlev *et al.*,[8] however, recent experimental studies[9] indicate that the low energy fusion cross section may be determined by pronounced single ^{12}C cluster resonance configurations that may dramatically change the present assumptions about the reaction rates. Similar to the cases of hydrogen and helium burning, carbon burning may be affected by the special nuclear structure configuration, the carbon cluster. The nucleosynthesis, energy production, and lifetime of these quiescent phases of stellar evolution are therefore closely correlated with the single particle and the cluster structure configurations of the associated nuclei.

In addition, the cluster structure of nuclei can have a dramatic impact in cataclysmic short-lived astrophysical environments. The thermonuclear runaway on accreting neutron stars, observed as X-ray bursts,[10] is driven in its initial phase by the break-out from the hot CNO cycles and the so-called αp-process, a sequence of (α,p) reactions by which CNO material is rapidly converted to nuclei in the calcium titanium range.[11] While traditionally the reaction rates have been estimated using statistical model predictions based on the assumption of high level density in the compound nuclei, recent experiments indicate that the rates are more likely determined by single α cluster states in the compound nuclei.[12] Type Ia supernova, considered as standard candles in cosmology, are ignited by the $^{12}C + ^{12}C$ fusion in the core of accreting white dwarf stars. A carbon cluster configuration of the compound nucleus ^{24}Mg may alter the ignition conditions for the supernova event. Similarly, superbursts in accreting neutron star systems have been interpreted as cataclysmic ignition of enhanced carbon material in the ashes of X-ray bursts.[13] The likelihood of such a superburst event would be strongly affected by a pronounced carbon cluster structure of the ^{24}Mg nucleus.

There is little direct work on the role of nuclear clusters in astrophysics. This paper will provide a summary of its different aspects, questions, and consequences it poses. The following section will focus on the microscopic description of nuclear clusters and their experimental signatures. This will be followed by a more detailed analysis of the way cluster configurations in nuclei determine the strength of specific reactions and determine the stellar burning pattern. This section will also set a number of experimental goals to address open questions in nuclear astrophysics that are associated with the cluster structure of the nucleus. Finally a number of specific examples and open questions for cluster driven nucleosynthesis patterns in nuclear astrophysics will be presented.

2. Cluster structure of nuclei

The formation of cluster structure in nuclei is an emergent phenomenon and requires a different paradigm than the single-particle or collective picture. Nucleons correlated in a cluster are on average closer to each other than to nucleons in a different configuration. This changes both the local and average spatial densities of the nucleus. This has been seen experimentally, where form factors have been measured for the highly α-clustered Hoyle state via inelastic electron scattering. This revealed that the Hoyle state is 3-4 times the extent of the ground state,[14] which implies the average density is less than that of the ground state. A phenomena in high-energy electron scattering called the European Muon Collaboration (EMC) effect and its relation to short-range correlations[15,16] is thought to depend on the local density of nucleons. The local density is different for a clustered nucleus such as ^{9}Be where the EMC effect is markedly different than what is expected from other nuclei that do not have strong cluster structure in their ground state.[17]

The resemblance of clusters to actual nuclei such as ^{4}He can result in the enhancement of nuclear reactions that involve these nuclei. This is important for understanding the energy and element production in stars as was mentioned in the preceding section. In this section, a general overview of cluster states and their general and microscopic description will be presented along with some theoretical models and specific experimental observables. Comprehensive overviews of clusters in nuclei can be found in a number of review papers.[18–21]

2.1. *Conditions for clustering*

The cluster structure comes from correlations arising from the many-body aspects of the nucleon-nucleon interaction. The fact that nuclei are composed of two types of particles, proton and neutrons, and that these particles are fermions and thus obey the Pauli-exclusion principle are essential in determining nuclear structure, including cluster structure. There are several conditions that are known to be important for cluster formation: nuclear density, deformation, and energy. How cluster formation specifically depends on other parameters such as spin, proton to neutron ratio, and mean field conditions is still in need of a more detailed and unified description. Experimental and theoretical work is progressing towards this goal.

The nuclear environment is important for cluster formation. Theoretical calculations show clusterization appears at one third the normal nuclear density,[22] which would be expected near the nuclear surface. It was also found that continuum states are important for the formation of cluster states.[23] Studies of α-transfer reactions indicate that α particles are transferred near the surface of heavier nuclei.[24] In all these cases, the role of the nuclear surface plays an important role in the formation of clusters.[25,26]

The most basic and prominent cluster observed in nuclei is the α cluster. The α particle or the ^{4}He nucleus has the largest binding energy per nucleon among low-mass nuclei so the nucleus can gain energy locally by forming α clusters. In addition, nucleons are fermions that obey the Pauli exclusion principle. The α particle has two protons and two neutrons that have total angular momentum equal to 0. This is the maximally correlated four nucleon system. Nuclei that are composed of an integer number of α particles, also known as α conjugate nuclei, should benefit from these correlations and have an enhanced stability. In fact, the abundance of elements in the solar system clearly shows a markedly higher abundance for elements

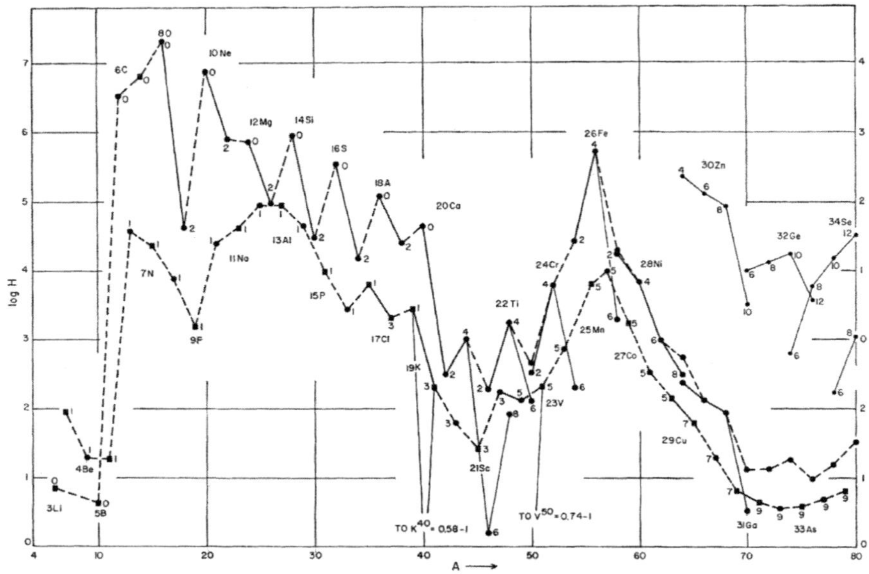

Fig. 1. Relative elemental abundances of light elements up to $A = 80$. The logarithm of the abundance is given as a function of mass number A where the abundance of Si is normalized to the value 6. The even and odd masses are plotted on different curves and the neutron excess is given next to each point. Reprinted figure with permission from H. Suess and H. Urey, Rev. Mod. Phys. 28, 53 (1956). Copyright 1956 by the American Physical Society.

that are mainly comprised of these α conjugate nuclei such as C, O, Ne, Mg, Si, S, Ca, and Ti as shown in Fig. 1.[27]

It has been found that α-cluster states appear near the α threshold for dissociating into α particles. Cluster states composed of larger clusters are also predicted to exist close to the associated breakup threshold. This is known as the Ikeda rule[28] and it is succinctly summarized in a graphical representation of where various clusters should appear as a function of energy (Fig. 2). For example, the Hoyle state is very close to the threshold for the dissociation of ^{12}C into three α particles. States composed of larger clusters are predicted to be found close to their associated thresholds as well such as the ^{12}C + ^{12}C cluster structure of ^{24}Mg and asymmetric clusters such as the ^{12}C + α structure in ^{16}O. The fact that these clusters are expected near thresholds make them important in astrophysical scenarios where most reactions happen near the threshold.

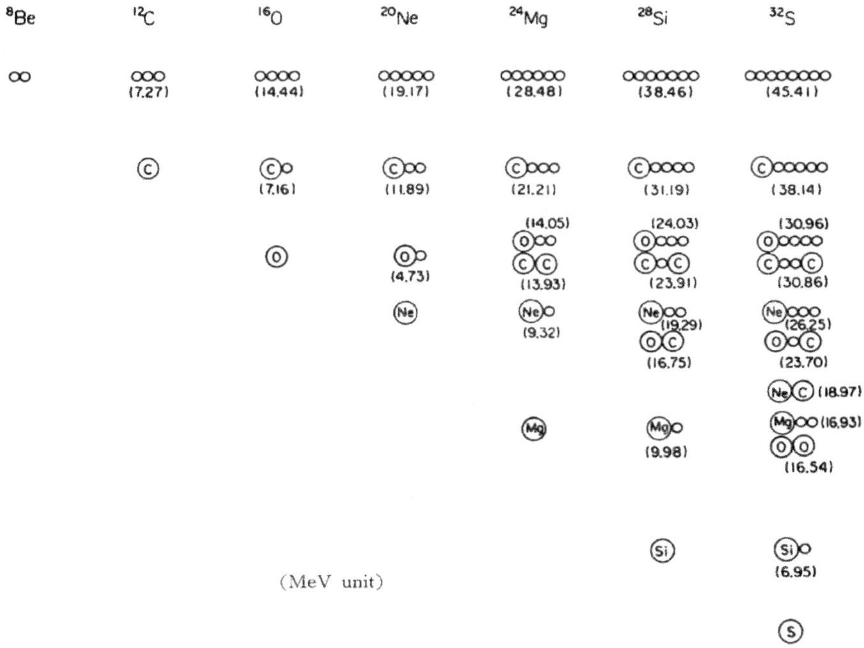

Fig. 2. Ikeda diagram for various cluster structures are shown with the energy of the associated breakup thresholds. Figure reprinted with permission from Horiuchi et al., Prog. Theor. Phys. Suppl. 52, 89 (1972).

Although the Ikeda rule predicts cluster structure near breakup thresholds, a degree of cluster structure is thought to exist in the ground state or at low excitation energy in some nuclei. Various ab-initio model calculations have shown there is a significant amount of clustering in the ground states of ^{8}Be[29] and ^{16}O.[30] Cluster states for neutron-rich nuclei may be expected at lower energies than the Ikeda rule suggests.[31] This shows that the details of the nuclear interaction and the proton to neutron ratios also affect the formation of clusters in addition to the excitation energy of these states. Ground state or low-energy cluster structure can significantly increase the cross section for direct capture reactions and enhance the non-resonant component of the corresponding reaction rates. Locating these cluster states for both stable and unstable nuclei is important for calculating nuclear reaction rates for nucleosynthesis.

2.2. *Nuclear molecules*

A rich array of structures are possible in nuclei that form α clusters as well as larger clusters such as ^{12}C and ^{16}O. As one goes to larger mass nuclei, states that resemble α clusters attached to a larger core can be seen. Additional nucleons outside of well formed clusters also add to the number of structures that can be formed, where the valence nucleons resemble electrons that are shared between two atoms. These clustering phenomena are collectively known as nuclear molecules[18] and play an important role in element formation.

2.2.1. *Core + α and larger clusters*

Early considerations of α clusters being preformed in heavy nuclei were give by Bethe in 1936 although several arguments were also presented against the case.[32] Despite these early objections, calculations and data point to α-cluster states existing in nuclei beyond ^{16}O, ^{20}Ne, and ^{24}Mg. One example is ^{32}S that is predicted to have a ^{28}Si core + α structure. Experimentally, a search for ^{28}Si + α cluster state bandheads in ^{32}S have been performed using the thick-target technique with elastic α scattering[33] and with a high energy beam looking for rotational bands to deduce moments of inertia.[34] A number of resonances were found in the elastic scattering experiment that were interpreted as fragmented α-cluster states that form a rotational structure. In the inelastic α-scattering experiment, candidates for the ^{28}Si + α bandhead were observed based on moments of inertia found for-parity-doublet bands that agree with a simple core + α system. Theoretical work in understanding the ^{28}Si + α structure including the breaking of the α clusters showed agreement with the rotational structure found by Lönnroth *et al.*[33,35] In ^{32}S, the cluster structures appear to be more fragmented and less pronounced than in lighter nuclei.

Going even heavier, the clusterization of the ^{44}Ti ground state as ^{40}Ca + α has been studied.[36,37] The α conjugate nucleus ^{44}Ti is a major product of the α process in the expanding shockfront of type II core collapse supernova events and is being produced via the ^{40}Ca(α,γ) radiative capture reaction. While most of the α process maintains statistical equilibrium, a modification of the reaction rate due to cluster effects may change these conditions and lead to modified freeze-out abundances for ^{44}Ti.[38] Similarly in ^{56}Ni, α clusterization is important for the α process in the expanding shockfont of core collapse supernovae where it is a similar phenomenon to that of ^{44}Ti and ^{40}Ca. The existence of α-clusters in ^{56}Ni as a gas structure has been proposed in a study of α breakup and was done using high-energy inelastic α scattering.[39] A high multiplicity for the α-particle breakup was observed that was not consistent with statistical models. This points to clusterized structure that would enhance the α breakup channels. Larger clusters and cluster structure in higher mass nuclei are relevant to cyclic burning cycles and certain clusters such as carbon play a prominent role. Clusters in larger mass nuclei such as Ca, Ti, and heavier are relevant for nucleosynthesis processes in explosive environments.[40,41]

The formation of α clusters with heavier cores continues to the extremes of heavy nuclei where evidence of a ^{208}Pb + α structure was found in ^{212}Po.[42] The unique finding of this study was that a cluster structure consisting of a ^{208}Pb core + α cluster could not be explained with the shell model, but was attributed to the oscillatory motion of the α cluster with respect to the core. This example shows that the correlations that favor α-cluster formation is not only important for light nuclei, but continue to exist in much heavier nuclei.

In addition to α clusters existing with a heavier core, nuclei can have prominent cluster structure consisting of larger clusters such as ^{12}C and ^{16}O. Where these structures are expected to exist is given by the Ikeda diagram, which includes the thresholds for breaking up into these larger clusters. Both symmetric clusters such as ^{12}C + ^{12}C and asymmetric clusters such as ^{12}C + ^{16}O can be formed.[43,44] Resonances for the ^{12}C + ^{12}C clusters are important for the interpretation of carbon burning in late stellar evolution[45] and for the ignition of superbursts[13,46] in the upper crusts of accreting neutron stars.

The description of a symmetric system that can form rotational bands is the ^{16}O + ^{16}O structure of ^{32}S that has been studied with Anti-symmetrized Molecular Dynamics (AMD) (see section 2.3.3) with a deformed basis where the fragmentation of the cluster structure was observed.[47] Cluster resonances in ^{16}O + ^{16}O may emerge during the fusion process at lower energies[48] and influence our present interpretation of oxygen burning environments.[49]

A common feature of nuclei with larger clusters is many of the levels found in these nuclei are less pure cluster states than for lighter α cluster nuclei. The cluster structure is found many times to be fragmented over several states,[50–52] which makes their study and impact on reaction rates more complex. Finding this impact will continue to be important for accurate nucleosynthesis reaction calculations.

2.2.2. *Non-α conjugate nuclei*

Clusters are also important for nuclei that are not α-conjugate nuclei, i.e. they have $N \neq Z$. In these cases, α-clusters can be helped to form with additional neutrons outside the α clusters that serve to "bind" the clusters together. The structures that can be formed from two or three α-clusters with one or more additional neutrons can be understood using two-center or three-center harmonic oscillator potentials, respectively. Examples of the effect of additional neutrons can been seen in the Be isotopes, a two-α center system. ^{8}Be is unstable against breakup into α particles, but ^{9}Be is stable and has a significantly clustered ground state. For ^{10}Be, structures can be seen that correspond to the valence neutrons arranging themselves in different molecular bond configurations also known as π and σ bonds. Going to 3-α+n cluster nuclei, an even larger number of possible configurations are possible as the 3-α clusters can arrange themselves into various geometrical configurations. A number of predictions for neutron-rich 3-α cluster carbon isotopes have been made that include exotic structures such as a linear-chain of α particles and triaxial configurations.[53–57] A review of α clusters with valence neutrons can be found in Horiuchi *et al.*[20]

The question still remains whether protons can also help with the stability of α cluster states in the same way as neutrons. There are a few experiments that have searched for cluster states on the proton-rich side of stability. One is an experiment to look for cluster states in the mirror nuclei ^{22}Mg and ^{22}Ne using resonant α scattering.[58] It was found a 3^- doublet that contains a large fraction of the α-cluster strength in ^{22}Ne was not seen in ^{22}Mg, but the reason for the non-observation was inconclusive. Another example is the proton-rich nucleus ^{10}C, which is predicted to have a 2-α + 2-p cluster structure. In a study of the breakup reactions of ^{10}C, a candidate for this structure was found at 4.70 MeV based on various possibilities for the shape of the clusters and role of the valence protons.[59,60]

The role of extra spectator nucleons in preserving or helping the formation of cluster states is important for reaction rates in nucleosynthesis both on the proton-rich and neutron-rich sides of stability as will be discussed in Sections 5 and 6. Data is especially lacking on the proton-rich side of stability. The study of isobaric analog states will be important for elucidating the role isospin plays in cluster formation.

Putting all of these clustering concepts together, we can see the richness of the structures that can be developed. An example is the structure of ^{22}Ne, where the α cluster, ^{16}O cluster, and valence neutrons all play a

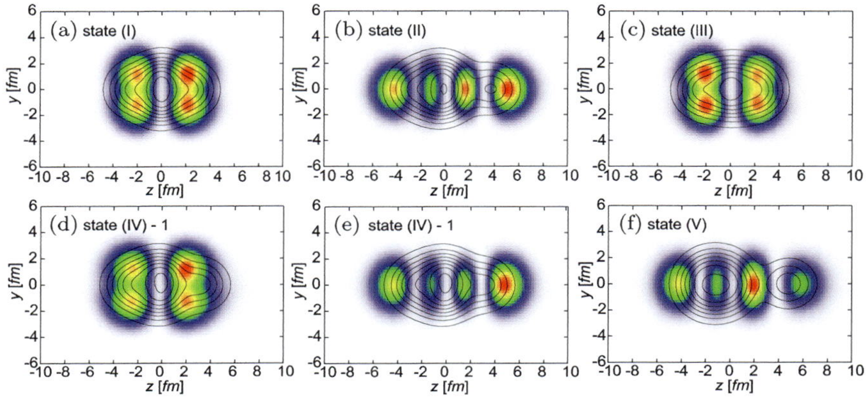

Fig. 3. Density distributions calculated for states in ^{22}Ne with the deformed-basis AMD. The states I-V are five different configurations based on constraints on the number of harmonic oscillator quanta used in the calculation. Contours show the shape of the ^{16}O + α core while the colors represent the density distribution of the valence neutrons. Reprinted figure with permission from M. Kimura, Phys. Rev. C 75, 034312 (2007). Copyright 2007 by the American Physical Society.

role. Calculations for the neutron-rich ^{22}Ne were also made using the AMD model that addresses both the neutron single-particle and ^{16}O + α cluster structure.[61] Calculated structures are shown in Fig. 3 where different states (I-V) are five different configurations based on the number of harmonic oscillator quanta used in the model (see Kimura[61] for more details). The contours show the core consisting of ^{16}O + α with the colors representing the density of the remaining two valence neutrons. The calculations show the extent of the deformation of the ^{16}O + α core as well the structure of the valence neutrons. The production of ^{22}Ne is important for neutron sources for the slow-neutron caption or s process in massive stars where the capture of an α particle onto ^{18}O plays an important role. This is covered in more detail in Sec. 6.1.

2.3. *Microscopic Cluster description*

A large number of models and frameworks have been used to understand the formation and the properties of cluster states. Some of these are phenomenological models that explicitly include cluster structure. Fundamentally, the nucleus is a collection of individual nucleons and many models start with this premise: independently moving nucleons in a mean-field potential. This and other methods link cluster structure to realistic nucleon interactions. This section is by no means comprehensive, but explores several models and calculations for understanding cluster structure for both α-conjugate nuclei and nuclear molecules that have extra nucleons in addition to clusters.

2.3.1. *Cluster models*

There are a number of models that start with assumed cluster structure and are very useful in obtaining cluster state properties including cluster separations and arrangements. The Block-Brink α-cluster model calculates properties of cluster states by using α clusters in a harmonic oscillator potential. The model represents an α cluster as a Gaussian wave-packet with a certain spatial width.[62] The Hamiltonian of the system is determined by the relative kinetic energy of the clusters and the mutual nuclear and Coulomb potentials between them. The total wavefunction is represented by a single Slater determinant of the α clusters and the energy of the system is minimized by varying the location and width of the clusters. This model has been applied to a number of nuclei including an early study of ^{16}O, which concluded that the ground state should have a tetrahedral arrangement.[63]

Another type of α-cluster model was proposed by Tohsaki *et al.*[64] where states in α-conjugate nuclei are described by a condensate of α particles. Using this Tohsaki-Horiuchi-Schuck-Röpke (THSR) model, states in ^{12}C including the Hoyle state and states in ^{16}O were described as analogs to α-cluster condensates. This description of α clusters can be applied to other α-conjugate nuclei as was done for the study of the ground state of ^{20}Ne.[65] A review of the THSR gas condensate model can be found in Funaki *et al.*[14]

2.3.2. *Mean-field and density functional calculations*

Many cluster properties can be understood from models using fermions in a harmonic oscillator mean-field as well as models using self-consistent mean-field methods. Harmonic oscillator mean-field models use two- and three-center models to represent the number of clusters present in a nucleus where the separation of the mean-field potentials can be varied. A single deformed harmonic oscillator mean-field is also used as deformation is closely related to the formation of clusters. The change of single-particles levels as a function of deformation can give rise to new shells that favor the formation of α clusters. An example using the deformed mean-field is given by Marsh[66] where states of ^{24}Mg were predicted to have prominent cluster structure. Detailed overviews of the use of harmonic oscillator mean-field models can be found in the reviews of Von Oertzen *et al.*[18] and Freer.[19]

Self-consistent mean-field models have been used to study the density distribution of nucleons in the nucleus and have found significantly clustered structures. Self-consistent methods such as Hartree-Fock assume the many-body state can be represented by a single Slater determinant and uses a variational method to minimize an energy surface. Simple relations can also be found to make predictions for the deformation of nuclei. As an example, the use of a zero-range interaction results in a prediction for ^{8}Be that gives an expectation of superdeformation, a result that is consistent with its interpretation as a cluster state.[19] Alternative methods for finding density distributions come from using the density matrix, which finds similarly clustered structures for the ground states of α-conjugate nuclei.[67]

Self-consistent mean-field methods are typically used for the ground states of nuclei, but can be used to describe excited state and rotational structures as well. For example, the rotation of α-conjugate nuclei, ^{16}O, ^{24}Mg, and ^{32}S up to the point of fission have been performed.[68] It is interesting to note that the density distributions show prominent cluster states especially for ^{24}Mg. ^{12}C + ^{12}C reactions studied with constrained Hartree-Fock show similar density distributions for ^{24}Mg.[69] Cranked Hartree-Fock calculations have also resulted in the rotation of unique clustered states such as the linear-chain of clusters in ^{24}Mg.[70]

Density functional theory has also been used to study the formation of clusters in nuclei. Density functional theory replaces the many-body wave function with a nucleon density functional, which gives the properties of the ground state of nuclei. Similar to self-consistent methods, the resulting nucleon densities can give insight into the parameters that are important for cluster formation. One example is by Ebran *et al.* that found more prominent cluster structures were realized from relativistic density functional calculations compared to ones using non-relativistic density functionals.[71] The reason was attributed to a deeper mean-field potential that is part of the relativistic density functional calculations. The effect of single-particle orbits and level degeneracy is also explored using the same framework.[72] A number of exotic cluster structures that include linear chains of clusters were found that correspond to local minima in the energy surface.

2.3.3. *Ab-initio models*

There are a number of approaches that do not assume a cluster structure *a priori*. One is Anti-symmetrized Molecular Dynamics (AMD) where individual nucleons instead of α clusters are represented by Gaussian wave packets.[73] An effective nucleon-nucleon interaction is used and nucleons are represented by a single Slater determinant. A variational method is used to minimize a potential energy surface. By looking at the density distribution of the particles, cluster structure can be identified. What makes AMD a powerful tool is one can observe the formation or dissolution of clusters based on model parameters. The explicit treatment of nucleons also allows one to see the role of valence nucleons in non-α conjugate nuclei where the spectator nucleons play an important role in cluster stability such as in the Be isotopes.

A closely related method is Fermionic Molecular Dynamics (FMD) where the width of the Gaussian wavepacket representing a nucleon as well as its spinor direction is treated as variational parameters.[74] FMD can be combined with a renormalization scheme such as the Unitary Commutator Method to use realistic nucleon-nucleon interactions. This combination has been used to study α-cluster states in light nuclei where calculated density distributions show strong cluster-like deformations in ^{9}Be, ^{20}Ne, and ^{26}Mg.[75] It was also found that the cluster structure of ^{20}Ne eventually erodes as extra neutrons are added as one approaches ^{26}Ne. These types of calculations show the strength of AMD and FMD approaches where both single nucleon and cluster structure can be probed within a single framework.

Another ab-initio method is the Green's Function Monte Carlo. Calculations using this method were performed for a number of $A = 8$ nuclei using the realistic interaction Argonne V18 as well as Urbana IX three-body interactions.[29] One of the nuclei included in the study ^{8}Be was found to have a ground state density distribution to be highly clustered. This is a significant result as cluster structure was not assumed in the calculation and the result agrees with the expectation of a 2-α structure ground state for ^{8}Be due to its primary decay mode of α breakup.

Similarity Renormalization Group is another example of using realistic nucleon-nucleon interactions including 3-nucleon forces. This technique was used to describe the structure of ^{12}C and ^{16}O and showed the importance of including 3-body forces although the description of excited states was only qualitative.[76] Lattice effective field theory has been used to understand the structure of the Hoyle state and its rotations[77] as well as the ground state of ^{16}O, which found its structure to be four α particles in a tetrahedral configuration,[30] which is the same conclusion reached using the Bloch-Brink α-cluster model.[63] The exact role of the details of realistic nucleon-nucleon interactions in cluster formation is still in need of investigation, but these examples show that progress in being made in this direction.

Several methods mentioned above have been combined to extend the ability to model cluster and molecular structure. Spherical shell model basis states are not well suited for the extended wavefunctions of cluster states, but hybrid models using no-core shell model wavefunctions for the individual clusters and combining these wavefunctions using the Resonating Group Method have been pursued to model clusters states that combines ab-initio and cluster features.[78] Another combination of methods is the use of AMD and self-consistent Hartree-Fock methods. This has the advantage of giving a better description of valence neutron wavefunctions in halo nuclei.[79] In this method, an AMD wavefunction is constructed. Then the single-particle motion is extracted from the AMD wavefunctions. Mean-field wavefunctions for nucleons are calculated and excited states can be constructed from particle-hole excitations. This is an extension of the AMD method by adding the ability to construct these states. This can be useful for the study of non-α conjugate nuclei and allows one to study the dynamics of the valence nucleons, which is especially important for loosely bound nucleons such as those in halo nuclei.

There are a large number of specific models that we have not mentioned, but this sample of models and calculations show that progress is being made in describing cluster structure and making direct connections between cluster structure and the nucleon-nucleon interactions. Prediction of cluster distributions can be important not only for the analysis of single specific reactions but can also provide a critical tool for generating statistical model predictions for a broad range of α-induced reaction processes, which require empirical corrections for matching the observed cross section data.[80] Nuclear cluster structure's impact on reaction rates can be dramatic and more robust models of nuclei will allow for a better understanding of element production especially for unstable nuclei.

2.4. *Observables related to cluster structure*

Prominent cluster structure in nuclei is manifested through several important observables. One observable is related to a bulk property of the nucleus, the mean charge radius. The larger spatial extent of cluster states will manifest as a larger mean charge radius, which is markedly different than a non-clustered nucleus. The $E0$ transition operator is directly proportional to the mean charge radius of nuclear states and so large monopole transition strengths can be used to identify cluster states.[81] One example is in the light nucleus ^{11}B where the excitation to the 8.65 MeV $3/2^-$ level in a (d, d') experiment resulted in a large measured monopole strength.[82] Fast-beam experiments performing inelastic α scattering that include measuring $E0$ transition strengths have be used to search for elongated cluster states in neutron-rich ^{16}C.[83–85]

A highly clustered state is expected to have a large deformation and there are several observables can be deduced from observed level structure and electromagnetic decays. Rotational bands arise from clustered states that rotate and moments of inertia can be deduced from the level spacing. There are predictions for particularly elongated structures such as linear α-chain states that can result in superdeformed or hyperdeformed bands, which correspond to 2:1 and 3:1 semi-major to semi-minor axis ratios, respectively. Systems containing larger clusters or a core + cluster configuration result in asymmetric structures. This gives rise to parity doublet bands and can be used to identify these type of asymmetric structures.[18] Two examples are ^{13}C and ^{16}O where asymmetric structures can be identified by these rotational bands.[43,55]

In addition to the formation of rotational bands, electromagnetic transitions should be enhanced between cluster states due to their large deformations. The measurement of electromagnetic transition rates and electric quadrupole moments give a strong constraint on nuclear models, especially with respect to nuclear shapes. Measured rates can be directly compared to the predictions of cluster models. Due to cluster states being close to the dissociation thresholds, particle emission or breakup channels can compete with γ transitions resulting in a relatively small γ branching ratio. This makes the observation of γ transitions difficult, but direct measurements have been made. A notable example is the measurement of the $4^+ \rightarrow 2^+$ γ transition strength in ^{8}Be.[86,87] This was done using high-efficiency BGO scintillators as γ detectors. The measured capture cross sections were compared with *ab-initio* model calculations and gives the observation of the electromagnetic transition resulting from the rotation of the highly clustered ^{8}Be. Observation and predictions for electromagnetic transitions in heavier Be isotopes is given in Von Oertzen *et al.*[18] where unusually long lifetimes can occur for transitions between different molecular cluster configurations. In heavier nuclei, enhanced $E1$ transitions can be an indicator for asymmetric cluster states as they have a non-uniform charge distribution.[88]

One of the most important observables are reaction cross sections, which includes the probability for a nucleus to breakup into smaller clusters. From the cross section, reduced α widths, spectroscopic factors, and transfer probabilities can be deduced where large values indicate the presence of cluster states. The amplitude of the α cluster configuration is reflected in the reduced reduced α width or spectroscopic factor. More details of this will be discussed in the Sec. 3. In the same way, a large α breakup probability points to enhanced cluster structure. This has also been observed for larger cluster breakup such as ^{24}Mg $\rightarrow$ ^{12}C + ^{12}C. By looking at the breakup probability, the fragmentation of cluster states can also be observed.[52,89]

In summary, it is of great importance to make progress both experimentally and theoretically in understanding the nature of clusterization in nuclei and relating its origins to its effects on nuclear reactions. From the fundamental to the applied, the goal is to relate the details of realistic nucleon-nucleon interactions to the manifestation of cluster structure and directly relate this to the production of elements through astrophysical nuclear reactions. The guidance of nuclear structure theory that can accurately predict cluster structure will be needed to understand in detail nucleosynthesis models and calculations.

3. Cluster structure and reaction rates

The role nuclear reactions play in stellar nucleosynthesis processes are determined by the stellar reaction rate which expresses the probability of a reaction to take place as a function of temperature. It is interesting to note that nuclear reactions associated with hydrogen burning are primarily facilitated and controlled by the single particle structure in the compound nuclei, while helium burning depends on the α cluster structure of the associated nuclei. Recent results indicate that a similar principle is observed in heavy ion burning processes, such as carbon burning, but is washed out by collective excitation components in the fusion process. Examples of that will be discussed in a later section; here we want to introduce first the nuclear reaction formalism that reflects the role of nuclear structure in quiescent and explosive stellar burning.

3.1. *The reaction rate formalism*

The nuclear reaction rate can be calculated from the reaction cross section $\sigma(E)$ by integration over the Maxwell-Boltzmann distribution of the interacting particles in a stellar environment of a temperature T. The reaction rate per interacting particle pair is given by:

$$N_A\langle\sigma\nu\rangle = \left(\frac{8}{\pi\mu}\right)^{\frac{1}{2}} \frac{N_A}{(kT)^{\frac{3}{2}}} \int_0^\infty \sigma(E)Ee^{-E/kT}dE \tag{1}$$

where μ is the reduced mass, N_A is Avogadro's number, and k is Boltzmann's constant. The reaction cross section $\sigma(E)$ and the its energy dependence are the key parameters for determining the reaction rate. The reaction contributions are determined by the underlying nuclear structure and reaction phenomena. In most cases, these include resonant reaction and nonresonant, direct reaction contributions.

3.1.1. *Resonant reaction contributions*

In cases where the reaction cross section is characterized by narrow resonances the expression can be simplified since the cross section can be expressed by a sum of single resonances described by the Breit-Wigner resonance function. For narrow resonances the integration over the Breit Wigner function allows one to express the reaction rate in a numerical form in terms of the resonance strength $\omega\gamma$ as:

$$N_A\langle\sigma\nu\rangle_R = 1.54{\cdot}10^5(\mu T_9)^{-3/2}{\cdot}\sum_i(\omega\gamma)_i \exp\left(\frac{-11.605E_{R,i}}{T_9}\right)\frac{cm^3}{sec\cdot mol} \tag{2}$$

where, μ is the reduced mass, T_9 is the temperature in GK, $(\omega\gamma)_i$ is the resonance strength of the ith resonance in eV and $E_{R,i}$ is the resonance energy in the center of mass frame of the ith resonance in MeV.

The resonance strengths are proportional to the probabilities of the production and decay channels, Γ_{in} and Γ_{ex}. In case of radiative α capture reactions these channels correspond to the α and γ partial widths Γ_α and Γ_γ, respectively. In this terminology the resonance strength is:

$$\omega\gamma_{(\alpha,\gamma)} = \frac{2J+1}{(2J_1+1)(2J_2+1)}\frac{\Gamma_\alpha\Gamma_\gamma}{\Gamma} \tag{3}$$

where, J represents the spin of the resonance and J_1 and J_2 represent the spin of the target and the α projectile, respectively. For resonances in the stellar energy range the α width is largely reduced because of the low penetrability through the Coulomb barrier therefore $\Gamma_\alpha \ll \Gamma_\gamma$ and with $(\Gamma = \Gamma_\alpha + \Gamma_\gamma)$ the resonance strength is directly proportional:

$$\omega\gamma_{(\alpha,\gamma)} \approx (2J+1)\Gamma_\alpha \tag{4}$$

The α partial width $\Gamma_\alpha = 2P_\alpha \cdot \gamma_\alpha^2$ is defined in terms of the α penetrability through the Coulomb barrier P_α and the reduced α width γ_α^2 . The reduced width is often expressed in terms of the Wigner limit and the dimensionless reduced width or spectroscopic α factor θ_α^2

$$\gamma_\alpha^2 = \theta_\alpha^2 \cdot \gamma_W^2 \tag{5}$$

with the Wigner limit defined as

$$\gamma_W^2 = \frac{3}{2}\frac{\hbar^2}{\mu a_c^2} \quad \text{or} \quad \frac{\hbar^2}{\mu a_c^2} \tag{6}$$

where μ is the reduced mass, and a_c is the channel radius. The reduced width or spectroscopic factor reflects the amplitude of the α cluster configuration in the wavefunction of the resonance state and is proportional to the asymptotic normalization coefficient (ANC) as defined by Mukhamedzhanow and Tribble 1999.[90] In view of this the strength of low energy single resonances in an α capture reaction is therefore directly correlated with the α cluster distribution in the compound nucleus above the α threshold.

3.1.2. *Nonresonant reaction contributions*

Nonresonant reaction components in low energy cross sections are often ill defined. They can include one-step direct-reaction components, the low energy tails of broad resonances at higher energies, as well as the high energy component of sub-threshold states tailing into the continuum. For considering non-resonant reaction contributions, traditionally the reaction cross section $\sigma(E)$ is expressed in terms of the astrophysical S-factor

$$S(E) = \sigma(E) \cdot E \cdot e^{2\pi\eta}. \tag{7}$$

The exponential term $e^{2\pi\eta}$ corrects, in a first-order-approximation, for the influence of the Coulomb-barrier on the cross section. Therefore $S(E)$ represents the nuclear transition strength in the reaction mechanism and the contribution of higher order orbital momentum components in the wave functions.

For nonresonant reaction contributions the reaction rate frequently scales with the S-factor, so that in a first approximation the S-factor can be considered to be constant $S_{eff}(E_0)$ over the stellar energy range or Gamow window E_0 and the integral in equation 1 has to be solved for the Maxwell Boltzmann exponential function only. In that case the nonresonant term of the reaction rate can be expressed by

$$N_A\langle\sigma\nu\rangle_{NR} = 7.83\cdot 10^9 \left(\frac{Z_1 Z_2}{\mu T_9^2}\right)^{1/3}$$
$$\cdot S_{eff}(E_0)\exp\left(4.248\frac{Z_1^2 Z_2^2\mu}{T_9}\right)^{1/3}\frac{cm^3}{sec\cdot mol}. \quad (8)$$

The kind of non-resonant contributions at stellar energies depend on the specific reaction mechanism, direct reaction component, and the tails of broad resonances or subthreshold states. In addition there are possible interference contributions between the various broad reaction components. The strength of the resonance tails are proportional to the α partial widths of the correspondent levels. For direct or one-step α capture or α transfer reactions the cross section can be described in the framework of a potential model approach. Such models are the Distorted Wave Born Approximation or DWBA for α transfer reactions and the Direct Capture model (DC) for radiative α capture reactions.

The differential cross section scales directly with the transition probability between the initial and final configuration, which is proportional to the α spectroscopic factor or ANC value of the final state. The S_{eff} factor can be expressed in terms of an orbital momentum dependent component S_ℓ that contains all the orbital momentum coupling other energy dependent components in the transition operator and the α spectroscopic factor of the final state.

$$S_{eff}(E_0) = S_\ell(E_0)\cdot\theta_\alpha^2 \quad (9)$$

often plays a critical role and the cross section scales with the α cluster strength of the final state. Also broad resonances scale with the strength of the α channel. Alpha clustering therefore dictates the reaction flow pattern in helium burning.

Equations 5 and 9 demonstrate that cluster configurations in compound and/or final nuclear states will be reflected in strong resonant and nonresonant reaction contributions and can be identified over a direct measurement of resonance strength or non-resonant reaction cross section, respectively.

On the other hand an indirect determination of these level parameters can be used for calculating the contributions these states have to the reaction mechanisms.

3.2. *Experimental methods and tools*

There is a number of different and often complementary experimental techniques to determine the impact of cluster states on the reaction cross section and reaction rates in stellar environments. The traditional approach is the direct measurement of the different contributions which in turn defines the strength of the cluster configuration. This approach is not always possible because of the Coulomb barrier that largely reduces the reaction cross section, making a direct measurement extremely challenging. In these cases indirect methods have to be utilized to determine and evaluate the role of cluster configurations in the reaction process.

3.2.1. *Direct Methods*

The classical approach to determine reaction rates for questions associated with stellar burning processes is the direct measurement of the contributing resonant and non-resonant processes to determine the resonance strength and the S-factor.

The strength of a narrow resonance $\omega\gamma$ is directly proportional to the maximum yield Y_{max} of a thick-target yield curve which is determined by integrating the resonance Breit-Wigner cross section over the target distribution.[91]

$$\omega\gamma = \frac{2\eta}{\lambda_{cm}^2} \frac{m_T}{m^T + m_p} \cdot Y_{max} \tag{10}$$

with η the stopping power of the projectile in the target material, λ_{cm} the wavelength of the incoming projectile in the center-of mass system, and m_T and m_p the mass of target nucleus and projectile respectively. Unbound resonant states with pronounced α cluster configurations are therefore characterized by strong resonance yields in α induced reactions.

For nonresonant processes the reaction cross section $\sigma(E)$ is directly proportional to the reaction yield $Y(E)$

$$\sigma(E) = \frac{\eta}{\Delta} \cdot Y(E) \tag{11}$$

with η being the stopping power of the projectile in the target material and Δ the target thickness. Again, a high nonresonant yield for a specific

α transfer or α capture transition correlate with a strong α cluster configuration of the final state.

The direct measurement of nuclear reactions of relevance for nuclear astrophysics considerations has the advantage that the critical resonant and non-resonant contributions can be directly studied with the appropriate experimental arrangements. The main disadvantage is, however, the relative low energy range at which these reactions need to be studied in the laboratory. This energy range is called the Gamow window and represents the effective energy range of the reaction cross section for a stellar temperature T; it corresponds to the integrand in Eq. 1. For a constant S-factor it is approximated by a Gaussian distribution around the mean energy E_0 in units [MeV]

$$E_0 = 0.122 \cdot (Z_1^2 Z_2^2 \mu \cdot T^2)^{1/3} \tag{12}$$

with Z_1 and Z_2 the charges of the interacting particles and μ the reduced mass; T represents the temperature in units [GK]. Using the same notation, the width ΔE of the Gamow range is given by

$$\Delta E = 0.236 \cdot (Z_1^2 Z_2^2 \mu \cdot T^5)^{1/6} \tag{13}$$

For a sharp low energy resonance nearby the Gamow range the conditions are different. For a resonance with a width Γ_{tot}, the effective energy range corresponds effectively to the resonance width. Using Equations 12 and 13 it can be quickly estimated that for stellar helium burning conditions with temperatures between 0.2 and 1.0 GK the effective nuclear energy ranges from 0.2 MeV to 2 MeV and needs to be explored. For explosive helium burning the temperatures can be considerably higher ranging from 0.5 GK to 3 GK. This translates into a Gamow range stretching from 0.5 MeV to 5.0 MeV center of mass energies for the associated reactions with radioactive particles.

At such low energy conditions corresponding to the Gamow window, the experimental yield is largely suppressed by the low penetrability through the Coulomb barrier and the cross sections for α induced radiative capture or transfer reactions are extremely small. This makes the direct measurements very challenging; reactions on stable particles are handicapped not only by the extremely low reaction yield, but also by the large count rates in the detectors due to background signals from cosmogenic and radiogenic processes. In addition to that beam induced background on target impurities may add additional background signals. These problems can be addressed by inverse kinematics experiments with direct counting of the reaction products or by experiments at deep underground accelerators where the cosmogenic background is largely reduced.[92] For experiments with radioactive particles, inverse kinematics is necessary, but the experiments are mainly handicapped by the low intensity of radioactive beams.

To compensate for that indirect studies of the compound nucleus offer an alternative approach complementing the efforts of direct measurement of the critical reaction.

3.2.2. *Indirect Methods*

As mentioned in the previous section, large α widths for reactions with α particles correspond to a large overlap in the structure of the α particle with the corresponding quantum states in the formed nucleus. Besides direct capture reactions, transfer and resonant scattering cross sections can be measured to deduce the α width.

The measurement of α transfer reactions such as (^{6}Li,d) or (^{7}Li,t) populating excited states in the final nucleus is a powerful method to identify cluster states. These measurements are performed at high energies to remove possible resonant contributions from the cross section. The observed differential cross section scales directly with the relative α spectroscopic factor ($\theta^2_{\alpha(rel)}$) and the predicted cross section for a single cluster transfer calculated in the framework of the Distorted Wave Born Approximation (DWBA) σ_{DWBA} is expressed by the following equation:

$$\frac{d\sigma_{exp}}{d\Omega} = \theta^2_{\alpha(rel)} N \sigma_{DWBA} \tag{14}$$

where, N is the normalization constant. For (^{6}Li,d) and (d,^{6}Li) reactions, N has been found [93] to be equal to 2.67 from a comparison of α-transfer and α-decay data. The $S_{\alpha(rel)}$ value determined using the above equation is traditionally scaled to the strength of a transition of a well known α cluster state to remove possible systematic uncertainties in the experimental arrangement and obtain the spectroscopic factor θ^2_α and the reduced α width using Eq. 5 for determining the α partial width Γ_α. In the case of unbound states, this information provides the resonance strength of possible cluster states and in case of bound final states this information helps to estimate the strength on direct α induced reaction components. A recent example of this approach is given by Talwar *et al.* (2016).[94]

ANC approach THM approach A modified approach is the determination of the asymptotic normalization coefficients (ANCs) using sub-Coulomb α-transfer reactions to probe near-threshold resonances. This approach promises a very valuable tool for studies of astrophysically important reaction rates since the results are claimed to be model independent. A recent study[95] the 5.79 MeV (1^-) state in ^{20}Ne using the ^{16}O(^{6}Li,d)^{20}Ne reaction at sub-Coulomb energies yielded good agreement with the results of a direct radiative capture measurement of the resonance strength within the given uncertainties.[96]

Another method to extract cross sections at low energy is is called the Trojan Horse method (THM).[97,98] In this method a suitable reaction is found to replace the reaction of interest as a surrogate process.[99] This allows one to directly mimic the reaction mechanism bypassing the hindrance of the Coulomb barrier that handicaps the direct measurement at low energies. For direct reactions, one also has modification of the cross section due to the effect of electron screening, which is not present in astrophysical environments. The THM uses a reaction $A + (b + x) \to b + c + C$ to study the reaction of interest $A + x \to c + C$. The Fermi momentum of the particle x contained within the particle $a = b + x$ can allow the reaction of $x + A$ to happen at very low energies even approaching zero energy. The THM uses the particle a, which is accelerated to energies above the Coulomb barrier. This allows for particle x to overcome the Coulomb barrier and then interact with particle A as a breaks up. By choosing the kinematics of the final particles, one can deduce the cross section of the reaction of interest, which happens as a subprocess. The particle x is considered quasi-free while particle b is not involved and is a spectator in the process. This condition can be satisfied if the spectator is detected near the maximum of its momentum distribution.[100] The THM can be used to deduce cross sections for both direct and resonant reactions.

For a list of examples of the Trojan Horse method see Tumino *et al.*[99] and references therein. Some of the reactions studied include ^{2}H(^{6}Li,α)^{4}He using the ^{6}Li(^{6}Li,$\alpha\alpha$)^{4}He three-body reaction,[101] ^{7}Li(p,α)^{4}He using ^{7}Li(d,$\alpha\alpha$)n,[102] and ^{10}B(p,α)^{7}Be using the ^{2}H(^{10}B,α^7Be)n reaction.

4. Use of radioactive beams

Reactions that occur with unstable nuclei require the use of radioactive beams. There are a number of reasons, first of all the fact that in most

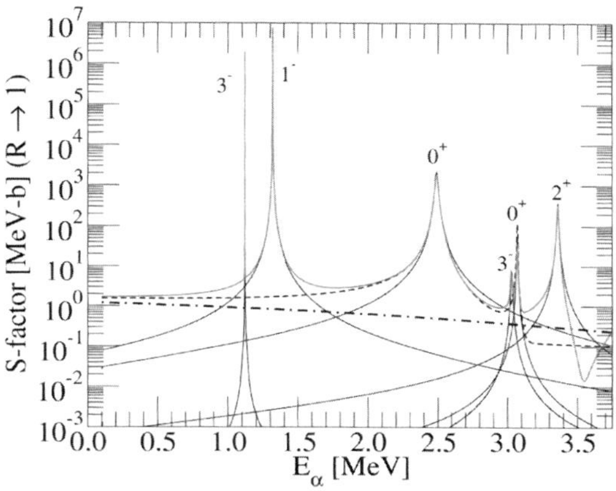

Fig. 4. The R-matrix prediction for the S-factor excitation curve of $^{16}O(\alpha,\gamma)^{20}Ne$ based on the experimental data of Constantini *et al.*.[96] The different resonances are indicated as well as the sum of the contributions from non-resonant E2 transitions to bound states in ^{20}Mg indicated in figure 5. Reprinted figure with permission from H. Constantini *et al.*, Phys. Rev. C 82, 035802 (2010). Copyright 2010 by the American Physical Society.

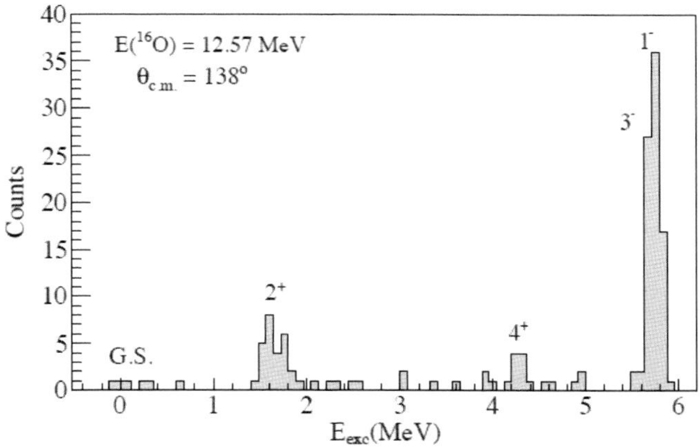

Fig. 5. The deuteron spectrum obtained in a $^{16}O(^{6}Li,d)^{20}Ne$ study populating bound and unbound states in ^{20}Ne up to 6 MeV excitation energy.[95] The two levels between 5.5 and 6.0 MeV excitation correspond to the resonance states observed in $^{16}O(\alpha,\gamma)^{20}Ne$ shown in Fig. 4 between 1.0 and 1.5 MeV α energy. Reprinted figure with permission from M. Avila *et al.*, Phys. Rev. C 90, 042801 (2014). Copyright 2014 by the American Physical Society.

cases the lifetimes are too short for producing a target and second because of the very large activity that would be generated by a radioactive target. Radioactive beams both proton- and neutron-rich and close and far from the valley of stability can be produced using a number of methods. For example, the open question of the nature of clusters in proton-rich nuclei is due to the lack of the number of reaction studies that have been performed requiring radioactive beams. Radioactive beams are produced as secondary reaction products from a reaction with a primary beam and stable target. The secondary reaction product of interest is usually filtered out from other unwanted reaction products as well as the primary beam and scattered target nuclei. Several facilities produce radioactive beams by using transfer reactions.[103] These beams are usually closer to the valley of stability and may range from several to 15 MeV/n depending on the capabilities of the accelerator producing the primary beam. The beams produced in this way can normally be used directly with detector setups, though in some cases degraders may be needed to bring the energy of the secondary beam down to the desired energies.

To reach nuclei that are far from the valley of stability, the fragmentation or fission of heavier nuclei, or multi-nucleon transfer can be used. These processes require higher energies and for studies relevant for astrophysics at low energies, the secondary beam of interest must be slowed down. To reach these low energies, the secondary beams can be degraded through the use of degrading foils or be stopped and reaccelerated to the appropriate energy.

Due to using the radioactive species in the reaction as the beam, reactions are performed in inverse kinematics using light-ion targets. Proton, deuteron, ^{3}He, and ^{4}He targets are naturally gases so pure targets require the use of a gas cell or a gas jet. Polyethylene and deuterated polyethylene targets can also be used for protons and deuterons. Radioactive beams have also been used to study fusion processes that can be enhanced or hindered for nuclei farther from the valley of stability. A range of experiments with radioactive beams from light to heavy nuclei used for fusion studies can be found in the review of Back *et al.*[104]

Various experimental setups aims to capture the relevant observables for a reaction, the differential cross section. Light and heavy particles can be detected through various means, a common technique is using Si detectors at fixed angles. In inverse kinematics, the light particles must almost always be detected to have sufficient energy resolution.

Examples of the use of radioactive beams to study reactions involving α particles are with neutron-rich light isotope beams such as ^{6}He, ^{8}He, ^{12}Be, ^{11}Li up to heavier beams that can be produced through fission or fragmentation reactions such as ^{132}Sn and ^{196}Pb. Unstable nuclei on the proton-rich side of stability are important for understanding the hot CNO cycles and the rp-process.

The availability of radioactive beams is much more limited than stable beams both in intensity and specific isotopes. This in conjunction with the typically low cross sections associated with low-energy astrophysical reactions present a substantial challenge, but improvements in current radioactive beam facilities and the next generation of radioactive beam facilities along with the development of higher efficiency detector setups and techniques will allow for the investigation of some of the reactions of interest mentioned in this article in the near future.

4.1. *Thick-target and Active-target Measurements*

Many of the astrophysically relevant cross sections have low cross sections so it is desirable to have techniques that can increase the luminosity of experiments. One way is to increase the thickness of targets by using the thick-target or active-target techniques. These methods allow for measurements to be done with modest beam intensities as is often the case with radioactive beams. Active-target detectors use a gas target with the ability to image a reaction within the target material allows for energy measurements to extend to relatively low energies. This gives the possibility to measure some astrophysical reactions involving light charged particles including α particles. Additionally, the increase in the effective areal density of the target yields a higher luminosity that can compensate for low beam intensities.

Thick-target experiments using inverse kinematics and light gas targets for resonant scattering experiments have shown to be useful. Target chambers were filled with He gas that serves as the target and Si detectors are used to detect the scattered charged particles. Some examples of measurements using this technique with unstable beams include ^{10}Be[105] and ^{18}Ne.[58,106]

The next-generation of active-target detectors are currently being developed or are starting their initial scientific campaigns. An overview of active-target detectors including some of the more recent projects can be found in Mittig *et al.* and Beceiro *et al.*[107,108]

Experiments with the active-target detector Multi-Sampling and Tracking Proportional Chamber (MSTPC) focused on the ^{8}Li(α,n) reaction, a reaction that has been discussed in the context of inhomogeneous Big Bang environments[109] but is also important as a trigger for light r-process nucleosynthesis in core-collapse supernova.[110] This reaction might be facilitated through cluster structure in the ^{12}Be compound nucleus as indicated by early experiments.[111] The MSTPC detector had auxiliary neutron detectors surrounding its gas target, which allowed it to measure the exclusive cross section. Although the measured cross sections in Mizoi *et al.* were later found to lower than other measurements,[112] which was attributed to systematics in calibrating neutron detector efficiency, the MSTPC active-target allowed for measurements to be performed with modest secondary beam rates of 1000-2000 pps.

Measurements with a He filled Multi-Sampling Ionization Chamber (MUSIC)[113] were performed using radioactive beams to measure cross sections relevant for carbon fusion with several carbon isotopes, which are important for understanding X-ray bursts.[114] The experiment shows the utility of the active-target technique even with a relatively low number of electronic channels. Another example of the use of radioactive beams to measure fusion cross sections was ^{10}Be + ^{40}Ar using the Prototype Active-Target Time Projection Chamber.[115] Both total fusion and fusion with charged-particle emission were measured with a ^{10}Be beam rate of about 100 pps. Although statistics were limited, this measurement shows that inclusive fusion cross sections can be precisely measured using low beam rates.

5. Nuclear cluster in α-conjugate nuclei and their role in light element nucleosynthesis patterns

As already summarized in the introduction, there is a close correlation between the stellar burning phases and the nuclear structure of the nuclei involved in the associated nucleosynthesis processes. This correlation manifests itself in many of the observational signatures of nucleosynthesis processes. The observed abundance distribution of the elements is directly correlated with the binding energies of the associated nuclei. This is clearly demonstrated in Fig. 1, which shows the high abundances of the light α-conjugate nuclei that are characterized by their large binding energy. Another notable example is the abundance distribution of isotopes above iron that are produced by either the slow neutron capture (s-process) or the rapid neutron capture (r-process) driven nucleosynthesis events that are responsible for the build-up of all heavy elements in our universe. For both processes the observed abundance distribution shows distinct features of elemental enrichment that are directly explained by the shell structure of the nucleus. While there are many more examples on the close correlation between nuclear structure and astrophysics phenomena, we will focus in the following sections on the specific case of cluster structure in nuclei and where it appears to influence the reaction pattern and reaction flow in the nucleosynthesis of light elements in quiescent and explosive stellar environments. A number of examples will be discussed in which such correlation has been observed. The following section will then focus on the role of nuclear clusters in the production of heavy elements in quiescent and explosive stellar burning events.

5.1. *Cyclic burning patterns in stellar hydrogen burning*

Stellar burning cycles are a well known phenomenon and provide a catalytic process for energy production in stars during the hydrogen burning phase. Cyclic burning is based on a sequence of four proton capture reactions interspaced by two β decay processes being initiated on an α conjugate nucleus. By this reaction sequence a new and heavier α-conjugate nucleus is formed in an excited α cluster configuration. This decays under emission of an α particle back to its initial configuration. The efficient operation of the cycle depends on the formation and function of an α cluster in its course.

The $T = 0$ α-conjugate nuclei such as ^{4}He, ^{12}C, ^{16}O, ^{20}Ne, etc. have a particularly strong binding energy, which is featured in the large abundances these isotopes show in the low mass range of the abundance distribution curve. The cluster structure is the reason for the emergence of the catalytic burning processes that characterize many of the cyclic reaction sequences in stellar and explosive hydrogen burning, such as the CNO cycles that generate the energy in massive main-sequence stars, the hot CNO cycles that determine the time structure of nova explosions,[116] the NeNa and MgAl cycles that characterize hydrogen shell burning during late star evolution as well as the thermonuclear runaway in Ne-novae, and finally the Sn-Te cycles that have been identified as an end-point of the rp-process.[117] Such cyclic processing has already been envisioned by v. Weizsäcker in 1936[118] where he suggested the α cluster as a possible catalyzing element for nucleosynthesis processes. In a subsequent paper[118] he modified the initial idea by postulating a carbon cycle, known today as the CNO cycle. During the same year Hans Bethe provided the mathematical formalism for quantifying these kinds of cyclic burning mechanisms.[119]

The reaction flow in the classical CNO cycle ^{12}C(p,γ)^{13}N($\beta^+\nu$)^{13}C (p, γ)^{14}N followed by the ^{14}N(p,γ)^{15}O($\beta^+\nu$)^{15}N(p,α)^{12}C sequence that closes the cycle depends sensitively on the α cluster configuration in the ^{16}O compound nucleus of the last reaction. The widths of these resonances correspond to the α width and the α-cluster structure of the states. The strong α cluster resonances in the 12.5 MeV range (above 300 keV center-of-mass energy for protons) exhibit a strong resonance strength $\omega\gamma_{(p,\alpha)}$ in the α channel compared to the γ channel $\omega\gamma_{(p,\gamma)}$

$$\frac{\omega\gamma_{(p,\alpha)}}{\omega\gamma_{(p,\gamma)}} = \frac{\Gamma_\alpha}{\Gamma_\gamma}. \tag{15}$$

This ensures that the (p,α) reaction channel dominates over the competing (p,γ) branch by more than three orders of magnitude.[120] The resonances can be seen in the S-factor for the ^{15}N(p,α_0)^{12}C reaction shown in Fig. 6. For this reason most of the CNO material remains within the first cycle. Likewise this is the case for the ^{20}Ne nucleus, that has a well known

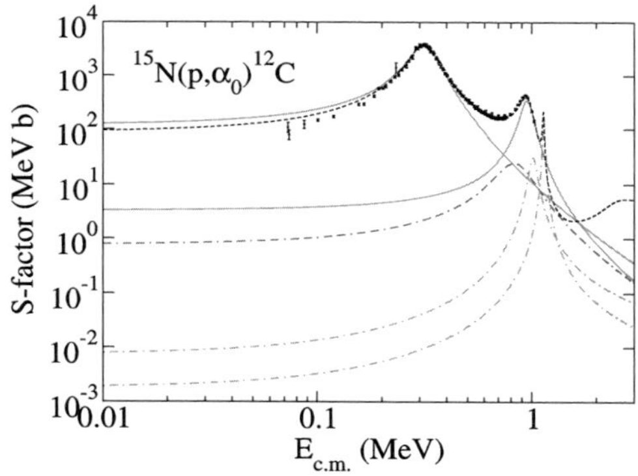

Fig. 6. The S-factor curve for the ^{15}N(p,α_0)^{12}C reaction is shown. The experimental data, given by the black points indicate two broad interfering natural parity resonance structures. The width of these resonances correspond to α width and indicate the α-cluster structure of the states. This pronounced α strengths allows the (p,α) reaction channel to dominate over the competing (p,γ) reaction branch. A full R-matrix fit is shown by the dashed black line, the resonance components of the 1^- levels by the solid red, the 2^+ levels by the dot-dash-dashed blue line, and the 3^- levels by the dot-dashed green line. Reprinted figure with permission from R. J. deBoer *et al.*, Phys. Rev. C 87, 015802 (2013). Copyright 2013 by the American Physical Society.

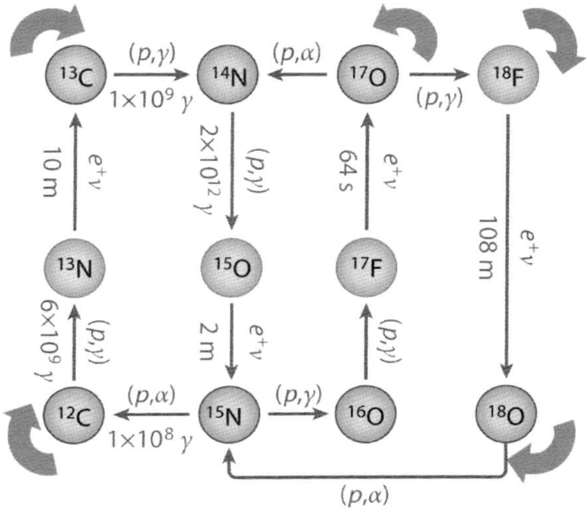

Fig. 7. The CNO cycle where the center row contains radioactive isotopes that contribute to the solar neutrino flux and in bottom row contains the nuclei that exhibit cluster configurations facilitating the closing (p,α) reactions with the exception of ^{15}N. A fourth cycle occurs via the ^{18}O(p,γ)^{19}F reaction but is closed via the strong ^{19}F(p,α)^{16}O process with the α-conjugate compound nucleus ^{20}Ne.

^{16}O+α cluster structure.[121] This ensures that the α decay channel of the ^{19}F(p,α)^{16}O dominates over the competing ^{19}F(p,γ)^{20}Ne radiative capture reaction prohibiting a possible CNO material leak via the latter reaction feeding the NeNa cycle[122] (Fig. 7).

Similar cycle structures characterize hydrogen burning at higher temperatures such as in hydrogen shell burning in massive stars and explosive hydrogen burning in Ne-novae;[123] these are known as the NeNa cycle, the MgAl cycle, and possibly the SiP cycle, that emerge as nucleosynthesis patterns at certain temperature and density conditions in hot hydrogen burning environments.[124] This depends on the α cluster configurations of the T=0 nuclei ^{24}Mg, ^{28}Si, and ^{32}S, respectively. The existence of unbound α cluster configurations in these nuclei would facilitate a cyclic burning pattern through (p,α) resonances. In the case of resonances with a predominantly single-particle structure a radiative capture driven reaction flow would continue towards higher masses. However at least up to ^{32}S the reaction rates indicate a cyclic burning pattern that may in principle continue towards a higher mass range due to the cluster structure in nuclei such as

^{36}Ar, ^{40}Ca, and ^{44}Ti. At the high temperatures necessary for cyclic burning processes in the mass range above Z = 16, the cycles are bypassed by the rapid-proton capture or rp-process that shifts the nucleosynthesis path off the line of stability towards more neutron deficient nuclei, bypassing the cyclic burning pattern.[124]

However, the cyclic pattern re-emerges in the higher mass range where the N = Z line runs close to the drip line between ^{60}Zn and ^{100}Sn. In high temperature conditions as expected for the thermonuclear runaway on accreting neutron stars[11] the rp-process path runs along the drip line and the $T = 0$ isotopes ^{60}Zn, ^{64}Ge, ^{68}Se, ^{72}Kr up to ^{100}Sn emerge as waiting points. Yet, the cyclic pattern does not develop because near the drip line the proton threshold is lower than the alpha threshold preventing the emission of α particles. Only above Z = 50 in the Sb Te range a sharp drop in α binding energy opens the α decay channel again and a cyclic reaction pattern reemerges as the SbTe cycle.[117] Yet, based on present predictions ^{103}Sb is unbound and ^{104}Te cannot be populated lightly. The cycle therefore shifts to the α unbound ^{107}Te nucleus via the reaction sequence ^{105}Sn(p,γ)^{106}Sb(p,γ)^{107}Te(γ,α)^{103}Sn, releasing one ^{4}He particle for refueling the triple-α-process. However, more studies are necessary to determine the masses of the associated nuclei and explore the specifics of the α cluster structure to make reliable predictions on its consequences.

5.2. *Cluster in stellar helium burning*

The impact of cluster structure becomes more obvious in stellar helium burning, which is facilitated through resonant and nonresonant processes whose strength is directly correlated with the α cluster configuration of the compound nucleus. The triple-alpha process is often described as the ultimate example for the role of cluster configurations in the build-up of heavy elements. The triple-alpha reaction initiates stellar helium burning though a three α particle fusion mechanism. The fusion of two ^{4}He nuclei forms the weakly bound ^{8}Be nucleus, which can be described as a two α cluster. A third α particle is captured on the short-lived ^{8}Be forming a three α cluster resonance state in ^{12}C, the so-called Hoyle state at 7.6 MeV excitation energy. Fred Hoyle had argued in 1954[125] that without such a state stellar helium burning and therefore the evolution of carbon-oxygen based life would not be possible. The level was observed experimentally shortly afterwards but only last year a calculation of the internal structure of the Hoyle state using an modified effective field theory approach clearly demonstrated the α cluster nature of this state[77] through which the production of heavier elements is facilitated.

Also the next step, the subsequent capture of an α particle on the ground state of ^{12}C, the $^{12}C(\alpha,\gamma)^{16}O$ reaction depends sensitively on the α cluster configuration of the ^{16}O compound nucleus. The process is facilitated by two reaction components: resonant contributions from α unbound states as well as states below the α threshold in ^{16}O and direct α capture reactions to bound states in ^{16}O. The strength of both components depend directly on the α cluster structure of the associated states. In particular for the ground state the nucleons are dominantly arranged in a tetrahedral configuration of alpha clusters.[30]

The cluster configuration in the higher excitation range of the ^{16}O nucleus was recently studied using a cluster-nucleon configuration interaction model with advanced realistic shell-model Hamiltonians in order to investigate the strength of the clustering phenomena in the harmonic oscillator basis.[126] This study provides a comprehensive description of the cluster structure of the ^{16}O nucleus up to a very high excitation range based on the ^{12}C ground state configuration. The study demonstrates the existence of pronounced cluster configurations in particular for the energy range near the threshold that are reflected in large spectroscopic factors for the natural parity resonance and subthreshold levels between 6.0 and 8.5 MeV.

For modeling the radiative capture cross section of $^{12}C(\alpha,\gamma)^{16}O$ a number of cluster models of varying levels of sophistication have been applied in the past to investigate the impact of these effects on the various transition components. This is in particular important for the E2 ground-state transition due to the fact that the sub-threshold 2^+ state at 6.92 MeV in ^{16}O represents a high degree of clustering.[127] Also the E2 direct capture to the ground state in ^{16}O is enhanced due to its pronounced cluster configuration. In the E1 component in particular the cluster contribution of the 7.12 MeV 1^- subthreshold state and its interference with the broad 9.59 MeV 1^- cluster resonance state does add to the reaction strength. These cluster configurations clearly determine the stellar reaction rate of $^{12}C(\alpha,\gamma)^{16}O$ and define the $^{12}C/^{16}O$ ratio in our universe.[128]

5.3. *Carbon cluster and stellar carbon burning*

While the role of α cluster configurations in stellar helium burning is well established, there is still a rather large uncertainty associated with the possibility of ^{12}C clusters in heavier nuclei. For a long time there have been discussions about the nature and role of resonance structures that characterize the low energy cross section of the ^{12}C+^{12}C fusion process. These structures have been identified as molecular ^{12}C+^{12}C configurations in the ^{24}Mg compound nucleus.[50] However, the reaction rate is calculated on the basis of an average cross section integrating over the molecular resonance components.[129] There have been indications, however on the possible existence of pronounced low energy resonances at 2.1 MeV center of mass energy[130] that can only be explained by strong ^{12}C cluster configurations of the corresponding state in ^{24}Mg. More-so, there have been predictions based on phenomenological considerations of explosive stellar events such as superbursts that suggest a strong ^{12}C+^{12}C cluster resonance around 1.5 MeV center of mass energy in ^{24}Mg that would drastically enhance the energy production and may provide a direct nuclear driver for the superburst phenomenon.[46] There has been so far no direct observational evidence for such a state. First direct ^{12}C+^{12}C fusion measurements seemed to indicate such a resonance but the signals have subsequently been identified as beam induced background caused by inelastically scattered ^{12}C nuclei on deuterium impurities in the target material. The deuterons trigger the secondary ^{12}C(d,p) producing a background flux of neutrons. These protons have the same energy as protons that may emerge from the ^{12}C(^{12}C,p)^{23}Na reaction at 1.5 MeV center of mass energy.[131] However recent studies using the Trojan Horse technique confirmed the cluster level at 2.1 MeV and also suggest the existence of the predicted state at 1.5 MeV center of mass energy for the fusion reaction.[9] Such a discovery of a low energy ^{12}C–^{12}C cluster state would indeed have significant impact on the reaction rate and an experimental confirmation through direct fusion studies would be of utmost importance.

6. Nuclear clusters as facilitators for heavy element production

While theoretical cluster model studies for nuclear astrophysics have focused on the prediction of $T = 0$ even-even nuclei, a number of interesting astrophysical phenomena are associated with the cluster configuration of $T = 1$ even nuclei above the α threshold. There are two dominant phenomena. The first case is the role of α cluster resonances in $T_z = 1$ nuclei that facilitate the stellar neutron sources for the slow neutron capture or s-process. The s-process is characterized by a series of neutron capture reactions and β-decays on medium mass nuclei in the stellar material[132] by which heavy elements are being produced. The s-process is, next to the r-process, the dominant reaction chain for the production of heavy elements and depends sensitively on the neutron flux provided by (α,n) reactions on light $T_z = +1$ nuclei. In turn the flux depends sensitively on the existence of α-cluster resonances in the energy range of stellar burning. The possible occurrence of pronounced cluster configurations therefore plays an important role in determining the neutron flux and the subsequent production of heavy elements in s-process environments.

The second case is about the role of α cluster resonances in $T_z = -1$ nuclei that facilitate the ignition and timescale of thermonuclear runaway processes via the αp process in accreting binary star systems. These are observed as X-ray bursts but may also play a role for nova explosions. The αp process is a sequence of (α,p) reactions followed by radiative proton capture that feed on the highly enriched ^{14}O isotope in the hot CNO cycles.[3] By this reaction sequence CNO masses are rapidly transferred into the Ca Ti region. This process correlates to the risetime of X-ray bursts systems, but depends critically on the resonance structure of the associated compound nuclei.

6.1. *T_z=+1 cluster nuclei as stellar neutron sources*

The reaction ^{22}Ne(α,n) is considered the main neutron source for the weak s-process in massive stars.[6] It also contributes substantially to the neutron production in thermal helium pulsing of Asymptotic Giant Branch (AGB) stars which are the site for the main s-process.[133] While ^{22}Ne is a relatively rare stable isotope of neon, in stellar helium burning environments it is being produced by α capture on ^{14}N, $^{14}\mathrm{N}(\alpha,\gamma)^{18}\mathrm{F}(\beta\nu)^{18}\mathrm{O}(\alpha,\gamma)^{22}\mathrm{Ne}$. ^{14}N is the main product of CNO burning in massive stars and is converted rapidly to ^{22}Ne at the onset of stellar helium burning. This conversion is facilitated by a strong α cluster state in ^{22}Ne which dominates the reaction rate for $^{18}\mathrm{O}(\alpha,\gamma)^{22}\mathrm{Ne}$ as a single resonance populated at 662 keV α energy.[134,135] There are also possible contributions from lower energy resonances. α transfer studies indicate the existence of two resonance levels at even lower energies[136] which were confirmed in direct radiative capture measurements.[134,135]

Cluster states near the α threshold in ^{26}Mg may have a drastic impact. The ^{22}Ne(α,n)^{25}Mg neutron source depends very much on the resonance

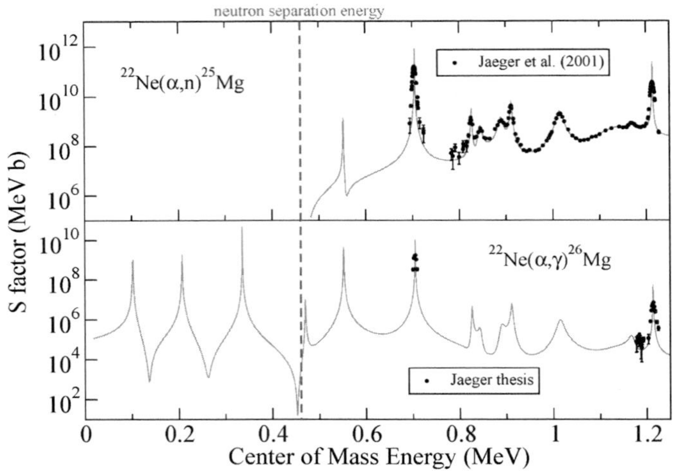

Fig. 8. The figure shows the S-factor excitation curves for the two competing reactions ^{22}Ne(α,n)^{25}Mg and ^{22}Ne(α,γ)^{26}Mg. Experimental data by Jäger *et al.*[137] as well as data from the thesis are shown.[138] The predictions at lower energies rely on R-matrix simulations based on ^{22}Ne(^{6}Li,d)^{26}Mg data shown in Fig. 9 and on ^{26}Mg(α,α) inelastic scattering data shown in Ref. 94. These predictions need to be verified experimentally.

at 702 keV center of mass energy that corresponds to a 1^- cluster state at 11.317 MeV in ^{26}Mg. This resonance has a comparable strength in the ^{22}Ne(α,γ)^{26}Mg reaction channel which underlines that this level only has a negligible single particle component.[94] The ^{22}Ne(α,n)^{25}Mg reaction, however, has a negative Q-value (−0.478 MeV) and hence, neutron production is only possible at higher temperatures that the stellar core reaches in its contraction phase towards the end of helium burning.[6] Prior to that the radiative capture reaction ^{22}Ne(α,γ)^{26}Mg gradually depletes the ^{22}Ne fuel depending on the strength of this reaction which in turn depends on the possibility of cluster states at lower energies. Indeed a recent study of ^{22}Ne(^{6}Li,d)^{26}Mg populating the excitation range of ^{26}Mg near and above the α threshold identified a number of natural parity levels with substantial strength that might correspond to additional cluster configurations above the alpha and below the neutron threshold.[94] Most notable is a 1^- state at 11.167 MeV which is about an order of magnitude stronger in the (α,γ) channel than in the competing (α,n) channel. This reduces the efficiency of the neutron source by roughly an order of magnitude.

Similar phenomena can be expected for the ^{26}Mg+α system, where α cluster states can be expected near the α threshold at 10.643 MeV in

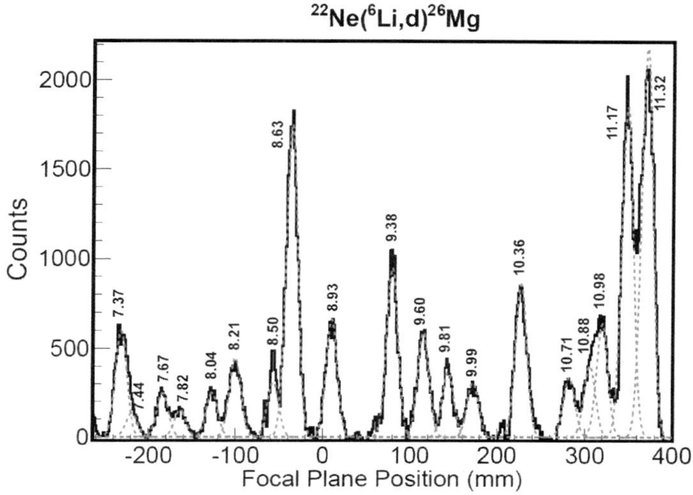

Fig. 9. Spectrum of the ^{22}Ne(^{6}Li,d)^{26}Mg reaction populating natural parity states in the ^{26}Mg nucleus below and above the alpha threshold at 10.615 MeV. The deuteron peaks are marked by the excitation energy of the final states in ^{26}Mg and several unbound states can be observed above the alpha threshold, which represent the resonances indicated in Fig. 8. Reprinted figure with permission from R. Talwar *et al.*, Phys. Rev. C 93, 055803 (2016). Copyright 2016 by the American Physical Society.

the ^{30}Si compound nucleus. There is only a limited amount of data available that could provide direct information on the existence of cluster resonances and their potential impact on the ^{26}Mg(α,n)^{29}Si and the competing ^{26}Mg(α,γ)^{30}Si reaction channels. Most of the published data on α induced reactions is limited to energies well above 2 MeV.[139,140] However, the α transfer reaction ^{26}Mg(^{6}Li,d)^{30}Si clearly populates a number of states near and above the α threshold indicating a pronounced cluster composition.[141] For some of the states α spectroscopic factors of up to 50% of the ground state strength were reported. However, to investigate the role of these states in helium burning scenarios requires more detailed study of the resonance strength in the ^{26}Mg(α,n)^{29}Si and the competing ^{26}Mg(α,γ)^{30}Si reaction channel.

6.2. *T_z=-1 cluster nuclei and the αp-process*

While cluster states in $T_z = +1$ nuclei can facilitate and control the stellar neutron production for the s-process, cluster configurations in the $T_z = -1$ isobars may facilitate thermonuclear runaway processes that occur in the

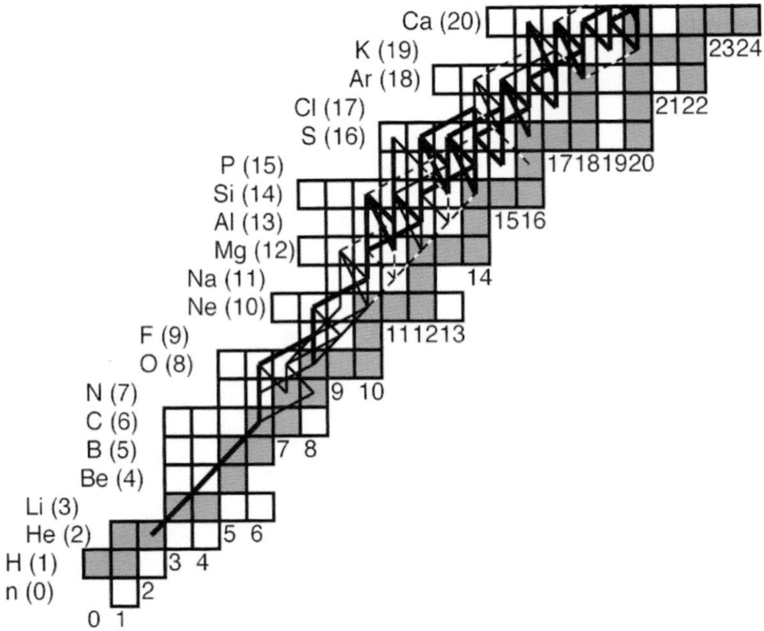

Fig. 10. The reaction path of the αp process driving the on-set of a thermonuclear runaway in the atmosphere of an accreting neutron star. The process is characterized by a sequence of (α, p) reactions on $T_z = -1$ nuclei followed by rapid proton capture converting CNO material towards heavier masses in the Ca Ti region.

outer atmosphere of accreting neutron stars. These phenomena are observed as X-ray bursts[10] that count as the most powerful thermonuclear explosions in our universe. X-ray bursts occur in close binary systems with a neutron star accreting hydrogen and helium rich material from the outer layers of his companion star. The material accumulates at highly electron-degenerate conditions in the atmosphere of the neutron star. Nuclear burning occurs primarily through the hot CNO cycles[122] enriching the material in radioactive ^{14}O and ^{15}O isotopes. The accreted layer is heated until a thermonuclear runaway occurs through the break-out from the first cycle via the ^{14}O(α,p)^{17}F and the ^{15}O(α, γ)^{19}Ne reactions triggering the αp-process. The strength of the out-flow depends on the resonance structure of the respective reactions, in particular in the first case on the α clustering of the ^{18}Ne compound nucleus.

Only limited information is available about the low energy cross sections of the ^{14}O(α,p)^{17}F reaction and is typically restrained to the study of the time-inverse reaction, ^{17}F(p,α)^{14}O[142] and the inelastic scattering process ^{17}F(p,p')^{17}F*.[143] No direct investigations of α transfer or capture reactions have been undertaken at low energies, due to the need for intense radioactive ^{14}O beams.However inelastic α scattering experiments in inverse kinematics at higher energies have been successful in demonstrating the α cluster configuration of ^{18}Ne at higher excitation energies above 8.0 MeV.[144] An expansion of the investigated range towards lower energies by a combination of inelastic α scattering and α transfer reactions is desirable but presently not feasible given the required ^{14}O beam intensities.

It has been argued by Kurath and Towner[145] that there is a close correlation between α-transfer and two-nucleon transfer reactions at higher energies when direct reaction components dominate. It was demonstrated that the amplitudes can be written as a sum of coupled two-neutron and two-proton spectroscopic amplitudes, illustrating that many properties of α-transfer reactions can be understood in terms of our experience with two-nucleon transfer reactions. This effect has been experimentally confirmed in the comparison of the ^{24}Mg(^{6}Li,d)^{28}Si reaction with the two-proton transfer ^{26}Mg(^{3}He,n)^{28}Si reaction.[146]

Based on this correlation ^{16}O(^{3}He,n)^{18}Ne two proton transfer reaction was used to probe the energy range near the α threshold in ^{18}Ne at 5.125 MeV.[147] A number of pronounced states near and above this threshold have been found at 5.1 MeV, 6.15 MeV, 6.3 MeV and 7.06 MeV excitation energy. The strength of the transitions to these states suggest an appreciable α strength which is reflected in the predicted strengths for the corresponding ^{14}O(α,p)^{17}F resonances. First direct measurements of this reaction have been attempted,[148] but have not reached low enough energies to allow a direct comparison.

The subsequent reaction flow breaks out of the mass range of the CNO cycles via the ^{18}Ne(α,p)^{21}Na. This reaction triggers a sequence of (α,p) reactions intersequenced by radiative proton capture processes along the $T_z = -1$ line, ^{22}Mg(α,p)^{25}Al, ^{26}Si(α,p)^{29}P, ^{30}S(α,p)^{33}Cl, ^{34}Ar(α,p)^{37}K, and ^{38}Ca(α,p)^{41}Sc, which is called the αp-process. This process rapidly converts the CNO material in the neutron star atmosphere into the Ca Ti mass range. The αp process has direct impact on the shape of the X-ray burst light curve[149] and the strength of these reactions provide critical information for the ignition and timescale of the event. Traditionally the αp process reaction rates have been calculated using statistical model theory.[11] However, a number of recent studies, primarily based on two-particle transfer reaction techniques suggest that the level density of natural parity states in the compound nuclei is substantially lower than required for a statistical treatment.[12] However, all studies suggest, based on the assumption of the correlation between two particle transfer and α particle transfer the existence of α cluster states in the respective compound nuclei.

The level structure in ^{22}Mg, the compound nucleus of the ^{18}Ne(α,p)^{21}Na reaction was studies by ^{24}Mg(p,t)^{22}Mg[150] that showed a number of pronounced states in the triton spectrum populating states in ^{22}Mg above the α threshold at 8.14 MeV. Several pronounced states are observed in the energy range above the threshold that correspond roughly to the astrophysical range of interest. Most notably are states at 8.18, 8.39, 8.52, 8.66, 8.78 as well as 8.93 MeV. Comparison with the levels in the mirror nucleus ^{22}Ne indicate that several of these states have an appreciable α strength. For a number of these levels the width is not known and additional work was done by analyzing the impact of these states in the framework of a Monte Carlo approach[151] adopting a Porter-Thomas distribution for the α strength leading to a mean value for the reduced α width Θ_α=0.03$\pm$0.01.[152] The uncertainty range based on these statistical assumptions ignores the possibility of a more pronounced α cluster configuration for these states. Direct measurements of these by α transfer reactions are necessary to address this systematic uncertainty factor not included in the Monte Carlo analysis.

Similar (p,t) two-neutron transfer studies have been performed for searching for possible resonances in (α,p) reactions along the αp process path. This includes the studies of the α-unbound states in the compound nuclei ^{26}Si,[153] ^{30}S,[154] and more recently for ^{38}Ca.[12] In all cases the (p,t) spectra are characterized by a limited number of α unbound states, candidates for α cluster configurations. The resonance strengths are directly proportional to the α strengths of these levels that are typically not known and are subject to assumptions. Adopting a low strength for the observed natural parity states as suggested by the assumption of a statistical distribution[152] suggests a reaction rate that is typically two order of magnitude lower than rates that are based on the assumption of a high level density implicit to the Hauser-Feshbach predictions. After identifying possible resonances in these reactions, the next step should be to provide realistic cluster model predictions for a strength distribution on the theoretical side and develop experimental techniques with radioactive beams for the experimental verification. In the mean time the rates will maintain an

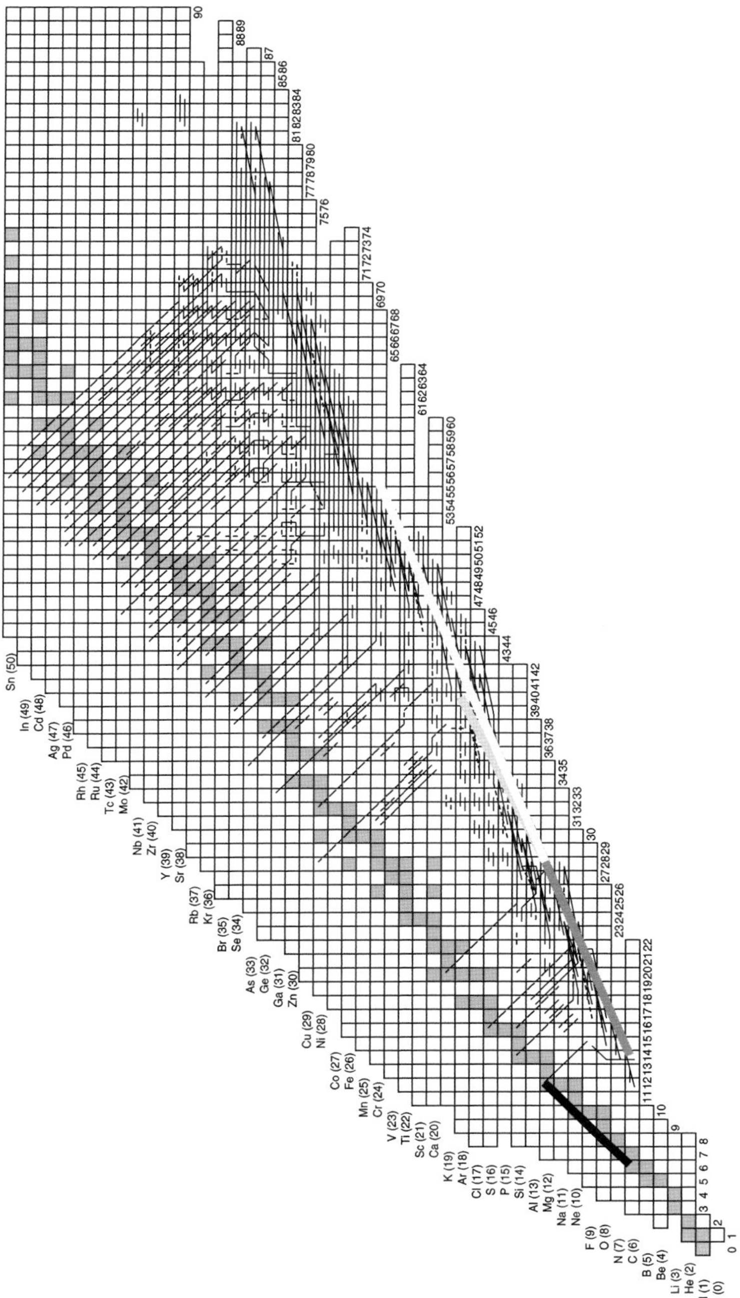

Fig. 11. The fat black lines indicate pycnonuclear fusion of neutron rich carbon, oxygen, neon, and magnesium isotopes releasing energy in the deep crust of an accreting neutron star. These nuclei are produced by electron capture and electron capture induced neutron emission from the ashes of the X-ray burst in the upper layers of the neutron star.

uncertainty spanning about two orders of magnitude due to the systematic uncertainty associated with the assumptions for the cluster configurations and α strength distributions in $T_z = -1$ even-even isotopes.

7. Conclusion

The cluster structure in nuclei has great impact on a number of stellar burning scenarios. In particular cluster states near the threshold may dominate the reaction flow and determine the reaction path in helium as well as in oxygen and carbon burning as outlined in this paper. Because of the difficulties associated with a direct experimental study of cluster states indirect methods probing the compound nucleus are becoming increasingly important to provide the necessary information for determining and calculating the various cluster contributions to the reaction rate.

In this paper we have primarily focused on the better known cases where clusters play an important role. In nuclear astrophysics the most obvious cases are the triple-α-process and the subsequent $^{12}C(\alpha,\gamma)^{16}O$ reaction but there are the lesser known cases where the clusters in nuclei play a role in multiple other environments, maintaining the cyclic structure of nucleosynthesis patterns in hydrogen burning, governing the production of neutrons for the s-process, and facilitating the energy release in carbon and oxygen burning through larger cluster configurations.

There are, however, other yet not recognized cases where the α cluster structure in light nuclei or even heavier nuclear cluster configurations may play an important role. A prime example is the "light" r-process that is expected to facilitate the production of medium mass nuclei (between Fe and Ag) in neutrino-driven winds following core-collapse supernova explosions. In these events, (α,n) reactions, are key to move matter towards the region of higher proton number, while producing a high flux of neutrons.[155] Other reactions of relevance are (α,p) or (α,γ) reactions that are associated with the νp-process in the neutrino driven wind.[156] The reaction rates used to model the reaction path and determine the predicted abundance distributions rely entirely on statistical model calculations.[157] These carry large uncertainties, but in particular for light nuclei the Hauser-Feshbach formula typically applied for these cases might not be applicable. In these or at least in some of these cases the reaction rates might be determined by single resonance cluster configurations and a closer look into these aspects seems advisable.

A further even more exotic example is the case of pycnonuclear reactions in the core of white dwarfs or the crust of neutron stars, where fusion reactions at extreme density conditions lead to sudden or continuous energy release given the environmental conditions. This is related to the ignition of type Ia supernovae explosions[158,159] as well as to the ignition of superbursts in the crust of accreting neutron stars.[13,46] The ignition depends on the $^{12}C+^{12}C$ fusion rate which might be characterized by the ^{12}C-^{12}C molecular structures in the ^{24}Mg compound nucleus as outlined above. However, pycnonuclear fusion may take place in even deeper layers of the neutron star where very neutron-rich even-even isotopes of carbon, neon, and magnesium are frozen out in a lattice structure.[8,160] Pycnonuclear fusion between these nuclei may serve as an internal nuclear energy source that prevents cooling of X-ray bursts transients.[161] The fusion mechanism at such extreme densities of 10^{12} to 10^{14} g/cm^3 is completely unknown and presently only approximated in the framework of a simple potential model approach.[129,162] First tests using radioactive carbon beams indicated good agreement between the predicted fusion cross sections and the experimentally studied ones.[114] However, these experiments were performed at relatively high energies, while pycnonuclear reactions between a lattice of nuclei take place at near zero energy. Since even-even nuclei appear to emerge as the most abundant ones in the pycnonuclear burning pattern, it is well possible that the actual reaction rates can be enhanced by molecular configurations in the respective fusion compound systems. This is a question which requires a better understanding and treatment of the fusion mechanism at extreme density and temperature conditions.

Acknowledgements

This research was supported by the National Science Foundation through Grant No. Phys-0758100, and the Joint Institute for Nuclear Astrophysics through Grant No. Phys-0822648 and PHY-1430152 (JINA Center for the Evolution of the Elements). The authors thank R. J. DeBoer, M. Beard from the University of Notre Dame, G. Rogachev from Texas A & M University, and M. Kimura from Hokkaido University for providing some of the figures used in this paper.

References

1. L. R. Hafstad and E. Teller, *Phys. Rev.* **54**, 681 (Nov 1938).

2. E. G. Adelberger, A. García, R. G. H. Robertson, K. A. Snover, A. B. Balantekin, K. Heeger, M. J. Ramsey-Musolf, D. Bemmerer, A. Junghans, C. A. Bertulani, J.-W. Chen, H. Costantini, P. Prati, M. Couder, E. Uberseder, M. Wiescher, R. Cyburt, B. Davids, S. J. Freedman, M. Gai, D. Gazit, L. Gialanella, G. Imbriani, U. Greife, M. Hass, W. C. Haxton, T. Itahashi, K. Kubodera, K. Langanke, D. Leitner, M. Leitner, P. Vetter, L. Winslow, L. E. Marcucci, T. Motobayashi, A. Mukhamedzhanov, R. E. Tribble, K. M. Nollett, F. M. Nunes, T.-S. Park, P. D. Parker, R. Schiavilla, E. C. Simpson, C. Spitaleri, F. Strieder, H.-P. Trautvetter, K. Suemmerer and S. Typel, *Rev. Mod. Phys.* **83**, 195 (Apr 2011).
3. M. Wiescher, J. Görres, E. Uberseder, G. Imbriani and M. Pignatari, *Annu. Rev. Nucl. Part. Sci.* **60**, 381 (Nov 2010).
4. M. Freer and H. Fynbo, *Prog. Part. Nucl. Phys.* **78**, 1 (Sep 2014).
5. W. A. Fowler, *Science* **226**, 922 (Nov 1984).
6. F. Kaeppeler, M. Wiescher, U. Giesen, J. Goerres, I. Baraffe, M. El Eid, C. M. Raiteri, M. Busso, R. Gallino, M. Limongi and A. Chieffi, *Astrophys. J.* **437**, 396 (Dec 1994).
7. D. A. Bromley, J. A. Kuehner and E. Almqvist, *Phys. Rev. Lett.* **4**, 365 (Apr 1960).
8. D. G. Yakovlev, M. Beard, L. R. Gasques and M. Wiescher, *Phys. Rev. C* **82**, 044609 (Oct 2010).
9. A. Tumino, C. Spitaleri, S. Cherubini, L. Guardo, M. Gulino, I. Indelicato, M. La Cognata, L. Lamia, R. Pizzone, S. Puglia, G. Rapisarda, S. Romano, M. Sergi and R. Spartá, *EPJ Web Conf.* **117**, 09004 (May 2016).
10. R. E. Taam, S. E. Woosley, T. A. Weaver and D. Q. Lamb, *Astrophys. J.* **413**, 324 (Aug 1993).
11. H. Schatz, A. Aprahamian, J. Görres, M. Wiescher, T. Rauscher, J. Rembges, F.-K. Thielemann, B. Pfeiffer, P. Möller, K.-L. Kratz, H. Herndl, B. Brown and H. Rebel, *Phys. Rep.* **294**, 167 (Feb 1998).
12. A. Long, T. Adachi, M. Beard, G. P. A. Berg, J. Carter, M. Couder, R. W. Fearick, S. V. Fortsch, J. Goerres, J. Mira, S. H. T. Murray, R. Neveling, J. A. Swartz, R. Talwar, I. T. Usman, M. Wiescher and J. J. Van Zyl, *submitted.*
13. E. F. Brown, *Astrophys. J.* **614**, L57 (Oct 2004).
14. Y. Funaki, A. Tohsaki, H. Horiuchi, P. Schuck and G. Röpke, *Eur. Phys. J. A* **28**, 259 (Jul 2006).
15. L. B. Weinstein, E. Piasetzky, D. W. Higinbotham, J. Gomez, O. Hen and R. Shneor, *Phys. Rev. Lett.* **106**, 052301 (Feb 2011).
16. N. Fomin, J. Arrington, R. Asaturyan, F. Benmokhtar, W. Boeglin, P. Bosted, A. Bruell, M. H. S. Bukhari, M. E. Christy, E. Chudakov, B. Clasie, S. H. Connell, M. M. Dalton, A. Daniel, D. B. Day, D. Dutta, R. Ent, L. El Fassi, H. Fenker, B. W. Filippone, K. Garrow, D. Gaskell, C. Hill, R. J. Holt, T. Horn, M. K. Jones, J. Jourdan, N. Kalantarians, C. E. Keppel, D. Kiselev, M. Kotulla, R. Lindgren, A. F. Lung, S. Malace, P. Markowitz, P. McKee, D. G. Meekins, H. Mkrtchyan, T. Navasardyan, G. Niculescu, A. K.

Opper, C. Perdrisat, D. H. Potterveld, V. Punjabi, X. Qian, P. E. Reimer, J. Roche, V. M. Rodriguez, O. Rondon, E. Schulte, J. Seely, E. Segbefia, K. Slifer, G. R. Smith, P. Solvignon, V. Tadevosyan, S. Tajima, L. Tang, G. Testa, R. Trojer, V. Tvaskis, W. F. Vulcan, C. Wasko, F. R. Wesselmann, S. A. Wood, J. Wright and X. Zheng, *Phys. Rev. Lett.* **108**, 092502 (Mar 2012).

17. J. Seely, a. Daniel, D. Gaskell, J. Arrington, N. Fomin, P. Solvignon, R. Asaturyan, F. Benmokhtar, W. Boeglin, B. Boillat, P. Bosted, a. Bruell, M. H. S. Bukhari, M. E. Christy, B. Clasie, S. Connell, M. M. Dalton, D. Day, J. Dunne, D. Dutta, L. El Fassi, R. Ent, H. Fenker, B. W. Filippone, H. Gao, C. Hill, R. J. Holt, T. Horn, E. Hungerford, M. K. Jones, J. Jourdan, N. Kalantarians, C. E. Keppel, D. Kiselev, M. Kotulla, C. Lee, a. F. Lung, S. Malace, D. G. Meekins, T. Mertens, H. Mkrtchyan, T. Navasardyan, G. Niculescu, I. Niculescu, H. Nomura, Y. Okayasu, a. K. Opper, C. Perdrisat, D. H. Potterveld, V. Punjabi, X. Qian, P. E. Reimer, J. Roche, V. M. Rodriguez, O. Rondon, E. Schulte, E. Segbefia, K. Slifer, G. R. Smith, V. Tadevosyan, S. Tajima, L. Tang, G. Testa, R. Trojer, V. Tvaskis, W. F. Vulcan, F. R. Wesselmann, S. a. Wood, J. Wright, L. Yuan and X. Zheng, *Phys. Rev. Lett.* **103**, 202301 (Nov 2009).
18. W. Von Oertzen, M. Freer and Y. Kanada Enyo, *Phys. Rep.* **432**, 43 (Sep 2006).
19. M. Freer, *Reports Prog. Phys.* **70**, 2149 (Dec 2007).
20. H. Horiuchi, K. Ikeda and K. Kat, *Prog. Theor. Phys. Suppl.* **192**, 1 (Aug 2012).
21. Y. Funaki, H. Horiuchi and A. Tohsaki, *Prog. Part. Nucl. Phys.* **82**, 78 (May 2015).
22. D. Brink and J. Castro, *Nucl. Phys. A* **216**, 109 (Nov 1973).
23. J. Okołowicz, M. Płoszajczak and W. Nazarewicz, *Prog. Theor. Phys. Suppl.* **196**, 230 (Oct 2012).
24. T. Fukui, Y. Taniguchi, T. Suhara, Y. Kanada-En'yo and K. Ogata, *Phys. Rev. C* **93**, 034606 (Mar 2016).
25. J. W. Clark and T.-P. Wang, *Ann. Phys. (N. Y).* **40**, 127 (Oct 1966).
26. G. Mueller and J. Clark, *Nucl. Phys. A* **155**, 561 (Oct 1970).
27. H. E. Suess and H. C. Urey, *Rev. Mod. Phys.* **28**, 53 (Jan 1956).
28. K. Ikeda, T. Marumori, R. Tamagaki and H. Tanaka, *Prog. Theor. Phys. Suppl.* **52**, 1 (Aug 1972).
29. R. B. Wiringa, S. C. Pieper, J. Carlson and V. R. Pandharipande, *Phys. Rev. C* **62**, 014001 (Jun 2000).
30. E. Epelbaum, H. Krebs, T. A. Lähde, D. Lee, U.-G. Meissner and G. Rupak, *Phys. Rev. Lett.* **112**, 102501 (Mar 2014).
31. H. Horiuchi, *J. Phys. Conf. Ser.* **436**, 012003 (Apr 2013).
32. H. A. Bethe and R. F. Bacher, *Rev. Mod. Phys.* **8**, 82 (Apr 1936).
33. T. Lönnroth, M. Norrby, V. Z. Goldberg, G. V. Rogachev, M. S. Golovkov, K. M. Källman, M. Lattuada, S. V. Perov, S. Romano, B. B. Skorodumov,

G. P. Tiourin, W. H. Trzaska, A. Tumino and A. N. Vorontsov, *Eur. Phys. J. A* **46**, 5 (Oct 2010).
34. M. Itoh, S. Kishi, H. Sakaguchi, H. Akimune, M. Fujiwara, U. Garg, K. Hara, H. Hashimoto, J. Hoffman, T. Kawabata, K. Kawase, T. Murakami, K. Nakanishi, B. Nayak, S. Terashima, M. Uchida, Y. Yasuda and M. Yosoi, *Phys. Rev. C* **88**, 064313 (Dec 2013).
35. Y. Yoshida, Y. Kanada-En'yo and F. Kobayashi, *Prog. Theor. Exp. Phys.* **2016**, 043D01 (Apr 2016).
36. K. Itonaga, *Prog. Theor. Phys.* **66**, 2103 (Dec 1981).
37. S. Ohkubo, Y. Hirabayashi and T. Sakuda, *Phys. Rev. C* **57**, 2760 (May 1998).
38. G. Magkotsios, F. X. Timmes and M. Wiescher, *Astrophys. J.* **741**, (Nov 2011) 78.
39. H. Akimune, M. Fujiwara, J. Gibelin, M. N. Harakeh, L. Achouri, S. Bagchi, B. Bastin, K. Boretzky, H. Bouzomita, L. Caceres, S. Damoy, F. Delaunay, B. Fernández-Domínguez, M. Caamano, U. Garg, G. F. Grinyer, N. Kalantar-Nayestanaki, O. Kamalou, E. Khan, a. Krasznahorkay, G. Lhoutellier, S. Lukyanov, K. Mazurek, M. Najafi, J. Pancin, Y. Penionzkhevich, L. Perrot, R. Raabe, C. E. Rigollet, T. Roger, S. Sambi, H. Savajols, M. Senoville, C. Stodel, L. Suen, J. C. Thomas, J. van de Walle and M. Vandebrouck, *J. Phys. Conf. Ser.* **436**, 012010 (Apr 2013).
40. G. Magkotsios, F. X. Timmes, A. L. Hungerford, C. L. Fryer, P. A. Young and M. Wiescher, *Astrophys. J. Suppl. Ser.* **191**, 66 (Nov 2010).
41. R. D. Hoffman, S. A. Sheets, J. T. Burke, N. D. Scielzo, T. Rauscher, E. B. Norman, S. Tumey, T. A. Brown, P. G. Grant, A. M. Hurst, L. Phair, M. A. Stoyer, T. Wooddy, J. L. Fisker and D. Bleuel, *Astrophys. J.* **715**, 1383 (Jun 2010).
42. A. Astier, P. Petkov, M.-G. Porquet, D. S. Delion and P. Schuck, *Phys. Rev. Lett.* **104**, 042701 (Jan 2010).
43. W. von Oertzen, *Eur. Phys. J. A* **11**, 403 (Oct 2001).
44. S. Ohkubo and K. Yamashita, *Phys. Lett. B* **578**, (2004) 304.
45. M. Pignatari, R. Hirschi, M. Wiescher, R. Gallino, M. Bennett, M. Beard, C. Fryer, F. Herwig, G. Rockefeller and F. X. Timmes, *Astrophys. J.* **762**, 31 (Jan 2013).
46. R. L. Cooper, A. W. Steiner and E. F. Brown, *Astrophys. J.* **702** (Sep 2009) 660.
47. M. Kimura and H. Horiuchi, *Phys. Rev. C* **69**, 051304 (May 2004).
48. A. Diaz-Torres, L. Gasques and M. Wiescher, *Phys. Lett. B* **652**, 255 (2007).
49. W. D. Arnett, *Astrophys. J.* **173**, 393 (Apr 1972).
50. K. A. Erb, R. R. Betts, S. K. Korotky, M. M. Hindi, P. P. Tung, M. W. Sachs, S. J. Willett and D. A. Bromley, *Phys. Rev. C* **22**, 507 (Aug 1980).
51. H. Voit, W. Galster, W. Treu, H. Fröhlich and P. Dück, *Phys. Lett. B* **67**, 399 (Apr 1977).
52. N. Curtis, N. M. Clarke, B. R. Fulton, S. J. Hall, M. J. Leddy, A. S. J. Murphy, J. S. Pople, R. P. Ward, W. N. Catford, G. J. Gyapong, S. M.

Singer, S. P. G. Chappell, S. P. Fox, C. D. Jones, D. L. Watson, W. D. M. Rae and P. M. Simmons, *Phys. Rev. C* **51**, 1554 (Mar 1995).
53. W. von Oertzen, *Zeitschrift für Phys. A Hadron. Nucl.* **357**, 355 (Apr 1997).
54. N. Itagaki, S. Okabe, K. Ikeda and I. Tanihata, *Phys. Rev. C* **64** (May 2001) 014301.
55. M. Milin and W. von Oertzen, *Eur. Phys. J. A* **14**, 295 (Jul 2002).
56. N. Itagaki, W. von Oertzen and S. Okabe, *Phys. Rev. C* **74** (Dec 2006) 067304.
57. T. Suhara and Y. Kanada-En'yo, *Phys. Rev. C* **82**, 044301 (Oct 2010).
58. V. Z. Goldberg, G. V. Rogachev, W. H. Trzaska, J. J. Kolata, A. Andreyev, C. Angulo, M. J. G. Borge, S. Cherubini, G. Chubarian, G. Crowley, P. Van Duppen, M. Gorska, M. Gulino, M. Huyse, P. Jesinger, K.-M. Källman, M. Lattuada, T. Lönnroth, M. Mutterer, R. Raabe, S. Romano, M. V. Rozhkov, B. B. Skorodumov, C. Spitaleri, O. Tengblad and A. Tumino, *Phys. Rev. C* **69**, 024602 (Feb 2004).
59. N. Curtis, N. L. Achouri, N. I. Ashwood, H. G. Bohlen, W. N. Catford, N. M. Clarke, M. Freer, P. J. Haigh, B. Laurent, N. A. Orr, N. P. Patterson, N. Soić, J. S. Thomas and V. Ziman, *Phys. Rev. C* **77** (Feb 2008) 021301.
60. N. Curtis, N. L. Achouri, N. I. Ashwood, H. G. Bohlen, W. N. Catford, N. M. Clarke, M. Freer, P. J. Haigh, B. Laurent, N. A. Orr, N. P. Patterson, N. Soić, J. S. Thomas and V. Ziman, *Phys. Rev. C* **82** (Aug 2010) 029907.
61. M. Kimura, *Phys. Rev. C* **75**, 034312 (Mar 2007).
62. D. Brink, No Title, in *Proc. Int. Sch. Phys. "Enrico Fermi"*, ed. C. Bloch (New York, Academic Press, Varenna, 1966), p. 247.
63. W. Bauhoff, H. Schultheis and R. Schultheis, *Phys. Rev. C* **29** (Mar 1984) 1046.
64. A. Tohsaki, H. Horiuchi, P. Schuck and G. Röpke, *Phys. Rev. Lett.* **87**, 192501 (Nov 2001).
65. B. Zhou, Z. Ren, C. Xu, Y. Funaki, T. Yamada, A. Tohsaki, H. Horiuchi, P. Schuck and G. Röpke, *Phys. Rev. C* **86**, 014301 (Jul 2012).
66. S. Marsh and W. Rae, *Phys. Lett. B* **180**, 185 (Nov 1986).
67. P.-G. Reinhard, J. A. Maruhn, A. S. Umar and V. E. Oberacker, *Phys. Rev. C* **83**, 034312 (Mar 2011).
68. H. Flocard, P. H. Heenen, S. J. Krieger and M. S. Weiss, *Prog. Theor. Phys.* **72**, 1000 (Nov 1984).
69. J. Cugnon, H. Doubre and H. Flocard, *Nucl. Phys. A* **331**, 213 (Nov 1979).
70. Y. Iwata, T. Ichikawa, N. Itagaki, J. A. Maruhn and T. Otsuka, *Phys. Rev. C* **92**, 011303 (Jul 2015).
71. J.-P. Ebran, E. Khan, T. Nikšić and D. Vretenar, *Nature* **487** (Jul 2012) 341.
72. J.-P. Ebran, E. Khan, T. Nikšić and D. Vretenar, *Phys. Rev. C* **90**, 054329 (Nov 2014).
73. Y. Kanada-En'yo and H. Horiuchi, *Prog. Theor. Phys. Suppl.* **142**, 205 (2001).
74. H. Feldmeier and J. Schnack, *Rev. Mod. Phys.* **72**, 655 (Jul 2000).

75. R. Roth, T. Neff, H. Hergert and H. Feldmeier, *Nucl. Phys. A* **745**, 3 (Nov 2004).
76. R. Roth, J. Langhammer, A. Calci, S. Binder and P. Navrátil, *Phys. Rev. Lett.* **107**, 072501 (Aug 2011).
77. E. Epelbaum, H. Krebs, T. A. Lähde, D. Lee and U.-G. Meißner, *Phys. Rev. Lett.* **109**, 252501 (Dec 2012).
78. S. Quaglioni, C. Romero-Redondo and P. Navrátil, *Phys. Rev. C* **88** (Sep 2013) 034320.
79. A. Doté, H. Horiuchi and Y. Kanada-En'yo, *Phys. Rev. C* **56**, 1844 (Oct 1997).
80. T. Rauscher, F.-K. Thielemann, J. Görres and M. Wiescher, *Nucl. Phys. A* **675**, 695 (Aug 2000).
81. T. Yamada, Y. Funaki, H. Horiuchi, K. Ikeda and A. Tohsaki, *Prog. Theor. Phys.* **120**, 1139 (Dec 2008).
82. T. Kawabata, H. Akimune, H. Fujita, Y. Fujita, M. Fujiwara, K. Hara, K. Hatanaka, M. Itoh, Y. Kanada-En'yo, S. Kishi, K. Nakanishi, H. Sakaguchi, Y. Shimbara, A. Tamii, S. Terashima, M. Uchida, T. Wakasa, Y. Yasuda, H. Yoshida and M. Yosoi, *Phys. Lett. B* **646**, 6 (Mar 2007).
83. P. J. Leask, L. Achouri, J. C. Angélique, B. Benoit, C. Bremner, W. N. Catford, R. Chapman, S. P. G. Chappell, N. M. Clarke, E. d. G. Brennard, L. Donadille, M. Freer, B. R. Fulton, S. Grévy, F. Hanappe, K. L. Jones, J. L. Lecouey, R. Lemmon, M. Lewitowicz, X. Liang, F. M. Marqués, M. P. Nicoli, A. Ninane, N. A. Orr, M. Shawcross, S. M. Singer, M. Smith, K. Spohr and L. Stuttgé, *J. Phys. G Nucl. Part. Phys.* **27**, B9 (Aug 2001).
84. B. J. Greenhalgh, B. R. Fulton, D. L. Watson, N. M. Clarke, L. Donadille, M. Freer, P. J. Leask, W. N. Catford, K. L. Jones and D. Mahboub, *Phys. Rev. C* **66**, 027302 (Aug 2002).
85. H. Otsu, M. Kurata-Nishimura, Y. Togano, Y. Shimizu, H. Sato, T. Isobe, H. Baba, M. Sasano, S. Takeuchi, P. Doornenbal, N. Fukuda, K. Kusaka, J. Ohnishi, K. Yoneda, T. Motobayashi, J. Zenihiro, J. Lee, H. Liu, H. Sakurai, T. Kobayashi, K. Sekiguchi, N. Chiga, T. Sumikama, K. Muto, Y. Wada, Y. Kondo, R. Minakata, S. Ogoshi, T. Nakamura, Y. Satou, J. Hwang, Y. Matsuda, T. Murakami, N. Nakasuka, T. Kawabata, J. Gibelin, S. Leblond, N. Kobayashi, M. Niikura, M. Matsushita, E. Y. Nikolskii, S. Sakaguchi, D. Beaumel, A. Navin, V. Lapoux and M. Itoh, *RIKEN Accel. Prog. Rep.* **47** (2014).
86. V. M. Datar, S. Kumar, D. R. Chakrabarty, V. Nanal, E. T. Mirgule, A. Mitra and H. H. Oza, *Phys. Rev. Lett.* **94**, 122502 (Apr 2005).
87. V. M. Datar, D. R. Chakrabarty, S. Kumar, V. Nanal, S. Pastore, R. B. Wiringa, S. P. Behera, A. Chatterjee, D. Jenkins, C. J. Lister, E. T. Mirgule, A. Mitra, R. G. Pillay, K. Ramachandran, O. J. Roberts, P. C. Rout, A. Shrivastava and P. Sugathan, *Phys. Rev. Lett.* **111**, 062502 (Aug 2013).
88. F. Iachello, *Phys. Lett. B* **160**, 1 (Oct 1985).
89. B. Fulton, S. Bennett, M. Freer, J. Murgatroyd, G. Gyapong, N. Jarvis, C. Jones, D. Watson, J. Brown, W. Rae, A. Smith and J. Lilley, *Phys. Lett. B* **267**, 325 (Sep 1991).

90. A. M. Mukhamedzhanov and R. E. Tribble, *Phys. Rev. C* **59**, 3418 (Jun 1999).
91. W. A. Fowler, C. C. Lauritsen and T. Lauritsen, *Rev. Mod. Phys.* **20**, 236 (Jan 1948).
92. M. Wiescher, F. Käppeler and K. Langanke, *Annu. Rev. Astron. Astrophys.* **50**, 165 (Sep 2012).
93. F. Milder, J. Jänecke and F. Becchetti, *Nucl. Phys. A* **276**, 72 (Jan 1977).
94. R. Talwar, T. Adachi, G. P. A. Berg, L. Bin, S. Bisterzo, M. Couder, R. J. DeBoer, X. Fang, H. Fujita, Y. Fujita, J. Görres, K. Hatanaka, T. Itoh, T. Kadoya, A. Long, K. Miki, D. Patel, M. Pignatari, Y. Shimbara, A. Tamii, M. Wiescher, T. Yamamoto and M. Yosoi, *Phys. Rev. C* **93** (May 2016) 055803.
95. M. L. Avila, G. V. Rogachev, E. Koshchiy, L. T. Baby, J. Belarge, K. W. Kemper, A. N. Kuchera and D. Santiago-Gonzalez, *Phys. Rev. C* **90** (Oct 2014) 042801.
96. H. Costantini, R. J. DeBoer, R. E. Azuma, M. Couder, J. Görres, J. W. Hammer, P. J. LeBlanc, H. Y. Lee, S. O'Brien, A. Palumbo, E. C. Simpson, E. Stech, W. Tan, E. Uberseder and M. Wiescher, *Phys. Rev. C* **82** (Sep 2010) 035802.
97. G. Baur, *Phys. Lett. B* **178**, 135 (Oct 1986).
98. A. M. Mukhamedzhanov, L. D. Blokhintsev, B. a. Brown, V. Burjan, S. Cherubini, C. a. Gagliardi, B. F. Irgaziev, V. Kroha, F. M. Nunes, F. Pirlepesov, R. G. Pizzone, S. Romano, C. Spitaleri, X. D. Tang, L. Trache, R. E. Tribble and A. Tumino, *Eur. Phys. J. A* **27**, 205 (Mar 2006).
99. A. Tumino, C. Spitaleri, S. Cherubini, M. L. Cognata, L. Lamia, R. G. Pizzone, S. M. R. Puglia, G. G. Rapisarda, S. Romano and M. L. Sergi, *J. Phys. Conf. Ser.* **205**, 012048 (Jan 2010).
100. S. Typel and H. H. Wolter, *Few-Body Syst.* **29**, 75 (Nov 2000).
101. C. Spitaleri, S. Typel, R. Pizzone, M. Aliotta, S. Blagus, M. Bogovac, S. Cherubini, P. Figuera, M. Lattuada, M. Milin, D. Miljanić, a. Musumarra, M. Pellegriti, D. Rendić, C. Rolfs, S. Romano, N. Soić, a. Tumino, H. Wolter and M. Zadro, *Phys. Rev. C* **63** (Apr 2001) 055801.
102. C. Spitaleri, M. Aliotta, S. Cherubini, M. Lattuada, D. Miljanić, S. Romano, N. Soic, M. Zadro and R. A. Zappalà, *Phys. Rev. C* **60** (Oct 1999) 055802.
103. S. Kubono, *Prog. Theor. Phys.* **96**, 275 (Aug 1996).
104. B. B. Back, H. Esbensen, C. L. Jiang and K. E. Rehm, *Rev. Mod. Phys.* **86** (Mar 2014) 317.
105. M. Freer, J. D. Malcolm, N. L. Achouri, N. I. Ashwood, D. W. Bardayan, S. M. Brown, W. N. Catford, K. A. Chipps, J. Cizewski, N. Curtis, K. L. Jones, T. Munoz-Britton, S. D. Pain, N. Soić, C. Wheldon, G. L. Wilson and V. A. Ziman, *Phys. Rev. C* **90**, 054324 (Nov 2014).
106. G. V. Rogachev, V. Z. Goldberg, T. Lönnroth, W. H. Trzaska, S. A. Fayans, K.-M. Källman, J. J. Kolata, M. Mutterer, M. V. Rozhkov and B. B. Skorodumov, *Phys. Rev. C* **64**, 051302 (Oct 2001).
107. W. Mittig, S. Beceiro-Novo, A. Fritsch, F. Abu-Nimeh, D. Bazin, T. Ahn, W. Lynch, F. Montes, A. Shore, D. Suzuki, N. Usher, J. Yurkon, J. Kolata, A.

Howard, A. Roberts, X. Tang and F. Becchetti, *Nucl. Instruments Methods Phys. Res. Sect. A Accel. Spectrometers, Detect. Assoc. Equip.* **784** (Nov 2015) 494.
108. S. Beceiro-Novo, T. Ahn, D. Bazin and W. Mittig, *Prog. Part. Nucl. Phys.* **84**, 124 (Jul 2015).
109. T. Rauscher, J. H. Applegate, J. J. Cowan, F.-K. Thielemann and M. Wiescher, *Astrophys. J.* **429**, 499 (Jul 1994).
110. Y. Mizoi, T. Fukuda, Y. Matsuyama, T. Miyachi, H. Miyatake, N. Aoi, N. Fukuda, M. Notani, Y. Watanabe, K. Yoneda, M. Ishihara, H. Sakurai and A. Yoshida, *Phys. Rev. C* **62**, 065801 (Nov 2000).
111. X. Gu, R. Boyd, M. Farrell, J. Kalen, C. Mitchell, J. Kolata, M. Belbot, K. Lamkin, K. Ashktorab, F. Becchetti, J. Brown, D. Roberts, K. Kimura, I. Tanihata, K. Yoshida and M. Islam, *Phys. Lett. B* **343** (Jan 1995) 31.
112. M. La Cognata and A. Del Zoppo, *Astrophys. J.* **736**, 148 (Aug 2011).
113. P. Carnelli, S. Almaraz-Calderon, K. Rehm, M. Albers, M. Alcorta, P. Bertone, B. Digiovine, H. Esbensen, J. Fernández Niello, D. Henderson, C. Jiang, J. Lai, S. Marley, O. Nusair, T. Palchan-Hazan, R. Pardo, M. Paul and C. Ugalde, *Nucl. Instruments Methods Phys. Res. Sect. A Accel. Spectrometers, Detect. Assoc. Equip.* **799**, 197 (Nov 2015).
114. P. F. F. Carnelli, S. Almaraz-Calderon, K. E. Rehm, M. Albers, M. Alcorta, P. F. Bertone, B. Digiovine, H. Esbensen, J. O. F. Niello, D. Henderson, C. L. Jiang, J. Lai, S. T. Marley, O. Nusair, T. Palchan-Hazan, R. C. Pardo, M. Paul and C. Ugalde, *Phys. Rev. Lett.* **112**, 192701 (May 2014).
115. J. Kolata, A. Howard, W. Mittig, T. Ahn, D. Bazin, F. Becchetti, S. Beceiro-Novo, Z. Chajecki, M. Febbrarro, A. Fritsch, W. Lynch, A. Roberts, A. Shore and R. Torres-Isea, *Nucl. Instruments Methods Phys. Res. Sect. A Accel. Spectrometers, Detect. Assoc. Equip.* **830**, 82 (Sep 2016).
116. S. Starrfield, J. W. Truran and W. M. Sparks, *Astrophys. J.* **226** (Nov 1978) 186.
117. H. Schatz, A. Aprahamian, V. Barnard, L. Bildsten, A. Cumming, M. Ouellette, T. Rauscher, F. K. Thielemann and M. Wiescher, *Phys. Rev. Lett.* **86** (Apr 2001) 3471.
118. C. F. von Weizsaecker, *Phys. Zeitschrift* **39** (1938) 633.
119. H. A. Bethe, *Phys. Rev.* **55**, 434 (Mar 1939).
120. P. J. LeBlanc, G. Imbriani, J. Görres, M. Junker, R. Azuma, M. Beard, D. Bemmerer, A. Best, C. Broggini, A. Caciolli, P. Corvisiero, H. Costantini, M. Couder, R. DeBoer, Z. Elekes, S. Falahat, A. Formicola, Z. Fülöp, G. Gervino, A. Guglielmetti, C. Gustavino, G. Gyürky, F. Käppeler, A. Kontos, R. Kuntz, H. Leiste, A. Lemut, Q. Li, B. Limata, M. Marta, C. Mazzocchi, R. Menegazzo, S. O'Brien, A. Palumbo, P. Prati, V. Roca, C. Rolfs, C. Rossi Alvarez, E. Somorjai, E. Stech, O. Straniero, F. Strieder, W. Tan, F. Terrasi, H. P. Trautvetter, E. Uberseder and M. Wiescher, *Phys. Rev. C* **82**, 055804 (Nov 2010).
121. Y. Kanada-En'yo and H. Horiuchi, *Prog. Theor. Phys.* **93**, 115 (Jan 1995).

122. M. Wiescher, J. Görres and H. Schatz, *J. Phys. G Nucl. Part. Phys.* **25**, R133 (Jun 1999).
123. J. W. Truran and M. Livio, *Astrophys. J.* **308**, 721 (Sep 1986).
124. L. van Wormer, J. Goerres, C. Iliadis, M. Wiescher and F.-K. Thielemann, *Astrophys. J.* **432**, 326 (Sep 1994).
125. F. Hoyle, *Astrophys. J. Suppl. Ser.* **1**, 121 (Sep 1954).
126. A. Volya and Y. M. Tchuvil'sky, *Phys. Rev. C* **91**, 044319 (Apr 2015).
127. G. Brown and A. Green, *Nucl. Phys.* **75**, 401 (Jan 1966).
128. R. J. de Boer and M. Wiescher, *submitted.*
129. L. R. Gasques, A. V. Afanasjev, M. Beard, J. Lubian, T. Neff, M. Wiescher and D. G. Yakovlev, *Phys. Rev. C* **76**, 045802 (Oct 2007).
130. T. Spillane, F. Raiola, C. Rolfs, D. Schürmann, F. Strieder, S. Zeng, H.-W. Becker, C. Bordeanu, L. Gialanella, M. Romano and J. Schweitzer, *Phys. Rev. Lett.* **98**, 122501 (Mar 2007).
131. J. Zickefoose, J. Schweitzer, T. Spillane, F. Strieder, H.-W. Becker, C. Rolfs, A. di Leva, M. De Cesare, N. De Cesare, F. Terrasi, L. Gialanella, D. Schurmann, Y. Guan, G. Imbriani and B. Limata, *"Proceedings 11th Symp. Nucl. Cosmos. 19-23 July 2010. Heidelb.* (2010).
132. F. Käppeler, R. Gallino, S. Bisterzo and W. Aoki, *Rev. Mod. Phys.* **83**, 157 (Apr 2011).
133. S. Bisterzo, C. Travaglio, R. Gallino, M. Wiescher and F. Käppeler, *Astrophys. J.* **787**, 10 (May 2014).
134. H. Trautvetter, M. Wiescher, K.-U. Kettner, C. Rolfs and J. Hammer, *Nucl. Phys. A* **297**, 489 (Mar 1978).
135. S. Dababneh, M. Heil, F. Käppeler, J. Görres, M. Wiescher, R. Reifarth and H. Leiste, *Phys. Rev. C* **68**, 025801 (Aug 2003).
136. U. Giesen, C. Browne, J. Görres, S. Graff, C. Iliadis, H.-P. Trautvetter, M. Wiescher, W. Harms, K. Kratz, B. Pfeiffer, R. Azuma, M. Buckby and J. King, *Nucl. Phys. A* **561**, 95 (Aug 1993).
137. M. Jaeger, R. Kunz, A. Mayer, J. W. Hammer, G. Staudt, K. L. Kratz and B. Pfeiffer, *Phys. Rev. Lett.* **87**, 202501 (Oct 2001).
138. M. Jaeger, No Title, PhD thesis, University of Stuttgart (2001).
139. J. Russell, W. Taylor, F. Dunnam and H. Van Rinsvelt, *Nucl. Phys. A* **187**, 449 (Jun 1972).
140. M. Anderson, L. Mitchell, M. Sevior and D. Sargood, *Nucl. Phys. A* **405** (Aug 1983) 170.
141. T. Tanabe, K. Haga, M. Yasue, K. Sato, K. Ogino, Y. Kadota, M. Tochi, K. Makino, T. Kitahara and T. Shiba, *Nucl. Phys. A* **399**, 241 (May 1983).
142. J. Blackmon, D. Bardayan, W. Bradfield-Smith, A. Champagne, A. Chen, T. Davinson, K. Hahn, R. Kozub, Z. Ma, P. Parker, G. Rajbaidya, R. Runkle, C. Rowland, A. Shotter, M. Smith, K. Swartz, D. Visser and P. Woods, *Nucl. Phys. A* **688**, 142 (May 2001).
143. J. J. He, J. Hu, S. W. Xu, Z. Q. Chen, X. Y. Zhang, J. S. Wang, H. W. Wang, W. D. Tian, X. Q. Yu, L. Y. Zhang, L. Li, Y. Y. Yang, P. Ma, X. H. Zhang, J. Su, E. T. Li, Z. G. Hu, Z. Y. Guo, X. Xu, X. H. Yuan, W. Lu, Y.

H. Yu, Y. D. Zang, S. W. Ye, R. P. Ye, J. D. Chen, S. L. Jin, C. M. Du, S. T. Wang, J. B. Ma, L. X. Liu, Z. Bai, X. Q. Li, X. G. Lei, Z. Y. Sun, Y. H. Zhang, X. H. Zhou and H. S. Xu, *Eur. Phys. J. A* **47**, 67 (May 2011).
144. C. Fu, V. Goldberg, G. Rogachev, G. Tabacaru, G. Chubarian, B. Skorodumov, M. McCleskey, Y. Zhai, T. Al-Abdullah, L. Trache and R. Tribble, *Phys. Rev. C* **77**, 064314 (Jun 2008).
145. D. Kurath and I. Towner, *Nucl. Phys. A* **222**, 1 (Apr 1974).
146. R. N. Boyd, D. Elmore, W. P. Alford, J. P. Draayer and R. A. Lindgren, *Phys. Rev. C* **14**, 4 (Jul 1976).
147. S. Almaraz-Calderon, W. P. Tan, A. Aprahamian, B. Bucher, A. Roberts, M. Wiescher, C. R. Brune, T. N. Massey, N. Özkan, R. T. Güray and H. Mach, *Phys. Rev. C* **86**, 025801 (Aug 2012).
148. A. Kim, N. H. Lee, M. H. Han, J. S. Yoo, K. I. Hahn, H. Yamaguchi, D. N. Binh, T. Hashimoto, S. Hayakawa, D. Kahl, T. Kawabata, Y. Kurihara, Y. Wakabayashi, S. Kubono, S. Choi, Y. K. Kwon, J. Y. Moon, H. S. Jung, C. S. Lee, T. Teranishi, S. Kato, T. Komatsubara, B. Guo, W. P. Liu, B. Wang and Y. Wang, *Phys. Rev. C* **92**, 035801 (Sep 2015).
149. J. L. Fisker, F.-K. Thielemann and M. Wiescher, *Astrophys. J.* **608** (Jun 2004) L61.
150. A. Matic, A. M. van den Berg, M. N. Harakeh, H. J. Wörtche, G. P. A. Berg, M. Couder, J. L. Fisker, J. Görres, P. LeBlanc, S. O'Brien, M. Wiescher, K. Fujita, K. Hatanaka, Y. Sakemi, Y. Shimizu, Y. Tameshige, A. Tamii, M. Yosoi, T. Adachi, Y. Fujita, Y. Shimbara, H. Fujita, T. Wakasa, P. O. Hess, B. A. Brown and H. Schatz, *Phys. Rev. C* **80** (Nov 2009) 055804.
151. P. Mohr, R. Longland and C. Iliadis, *Phys. Rev. C* **90**, 065806 (Dec 2014).
152. I. Pogrebnyak, C. Howard, C. Iliadis, R. Longland and G. E. Mitchell, *Phys. Rev. C* **88**, 015808 (Jul 2013).
153. A. Matic, A. M. van den Berg, M. N. Harakeh, H. J. Wörtche, M. Beard, G. P. A. Berg, J. Görres, P. LeBlanc, S. O'Brien, M. Wiescher, K. Fujita, K. Hatanaka, Y. Sakemi, Y. Shimizu, Y. Tameshige, A. Tamii, M. Yosoi, T. Adachi, Y. Fujita, Y. Shimbara, H. Fujita, T. Wakasa, J. P. Greene, R. Crowter and H. Schatz, *Phys. Rev. C* **84**, 025801 (Aug 2011).
154. S. Almaraz-Calderon, W. P. Tan, A. Aprahamian, M. Beard, G. P. A. Berg, B. Bucher, M. Couder, J. Görres, S. O'Brien, D. Patel, A. Roberts, K. Sault, M. Wiescher, C. R. Brune, T. N. Massey, K. Fujita, K. Hatanaka, D. Ishiwaka, H. Matsubara, H. Okamura, H. J. Ong, Y. Sakemi, Y. Shimizu, T. Suzuki, Y. Tameshige, A. Tamii, J. Zenihiro, T. Kubo, Y. Namiki, Y. Ohkuma, Y. Shimbara, S. Suzuki, R. Watanabe, R. Yamada, T. Adachi, Y. Fujita, H. Fujita, M. Dozono and T. Wakasa, *Phys. Rev. C* **86** (Dec 2012) 065805.
155. A. Arcones and F. Montes, *Astrophys. J.* **731**, 5 (Apr 2011).
156. S. Wanajo, H.-T. Janka and S. Kubono, *Astrophys. J.* **729** (Mar 2011) 46.
157. J. Pereira and F. Montes, *Phys. Rev. C* **93**, 034611 (Mar 2016).
158. E. Bravo, L. Piersanti, I. Domínguez, O. Straniero, J. Isern, J. A. Escartin and J. A. Escartin, *Astron. Astrophys.* **535** (Nov 2011) A114.

159. E. Bravo and G. Martínez-Pinedo, *Phys. Rev. C* **85**, 055805 (May 2012).
160. P. Haensel, J. L. Zdunik and J. L. Zdunik, *Astron. Astrophys.* **404** (Jun 2003) L33.
161. E. F. Brown, *Astrophys. J.* **531**, 988 (Mar 2000).
162. M. Beard, A. Afanasjev, L. Chamon, L. Gasques, M. Wiescher and D. Yakovlev, *At. Data Nucl. Data Tables* **96** (2010) 541.

CHAPTER 9

PERSISTENCE OF CLUSTERING AT HIGH EXCITATION ENERGY: CLUES FROM ^{24}Mg

F. Gulminelli

Normandie Univ, ENSICAEN, UNICAEN, CNRS/IN2P3, LPC Caen, 14000 Caen, France

L. Morelli, M. Bruno, and M. D'Agostino

Dipartimento di Fisica ed Astronomia dell'Università and INFN, Bologna, Italy

G. Baiocco

Dipartimento di Fisica dell'Università, Pavia, Italy

The decay of highly excited states of ^{24}Mg is studied in fusion evaporation events completely detected in charge in the reactions ^{12}C+^{12}C and ^{14}N+^{10}B at 95 and 80 MeV incident energy, respectively. A detailed comparison with a dedicated statistical model reveals that the dominant reaction mechanism is compound nucleus formation and decay. However, in both reactions, a discrepancy with statistical expectations is found for α particles detected in coincidence with Carbon, Oxygen and Neon residues. The comparison between the two reactions shows that this discrepancy is only partly explained by an entrance channel effect. The deviation between the measured and calculated branching ratios for the channels involving multiple α particle emission shows that α decay is strongly favored in evaporation channels leading to even-Z residues. This suggests that cluster correlations in ^{24}Mg are preserved at high excitation energy.

1. Introduction

Since the first heuristic proposition of α-chains as possible building blocks of even-even nuclei in the late sixties,[1] the subject of α-clustering has been a central issue in nuclear physics.[2–6]

On the theoretical side, low density nuclear matter at zero and finite temperature is well known to be unstable against α-particle formation.[7] Concerning finite nuclei, highly sophisticated ab-initio calculations have shown pronounced cluster features in the ground state of a large number of light nuclei.[8] Evidences also exist that excited states of ^{12}C and ^{16}O can be modelled assuming a weakly interacting gas of almost free α-particles.[9] In addition, in recent years, different constrained density functional approaches have consistently found clear α-cluster correlations in all light and medium-heavy even-even nuclei at excitation energies around the threshold of breakup into constituent clusters.[10–13]

Concerning experimental research, rotational bands consistent with α-cluster structures have been identified in different even-even light nuclei and shown to persist even along their isotopic chains.[3] Exotic non-statistical decays of these correlated states have been evidenced in the recent literature.[14]

A natural extension of the concept of nuclear clusters concerns nuclear molecules. Early studies dating back from the '70s have observed several interesting resonances in the ^{12}C+^{12}C reaction in the inelastic[15] and α-transfer channels,[16] corresponding to resonant structures persisting in the ^{24}Mg system at least up to around 50 MeV excitation energy. Concerning the α-transfer channel, experimental results have been reproduced by coupled cluster calculations[17] where the cross section is dominated by a four-cluster $(\alpha+\alpha)+(\alpha+^{12}\text{C})$ state of highly excited ^{24}Mg around 30 MeV.

In our recent work,[18–22] we have analyzed in detail ^{12}C+^{12}C central and peripheral reactions at 95 MeV and systematically compared a large number of exclusive experimental observables to the results of a dedicated Monte Carlo Hauser-Feshbach code (HFℓ). The main result of these works is that the globality of the reaction observables can be reasonably well explained in the framework of compound nucleus sequential statistical decay of ^{24}Mg (^{12}C) for the central (peripheral) collisions, including the decay kinematics of the Hoyle state.[23] A notable exception to this statement is given, for central events, by the branching ratio of the decay towards the different open multiple-α channels, which are systematically underestimated by the statistical model.

This result points towards a persistence of α-cluster correlations in ^{24}Mg at the excitation energy of the compound $E^*(^{24}\text{Mg}) = 62.4$ MeV, that is well above the threshold for multiple-α decay. This is a surprising result, as a pure statistical behaviour might be expected due to the extremely high number of available states at such high excitation. Alternatively, the increase of α correlations could be due to a cluster structure for the daughter nucleus ^{20}Ne which, according to the HFℓ calculations, should be excited at $E^*(^{20}\text{Ne}) \approx 40$ MeV. This value is still much higher than the energy threshold for 5-α dissociation $E^*_{theo}(^{20}\text{Ne}) \approx 16.74$ MeV where cluster states are theoretically expected,[11,13] but cluster structures at energies as high as 30 MeV have been already reported in the literature.[17]

A possible clue to understand this remarkable persistence of α clusters at high excitation energy could be a specific state selectivity due to the entrance channel of the reaction. A model independent answer to this question can in principle be obtained by comparing two different reactions corresponding to the formation of the same compound nucleus. A perfect mapping is extremely hard because it is virtually impossible to gate on the compound spin at those high excitation energies, but very useful indications can be obtained comparing two compound reactions corresponding to the same mass, charge and excitation energy, measured with the same experimental apparatus. The comparison between compound nucleus events for ^{12}C+^{12}C at 95 MeV, and ^{14}N+^{10}B at 80 MeV[24] have revealed that the entrance channel has a sizeable influence on the inclusive α multiplicities and spectra, but a very limited effect on the observed α correlations, which appear essentially linked to the structure of the hot Mg source.

The experiments were performed at *Laboratori Nazionali di Legnaro* with beams from the Tandem-Alpi complex, in the framework of the INFN NUCL-EX collaboration.[a] The experimental setup[25] comprises the GARFIELD [26] apparatus and the Ring-Counter (RCo) annular detector,[27] fully equipped with digital electronics.[28] The combination of the two devices allows a nearly-4π coverage of the solid angle, which, combined with a high granularity, allows to measure the charge, the energy and the emission angles of nearly all the charged reaction products, with an excellent discrimination of the different reaction mechanisms. They also provide information on the mass of the emitted charged products in a wide range of particle energy and type.

A schematic sketch of the apparatus is shown in Fig. 1. Further details on the apparatus and the experimental conditions can be founds in Refs. 23, 25,26.

---

[a]M. Bruno, M. D'Agostino, L. Morelli (Bologna); G. Baiocco (Pavia); M. Cinausero, F. Gramegna, T. Marchi (LNL); D. Fabris (Padova): S. Barlini, M. Bini, G. Casini, A. Olmi, G. Pasquali, S. Piantelli, S. Valdrè (Firenze).

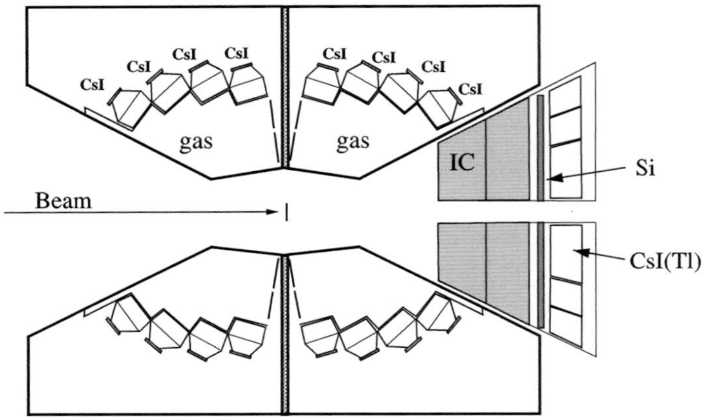

Fig. 1. Vertical cross section of the GARFIELD+RCo apparatus, azimuthally symmetric around the beam axis. Figure taken from Ref. 22.

In this contribution we will specifically concentrate on the analysis of the channel showing the most important deviation with respect to a standard statistical behavior, namely the $^{16-x}$O-α-α-xn channel.

We will show that two different features contribute to the observed deviation. On one side, the α transfer channel is shown to be favored in the ^{12}C+^{12}C reaction, leading to a high energy component in the alpha spectrum. Once this component, corresponding to low dissipated energy, is removed from the sample, an extra population of multiple alpha channels is still observed with respect to statistical expectation, independent of the entrance channel of the reaction. This alpha excess is consistent with the population of the recently predicted[12,29] α-^{16}O-α linear chain in the excited spectrum of ^{24}Mg. Evidences will be presented that this clustered configuration persists at high excitation energies, above the neutron separation energy for ^{16}O.

2. The Hauser-Feshbach code

A possible experimental evidence of clustering at high excitation energy is given by the measurement of observable distributions which are not compatible with a structureless decay dominated by pure statistical laws.

When a very huge number of channels is open, such deviations appear as peaks over a smooth background behavior. However, when the number of channels is limited by the available energy, angular momentum and small number of bodies, conservation laws constitute stringent conditions that deeply affect the kinematic correlations. In such a situation, extracting cluster correlations requires optimized predictions of the statistical background. For this reason, we have introduced a dedicated Monte Carlo Hauser-Feshbach code (HFℓ) which is specifically optimized on reactions with light ions.

In this section we give the main features of the code. For further details, see Ref. 19. The evaporation of light particles is treated with the standard Hauser-Feshbach (HF) formalism of CN decay, with n, p, d, t, ^{3}He, α particles and ^{6}Li, ^{7}Li emission channels included. The expression for the decay width in channel ξ for a hot nucleus (A, Z) excited in a state of energy E^* and angular momentum J reads:

$$\Gamma_\xi = \frac{1}{2\pi\rho_C}\int_0^{E^*-Q} d\epsilon_\xi \sum_{J_D} \sum_{j=|J-J_d|}^{J+J_D} \sum_{\ell=|j-s_p|}^{j+s_p} T^J_{j,s_p}(\epsilon_\xi)\rho_D \tag{1}$$

where ϵ_ξ is the relative kinetic energy of the decay products (the daughter nucleus, labeled by D, and the evaporated particle, labeled by p); Q is the decay Q-value; J_d, s_p and ℓ are the angular momentum of the daughter nucleus, the spin of the evaporated particle and the orbital angular momentum of the decay, respectively and summations include all angular momentum couplings between the initial and final states; T is the transmission coefficient; $\rho_C(E^*, J)$ and $\rho_D(E^* - Q - \epsilon_\xi, J_d)$ are the nuclear level density of the decaying and of the residual nucleus, respectively. The widths Γ_i are calculated for all possible decay channels and the Branching Ratio (BR) associated with a specific channel ξ is obtained as the ratio between Γ_ξ and the total decay width for the hot nucleus, $BR(\xi) = \Gamma_\xi / \sum_i \Gamma_i$. This decay probability constitutes the main ingredient of the Monte Carlo simulation: in the statistical code, starting from a given compound configuration defined by (A, Z, J, E^*), the decay pattern is calculated with the Monte-Carlo technique as a sequence of two-body decays governed by the emission probability given by (1).

In the case of the very light compound nucleus studied in this work, simple analytical expressions can be safely employed for the transmission coefficients. In our code we have adopted the empirical work of Ref 30:

$$T_\ell(\epsilon_\xi) = \frac{1}{1 + \exp\left(\frac{V_b - \epsilon_\xi}{\delta \cdot V_b}\right)} \tag{2}$$

where the barrier V_b is a sum of a Coulomb and a centrifugal term depending on ℓ, hence on all coupled angular momenta, see (1). Its full expression reads:

$$V_b = \frac{1.44}{r_Z} \frac{Z_p(Z - Z_p)}{(A - A_p)^{1/3} + A_p^{1/3}} + \frac{\hbar^2 \ell(\ell+1)}{2r_Z^2} \frac{\frac{A}{A_p(A-A_p)}}{\left[(A - A_p)^{1/3} + A_p^{1/3}\right]^2}$$

The two free parameters δ and r_Z were optimized to reproduce the decay of discrete resonances.[30] They depend on the charge Z_p of the evaporated particle, and δ also depends on whether the emission takes place in the sub- or above-barrier region.

The angular momentum distribution of the compound nucleus was taken from Refs. 31,32, where the reproduction of the systematics of the fusion cross section gives maximum values $J_{0\ max} = 18(15)\ \hbar$ for the ^{12}C+^{12}C (^{14}N+^{10}B) reaction, respectively. In same way the diffuseness parameter is for both cases $\Delta J = 2$,[32,33] and these values will be employed in the results presented in the following. A detailed study on the optimization of the angular momentum distribution will be presented elsewhere.

Concerning the kinematics of the decay with angular momentum, we have adopted the semi-classical approach proposed by the GEMINI code.[34] Angular momenta are considered as classical vectors, and j_a and j_b are coupled under the assumption of equiprobability for the module of their sum j_s between $|j_a - j_b| \leq j_s \leq |j_a + j_b|$. Once the decay channel has been selected, the angular momentum J_D is obtained through a maximization of $\rho_D(J)$ as a function of J. Decay Q-values are calculated from experimental binding energies.[35] Finally, a special effort has been devoted to the implementation of the level density model. In particular, information on measured excited levels from the online archive NUDAT2[36] has been explicitly and coherently included in the decay calculation.

2.1. *The level density model*

The back-shifted Fermi gas model (BSFG), with the level density parameter and the pairing backshift left as free fit parameters, is known to be a phenomenological approach well suited to reproduce the many-body correlated nuclear level density: pairing effects are included through the backshift Δ_p, and all correlations are taken into account in the renormalization of the parameter $a(E^*)$.

In Ref. 37 level density parameters for the BSFG model have been determined for a large set of nuclei (310 nuclei between ^{18}F and ^{251}Cf), by the fit of complete level schemes at low excitation energy and s-wave neutron resonance spacings at the neutron binding energy.

The adopted expression for $\rho(E^*)$ (after integration on angular momentum J and parity π) reads:

$$\rho(E^*) = \frac{\exp[2\sqrt{a(E^* - E_2)}]}{12\sqrt{2}\sigma a^{1/4}(E^* - E_2)^{5/4}} \tag{3}$$

where σ is the spin cut-off parameter:

$$\sigma^2 = 0.0146A^{5/3}\frac{1 + \sqrt{1 + 4a(E^* - E_2)}}{2a} \tag{4}$$

The energy backshift $\Delta_p = E_2$ is left as the first free parameter in the data fitting. The second fit parameter is the asymptotic value $\tilde{a}$ of the following functional form for $a(E^*, Z, N)$:[38]

$$a = \tilde{a}\left[1 + \frac{S(Z,N) - \delta E_p}{E^* - E_2}\left(1 - e^{-0.06(E^* - E_2)}\right)\right] \tag{5}$$

where $S(Z,N) = M_{exp}(Z,N) - M_{LD}(Z,N)$ is a shell correction term, M_{exp} and M_{LD} being respectively the experimental mass and the mass calculated with a macroscopic liquid drop formula for the binding energy not including any pairing or shell corrections. δE_p is a pairing term expressed in terms of the deuteron separation energy. Full details on the parameter definition and fit procedure can be found in Ref. 37. As a final result, analytic formulas for E_2 and $\tilde{a}$ as a function of tabulated nuclear properties are given.

With such formulas for the calculation of the level density parameters, the model of (3) allows for a very good reproduction of experimental distributions of measured levels in the mass region of interest for the present work. Two selected examples are given in Fig. 2. The isotope ^{20}Ne belongs to the fitted data set, and the good agreement between the line and the histogram shows the quality of the fit procedure of Ref. 37 Concerning ^{16}O, the values of the parameters are an extrapolation of the formulas proposed in Ref. 37 out of the fitted data set; from the figure it is also clear that Eq. (3) can be considered reliable also for nuclei whose level density has not been directly optimized. A similar agreement is observed for all the other particle-stable isotopes in the mass region of interest for the present study.

Still, numerical values for the pairing backshift and for the asymptotic limit of $a(E^*)$ with increasing excitation energy obtained through

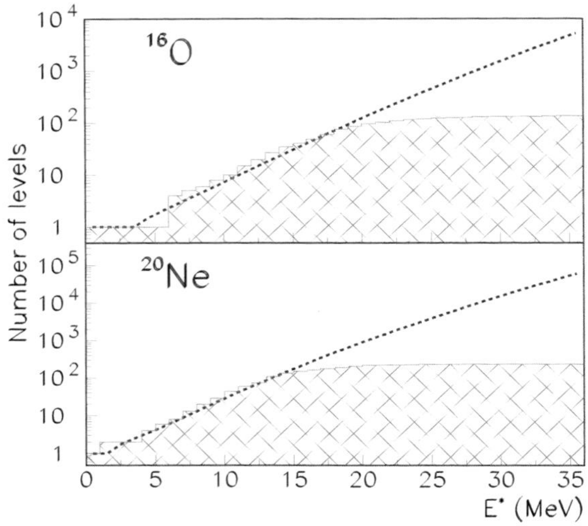

Fig. 2. Comparison between the cumulative number of levels given by (3) (lines), and the cumulative counting of experimentally measured levels from the NUDAT2 archive[36] (histograms).

this approach are to be considered reliable only up to $E^*/A \approx 1$ MeV for $A \approx 20$ nuclei. In particular, it is found that the values of the level density parameter needed to reproduce the information on discrete levels are usually lower than the ones coming from higher-energy constraints, through the reproduction of data for fusion-evaporation or evaporation-after fragmentation studies ($E^*/A \approx 2 \div 3$ MeV). A functional form giving a good reproduction of evaporation spectra at very high excitation energy was proposed in Ref. 39:

$$a_\infty = \frac{A}{14.6}\left(1 + \frac{3.114}{A^{1/3}} + \frac{5.626}{A^{2/3}}\right) \tag{6}$$

To correctly reproduce at the same time the low- and high-energy experimental constraints, we have adopted a functional form for the level density parameter that gives a continuous interpolation between Eq. (5) and Eq. (6):

$$a(E^*, A) = \begin{cases} a_D = (5) & if \quad E^* \le E_m + E_2 \\ a_C = \alpha \exp[-\beta(E^* - E_2)^2] + a_\infty & if \quad E^* > E_m + E_2 \end{cases} \tag{7}$$

The choice of a rapid (exponential) increase is imposed by the fact that the asymptotic value (6) is connected to the opening of the break-up or multifragmentation channels, which is a sharp threshold phenomenon. The α and β parameters are determined by the matching conditions between the low-energy and high-energy regime: $a_D(E_m, A) = a_C(E_m, A)$ and $a_C(E_l, A) = a_\infty \pm 10\%$.

Here, E_l represents the limiting energy at which the break-up or fragmentation regime is attained, while E_m is the excitation energy marking the transition between the discrete and the continuum part of the spectrum. This latter quantity is of the order of $E_m \approx 10$ MeV, coherently with the value of the critical energy for the damping of pairing effects in Ref. 40. In the case of light nuclei for which a large set of measured levels is available, this value well corresponds to the excitation energy maximizing the number of levels in bins of E^*. Above E_m the experimental information is too poor to consider the set of resolved levels exhaustive of the nuclear level density, due to the physical emergence of the continuum.

The limiting energy E_l governs the rapidity of the variation of $a(E^*)$ above E_m. As an example of the overall functional form resulting for the level density parameter, in Fig. 3 we plot $a(E^*, A)$ for ^{20}Ne, for two different choices of E_l. It is interesting to remark that the fiducial value for a in the GEMINI code for ^{20}Ne is $a(E^*) = 2.75$ MeV^{-1}, which does not fulfill

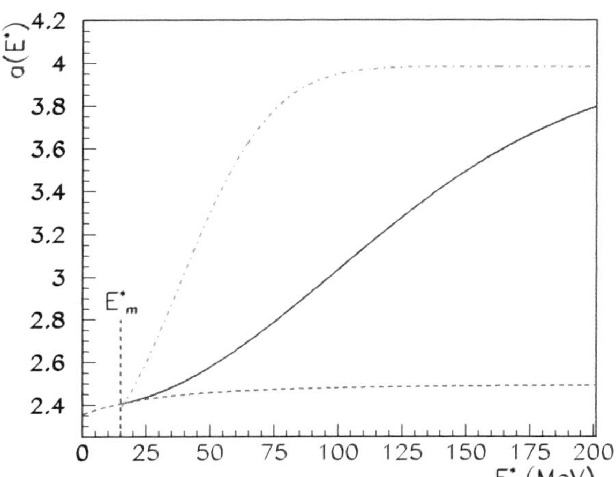

Fig. 3. Level density parameter calculation for ^{20}Ne. Blue dashed line: (5). Black solid and red dot-dashed lines: (7) with $E_l = 8$ and 3 AMeV, respectively.

neither the low nor the high energy experimental constraints for such a light nucleus. In our model, the E_l parameter is adjusted on the proton and α spectra. The best reproduction of data is obtained with the choice $E_l = 3$,[18,20] corresponding to $a \approx 3.5$ MeV^{-1} for $E^*/A = 3$ MeV. This underlines the importance of a dedicated model for the evaporation of very light nuclei.

The other important characteristics of light nuclei spectrum is the persistence of discrete levels at relatively high excitation energy (E_m). When the emitted particle leaves the daughter nucleus at an excitation energy $E_D^* < E_m$, the excitation energy is considered as a discrete variable, and one of the tabulated levels[36] of the daughter discrete spectrum is populated. The level is chosen according to the Breit-Wigner distribution of the discrete levels considering their respective widths, including the full spectroscopic information of Ref. 36, and the particle kinetic energy is adjusted if necessary to ensure energy conservation. When a particle bound level is populated, the subsequent decay is assumed to be due to a single γ emission to the corresponding ground state. If the daughter excitation energy E_d^* is greater than E_m, the spectroscopic information is not sufficient to fully constrain the spin and energy of the daughter nucleus. If measured excited states exist, they are populated with a probability given by the ratio between the measured level density from discrete states and the total level density including the continuum states and given by Eq. (3) with a given by Eq. (7). If no levels are known, the emission is assumed to take place in the continuum.

To summarize, the HFℓ is a standard Hauser-Feshbach Monte-Carlo code where however a special care was taken to include all the available experimental data concerning compound decay for light nuclei. The remaining free parameters were optimized on the $^{12}C+^{12}C$ data set,[18,20] and shown to lead to an excellent reproduction of a large set of inclusive and exclusive observables both for the $^{12}C+^{12}C$[18,20,22] and for the $^{14}N+^{10}B$[24] reaction.

These observables include charge and mass distributions, velocity spectra of residues, angular and energy distributions of light particles in coincidence with evaporation residues, isotopic distributions, and a large variety of correlation functions .[23] Because of that, we believe that deviations from the HFℓ results can unambiguously indicate non-statistical effects in the decay pattern, and specifically α-structure effects in the high-lying density of states.

3. α emission and clustering effects

For the analysis of compound nucleus decay, only complete events are analyzed, where the totality of the charge Z_{tot}=12 and at least 95% of the incoming parallel momentum is detected. These events correspond to the detection of a single residue in the forward direction (RCo) and one or more light particles at larger angles (GARFIELD). Due to the kinematics of the reaction and the detection geometry, this condition is enough to exclude from the data set most of the binary events associated to the presence of a quasi-target.[20] This is shown by the excellent reproduction of a large set of observables by the statistical calculation, as discussed above.

3.1. *Kinematic observables*

If we limit ourselves to kinematical observables, the only important deviation from a statistical behavior concerns decay channels with α particles detected in coincidence with an Oxygen fragment in the $^{12}C+^{12}C$ reaction. This is shown in Fig. 4, which presents energy spectra of protons and α particles detected in coincidence with a residue of a given atomic number.

We can see that with the exception of the $\alpha - O$ coincidence for the ^{12}C beam, particle energy spectra are very well reproduced by the statistical model, which in particular gives strong confidence to our level density model (see Section 2).

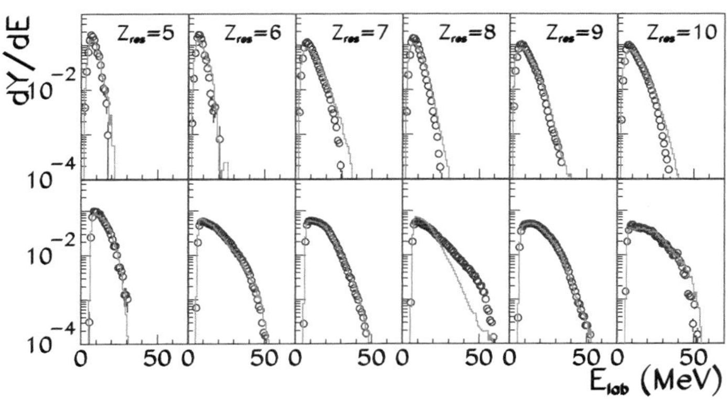

Fig. 4. Proton (upper part) and α (lower part) laboratory energy spectra in complete $^{12}C+^{12}C$ events detected in coincidence with a residue of charge Z_{res}, indicated in each column. Data (symbols) are compared to model calculations (lines). All spectra are normalized to unitary area. Figure taken from Ref. 20

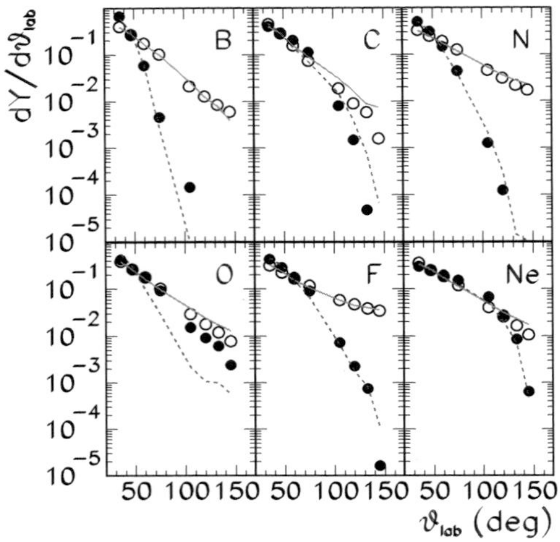

Fig. 5. Proton (open symbols) and α (full symbols) angular distributions in the laboratory frame, detected in complete $^{12}C+^{12}C$ events in coincidence with the indicated residues. Data are compared to model calculations. Red solid lines correspond to protons, blue dashed lines to α's. Experimental and calculated distributions are normalized to unitary area. Figure taken from Ref. 20

The angular distributions of protons and α particles in coincidence with each residue for the same reaction is shown in Fig. 5. The good agreement among data and model predictions as far as proton distributions are concerned is confirmed. Again, a large discrepancy is evident for α particles at backward laboratory angles detected in coincidence with an Oxygen fragment.

The well-known peculiarity of the $^{12}C+^{12}C$ fusion-evaporation reaction is that both the entrance channel and the compound nucleus have potentially α structure correlations. In particular, inelastic and α transfer reactions could be mixed to fusion-evaporation events, thus explaining the failure of the statistical model to correctly reproduce the Oxygen channel.

The distinction between direct and compound reactions is not completely well defined. Indeed the physical processes are continuous and the compound ^{24}Mg could be in a quasi-molecular state reminiscent of the entrance channel, even if it is the source of α emission. A discrimination between entrance channel effects and compound effects can be done by comparing the decay of the ^{24}Mg formed in two different entrance channels leading to the same excitation energy.

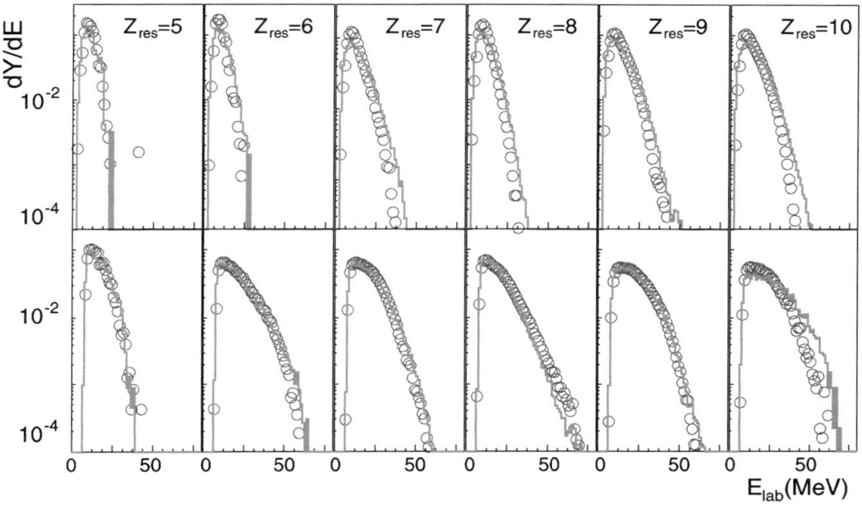

Fig. 6. Proton (upper part) and α (lower part) laboratory energy spectra in complete $^{14}N+^{10}B$ events detected in coincidence with a residue of charge Z_{res}, indicated in each column. Data (symbols) are compared to model calculations (lines). All spectra are normalized to unitary area.

The spectra and angular distributions obtained from the reaction $^{14}N+^{10}B$ are displayed in Figs. 6 and 7. The distributions look very close in the two reactions, and the agreement with the statistical model is also globally very good. Two notable differences between the two data-sets however appear. Concerning the Ne channel, we can see from the α energy spectra that some non-compound reactions affect the $^{14}N+^{10}B$ sample. Concerning the O channel we are especially focussing on, we can see that indeed the large discrepancy observed in the Oxygen channel is considerably reduced in this second data set, even if a discrepancy is still apparent in the angular distribution. This points towards an incomplete memory loss in the $^{12}C+^{12}C$ entrance channel or to a selective excitation of specific quasi-molecular states.

To further explore this hypothesis, we study the $\alpha-$channels as a function of the dissipated energy. We introduce an estimate of the dissipation by the event-by-event quantity:

$$Q_{kin} = E_{kin} - E_{beam} = \sum_{i=1}^{N} E_i - E_{beam} \tag{8}$$

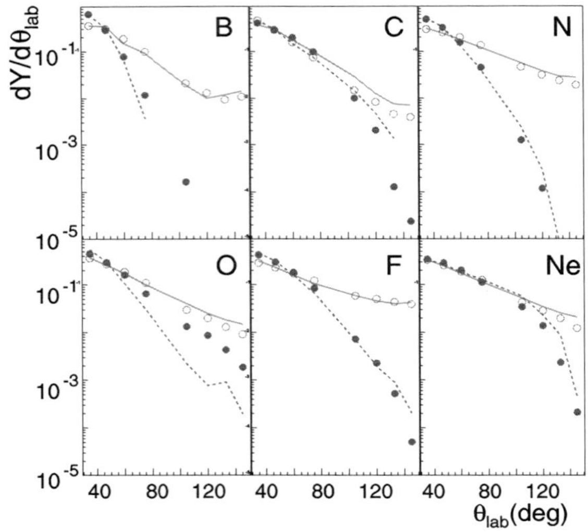

Fig. 7. Proton (open symbols) and α (full symbols) angular distributions in the laboratory frame, detected in complete ^{14}N+^{10}B events in coincidence with the indicated residues. Data are compared to model calculations. Red solid lines correspond to protons, blue dashed lines to α's. Experimental and calculated distributions are normalized to unitary area.

where E_i is the energy of the particle in the laboratory frame and the sum extends to the N particles or fragments that are detected in coincidence, and exhausts the total charge $Z_{tot} = 12$. The quantity Q_{kin} corresponds to the real Q-value of the reaction, given by the mass balance between the initial and final state, provided that the whole mass is collected in the outgoing channel (that is, in the absence of neutron emission).

Figure 8 displays the experimental (symbols) and theoretical (lines) Q_{kin} distributions in the channel O-α-α, for the two reactions.

We can see that the theoretical and experimental spectra show a common structure, with narrow peaks at high Q_{kin} values ($Q_>$ region), and a broader region extending up to an important amount of missing energy ($Q_<$ region). The highest Q_{kin} peak at $Q_{kin} \approx 0$ $(15) MeV$ for ^{12}C+^{12}C (^{14}N+^{10}B) corresponds to the ground state of ^{16}O. In the statistical calculation this peak is obtained when the last-step α emission from the ^{24}Mg* compound nucleus leaves the residue directly in its ground state. The lower Q_{kin} peaks at $Q_{kin} \approx -6.5$ $(8.5) MeV$ for ^{12}C+^{12}C (^{14}N+^{10}B) correspond to the population of one of the particle bound excited states of ^{16}O, which further γ-decay to the respective ground state. The limited energy

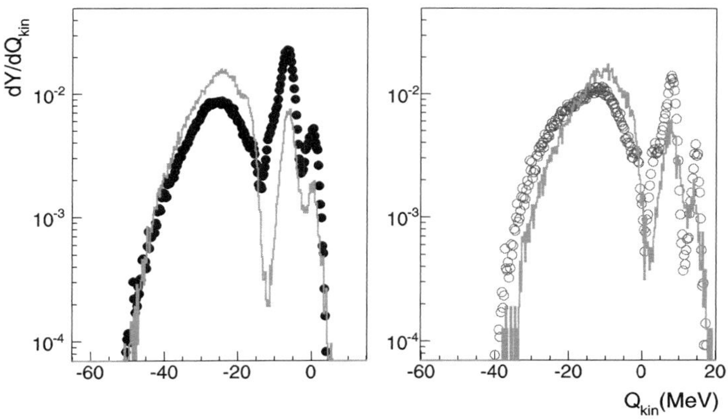

Fig. 8. Q_{kin} distributions in complete events when an Oxygen residue is detected in coincidence with two $\alpha-$particles for the $^{12}C+^{12}C$ (left side) and $^{14}N+^{10}B$ (right side). Data (symbols) are compared to HFℓ calculations (red lines). The dashed lines indicate the threshold Q-value for neutron emission that we have adopted to separate more dissipative from less dissipative events. All distributions are normalized to unitary area.

resolution in the Q_{kin} reconstruction, which broadens the peaks, does not allow to further separate the discrete levels. Starting from the threshold values for neutron emission ($Q_{kin} = -15.8$ or $-0.86\ MeV$ respectively for $^{12}C+^{12}C$ and $^{14}N+^{10}B$) a broader distribution is observed due to neutron emission, and the amount of missing energy corresponds to the undetected neutron kinetic energy.

We can remark that a much higher percentage of events populates the $Q_>$ region in the experimental sample with respect to model predictions for the $^{12}C+^{12}C$ experiment, while the discrepancy is reduced for $^{14}N+^{10}B$. This larger experimental branching ratio in the $Q_>$ region is an indication of the presence of direct reactions or transient quasi-molecular states, in competition with standard fusion-evaporation.

3.2. *Branching ratio for α emission*

However, the observed deviation with respect to the statistical calculation cannot be entirely attributed to direct reactions. To demonstrate this point, we introduce a variable quantifying the experimental branching ratio excess for α emission:

$$R_{clus}(Z) = \frac{Y_{exp}(Z; n_Z\alpha)}{Y_{exp}(Z)} - \frac{Y_{HF\ell}(Z; n_Z\alpha)}{Y_{HF\ell}(Z)} \qquad (9)$$

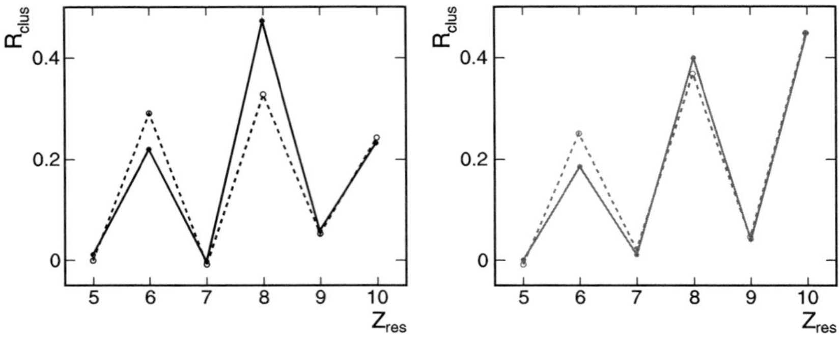

Fig. 9. Branching ratio excess for α decay as a function of the atomic number of the final residue for the ^{12}C+^{12}C (left side) and ^{14}N+^{10}B (right side). Full symbols: all events. Open symbols: more dissipative events only. Lines are drawn to guide the eye. All distributions are normalized to the unitary area.

Table 1. Branching ratio for channels with even residue in experimental (e suffix) ^{14}N + ^{10}B and ^{12}C + ^{12}C data compared with HFℓ predictions (t suffix). Errors on the experimental values (about 5%) take into account both the statistical error and the possible ^{3}He-α contamination.

Z_{res}	channel	BR_e (N+B)	BR_t (N+B)	BR_e (C+C)	BR_t (C+C)
6	C+3α + xn	96%	77%	98%	78%
8	O+2α + xn	56%	15%	63%	15%
10	Ne+α + xn	47%	3%	26%	3%

Here $Y(Z; n_Z\alpha)$ $(Y(Z))$ indicate coincident (inclusive) yields; $n_Z\alpha$ is the (nearest integer) maximum α multiplicity associated with the residue of charge Z $(n_Z\alpha = (12 - Z)/2)$. The subscripts *exp* and *HFℓ* refer to experimental data and model predictions, respectively.

The extra probability of α emission defined by (9) is plotted for the two reactions in Fig. 9. The same information is given more quantitatively in Table 1.

We can see that the evaporation chains leading to a final even Z residue systematically show a preferential α decay. Remarkably, the same qualitative effect is observed for the less dissipative $Q_>$ and more dissipative $Q_<$ events, showing that the alpha excess is not associated to direct reactions. Even more important, the same trend is observed for the two reactions, pointing towards the presence of residual α structure correlations in the excited ^{24}Mg or in its daughter nucleus ^{20}Ne.

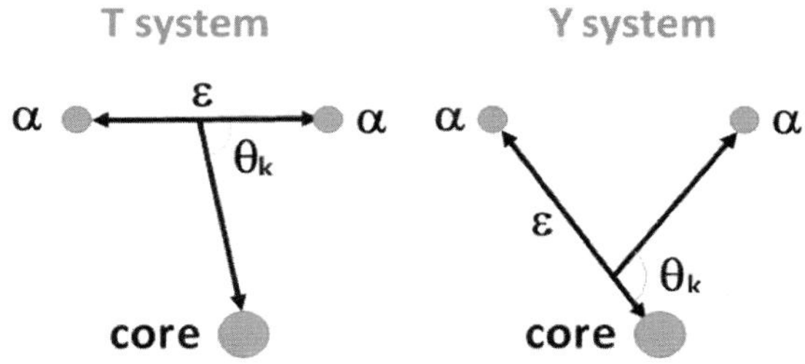

Fig. 10. Schematic representation of T and Y Jacobi coordinates for a three body system.

3.3. *Characterizing the α-O-α channel*

To better understand the physical processes leading to the α-α-^{A}O channels, we turn to study in greater detail the topology of these three-body coincidence events as revealed by their kinematic correlations.

Dealing with three-body systems, a useful representation can be obtained making use of Jacobi energy-angular correlations, described in detail in.[41]

Since in the α-α-O channel two identical particles are present, two Jacobi systems can be used, T and Y. These systems are sketched in Fig. 10.

We limit ourselves to the study of the correlations between the two $\alpha-$particles in the T-system, where the core is the Oxygen. Here, one of the coordinates is the cosinus of the relative angle θ_k of the Oxygen residue momentum $\vec{k}_O$ and the α-α relative momentum $\vec{k}_{\alpha-\alpha}$:[42] $cos(\theta_k) = (\vec{k}_O \cdot \vec{k}_{\alpha-\alpha})/(k_O\ k_{\alpha-\alpha})$.
The other coordinate is $\epsilon = E_{\alpha-\alpha}/E_T$, i.e. the ratio of the relative energy between the two α's to the total available energy. In our case, for more dissipative $Q_<$ events the undetected neutron prevents the calculation of the normalized relative energy ϵ, but $cos(\theta_k)$ can still be used.

Figure 11 shows the $cos(\theta_k)$ (upper panel) and the α-α relative energy (lower panel) distributions for the ^{12}C+^{12}C system as a function of Q_{kin} (defined in (8)).

We can recognize that the two dissipation event classes $Q_<$ and $Q_>$ defined in section 3.1 correspond to different patterns of relative α-α motion. Non-dissipative events are characterized by a relative energy that can overcome 80% of the total available energy, and a back-to-back emission with respect to the Oxygen residue. More dissipative events correspond to particles closer in momentum space, consistent with an increased memory

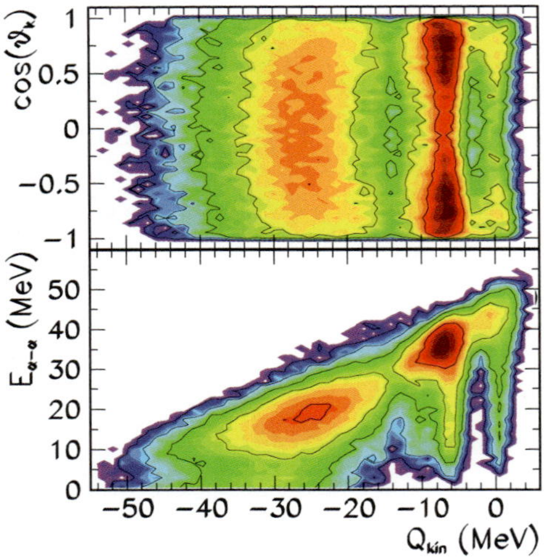

Fig. 11. Logarithmic contour plot of the correlation between $cos(\theta_k)$ (upper panel) and the α-α relative energy (lower panel) as a function of Q_{kin} (8) in O-α-α complete events for the $^{12}C+^{12}C$ reaction. Figure taken from Ref. 21

loss of the entrance channel. For these events, the $cos(\theta_k)$ angular distribution is essentially flat.

The energy and angular correlations for the two classes of dissipation are separately shown in Fig. 12, where data are compared to model predictions.

As far as the relative $\alpha-\alpha$ energy is concerned (middle panel), two contributions are visible in the distribution associated with dissipative events (left side). The small peak at low relative energies suggests that a part of the observed 2-α emission might be associated with a correlated Be state. It is important however to point out that events associated with the formation of discrete *Be* levels do not represent more than 3% of the experimental yield in the more dissipative $^{A}O-2\alpha$ channel, and can not be responsible for the global excess in the branching ratio.[21]

The distribution of low dissipation events (right side) is also strongly asymmetric. However, as we can see from the bidimensional representation Fig. 11, the low energy shoulder does not correspond to the population of a specific excited state, but only reflects the kinematic focussing of the decay to the particle bound states of ^{16}O. These shapes are very well reproduced by the statistical model.

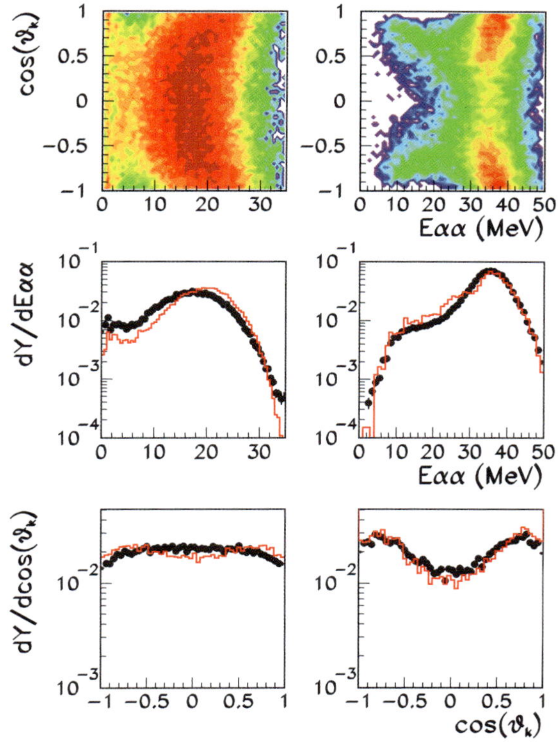

Fig. 12. Left panels correspond to more dissipative and right panels to less dissipative O-α-α complete events for the $^{12}C+^{12}C$ reaction. Upper panels: Logarithmic contour plot of the correlation between $cos(\theta_k)$ and the α-α relative energy. Middle panels: projections of the α-α relative energy. Lower panels: projections of the angular correlation $cos(\theta_k)$. Data (symbols) are compared to HFℓ calculations (lines). All distributions are normalized to the unitary area. Figure taken from Ref. 21

Coming to the angular correlation of Fig. 12, in the case of a simultaneous three-body event, it is expected that $cos(\theta_k) = \pm 1$ should be depleted due to the Coulomb repulsion.[41] Therefore, the observed flat distribution associated with $Q_<$ appears compatible with sequential emission. This is confirmed by the fact that the shape of the distribution is well reproduced by the HFℓ calculations. The sequential statistical model correctly reproduces also the forward-backward peaked distribution of $Q_>$ events.

However we recall that the branching ratio towards two-α decay associated with the oxygen residue shows an important excess in the experiment compared to the HFℓ calculation for both classes of dissipation. For this

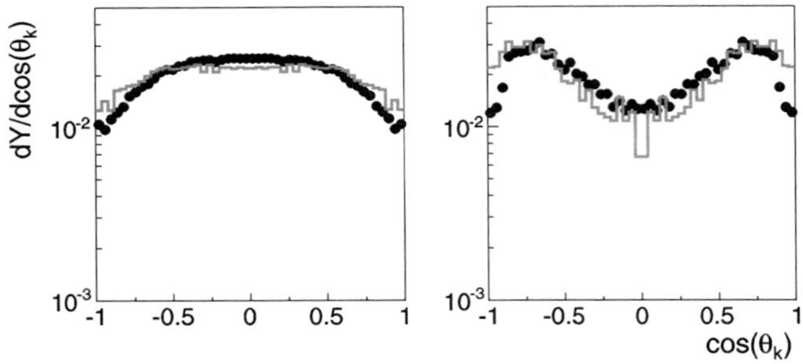

Fig. 13. Distribution of the Jacobi relative angle $cos(\theta_k)$ for more dissipative ($Q_<$, left) and less dissipative ($Q_>$, right) O-α-α complete events for the ^{14}N+^{10}B reaction. Data (symbols) are compared to HFℓ calculations (lines). All distributions are normalized to the unitary area.

reason, we can not interpret the good reproduction of the $cos(\theta_k)$ distribution by the model as a proof of standard statistical decay in the data, but only as a proof that the decay of the excited state is compatible with a sequential α emissions.

In particular, concerning the $Q_>$ events, it is important to notice that the kinematical focussing is compatible with the recently predicted[12,29] α-^{16}O-α linear chain in the excited spectrum of ^{24}Mg. The alternative picture of an incomplete fusion, with two α particles left over from the two α-clustered collision partners could also be invoked.

However is difficult to retain this interpretation since the decay pattern is the same independent of the entrance channel. This is demonstrated in Fig. 13, which shows the $cos(\theta_k)$ distribution for the system ^{14}N+^{10}B. As we discussed in Section 3.1, for this system an important contribution of direct α-^{16}O-α reactions can be safely excluded. We can see that the same back-to-back emission pattern for the alphas is observed in the reaction, which again is reasonably well reproduced by the model. This shows that the kinematical focussing cannot be attributed to entrance channel memory effects.

Concerning the most dissipative (^{A}O$-\alpha-\alpha$) events, the kinematical correlation is blurred by the undetected neutron(s). However, with the help of the model where this kinematical effect is fully accounted for, the $cos(\theta_k)$ distribution can be further exploited to gather further insight in the physical process at play.

This is shown in Figure 14, where we compare the experimental distribution with the HFℓ predictions corresponding to different evaporation

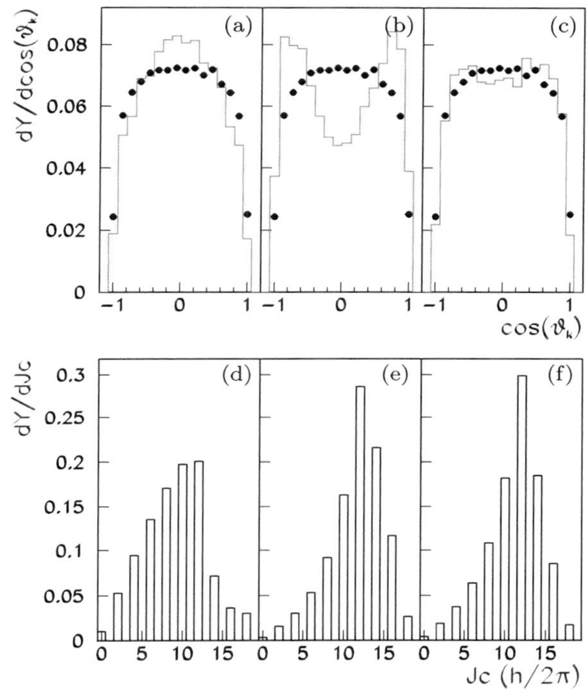

Fig. 14. Analysis of dissipative ^{15}O+2α complete events for the ^{12}C+^{12}C reaction. Upper part: the experimental (dots) $cos(\theta_k)$ distribution is compared to the calculated ones (red lines) according to different emission sequences (see text). Lower part: Compound angular momentum distribution in the unfiltered HFℓ calculations. Panel (a) and (d): n is the first emitted particle. Panel (b) and (e): n is the second emitted particle. Panel (c) and (f): n is the third emitted particle. All distributions are normalized to unitary area. Figure taken from Ref. 21

sequences. The corresponding angular momentum distributions of the initial ^{24}Mg compound are also displayed in bottom panels of the same figure.

When the neutron is the first particle emitted from ^{24}Mg (panel a) and the two α's correspond to the last two steps of the de-excitation chain, the distribution is single-peaked. Indeed the evaporation sequence corresponds to the lowest average angular momentum for the initial compound (panel d), and moreover the angular momentum available for the first α emission is reduced by the first step of the de-excitation. Conversely, if the two α's are the first and last emitted particles, with the neutron being emitted in the intermediate step (panel b), the distribution shows two clear peaks. This is consistent with two sequential emissions from a high J source and a low J

source. An intermediate situation is observed in panel c), which displays the $cos(\theta_k)$ distribution obtained when two successive α's are emitted from the excited ^{24}Mg, and the residual ^{16}O is excited above the neutron separation energy, leading to neutron emission in the last evaporation step. In this case the initial angular distribution is compatible with the one of panel b), but the angular momentum of the ^{20}Ne is not negligible. As a consequence, the second emission leads to a modification of the direction of $\vec{k}_{\alpha-\alpha}$ with respect to the first α emission angle, and a smoothing of the two-peaked distribution.

From Figure 14 we can conclude that the measured distributions are closer to the kinematical configuration expected for the $\alpha-\alpha$−n sequence. In turn, this implies that the preferential α emission observed in the data can be attributed to the excited state of ^{24}Mg and/or ^{20}Ne. In particular, this finding appears consistent with a persistence at high excitation energy of the excited ^{24}Mg state as a clustered α-^{16}O-α linear chain, recently predicted.[12,29]

4. Conclusions

Two different reactions leading to a ^{24}Mg compound nucleus at the same $E^* = 62$ MeV excitation energy have been studied with the GARFIELD+ RCo experimental setup at LNL-INFN, namely ^{12}C+^{12}C at 95 MeV and and ^{14}N+^{10}B at 80 MeV beam energy. Events completely detected in charge ($Z_{tot} = 12$) have been selected and compared to Hauser Feshbach statistical model calculations performed with the dedicated code HFℓ.

Some clear deviations from a statistical behavior in the decay have been found. These deviations concern an anomalously high branching ratio towards multiple (two or three) α emission. In this contribution, we have concentrated on the 2 α−particle exit channel.

Two different phenomena can explain the extra yield observed. First, an extra experimental cross section for the three-body less dissipative ^{16}O−α−α decay channel has been attributed to a non-negligible contribution, in the ^{12}C+^{12}C sample, of α transfer reactions or transient quasi-molecular states.

This reaction mechanism mixing is not the unique source of discrepancy with the HFℓ predictions. Despite these low dissipation events have been excluded and all the kinematic characteristics were reproduced, the branching ratios of the multiple $\alpha-$decay channels were found to be still largely underestimated by the calculation.

A detailed analysis of the multiple $\alpha-$particle correlations in these channels indicates a sequential process with a clear hierarchy in the emission sequence. The highest probability is associated with the first chance emission of an $\alpha-$particle from a highly excited ^{24}Mg compound, leaving the daughter ^{20}Ne nucleus still well above its particle emission threshold. Then, the excited ^{20}Ne preferentially emits another $\alpha-$particle, leaving an Oxygen final evaporation residue. Neutron emission was not directly measured, but it was attributed to the last emission step, according to kinematical correlations. As a general conclusion, the persistence of cluster structures for ^{24}Mg and/or its daughter nucleus ^{20}Ne, at excitation energies well above the energy threshold for full disintegration into $\alpha-$particles, can be inferred.

Acknowledgments

Thanks are due to the accelerator staff of Legnaro Laboratories for having provided good quality beams. This work was supported in part by the European Funds for Large Scale Facilities (Seventh Framework Program ENSAR 262010), by grants of Alma Mater Studiorum (Bologna University) and by grants of Italian Ministry of Education, University and Research under contract PRIN 2010-2011.

References

1. K. Ikeda, N. Tagikawa and H. Horiuchi, *Prog. Theor. Phys. Suppl.* **E68**, 464 (1968).
2. S. J. Sanders, A. Szanto de Toledo and C. Beck, *Phys. Rep.* **311**, 487–551 (1999).
3. M. Freer, *Rep. on Prog. in Phys.* **70**, 2149 (2007).
4. C. Beck (ed.), *Clusters in Nuclei - Vol.1* Lecture Notes in Physics **818**, (2010).
5. C. Beck (ed.), *Clusters in Nuclei - Vol.2* Lecture Notes in Physics **848**, (2012).
6. C. Beck (ed.), *Clusters in Nuclei - Vol.3* Lecture Notes in Physics **875**, (2014).
7. G. Roepke, *Phys. Rev. C* **92**, 054001 (2015).
8. R. B. Wiringa, S. C. Pieper, J. Carlson and V. R. Pandharipande, *Phys. Rev. C* **62**, 014001 (2000); Y. Kanada-Enyo and H. Horiuchi, *Prog. Theor. Phys.*

Suppl. **142**, 205 (2001); T. Neff and H. Feldmeier ,*Nucl. Phys. A* **738**, 357 (2004).
9. A. Tohsaki, H. Horiuchi, P. Schuck and G. Rpke, *Phys. Rev. Lett.* **87**, 192501 (2001); Y. Funaki, T. Yamada, H. Horiuchi, G. Rpke, P. Schuck and A. Tohsaki, *Phys. Rev. Lett.* **101**, 082502 (2008).
10. H. Takemoto, *et al.*, *Phys. Rev. C* **69**, 035802 (2004).
11. J. A. Maruhn, M. Kimura, S. Schramm, P. G. Reinhard, H. Horiuchi and A. Tohsaki, *Phys. Rev. C* **74**, 044311 (2006); P. G. Reinhard, J. A. Maruhn, A. S. Umar and V. E. Oberacker, *Phys. Rev. C* **83**, 034312 (2011).
12. M. Girod and P. Schuck, *Phys. Rev. Lett.* **111**, 132503 (2013).
13. J. P. Ebran, E. Khan, T. Niksic and D. Vretenar, *Nature* **487**, 341 (2012); *Phys. Rev. C* **89**, 031303(R) (2014).
14. E. Costanzo *et al.*, *Phys. Rev. C* **44**, 111 (1991); F. Grenier *et al.*, *Nucl. Phys. A* **811**, 233 (2008); Ad. R. Raduta *et al.*, *Phys. Lett. B* **705**, 65 (2011); Tz. Kokalova, N. Itagaki, W. von Oertzen and C. Wheldon, *Phys. Rev. Lett.* **96**, 192502 (2006); W. von Oertzen *et al.*, *Eur. Phys. Journ. A* **36**, 279 (2006); A. Di Pietro *et al.* *Journal of Physics: Conference Series* **366**, 012013 (2012); E. Vardaci *et al.*, *Journal of Physics: Conference Series* **436**, 012054 (2013) and references quoted therein; B. Borderie, *et al.*, *Phys. Lett. B*, in press (2016).
15. E. R. Cosman *et al.*, *Phys. Rev. Lett.* **35**, 265 (1975); C. A. Bremner *et al.*, *Phys. Rev. C* **66**, 034605 (2002); T. M. Cormier, C. M. Jachcinski, G. M. Berkowitz, P. Braun-Munzinger, P. M. Cormier, M. Gai, J. W. Harris, J. Barrette and H. E. Wegner, *Phys. Rev. Lett.* **40**, 924 (1978); A. Morsad, F. Haas, C. Beck and R. M. Freeman, *Zeit. Phys. A* **338**, 61 (1991).
16. A. H. Wuosmaa *et al.*, *Phys. Rev. Lett.* **68**, 1295 (1992); M. Aliotta *et al.*, *Zeit. Phys. A* **353**, 43 (1995); R. A. Le Marechal *et al.*, *Phys. Rev. C* **55**, 1881 (1997).
17. M. Takashina *et al.*, *Nucl. Phys. A* **738**, 352 (2004).
18. G. Baiocco *et al.*, *Phys. Rev. C* **87**, 054614 (2013).
19. G. Baiocco, *PhD Thesis, Università di Bologna and Université de Caen Basse-Normandie* (2012); http://amsdottorato.cib.unibo.it/4295/.
20. L. Morelli *et al.*, *J. Phys. G* **41**, 075107 (2014).
21. L. Morelli *et al.*, *J. Phys. G* **41**, 075108 (2014).
22. L. Morelli *et al.*, *J. Phys. G* **43**, 045110 (2016).
23. L. Morelli *et al.*, contribution to this topical volume.
24. L. Morelli *et al.*, *Proceedings of Science X LASNPA* 62 (2013); L. Morelli *et al.*, *EPJ Web of Conferences* **vol. 66**, 03064 (2014); http://dx.doi.org/10.1051/epjconf/20146603064; http://pos.sissa.it/archive/conferences/194/062/X{\%}20LASNPA_062.pdf; L.Morelli *et al.*, *EPJ Web of Conferences* **vol.88**, 26 (2015); http://dx.doi.org/10.1051/epjconf/20158800026.
25. M. Bruno *et al.*, *Eur. Phys. Journ. A* **49**, 128 (2013).
26. F. Gramegna *et al.*, *Nucl. Instr. And Meth. A* **389**, 474 (1997); F. Gramegna *et al.*, 2004 IEEE Nucl. Science Symposium, Rome, 16–22 October (2004).
27. A. Moroni *et al.*, *Nucl. Instr. And Meth. A* **556**, 516 (2006).
28. G. Pasquali *et al.*, *Nucl. Instr. And Meth. A* **570**, 126 (2007).

29. T. Ichikawa *et al.*, *Phys. Rev. Lett.* **109**, 232503 (2012).
30. Z. Chen and C. K. Gelbke C K *Phys. Rev. C* **38**, 2630 (1988).
31. M. E. Ortiz *et al.*, *Phys. Rev. C* **25**, 1436 (1982).
32. D. Mahboub *et al.*, *Phys. Rev. C* **69**, 034616 (2004).
33. O. B. Tarasov and D. Bazin *Nucl. Instr. And Meth. B* **204**, 174 (2003).
34. http://www.chemistry.wustl.edu/~rc/gemini++/
35. http://www.nndc.bnl.gov/amdc/
36. http://www.nndc.bnl.gov/nudat2/
37. T. Von Egidy and D. Bucurescu, *Phys. Rev. C* **72**, 044311 (2005); T. Von Egidy and D. Bucurescu, *Phys. Rev. C* **73**, 049901 (2006).
38. A. V. Ignatyuk *et al.*, *Sov. J. Nucl. Phys.* **21**, 255 (1975); A. S. Iljinov *et al.*, *Nucl. Phys. A* **543**, 450 (1979).
39. J. Toke and W. J. Swiatecki, *Nucl. Phys. A* **372**, 141 (1981).
40. M. V. Ricciardi *et al.*, *Nucl. Phys. A* **733**, 299 (2004).
41. M. Pfützner, M. Karny, L. V. Grigorenko and K. Riisager, *Rev. Mod. Phys.* **84**, 567 (2012); L. V. Grigorenko *et al.*, *Phys. Rev. C* **80**, 034602 (2009); I. A. Egorova *et al.*, *Phys. Rev. Lett.* **109**, 202502 (2012); R. Charity *et al.*, *Phys. Rev. C* **84**, 014320 (2011).
42. G. G. Ohlsen, *Nucl. Instr. And Meth.* **37**, 240 (1965).

CHAPTER 10

PARTICLE-PARTICLE CORRELATIONS: A TOOL FOR INVESTIGATING EXCITED STATES AND CLUSTERING EFFECTS IN THE DECAY OF EXCITED NUCLEI

L. Morelli, M. Bruno and M. D'Agostino

Dipartimento di Fisica ed Astronomia dell'Università and INFN, Bologna, Italy

G. Baiocco

Dipartimento di Fisica dell'Università, Pavia, Italy

F. Gulminelli

Normandie Univ, ENSICAEN, UNICAEN, CNRS/IN2P3, LPC Caen, 14000 Caen, France

T. Marchi*

INFN, Laboratori Nazionali di Legnaro, Legnaro, Italy

S. Barlini

Dipartimento di Fisica dell'Università and INFN, Firenze, Italy

In this contribution we present a selection of results from experiments performed at INFN Laboratori Nazionali di Legnaro exploiting the 4π GARFIELD + RCo apparatus. The investigated reactions are the medium-heavy ^{32}S + ^{58,64}Ni, and the light ^{12}C + ^{12}C and ^{14}N + ^{10}B systems. For the first set of reactions, we extracted information on fragments in the last but one

*Present address:KU Leuven, Instituut voor Kern-en Stralingsfysica, B-3001 Leuven, Belgium

stage of the compound nucleus decay using particle-particle and fragment-fragment correlation functions. In particular, the issue of odd-even staggering was addressed. We show that the distribution of excited fragments displays odd-even effects, but reversed with respect to the asymptotic distributions. The conclusion is that the staggering cannot be only due to the pairing effect of nuclear masses, but it is also influenced by pairing and possibly isospin effects in the level density. In the light nuclei reactions the possible α-cluster states in even-even nuclei were investigated through particle-particle correlations. Thanks to this powerful technique we were able to reconstruct the particle emission sequence, thus showing that multiple α particle emission from the decay of $^{24}Mg^{*}$ and $^{12}C^{*}$ can be understood as a sequential process with an important contribution of doorway Be states. Other chosen applications of the correlation function technique are also briefly reviewed, including measurement of the size, emission time, temperature and symmetry energy of excited nuclear sources.

1. Introduction

Complex heavy-ion reactions allow producing transient nuclear systems excited in the continuum, well above the particle and fragment emission threshold. Therefore these reactions give a unique opportunity to experimentally probe nuclear systems in extreme conditions of density, temperature and isospin ratio. Because of the extremely short reaction times in the femtosecond range, the properties of these excited nuclei cannot be measured directly but only through the decay products of the reaction. The problem is that the exit channel of these reactions produces a large number of particles and fragments that typically cannot be fully detected on an event-by-event basis, which makes the experimental analysis extremely challenging. On the experimental side, complex detectors based on a (quasi)-4π geometry, with high granularity and low detection and energy thresholds, have been built. In this contribution we will concentrate on one of these last generation detection systems, the GARFIELD-RCo multi-detector.[1] From the viewpoint of data analysis, it is clear that complex observables have also to be defined, in order to disentangle the many-body information of the exit channel. Indeed inclusive observables in complex nuclear reactions can lead to conflicting interpretations, since they can typically be reproduced by different models. It is therefore important to extend the comparison to exclusive observables. In particular particle-particle and particle-fragment correlations are an ideal tool to study properties of decaying excited nuclei, and can be used to extract detailed information both on the property of the excited source, and on the mechanism of the decay.

In this contribution we will review some of the numerous applications of the correlation function technique. Most of the presented results were obtained in two different measurement campaigns, performed at the LNL tandem in Legnaro.

The plan of the paper is as follows. Sections 2 and 3 briefly recall the main characteristics of the detection system, and the data selection criteria adopted to select quasi-complete detected events, as well as to disentangle the different reaction mechanisms in the measured data sets. Section 4 is devoted to the analysis of the experimental results. The correlation function technique is first introduced in Section 4.1, and then applied to different exclusive observables. Particle-fragment relative energy correlation functions are shown in Section 4.1.1 to give quantitative information on the population of excited states prior to their particle decay. The importance of pairing effects in the nuclear level density is demonstrated. Particle-particle relative energy correlations can be used both to determine the size and lifetime of excited nuclear sources (Section 4.1.2), and to characterize the evaporation sequence of the decay (Section 4.2). In particular, the coupling with the Dalitz plot analysis allows assessing in a model-independent way the simultaneous or sequential character of the decay, the doorway state population, and also to detect exotic structures as linear cluster chains. This interesting application is exploited in Sections 4.2.1 and 4.2.2, which analyze the decay of the Hoyle state excited in peripheral $^{12}C+^{12}C$, and of the exotic 6-α decay of ^{24}Mg obtained in fusion events. In both cases we show that the data are compatible with an essentially sequential decay, with an important contribution of doorway Be states. Finally, correlations between isotopes yields are analyzed in Section 4.3. They can be used to determine fundamental properties of the excited nuclei, such as their temperature (Section 4.3.1), density, and symmetry energy (Section 4.3.2). Conclusions are drawn in Section 5, where summary and future perspectives are briefly given.

2. The measurements

The measurements have been performed at *Laboratori Nazionali di Legnaro* with beams from the Tandem-Alpi complex, in the framework of the INFN NUCL-EX collaboration.[a] The aim of these measurements consisted

[a]M. Bruno, M. D'Agostino, L. Morelli (Bologna); G. Baiocco (Pavia); M. Cinausero, F. Gramegna, T. Marchi (LNL); D. Fabris (Padova): S. Barlini, M. Bini, G. Casini, A. Olmi, G. Pasquali, S. Piantelli, S. Valdrè (Firenze).

Fig. 1. The GARFIELD forward chamber open for maintenance.

in investigating reaction mechanisms and decay of excited nuclei formed in nuclear reactions, by detecting charged particle characteristics.

In this paper we will consider experiments with ^{32}S beam at 14.5 AMeV interacting with ^{58}Ni and ^{64}Ni targets and with light ion beams such as ^{12}C at 95 MeV onto a ^{12}C target and ^{14}N at 80.7 MeV onto a ^{10}B target. The apparatus used is a combination of GARFIELD and RCo detectors, which has been extensively described in Ref. 1. GARFIELD is made by two drift chambers filled with CF_4 gas at a pressure of 50 mbar. The forward gas chamber is divided in 24 sectors (see Fig. 1), each one divided in two subsectors (covering 7.5° each in azimuthal angle) and in four polar angle regions by four CsI(Tl) scintillators covering approximately 14°. In this way the forward chamber allows the definition of 192 ΔE-E telescopes. The electrodes of the chamber consist in micro-strips arranged in such a way to allow for electron multiplication, being able to detect very small energy loss. Threshold identifications as low as $0.8 \div 1$ MeV are therefore possible. The backward chamber has an azimuthal aperture of 45° and is divided in 21 sectors with the same characteristics as in the forward chamber, allowing for 168 ΔE-E telescopes. ΔE-E telescopes identify particles and fragments in charge. Charge and mass of light particles (up to Z = 3) are identified thanks to the Fast-Slow Pulse Shape Analysis (PSA) of CsI(Tl) signals.[2]

Fig. 2. The Ring Counter ready for operations.

The RCo (Ring Counter) consists in a Ionization Chamber (IC) divided in 8 azimuthal sectors, each one followed by a 8-strip 300 μm thick silicon detector (Si) and 6 CsI(Tl) scintillators (see Fig. 2), defining 128 pseudo-telescopes. Charge identification can be obtained by ΔE-E technique in IC-Si or Si-CsI telescopes or through PSA in reverse mounted Silicons.[1] Masses of light particles and fragments can be obtained in Si-CsI ΔE-E analysis, by Fast-Slow Pulse PSA in CsI(Tl) or (to some extent) by PSA in Silicon detectors.[3]

3. The data selection

3.1. $^{32}S + ^{58}Ni$ *and* $^{32}S + ^{64}Ni$ *reactions*

We exploited the method of "shape analysis"[4] to separate the most peripheral and the most central reactions. We have built the momentum tensor for all the emitted fragments ($Z \geq 3$), extracting eigenvectors and eigenvalues. The event shape is an ellipsoid and it is possible to define the flow angle, θ_{flow}, i.e., the angle between the eigenvector corresponding to the largest eigenvalue and the beam direction. The correlation between the flow angle and the total detected charge allows for selecting central and peripheral events. A further selection for these reactions consists in the condition that at least 50% of the total incoming momentum parallel to the beam is detected.

These criteria allow to select peripheral events with $Z_{tot} \leq 25$, together with a further condition on $\cos(\theta_{flow}) \geq 0.77$ to avoid any contamination by

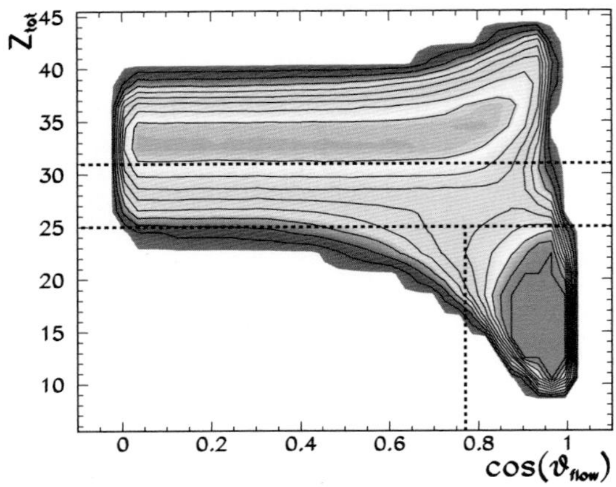

Fig. 3. Correlation of the total detected charge to cosinus of the flow angle for the ^{32}S + ^{58}Ni reaction at 14.5 AMeV.

non-peripheral events. With these selections we can observe that peripheral events are characterized by a detected charge close to the one of the projectile and a flow angle peaked in the forward direction, as it is shown in Fig. 3. Central events are selected with the condition of detecting at least 70% of the total charge, i.e., $Z_{tot} \geq 31$. They are characterized by a flow angle nearly uniformly distributed over the whole angular range, indicating an approximately isotropic distribution of the emitted fragments. With these selections for central events one finds that there is always a relatively large evaporation residue and a high charged particle multiplicity.[5]

3.2. ^{12}C+ ^{12}C *and* ^{14}N+ ^{10}B *reactions*

A selection of the same type as the one for the S + Ni reactions cannot be performed for these light nuclei reactions, since the multiplicities of charged particles are much lower than in the previous case, and multi-dimensional variables like the flow angle are not suited.

Central events have been selected allowing for the detection of the evaporation residue in the forward direction (RCo) and one or more light particles at large θ angles (GARFIELD). The evaporation residues are characterized by a velocity close to the centre of mass velocity of the reaction.[6] A two-dimensional total detected charge *vs.* total longitudinal momentum can be built. The corresponding plot is shown in Fig. 4. The fusion-evaporation events are located in the upper right region of Fig. 4.

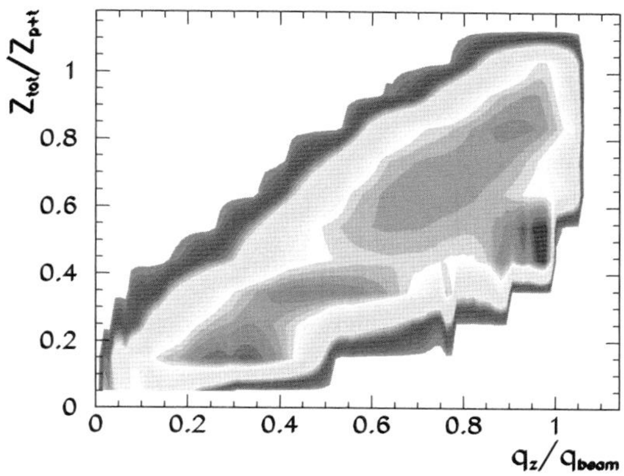

Fig. 4. Correlation between the total detected charge and the total detected momentum along the beam axis for the ^{12}C-^{12}C reaction at 95 MeV.

Due to the large number of events where the total charge is completely detected, a very stringent selection can be made, keeping only events where the total incoming charge ($Z_{tot} = 12$) and at least 95% of the incoming momentum have been measured.

A subset of peripheral events, corresponding to inelastic reactions $^{12}C + {}^{12}C \rightarrow {}^{12}C^* + {}^{12}C$ was also selected, allowing for the detection of three α-particles in the forward cone and no charged particles in the remaining solid angle. In this way we were able to select channels where the quasi-projectile and the quasi-target were in a well defined single quantum state. In particular we have analyzed events where the quasi-target was left in the ground state and the quasi-projectile was excited in the 7.65 MeV Hoyle[7] or in the 9.64 MeV state.[8]

4. Correlations

When comparing experimental results with model predictions it often happens that agreements or discrepancies hardly provide a conclusive validation of the model, especially when different models describe the experimental results with a similar degree of accuracy. This is often the case when inclusive observables are analyzed, such as, in heavy ion reactions, energy or angular distributions of a given light particle or fragment. A deeper insight into the mechanism of a reaction can be obtained exploiting exclusive variables or correlations among observables. Different techniques can be used to emphasize the adequacy of a model to explain physical results, from simple two dimensional plots to sophisticated correlation analyses. In the following we will illustrate some of the methods that have been used to characterize the reactions under study. Three different techniques will be exploited in the next subsections, namely

(1) relative energy particle-fragment correlation functions, which allow studying the properties of the excited fragments in their last stage of de-excitation;
(2) Dalitz plots, coupled to the particle-particle correlation function technique, for the characterization of the properties of three-body systems, and their decay mechanism;
(3) correlations among isotopes emitted in the decay stage of an excited nucleus, allowing to extract information on the temperature and to shed light on the symmetry energy coefficient C_{sym}.

In Sections 4.1, 4.2 and 4.3 we will show selected results obtained for the reactions described in Sec. 2.

4.1. *The correlation function technique*

The correlation function technique, originally introduced in astronomy to measure the size of distant objects,[9] has been used since the nineties as a powerful tool to investigate size and decay time emission of sources obtained in nucleus-nucleus collisions.[10,11] It has been applied in a large variety of nuclear reactions and different decay mechanisms.

Starting from $\vec{p}_1$ and $\vec{p}_2$, momenta of the two particles 1 and 2, one builds the relative energy $E_{rel} = \frac{(\vec{p}_1 - \vec{p}_2)^2}{2\mu}$, where μ is the reduced mass of the two-body system, and calculates the yield Y_{12} in a given bin of relative energy. The correlation function is defined as:

$$1 + R(E_{rel}) = \frac{\sum Y_{12}(\vec{p}_1, \vec{p}_2)}{\sum Y_1(\vec{p}_1) Y_2(\vec{p}_2)} \tag{1}$$

where Y_{12} is the two-particle coincidence yield of a given pair of particles and the $Y_i(\vec{p}_i)$ are the single particle yields for the two particles. Following the event mixing method, these quantities are generally extracted from two particles belonging to different events. In this way, with respect to Y_{12} one gets rid of the two single particle phase space.

When dealing with charged particles, it is necessary to consider the modifications of the two particle phase space by the long range Coulomb and short range nuclear interactions and to disentangle such contributions.

We can write:

$$1 + R(E_{rel}) = 1 + R_{Coul}(E_{rel}) + R_{nucl}(E_{rel}) \tag{2}$$

where $R_{nucl}(E_{rel})$ can be expressed in a Breit-Wigner formalism.[5] As suggested in Refs. 12 and 13 an empirical expression of the Coulomb contribution is given by:

$$1 + R_{Coul}(E_{rel}) = 1 - exp[-(E_{rel}/E_C)^{\gamma}] \tag{3}$$

which vanishes at zero relative energy and reaches unity at large relative energy.

A fitting procedure is performed on experimental data with the Coulomb background parameters E_C and γ, the nuclear ones and the weight of different contributions of excited states as free parameters. In this way the weight of different excited levels of the source can be determined.

Examples of typical correlation functions and of the resulting fits are shown in Fig. 5 for the deuteron-alpha pair looking for ^{6}Li* excited

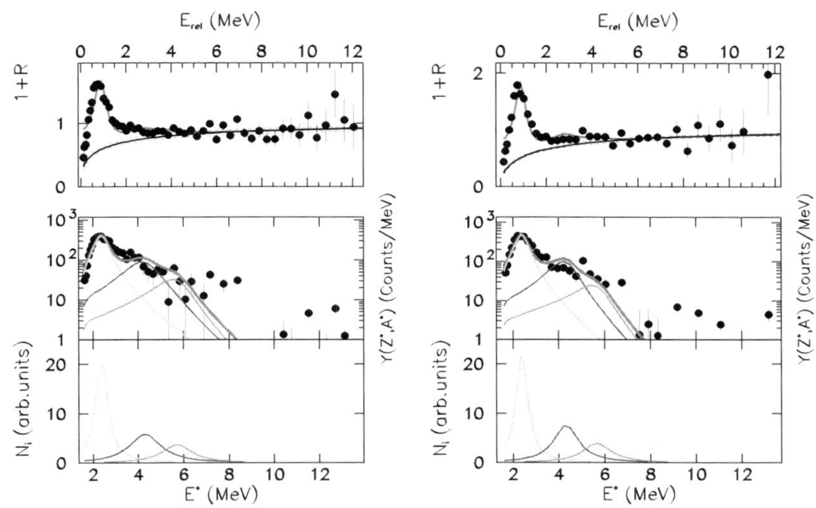

Fig. 5. d+α correlation functions obtained for peripheral (left) and central (right) ^{32}S + ^{58}Ni collisions at 14.5 AMeV. Upper panels: experimental correlation (symbols) and theoretical expectation from the known excited state information of ^{6}Li (red lines). The average Coulomb background (see text) is given by a full black line. The dashed lines give the uncertainty on the Coulomb fit. Middle panels: deduced population of ^{6}Li excited states. Both the population of the different states and the total population are given. Lower panels: Breit Wigner distributions corresponding to the states presented in the middle panels.

state population in peripheral (left panels) and in central (right panels) collisions. Primary yields are calculated by multiplying the nuclear contribution $R - R_{Coul}$ of the correlation function for the uncorrelated yield: $Y_{12}(E^*) = (R(E^*) - R_{Coul}(E^*)) \sum_E Y_1 Y_2$.[12–15] This experimentally reconstructed primary population is shown by full symbols in the middle panels of Fig. 5, together with the contributions from the different parent excited levels, shown as lines. The bottom panels of Fig. 5 additionally show the extracted Breit-Wigner distribution of the different populated $^6Li^*$ excited states.

Exploiting this technique, one can study the population of fragments excited well above the particle separation threshold, prior to their decay. The results obtained with this procedure can be used for different purposes:

- in heavy ion reaction at or close to Fermi energies the goal is to reconstruct from the final fragments the partition at the freeze-out stage. This allows for comparison with predictions of Statistical multifragmentation models, such as SMM[16] or MMM[17];
- to investigate the size and emission time of nuclear systems excited in the continuum;

In the following we will present a selection of the results obtained by the NUCL-EX collaboration on these different issues.

4.1.1. *Primary fragment reconstruction*

The fragment-fragment correlation functions have been used mainly to reconstruct primary fragment distributions. The results shown in Fig. 5 refer to peripheral (left panels) and central (right panels) $^{32}S + ^{58}Ni$ collisions. The panels show the reconstruction of the $^6Li^*$, with results similar for the two sets of data. As it can be seen in the left middle panel one can recognize the 2.186 MeV and 4.31 MeV levels corresponding roughly to 2/3 and 1/4 of the overall population of the $^6Li^*$ excited states. The analysis of central collisions (right part) leads to similar results. It has to be noted that the contribution of the excited states extends relatively far from the peaks. In addition, these are not symmetric, due to the phase space and the suppression factors.[14] However the extracted Breit-Wigner distributions, once corrected with these deformation factors, show nearly symmetric shapes as expected. This is shown in the lower panels of Fig. 5.

Other examples of particle-fragment correlation functions are shown in Fig. 6, for the total set of data (central + peripheral collisions). We can see

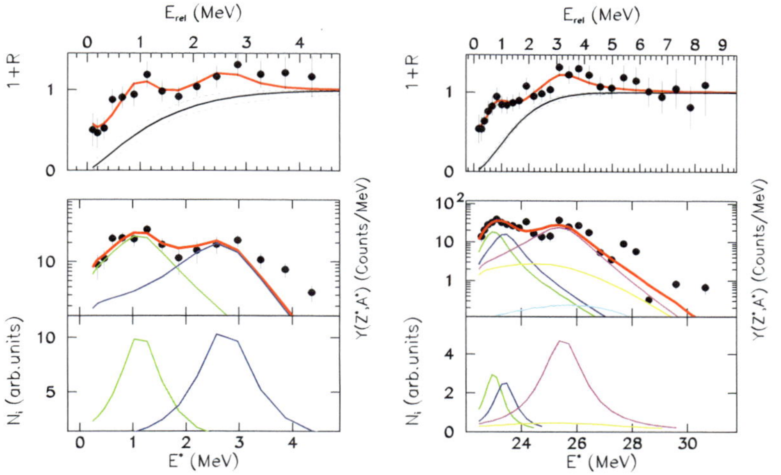

Fig. 6. Total p+^{7}Be (left) for ^{32}S + ^{64}Ni and d+^{6}Li (right) correlation functions for ^{32}S + ^{64}Ni collisions at 14.5 AMeV. Upper panels: experimental correlation (symbols) and theoretical expectation from the known excited state information of ^{8}Be* and ^{8}B*, respectively (red lines). The average Coulomb background (see text) is given by a full black line. The dashed lines give the uncertainty on the Coulomb fit. Middle panels: deduced population of ^{8}B and ^{8}Be excited states. Both the population of the different states and the total population are given. Lower panels: Breit Wigner distributions corresponding to the states presented in the middle panels.

that again an excellent reproduction of experimental correlation functions can be obtained, and this is true for all the other correlation functions.[14]

For the d+^{6}Li correlation function (right panels of Fig. 6), the ^{8}Be* excitation energy is above the threshold for the ^{6}Li decay in α + d. This means that a part of the ^{8}Be* population can be found in three-body α-d-d events. It would be interesting to perform a three-body decay correlation α + d + d to evaluate the total primary ^{8}Be* population. Unfortunately the statistics does not allow for this type of correlations and therefore we can only study what happens in the last stage of decay.

It is important to remark that the fit parameters determine the overall quality of the correlation function, and are not specific to the different excited states. These latter are fully determined by their spectroscopic information (spin, energy and width) which are experimentally known. This means that the determination of the excited state population is reliable even when no peak is visible in the correlation function, as it is the case for the broad resonance around 26 MeV in ^{8}Be* in Fig. 6.

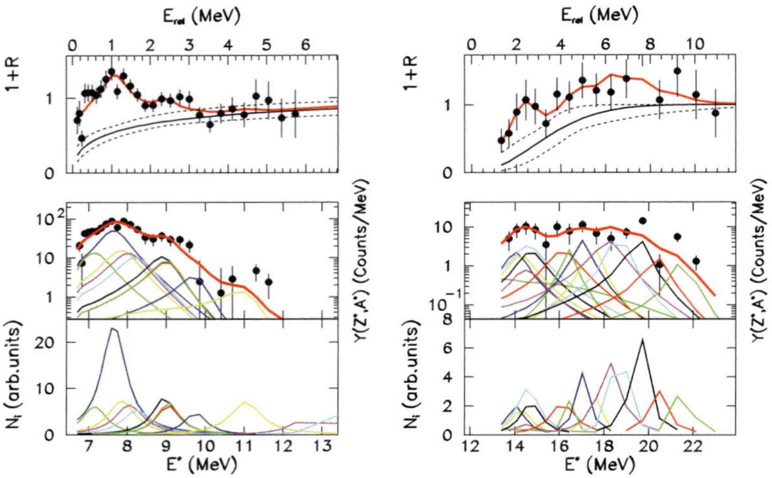

Fig. 7. Total p+^{9}Be (left) and ^{3}He+^{12}C (right) correlation functions measured in ^{32}S + ^{58}Ni collisions at 14.5 AMeV. Upper panels: experimental correlation (symbols) and theoretical expectation from the known excited state information of ^{10}B* and ^{15}O*, respectively (red line). The average Coulomb background (see text) is given by a full black line. The dashed lines give the uncertainty on the Coulomb fit. Middle panels: deduced population of ^{10}B and ^{15}O excited states. Both the population of the different states and the total population are given. Lower panels: Breit Wigner distributions corresponding to the states presented in the middle panels.

In the left panels of Fig. 6, the p + ^{7}Be correlation function is shown for the reaction ^{32}S + ^{64}Ni, as another example of two levels contributing to the excited ^{8}B*.

Finally in Fig. 7 we show two cases where several levels are involved. In the left panels of Fig. 7 the dominant contribution is given by a ^{10}B* level around 7.8 MeV, whereas in the right part of the same figure all the levels contribute with more or less the same strength to the excited states of ^{15}O*. Again, the Breit-Wigner fit allows to extract from any of these correlation functions the relative population of primary fragments, with a reliable separation of the different excited states even when they are not experimentally resolved.

The relative energy correlation function technique can be used to extract the "excited levels" temperature of the decaying system.[12,13] The weight of the excited levels, shown in Figs. 5–7 and in Ref. 14, have been also used to investigate the origin of the odd-even staggering effect on the charge distribution of fragments produced in the reactions ^{32}S + ^{58,64}Ni.

The staggering effect has been found in several reactions.[18] A possible interpretation was given in Ref. 19. Based on comparison to the evaporation ABLA model, the staggering was tentatively attributed to the difference in the proton and neutron separation energies in the last stage of the decay. In Ref. 14 the "warm" fragment distribution, i.e. the distribution of fragments in the last but one step of the decay sequence, has been reconstructed. As a result, the presence of the staggering effect was evidenced also for the excited fragment distribution, even if in opposite direction with respect to the distribution of final fragments. The *counter – staggering* of fragments at the last but one step indicates that not only the last decay step is responsible for the odd-even effect, but the effect concerns the whole evaporation chain. This also leads to the conclusion that staggering effects cannot be attributed only to the pairing term in the nuclear mass, but also to the interplay of pairing and isospin effects in the level density.

4.1.2. *Size and emission time of hot decaying system*

We report in the following selected results for central events of the reaction Au + Au at 35A MeV,[22] with the purpose of showing another use of the fragment-fragment correlation functions. The decay of an equilibrated excited nuclear system can be treated statistically. The assumptions of statistical codes can be a sequence of binary decays[20] or a simultaneous process through the freeze-out stage.[16,17] The comparison of experimental data to model predictions can be a way to evaluate the size of the source and the emission time of fragments.

A correlation function-based, model-independent method to assess the same information is based on the final-state interactions between emitted fragments. Indeed one can define the reduced velocity correlation function, similarly to Eq. (1), as:

$$1 + R(v_{red}) = C\frac{N_{corr}(v_{red})}{N_{uncorr}(v_{red})} \tag{4}$$

with

$$v_{red} = \frac{\vec{p}_1/m_1 - \vec{p}_2/m_2}{\sqrt{Z_1 + Z_2}}. \tag{5}$$

The background yield N_{uncorr} is built with fragments from different events.[11]

The Coulomb repulsion leads to a suppression of the two-fragment correlation function at small relative velocities. The width of this Coulomb

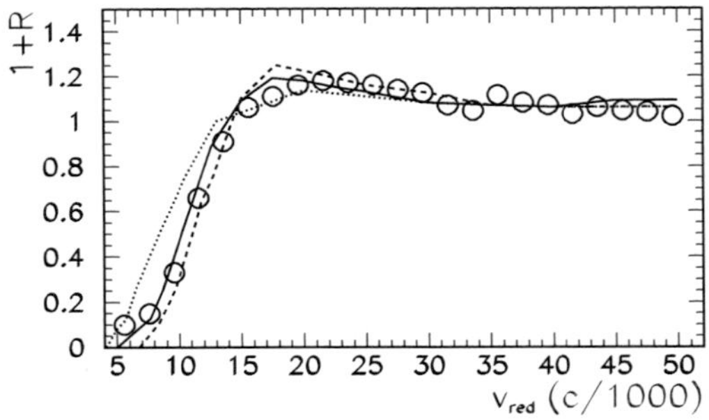

Fig. 8. Two-fragment correlation functions ($3 \leq Z_F \leq 20$) for the reaction ^{192}Au + ^{192}Au at 35A MeV measured by the MULTICS-Miniball collaboration.[22] Open symbols show experimental data. The solid, dashed and dotted lines are the the many-body trajectory predictions for τ = 85,50,150 fm/c, respectively.

hole increases with decreasing time between the emission of two subsequent fragments. The experimental results are then compared to a many body trajectory simulation,[21] starting from a source of radius R_S, mass M_S and charge Z_S. The fragments are emitted sequentially, and their characteristics are obtained by random sampling the experimental yields.

Different results are obtained with different emission times τ, with probability distributions $P(t) \propto \exp(-t/\tau)$. The Coulomb trajectory calculations for fragments, i.e. the velocities of the emitted fragments in the Coulomb field of the source, are performed randomly sampling the momenta distributions.

As an example of this technique, we show in Fig. 8 the result for the two fragment correlation function measured in central events of the reaction Au + Au at 35A MeV.[22] The experimental data are compatible with very short lifetimes ($\approx$ 85 fm/c) at the limits of the sensitivity of the model, confirming in this case a nearly instantaneous decay of the excited system. Results for other reactions show values of the lifetime more compatible with sequential decay.[23]

4.2. *Decay mechanisms of cluster excited states*

Another interesting application of correlation function is the study of the decay mechanism of excited cluster states. α-cluster correlations are thought to be ubiquitous structures in a large number of light and medium-heavy even-even nuclei at excitation energies around the threshold of breakup into constituent clusters.[24–27] From the experimental viewpoint, rotational bands consistent with α-cluster structures have been identified in different even-even light nuclei and shown to persist even along their isotopic chains.[28] Exotic non-statistical decays of these correlated states have been evidenced in the recent literature.[29] In this framework, it is interesting to understand if the final decay to α-particles can be considered as a simultaneous decay,[30] or if it proceeds through doorway Be states, as it is generally thought to be the case for the famous Hoyle state[31] in ^{12}C.

The ^{12}C + ^{12}C reaction, where both the collision partners and the fused system have well-known cluster structures, is an ideal ground for such a study. As explained in Section 3.2, we have selected both peripheral events leading to the production of a ^{12}C* source at low excitation energy, and fusion-evaporation events corresponding to the formation of a ^{24}Mg* at a high excitation energy of several tens of MeV. In both cases, an enhanced probability of multiple alpha emission was observed, with respect to the prediction of a dedicated Hauser-Feshbach code. The characterization of clustered states at high excitation energy in ^{24}Mg* is detailed in another contribution to this volume.[32] Specifically, we show indications of the survival of the α-^{16}O-α linear chain structure at excitation energies above the threshold of neutron emission of ^{16}O. Here we concentrate on the mechanism of decay into α-particles, both in the case of the Hoyle state, and in the exotic events where the ^{24}Mg* compound nucleus decays into six α-particles.

4.2.1. *Investigating the $^{12}C^*$ Hoyle state*

As it is well known, the Hoyle state of ^{12}C is of great importance both for nuclear structure and astrophysics. In particular the decay mechanism of this state is crucial for determining the abundance of ^{12}C in the primordial universe. The results in the literature are contradictory; most of them[31] show that non sequential mechanisms are negligible, whereas some of them[33] indicate a sizeable contribution of direct processes. Since the second type of results has been obtained for heavy ion reactions at Fermi energies, it is possible that the excitation mechanism of the state has an influence on the branching ratios of the decay. To explore this possibility, we have analyzed two different set of events.[8] The first one corresponds to peripheral two-body reactions where the quasi projectile is excited to the Hoyle state and the quasi-target is left in the ground state. The second class of events corresponds to central events where the excited Carbon is formed in the decay chain of ^{24}Mg*. Even if the system is relatively light,

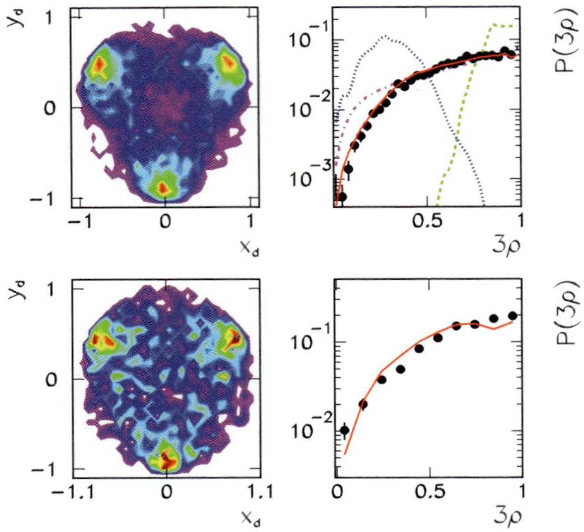

Fig. 9. Left panels: Dalitz plot of the energies of the three α-particles emitted from the 12C* Hoyle state in $^{12}C + ^{12}C$ reactions at 95 MeV. Right panels: Radial projections of the plots. Peripheral (central) collisions are shown in the upper (lower) panels. The lines represent predictions of HFℓ sequential decay (continuous) and for direct mechanisms DDE (dotted), DDL (dashed) and DDΦ (dot-dashed) (see text).

the presence of some in medium effects could indicate a deviation from a nearly pure sequential mechanism.

No sizeable difference has been found between the two sets of data,[8] even if a quantitative analysis has been possible only for peripheral events, due to the lack of statistics. As an example we show in Fig. 9 one of the correlations used for the analysis, i.e. the energy Dalitz plot of the three α-particles emitted by the $^{12}C^*$ and the radial projection of the plot.

The upper right panel of Fig. 9 shows the comparison of peripheral data to predictions of a sequential decay obtained with a dedicated Hauser-Feshbach code (HFℓ, see Ref. 32), and to three different direct decay mechanisms. These latter correspond to: three particles with equal energies (DDE), a linear α chain (DDL), and a uniform population of the phase space (DDΦ).

The upper limit of the direct decay mechanism resulted (1.1 ± 0.8) %, mainly due to the DDL contribution.

Our data are thus compatible with a sequential decay for this state. It would be interesting to extend this analysis to a heavier source in order

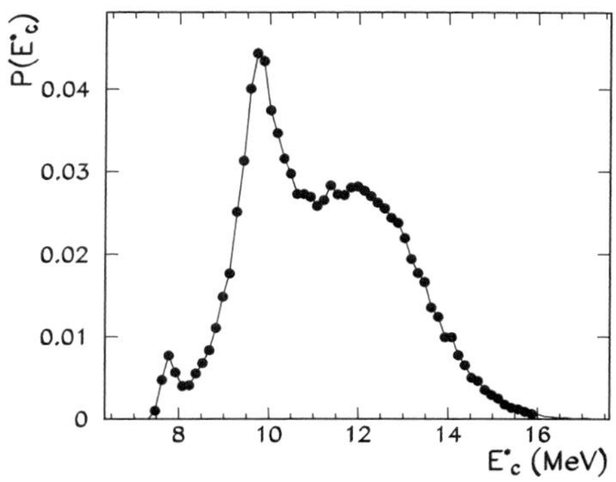

Fig. 10. Total energy distribution in the center of mass of the three lowest energy α particles detected in six-α events. Central $^{12}C + ^{12}C$ reactions at 95 MeV are considered.

to understand if in-medium effects can alter the observed sequential decay mechanism of the Hoyle state.

4.2.2. *Decay channels with six α-particles in the final state*

An important decay channel is the one with six α-particles in the final state. We have then selected the three α-particles out of the six, corresponding to the minimum total energy value. The energy distribution of these particles is compatible with an excited Carbon where the two lowest particle emitting states at 7.65 MeV and 9.64 MeV are clearly visible, as shown in Fig. 10. The high energy shoulder is a convolution of levels around 11−12 MeV. We can safely refer to these particles as to a (reconstructed) excited Carbon.

For these events, we have investigated the correlation between the three out of the six α-particles identified as reconstructed Carbon, and the set of the three remaining α-particles, hereafter referred to as Pseudo Carbon (PC).

If the mechanism of the decaying $^{24}Mg^*$ is a sequential statistical evaporation, the reconstructed Carbon represents the residue of the de-excitation, after three sequential α emissions from the compound nucleus. The Carbon then evaporates an α-particle and a ^8Be is formed. In this scenario the three α-particles of the Pseudo Carbon should not show any correlation.

This is not the case for the data. In Fig. 11 correlation functions are shown for α-particles belonging to the reconstructed Carbon (left panel) and to the PC (right panel). In both panels peaks corresponding to the ground state of the ^8Be are present. For the PC case also the first $^8Be^*$

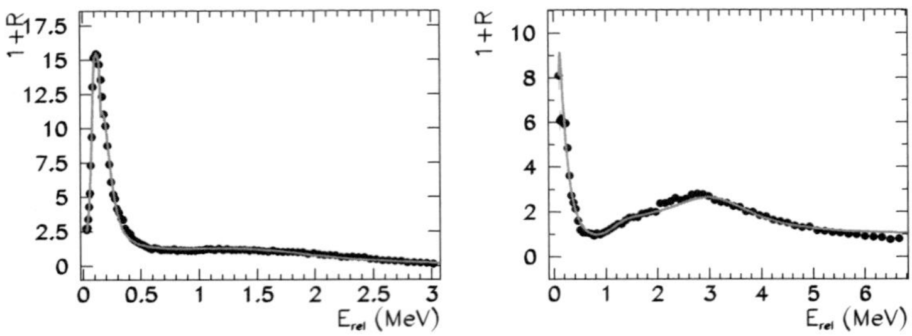

Fig. 11. $\alpha - \alpha$ correlation functions for Carbon (left panel) and PC (right panel) (see text). Dots are experimental results, continuous lines the fir with Eq. (2).

excited state at 3.03 MeV can be observed. The lines have been obtained by a fit with Eq. (2). We turn to investigate if the observed ^{8}Be's are emitted in the same event. Guided by the Breit-Wigner fits, we required a relative energy of alpha-particles smaller than 0.5 MeV for the reconstructed Carbon ($^8\text{Be}_{gs}$) and smaller than 6 MeV ($^8\text{Be}_{gs}$, $^8\text{Be}^*$) for the PC case. With these conditions, two ^{8}Be resulted in coincidence approximately in the 40% of the six alpha-particle events.

To summarize the results, in competition with the sequential emission of six α-particles the following channels are possible:

- $^{24}\text{Mg}^* \rightarrow\ ^{12}\text{C}^* +\ ^{12}\text{PC}^* \rightarrow\ ^8\text{Be} + \alpha +\ ^8\text{Be} + \alpha$
- $^{24}\text{Mg}^* \rightarrow\ ^{20}\text{Ne}^* + \alpha \rightarrow\ ^{12}\text{C}^* +\ ^8\text{Be} + \alpha \rightarrow\ ^8\text{Be} + \alpha +\ ^8\text{Be} + \alpha$
- $^{24}\text{Mg}^* \rightarrow\ ^{16}\text{O}^* +\ ^8\text{Be} \rightarrow\ ^{12}\text{C}^* +\ ^8\text{Be} + \alpha \rightarrow\ ^8\text{Be} + \alpha +\ ^8\text{Be} + \alpha$

where, as mentioned, the two ^{8}Be can be found either both in the ground state, or one in the ground state and the other in an excited state. Further analysis and experimental data are needed to quantify the weight of these decay chains and link these results to clustering effects in the $^{24}\text{Mg}^*$ or in the $^{12}\text{C}^*$ states.

4.3. *Correlations among isotopes*

In experiments where one can get information on the yield of isotopes emitted in the decay of an excited nucleus, isotope correlations can be built. In particular, the yield of the measured isotopes can give information on:

- the temperature of the decaying excited system, from the ratio of fragment populations measured in a given reaction;

- the scaling behaviour and the symmetry coefficient term C_{sym}, from the analysis of the ratio of isotopes coming from neutron-rich and neutron-poor systems excited in different reactions.

These two applications will be briefly reviewed in the following subsections.

4.3.1. *Temperature of the decaying system*

A method for extracting the temperature of hot excited systems has been developed in Ref. 34 and exploits correlations between isotopes emitted in the decay stage. The method consists in the calculation of the double ratio of isotopes in the ground state. Some assumptions are needed: a single emission source should be present, neutrons and fragments should be in thermal and chemical equilibrium described by the Maxwell-Boltzmann statistics, and the yield of fragments should be proportional to the source density at the emission time. This latter hypothesis implies a simultaneous decay, which can be experimentally validated through the correlation function technique of Section 4.1.2.

Within these hypotheses, the ratio R of the yields Y of four isotopes is given by:

$$R = \frac{Y(A_1, Z_1)/Y(A_1 + 1, Z_1)}{Y(A_2, Z_2)/Y(A_2 + 1, Z_2)} = \frac{e^{B/T}}{a} \tag{6}$$

where a is constant and

$$B = BE(A_1, Z_1) - BE(A_1 + 1, Z_1) - BE(A_2, Z_2) + BE(A_2 + 1, Z_2) \tag{7}$$

with BE(A,Z) binding energy of an isotope of mass A and charge Z.

In Fig. 12 the temperatures extracted with this method are shown, together with the values corrected for the secondary decays.[13,36,37] The fluctuations from the different thermometers should be interpreted as an error bar on the thermodynamical temperature, due to an imperfect event selection or secondary decay correction. The mean value of the temperature is T = 3.20 ± 0.05 for the ^{32}S + ^{58}Ni system and T = 3.27 ± 0.005 MeV for the ^{32}S + ^{64}Ni one. We can see that close values for the temperatures are extracted for the neutron-rich and neutron-poor system. This indicates that the temperature of an excited system formed in central collisions does not appreciably depend on the isospin.

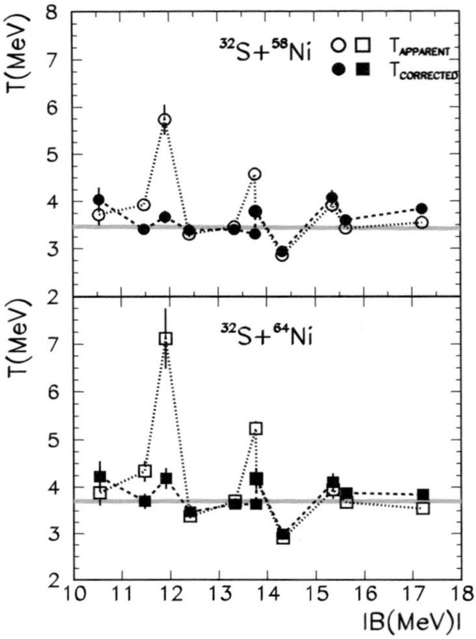

Fig. 12. Temperature extracted from different isotopic ratios as a function of the binding energy difference Eq. (7), in central $^{32}S + ^{58}Ni$ (upper part) and $^{32}S + ^{64}Ni$ (lower part) reactions. Empty symbols refer to rough data, while full symbols include corrections from secondary decay. Figure taken from Ref. 35.

4.3.2. *The isoscaling and the symmetry energy*

Obtaining information on the symmetry energy of an excited nucleus is one of the challenging item studied in recent heavy ion physics literature. One possible method is based on the measurement of isotope yields in two reactions involving nuclei of similar masses and different N/Z ratios.[38] Both statistical and dynamical models[39] indicate that the isotopic composition of fragments is sensitive to the symmetry energy and its surface and density dependence. In the grand canonical model, the ratio of the primary fragment yield depends exponentially on the neutron and proton number, which is referred to as isoscaling. We can write:

$$R_{21} = \frac{Y_1(N,Z)}{Y_2(N,Z)} = Ce^{\alpha N + \beta Z} \tag{8}$$

with Y_1 and Y_2 as the yield of the same (N,Z) fragment emitted by the neutron rich and neutron poor systems, respectively. C is a normalization constant, and α and β are the isoscaling parameters. The first parameter can be related to the symmetry energy coefficient:

$$\alpha = \frac{4C_{sym}}{T}\left[\frac{Z_1^2}{A_1^2} - \frac{Z_2^2}{A_2^2}\right], \tag{9}$$

where Z_1 (Z_2) and A_1 (A_2) are charge and mass of the neutron rich (poor) fragmenting system.

According to Eq. (9), an estimate of the C_{sym} coefficient can be obtained by the measurement of the temperature of the system and the neutron to proton ratio of the emitted fragments.[39]

The yields appearing in Eq. (8) refer to fragments at the freeze-out stage, and therefore data have to be corrected for secondary decay, similarly to what is done for extracting temperatures (see Section 4.3.1). To check the validity of the isoscaling formalism, the values of R_{21} of Eq. (8) have to be plotted vs. the isotope neutron and proton numbers. The result for central $^{32}S + ^{58,64}Ni$ collisions at 14.5 AMeV is presented in Fig. 13. We can

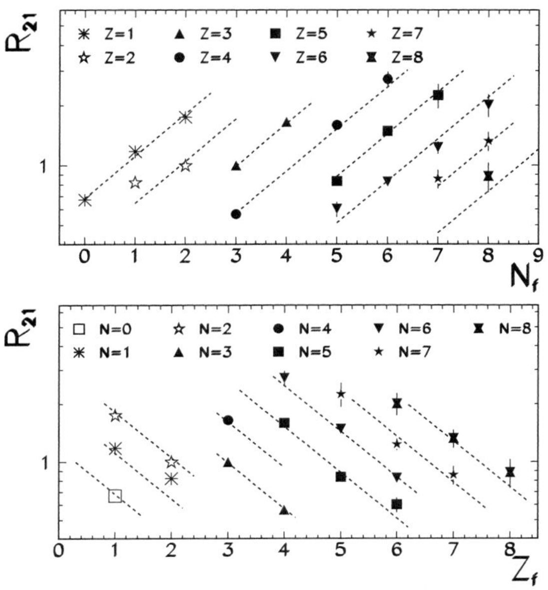

Fig. 13. Isotopic ratios defined by Eq. (8) for different isotopes as a function of the neutron number (upper panel), and for different isotones as a function of the proton number (lower panel) in central $^{32}S + ^{58,64}Ni$ reactions. Figure taken from Ref. 35.

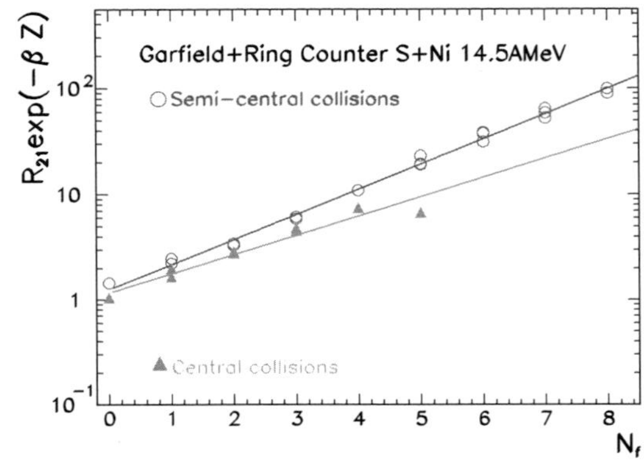

Fig. 14. Isotopic ratios defined by Eq. (8) for different isotopes as a function of the neutron number in ^{32}S + $^{58,64}Ni$ reactions. Open circles: semi-central collisions. Filled triangles: central events.

observe a very clear scaling, which gives confidence in the physical meaning of the C_{sym} parameter extracted in the analysis.

A value of $\alpha = 0.42 \pm 0.04$ is obtained, in agreement with previous measurements.[40–42] The results obtained for central events are compared to the ones for semi-central events in Fig. 14. In this last case the extracted α value is $\alpha = 0.55$, consistent with the lower temperatures achieved in semi-central collisions.[41]

Finally calculating the value of C_{sym} from the α parameter, one obtains 12 and 14 MeV for central and semi-central events, respectively. These values are very different from $C_{sym} \approx 25$ MeV expected for saturated matter at zero temperature. Again, such low values of C_{sym} are consistent with previous experiments.[40] From the theoretical point of view, these values can be understood as an effect of the low density associated to the excited nuclear sources, both because of the surface contribution and because of the expansion induced by the high excitation energy.[43]

5. Conclusions and perspectives

In this paper we have reviewed different applications of correlation function techniques in heavy ion collisions. Results were shown for different sets of data measured by the GARFIELD-RCo collaboration at the LNL laboratory. Four reactions were considered, namely ^{32}S + $^{58,64}Ni$ at 14.5 AMeV, ^{12}C + ^{12}C at 95 MeV and ^{14}N + ^{10}B at 80.7 MeV. For all the considered reactions, the quasi-complete detection of the reaction products allows disentangling the different reaction mechanisms. Specifically, central collisions leading to complete or quasi-complete fusion were identified both for the medium-heavy and for the light systems.

In the case of the light systems, the complete charge detection allows a precise identification of the emission source as ^{24}Mg, at an excitation energy $E^* = 62.4$ MeV. We have especially focussed on multiple α emission channels, thought to bear information on exotic α-cluster states which could be populated at high excitation energy.[32] For the exotic decay channel of a highly excited ^{24}Mg into six α particles, the correlation function technique shows that the decay is essentially sequential, due to an important contribution of intermediate Be states. In the case of the medium-heavy systems, the excitation energy of the fused source of the order of 3.3 AMeV is sufficiently high to lead to the opening of the multifragmentation channel. Excited light fragments are emitted from the multi-fragmenting source, and subsequently decay to their ground state through evaporation. For these reactions, the correlation function technique allows extracting both information on the multi-fragmenting source, and on the secondary evaporation chain. The multi-fragmenting source was characterized in terms of size, lifetime, temperature, density and symmetry energy. Concerning the evaporation chain, we have especially focussed on the physical origin of the observed staggering in the fragment distributions. We have shown that such effects originate from a complex interplay between pairing effects in the nuclear masses during the last evaporation step, and in the level density during the previous step of the desexcitation chain.

Peripheral collisions associated to the decay of a quasi-projectile source could also be identified in both reaction sets. For the medium-heavy system, an exact (N,Z) identification of the decaying source is not possible, but the correlation analysis allows studying the qualitative evolution with mass number and excitation energy of the nuclear temperature and symmetry energy. In the case of the light systems, the complete reconstruction of the reaction kinematics allows identifying the specific ^{12}C decay, without any contamination from other direct reactions. In particular the decay of the Hoyle state could be studied with the correlation function technique. A dominant sequential mechanism is observed, even if a small contribution of a linear α chain cannot be excluded. The results presented in this contribution give only a small snapshot of the rich and detailed information that can be extracted from multi-detectors on heavy-ion collisions. With these experimental resources, a complete study of nuclear thermodynamics is potentially at hand. This includes the characterization of nuclear temperature and symmetry energy as a function of the mass, charge, excitation energy, and isospin content of evaporation and multifragmentation systems. A detailed study of the systematics of the nuclear level density as a function of mass and isospin can also be achieved through the reconstruction of excited resonances prior to their particle decay. The persistence of α-clustered states at high excitation energy in light and medium-heavy even-even nuclei can be assessed, as well as the geometrical structure of these states. This challenging program requires a systematic study of a large set of reactions throughout the nuclear chart, and a complete detection of the reaction products both in mass and charge. Moreover, disentangling entrance channel effects in a fully reliable way demands not only a complete or-quasi complete detection, but also the availability of different projectile and target combinations to access the same excited nucleus through different reaction mechanisms.

The developement of new generation multi-detectors[44] and rare ion beam facilities in different laboratories (RIKEN, SPES, SPIRAL2, RAON, FAIR) are essential elements for the future success of this ambitious programme.

Acknowledgment

A special thank to the outstanding technical work of Mr. Andrea Zucchini, for the manufacturing of the mechanical support of detectors. Thanks are due to the accelerator staff of Legnaro Laboratories for having provided good quality beams. This work was supported in part by the European Funds for Large Scale Facilities (Seventh Framework Program ENSAR 262010), by grants of Alma Mater Studiorum (Bologna University) and by grants of Italian Ministry of Education, University and Research under contract PRIN 2010-2011.

References

1. F. Gramegna *et al.*, *Nucl. Instr. and Meth. A* **389**, 474 (1997); A. Moroni *et al.*, *Nucl. Instr. and Meth. A* **556**, 516 (2006); M. Bruno *et al.*, *Eur. Phys J A* **49**, 128 (2012).
2. L. Morelli *et al.*, *Nucl. Instr. and Meth. A* **620**, 305 (2010).
3. L. Bardelli *et al.*, *Nucl. Instr. and Meth. A* **654**, 272 (2011)
4. J. Cugnon, D. L'Hôte *Nucl. Phys. A* **397**, 519 (1983).
5. M. D'Agostino *et al.*, *Nucl. Phys. A* **861**, 47 (2011).

6. L. Morelli *et al.*, *J. of Phys. G* **41**, 075108 (2014).
7. F. Hoyle *The Astrophysical Journal, Supplement Series* **1**, 12 (1954).
8. L. Morelli *et al.*, *J. of Phys. G* **43**, 045110 (2016).
9. R. Hanbury-Brown and R. Q. Twiss, *Philos. Mag* **45**, 663 (1954); R. Hanbury-Brown, R. Q. Twiss, *Nature (London)* **177**, 27 (1956); R. Hanbury-Brown, R. Q. Twiss, *Nature (London)* **178**, 1046 (1956).
10. D. H. Boal, C. K. Gelbke, B. K. Jennings *Rev. Mod. Phys.* **62**, 553 (1990).
11. M. A. Lisa *et al.*, *Ann Rev. Nucl. Part. Sci.* **55**, 337 (2005).
12. J. Pochodzalla *et al.*, *Phys. Rev. C* **35**, 1695 (1987).
13. T. K. Nayak *et al.*, *Phys. Rev. C* **45**, 132 (1992).
14. M. D'Agostino *et al.*, *Nucl. Phys. A* **875**, 139 (2012).
15. W. P. Tan *et al.*, *Phys. Rev. C* **69**, 061304 (2004).
16. J. P. Bondorf *et al.*, *Nucl. Phys. A* **444**, 460 (1986).
17. Al. H. Raduta and Ad. R. Raduta *Phys. Rev. C* **55**, 1344 (1997); Al. H. Raduta and Ad. R. Raduta *Phys. Rev. C* **56**, 054610 (2002).
18. L. B. Yang *et al.*, *Phys. Rev. C* **60**, 041602 (1999) and reference quoted therein; E. Geraci *et al.*, *Nucl. Phys. A* **732**, 173 (2004); E. Bonnet *et al.*, *Int. J. Mod. Phys. E* **17**, 2359 (2009); G. Ademard *et al.*, *Phys. Rev. C* **83**, 054619 (2011); I. Lombardo *et al.*, *Phys. Rev. C* **84**, 024613 (2011); S. Piantelli *et al.*, *Phys. Rev. C* **88**, 064607 (2013).
19. M. V. Ricciardi *et al.*, *Nucl. Phys. A* **733**, 299 (2004).
20. R. J. Charity, *Phys. Rev. C* **82**, 014610 (2010).
21. T. Glasmacher *et al.*, *Phys. Rev. C* **50**, 952 (1994).
22. M. D'Agostino *et al.*, *Phys. Lett. B* **368**, 259 (1996).
23. D. R. Bowman *et al.*, *Phys. Rev. C* **52**, 818 (1995).
24. H. Takemoto, *et al.*, *Phys. Rev. C* **69**, 035802 (2004).
25. J. A. Maruhn, M. Kimura, S. Schramm, P. G. Reinhard, H. Horiuchi and A. Tohsaki *Phys. Rev. C* **74**, 044311 (2006); P. G. Reinhard, J. A. Maruhn, A. S. Umar and V. E. Oberacker *Phys. Rev. C* **83**, 034312 (2011).
26. M. Girod and P. Schuck *Phys. Rev. Lett.* **111**, 132503 (2013).
27. J. P. Ebran, E. Khan, T. Niksic and D. Vretenar *Nature* **487**, 341 (2012); *Phys. Rev. C* **89**, 031303(R) (2014).
28. M. Freer *Rep. on Prog. in Phys.* **70**, 2149 (2007).
29. E. Costanzo *et al.*, *Phys. Rev. C* **44**, 111 (1991); F. Grenier *et al.*, *Nucl. Phys. A* **811**, 233 (2008); Ad. R. Raduta *et al.*, *Phys. Lett. B* **705**, 65 (2011); Tz. Kokalova, N. Itagaki, W. von Oertzen and C. Wheldon *Phys. Rev. Lett.* **96**, 192502 (2006); W. von Oertzen *et al.*, *Eur. Phys. Journ. A* **36**, 279 (2006); A. Di Pietro *et al.*, *Journal of Physics: Conference Series* **366**, 012013 (2012); E. Vardaci *et al.*, *Journal of Physics: Conference Series* **436**, 012054 (2013) and references quoted therein.
30. B. Borderie *et al.*, *Phys. Lett. B*, in press (2016).
31. M. Freer *et al.*, *Phys. Rev. C* **49**, R1751 (1994); T. K. Rana *et al.*, *Proc DAE Symp. Nucl. Phys.* **56**, 492 (2011); J. Manfredi *et al.*, *Phys. Rev. C* **85**, 037603 (2012); O. S. Kirsebom *et al.*, *Phys. Rev. Lett.* **108**, 202501 (2012); T. K. Rana *et al.*, *Phys. Rev. C* **88**, 021601 (2013); M. Itoh *et al.*, *Phys. Rev.*

Lett. **113**, 102501 (2014); M. Freer and H. O. U. Fynbo 2014 *Progr. Part. Nucl. Phys* **78**, 1 (2014).

32. F. Gulminelli *et al.*, contribution to this volume.
33. Ad. R. Raduta *et al.*, *Phys. Lett. B* **705**, 65 (2011); L. Quattrocchi *et al.*, *Contribution to Nucleus nucleus 2015 NN2015, Catania, Italy J. Phys.:Conference Series* (to be published).
34. S. Albergo *et al.*, *Nuovo Cim. A* **89**, 1 (1985).
35. U. Abbondanno *et al.*, *LNL Ann. Rep.* p.24 (2005) www.lnl.infn.it/annrep/read_ar/2005/index_intro.htm
36. F. Zhu *et al.*, emphPhys. Rev. C **52**, 784 (1995).
37. M. B. Tsang *et al.*, *Phys. Rev. C* **64**, 054615 (2001).
38. A. Ono *et al.*, *Phys. Rev. C* **68**, 051601 (2003); A. S. Botvina *et al.*, *Phys. Rev. C* **65**, 044610 (2002).
39. H. S. Xu *et al.*, *Phys. Rev. Lett.* **85**, 716 (2000).
40. G. A. Soliutis *et al.*, emphPhys. Rev. C **73**, 024606 (2006); D. V. Shetty *et al.*, emphPhys. Rev. C **70**, 011601 (2007); A. Le Fevre *et al.*, *Phys. Rev. Lett.* **94**, 162701 (2005); D. Henzlova *et al.*, *J. of Phys. G* **37**, 85010 (2010).
41. E. Geraci *et al.*, *Eur. Phys. J. Special Topics* **150**, 21 (2007).
42. E. Geraci *et al.*, *Nucl. Phys. A* **734**, 524 (2004).
43. Ad. R. Raduta and F. Gulminelli *Phys. Rev. C* **75**, 024605 (2007).
44. R. Bougault *et al.*, *Eur. Phys. Journ. A* **50**, 47 (2014).

Part 4

Cluster Radioactivity/Fission and SHE

CHAPTER 11

NUCLEAR SIZE ISOMERS: THE EXCITED STATES OF LIGHT NUCLEI WITH CLUSTER STRUCTURE AND NONSTANDARD SIZES

A. A. Ogloblin,[*,¶] A. N. Danilov,[*] A. S. Demyanova,[*]
S. A. Goncharov[†] T. L. Belyaeva[‡] and W. Trzaska[§]

[*] *NRC Kurchatov Institute, Moscow RU-123182, Russia*
[†] *Lomonosov Moscow State University, Moscow, Russia*
[‡] *Universidad Autónoma del Estado de México, C. P. 50000, Toluca, México*
[§] *JYFL, Jyvaskyla University, Jyvaskyla, Finland*
[¶] *ogloblina@bk.ru*

Recently the methods of determining the radii of short-lived nuclear excited states (the modified diffraction model, methods using inelastic nuclear rainbow scattering and asymptotic normalization coefficients) were developed. Though model dependent they allowed getting evidence of the existence of nuclear excited states with abnormal radii. Two groups of such "size isomers" were identified: those with neutron halos and some alpha — cluster states. Most of these exotic structures were observed in stable nuclei ^{9}Be, ^{11}B, ^{12}C, ^{13}C in experiments with stable beams. The obtained results provide critical tests to some modern cluster theories, e.g., alpha particle condensation.

1. Introduction

The size of a nucleus defined by the radius of its nucleon (proton and neutron) density distribution and the proton charge distribution is one of the most fundamental and important nuclear characteristics. Nuclear radius determines the basic properties of nuclei and is a consequence of the fundamental features of the strong interaction. Even a moderate deviation from its standard value may be connected with a radical change of nuclear structure.

To date, there exist a number of reliable methods used to measure the radii of nuclei in their ground states (g.s.), among which the most precise and widespread is the elastic scattering of electrons on stable or long-lived targets.[1] More sophisticated methods, as, for instant, the laser spectroscopy, are used to measure the radii of nuclei in their excited states with half-lives of $10^{-8} - 10^{-9}$ and more. Until recently, an estimation of nuclear radii in the short-lived states (with $\tau_{1/2} < 10^{-12}$ s) decaying by emission of nucleons or clusters was possible only by indirect methods like fitting form factors to the data in inelastic electron scattering (see, e.g., Refs. 2 and 3). Nevertheless, direct experimental methods of measuring the radii of the short-lived nuclear states are now in progress.[4,5]

Now it is well recognized that unusual density distributions appear not only in the ground states of exotic radioactive nuclei, but in the excited states of some stable nuclei, located closely and above the particle-emission threshold. Note that as long as in 1959 Alfred Baz had predicted that nuclear states located near the neutron emission thresholds should possess enhanced radii.[6] Giant resonances, the excited states of exotic nuclei in the region of the drip lines, some quasimolecular and cluster states represent numerous examples of the states located near and above the particle emission thresholds. The nuclei which exited states above the neutron emission threshold manifest neutron halo properties are of special interest.

In the recent years, exactly these types of nuclear states with non-standard nucleon density distribution have attracted plenty of attention of nuclear physics community. The investigation of these properties can stimulate the observation of novel phenomena, the development of nuclear structure theory, and a progress of nuclear astrophysics.

In this review, we summarize the results of measuring and analysis of the radii of the short-lived excited states by applying three direct methods: a recently developed modified diffraction model (MDM), the inelastic nuclear rainbow scattering method (INRS) and the asymptotic normalization coefficients method (ANC).

We focus our attention on two types of nuclear structure with unusual properties: the excited states of light nuclei possessing α-cluster structure and neutron halos. Our analysis of some nuclear reactions provided evidence that there exist several excited states of ^{9}Be, ^{11}Be, ^{11}B, ^{12}C, ^{13}C whose radii exceed the radii of their ground states by ~20–30%. We specified these dilute states by name *nuclear size isomers.*[7]

2. Methods of determining the radii of nuclei in the short-lived excited states

Let us start with a short description of the three direct though model-dependent methods which have been proposed during last decade. The

MDM, INRS, and ANC are based on different concepts and are supplementary to each other: the first two allow us to study the excited dilute states in the target nucleus, while the ANC method deals with nuclei produced by transfer reactions.

2.1. *Modified diffraction model*

A diffraction scattering model,[8] a fairly rough approximation for calculating the differential cross sections, is quite adequate to determine nuclear radii from experimental data, because it operates only with a single parameter having the dimension of length, the diffraction radius R_{dif}, which is directly determined from the positions of the minima (maxima) of the experimental angular distributions.

Differential cross sections in this model are described by the squares of the cylindrical Bessel functions of various orders from the argument proportional to R_{dif}, which depends on both nuclear structure and interaction dynamics. If a diffraction radius is the same in the elastic and inelastic scattering, then the so-called Blair phase rules establish certain phase relations between the angular distributions of the elastic and inelastic scattering.[8] Because these relations are fulfilled very well in many cases of inelastic scattering with excitation of the lowest members of the rotational bands, the deviations from them can serve as an indicator of changing the radius of an excited state.

We have developed a modified diffraction model,[4,5,7,9] the main assumption of which is that the rms radius $<R^*>$ of a nucleus in the excited state is determined by the difference of diffraction radii of the excited and the ground states using the following relations

$$R_{\text{dif}}(\text{el}) = <R_0> + <R_1> + \Delta_{\text{el}}, \tag{1}$$

$$R_{\text{dif}}(\text{in}) = <R^*> + <R_1> + \Delta_{\text{in}}, \tag{2}$$

$$<R^*> = <R_0> + [R_{\text{dif}}(\text{in}) - R_{\text{dif}}(\text{el})] + (\Delta_{\text{el}} - \Delta_{\text{in}}). \tag{3}$$

Here $<R_0>$ and $<R_1>$ are the rms matter radii of the target and projectile nuclei in their ground states, which are assumed to be known.

The diffraction radii $R_{\text{dif}}(\text{in})$ and $R_{\text{dif}}(\text{el})$ are determined by finding out the positions of the minima and maxima of the experimental angular distributions of the inelastic and elastic scattering, respectively. Parameters Δ_{el} and Δ_{in} represent the difference between the diffraction radii and the sum of "true" (e.g., rms) radii of the nuclear states involved in the reactions;

and thus they determine the distance to which the colliding nuclei must approach to evoke a diffraction. Values of Δ include all the structural and dynamic effects that are neglected in the diffraction model. For the elastic scattering, the $\Delta_{\rm el}$ value can be determined directly from experimental data. Energy dependence of $\Delta_{\rm el}$ in the c.m. system for various combinations of colliding nuclei was shown in Fig. 5 of Ref. 5.

A common feature of $\Delta_{\rm el}$ values is their monotonic decrease with energy, meaning that a diffraction occurs at a stronger overlap of nuclear densities as the energy increases. Such behavior correlates with the known fact that the diffraction radius is close to the strong absorption radius. We also note that for the projectiles with lower average densities (deuterons, ^{6}Li), a diffraction takes place at a smaller distance between the colliding nuclei.

To determine the radius of a nucleus in the excited state, we suggest that $\Delta_{\rm in}$ and $\Delta_{\rm el}$ are similar in their main aspects. Thus, the rms radius $<R^*>$ of the excited state can be determined by the rms radius $<R_0>$ of the ground state plus a difference between the diffraction radii of the excited and the ground states. Though a hypothesis that $\Delta_{\rm in} = \Delta_{\rm el}$ cannot be strictly justified, we have empirically tested it in Ref. 5, and found the proximity of the mentioned diffraction radii difference obtained from the scattering of different projectiles with very different density distributions (^{3}He, ^{4}He, ^{6}Li, ^{12}C) on ^{12}C. We have also showed that the value of a difference $[R_{\rm dif}\ ({\rm in}) - R_{\rm dif}({\rm el})]$ in Eq. (3) does not depend either on energy or the projectile type (Fig. 1).

As one can see from Fig. 1, the radii of the g.s. and 4.44 MeV states coincide, whereas those of the 7.65 and 9.64 MeV states are enhanced.

In the following sections, we will show other examples of applying the MDM and its compatibility with other direct methods.

2.2. *Inelastic nuclear rainbow scattering method*

Ohkubo and Hirabayashi in Ref. 10 have showed that the nuclear rainbow inelastic scattering can provide information on the sizes of excited nuclear states. The main idea of the method is based on the conjecture that a trajectory of a particle scattered on a nucleus with the larger radius is longer than that for the more compact nucleus. Accordingly, the deflection angle (the Airy angle) should depend on a nuclear size. From the analysis of the elastic and inelastic scattering of ^{3}He on ^{12}C populating the ground 0_1^+ state and the excited 0_2^+ state at E_x = 7.65 MeV (the Hoyle state)

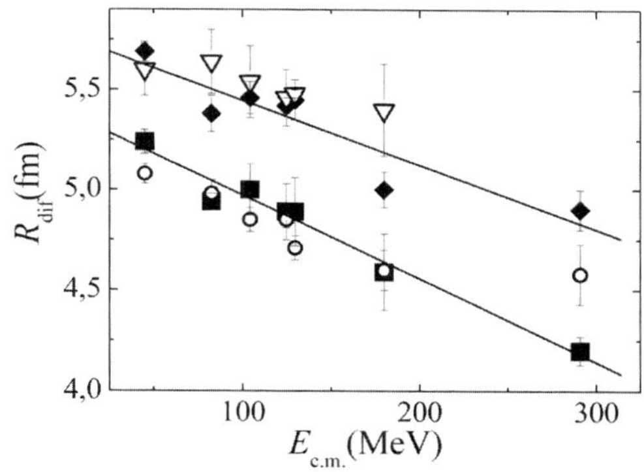

Fig. 1. Energy dependence of the diffraction radii for the $\alpha + {}^{12}C$ system determined from elastic scattering (filled squares) and inelastic scattering to the 4.44-MeV 2^+ state (open circles), the 7.65-MeV 0^+ state (filled rhombuses), and the 9.64-MeV 3^- state (open triangles).

in ${}^{12}C$, they have found that the predicted shift of nuclear rainbow (Airy) minima to the larger angles in inelastic scattering in comparison with elastic scattering really exist; that is, the excited state has an enhanced radius.

We have developed a model[4,9,11] based on the cited idea and used the explicit dependence of $\Theta_{\min}$ on the radius of nucleus, which was determined empirically. First, we constructed the systematic of the energy dependences of the positions of Airy minima in elastic α-particle scattering and found that the positions of Airy minima are located along the straight lines in the *Angle* vs. *Inverse energy* plane (Fig. 2). These lines can be reduced to an unique one by multiplying to a scaling factor proportional to $A^{2/3}$. Based on the analysis of the data on the elastic α-scattering, we have deduced an empirical relation

$$\Theta_{\min}(A) = C(A)A^{2/3}/E_{c.m.} + b(A) = C(A)R^2/E_{c.m.} + b(A), \qquad (4)$$

which connects the angle of the first Airy minimum and the radius of a nucleus. In Eq. (4), coefficients $C(A)$ and $b(A)$ depend on the mass number A of the target.

Assuming that the same type of relation is also valid for inelastic scattering and suggesting that the first half of the trajectory lies in the initial state and second one in the excited state, we obtained the relation

$$\frac{R_f}{R_i} = \left[\left(1 - \frac{Q}{E}\right) \times \left(\frac{2E_{cm}}{C}\Delta\Theta + 1\right)\right]^{1/2} \qquad (5)$$

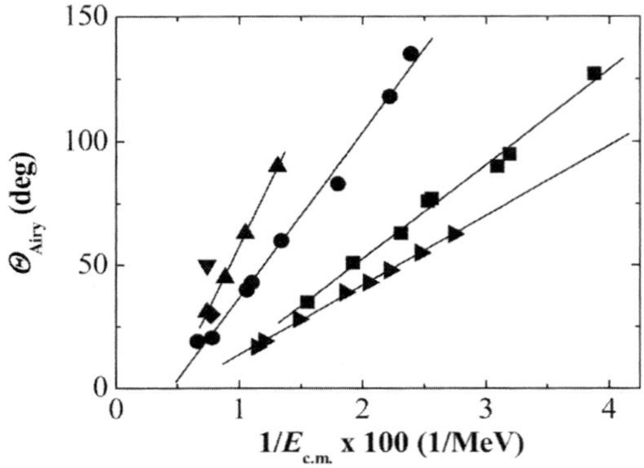

Fig. 2. Airy minima from α-particle scattering for different targets as function of inverse center-of-mass energy $1/E_{c.m.}$ Symbols ▶, ■, •, ◆, ▲, and ▼ correspond to the targets ^{12}C, ^{16}O, ^{40}Ca, ^{58}Ni, ^{90}Zr, and ^{208}Pb.

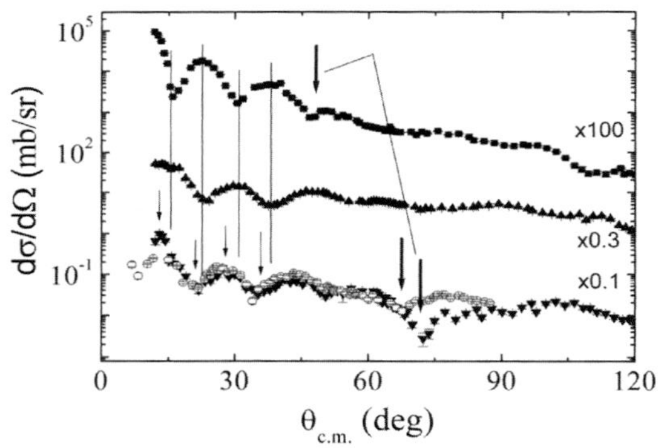

Fig. 3. Differential cross sections of the $\alpha + {}^{12}C$ scattering. The upper and medium curves denote the elastic scattering and inelastic scattering to the 2^+, 4.44 MeV at $E_\alpha = 60$ MeV[12]. The lower curves refer to the inelastic scattering exciting the 0_2^+, 7.65 MeV Hoyle state at $E_\alpha = 60$ MeV (filled triangles) and 65 MeV (open circles). Thin arrows denote the diffraction minima (maxima), thick ones show the Airy minima.

It connects the radius of an excited state with the difference $\Delta\Theta$ of the Airy minima angles in inelastic and elastic scattering differential cross-sections. Some examples of application of (5) are given in Refs. 4 and 9.

In Fig. 3 we show the shift of the Airy minimum in the inelastic α-particle scattering to the Hoyle state of ^{12}C in comparison with the elastic scattering. In the lowest curve, the Airy minimum is shifted to the larger angles relatively the upper curve indicating the enhancement of the radius of ^{12}C in the Hoyle state. The predicted shift of Airy minima with energy (from 60 to 65 MeV) is also observed. The diffraction radii are shifted to smaller angles as expected.

Application of the INRS method to the 7.65-MeV Hoyle state of ^{12}C gave its radius value to be a factor of $1.2-1.3$ larger than that of ^{12}C in the ground state. This result is in excellent agreement with the values obtained by the MDM analysis.

2.3. *Method of asymptotic normalization coefficients*

In addition to the MDM and the INRS methods that partly deal with empirical systematics, the ANC theory developed especially for peripheral reactions[13,14] can be adequately applied for describing the properties of the short-lived excited states. In particular, the ANC theory was used for determining the radii of the nuclei in the excited states produced in the nucleon transfer reactions, in which the neutron halos are formed.[15–18]

It is well known that many transfer reactions with light nuclei are actually sensitive only to a tail of the overlap function, i.e., they are peripheral. In accordance with a general definition, the ANCs represent the asymptotic behavior of the radial part $\phi_{Av,lj}\mathrm{g}(r)$ of the overlap function for the virtual cluster vertex $B \to A + v$. Numerous studies have shown (see, e.g., Refs. 13 and 14 and references therein) that for the peripheral direct nuclear reactions, the experimental ANC

$$C^{\mathrm{expt}}_{Av,lj} = [S^{\mathrm{expt}}_{lj}(B \to A + v)]^{1/2} b_{Av,lj} \tag{6}$$

is almost constant or only weakly depends on the model parameters, contrary to the empirical values of spectroscopic factors (SF) $S^{\mathrm{expt}}_{lj}(B \to A+v)$ and the single-particle (sp) ANC $b_{Av,lj}$ separately. The persistence of $C^{\mathrm{expt}}_{Av,lj}$ in a number of cross-section calculations with different sp wave functions is considered as a criterion of peripherality of this reaction.

In fact, the overlap and sp wave functions $\phi_{Av,lj}(r)$ and $u_{Av,lj}(r)$ may have quite different forms in the inner region of a nucleus, and experimental spectroscopic factors, determined as a normalization of the calculated cross section to the experimental data, may differ from the true SF defined as a square norm of the overlap function. Nevertheless, for a peripheral reaction, the experimental ANC, $C^{\mathrm{expt}}_{Av,lj}$, can be well determined and can be quite similar to the real ANC $C_{Av,lj}$.

The aim of an ANC analysis of the (d, p) reaction is to answer the question of whether there exist nuclear states that are formed due to the peripheral neutron-transfer reactions, that is, the reaction where the valence neutron is distant from the core approximately by the channel radius R_N or a larger distance. Thus, analyzing the differential cross section of the peripheral (d,p) reaction, one can determine parameters of the state "core plus neutron", the radius of the "orbit" of the transferred neutron in the final state and the weight of the asymptotic part of the wave function in the sp approximation.

Our recent studies allowed us to perform a critical test of all three models by determining the radius of the 3.09-MeV $1/2_1^+$ state of ^{13}C. This state is located 1.86 MeV below the $^{12}C + n$ threshold and is associated with a configuration where the last neutron occupies the $2s_{1/2}$ shell-model orbital. It was predicted in Refs. 15, 19 and 20 that a neutron halo can be formed in this state.

Comparison of the results obtained by the MDM, ANC, and INRS analysis, as well as the theoretical calculations within the orthogonality condition model (OCM) is presented in Table 1. First of all, the analysis shows that the 3.09-MeV $1/2_1^+$ state has an enhanced radius with respect to that of the ground state ($<R_0> = 2.33 \pm 0.03$ fm). This abnormally large value is connected with a high probability (≈70%) for the last neutron to be outside the range of the interaction radius $R_N = 5$ fm.[15,18] So, the existence of a neutron halo in this excited state is confirmed by several independent methods, in good agreement with the OCM predictions and the rough estimates made Refs. 21, 22.

The similarity of the results obtained by the ANC, MDM and INRS methods (see Table 1) confirms the reliability and adequacy of all three methods, at least at the energies of several tens of MeV. At higher energies the results could be distorted by some mechanisms differing from diffraction.

Table 1. Summary of rms matter and halo radii for the 3.09-MeV $1/2^+ \leftarrow$ state of ^{13}C.

Method	$R_{\rm rms}$ (fm)	$R_{\rm h}$ (fm)
MDM [21]	2.74 ± 0.06	5.88 ± 0.40
MDM [23]	2.92 ± 0.07	6.99 ± 0.41
ANC [15]	2.62 ± 0.20	5.04 ± 0.75
ANC [18]	2.72 ± 0.10	5.72 ± 0.16
INRS [23]	3.0 ± 0.1	7.4 ± 0.6
OCM [24]	2.68	5.47
R(^{12}C)+$\hbar(\mu\varepsilon)^{-1/2}$	2.7	

In order to relate the one-neutron halo radius $R_h = <r_n^2>^{1/2}$ and the rms $R_{\rm rms}$ radii of the core A and the (A + 1) system we used the two-body relationship proposed by Tostevin and Al-Khalili in Ref. 25:

$$(\mathrm{A}+1)[R_{\rm rms}(\mathrm{A}+1)]^2 = \mathrm{A}[R_{\rm rms}(\mathrm{A})]^2 + [\mathrm{A}/(\mathrm{A}+1)]R_{\rm h}^2 \quad (7)$$

For MDM and INRS the rms radii are "initial" values and the halo radii are calculated by Eq. (7).

3. Neutron halos in the excited states of light nuclei

A discovery of neutron halos in some neutron-rich radioactive light nuclei located close to the neutron drip line[26,27] was one of the most remarkable discoveries in nuclear physics from the end of the past century. Nuclei possessing neutron halos were called *exotic nuclei*,[28,29] and this term is now applied to some other nuclei having unusual properties.

Soon after the discovery of neutron halos, there appeared ideas about their possible existence not only in the ground states, but in the excited states as well including those of stable nuclei. For the first time such structure was predicted[19,20] for the 3.09-MeV state of ^{13}C and observed in Ref. 15 (some new results[18] are discussed in the previous Section). Up to now halos in the excited states were observed also in ^{12}B ($1_1^-, E_x = 2.62$ MeV) (for more discussion see Refs. 7, 21, 30, 31. All these halos possess properties which were considered to be immanent to them: (1) location of the corresponding states below neutron emission thresholds and close to them and (2) occupation of s-orbit by the valence neutron. In this paper we will discuss only new types of halos, namely those in continuum and with a neutron in a p — state. Such structures were observed in ^{11}Be and ^{9}Be.[18,21,22,32]

The ^{11}Be nucleus in the ground $1/2^+$ state is the most striking example of a nucleus with one-neutron halo.[15,18,29,33] For the two bound states of this nucleus, the $1/2^+$ g.s. and the first 0.32 MeV $1/2_1^-$ excited state, one can assume a single-particle structure of the ^{10}Be core in the 0^+ g.s. and the last neutron in the $2s_{1/2}$ and $1p_{1/2}$ sp orbitals, correspondingly. In Ref. 18, we showed that the halo radius of ^{11}Be in the $1/2^+$ g.s. is very large, $R_h = 8.0 \pm 0.2$ fm that agrees with the result obtained in recent measurements of the charge radii of the light halo nuclei.[33,34] This value corresponds to the contribution of the asymptotic part of the wave function of the last neutron of about 98%.

Also in our ANC analysis, we found that ^{11}Be in the 0.32 MeV $1/2^-_1$ excited state can possess a neutron halo as well. The halo radius occurred to be $R_h = 4.65 \pm 0.20$ fm. Much smaller halo radius in comparison with the g.s. is consequence of the fact that the valence neutron has an angular moment $L = 1$. The probability of the last neutron to be outside the range of the interaction radius $R_N = 5$ fm is larger than 76%.

Another interesting feature of ^{11}Be is the existence of the K $= 1/2^+$ rotational band based on its ground state. The quasielastic and inelastic ^{11}Be + ^{12}C scattering leading to formation of the g.s. and the $5/2^+$ and $3/2^+$ unbound excited states belonging to this band in ^{11}Be (Fig. 6) has been measured in Refs. 35 and 36; and these scattering data were analyzing by the MDM.[21,22] It was found that the diffraction radii for the all three members of the ^{11}Be rotational band are quite similar and sufficiently enlarged. This means that the neutron halo exists in ^{11}Be not only in the bound states (g.s. and the first excited $1/2^-_1$ state), but in the unbound excited states, members of the ^{11}Be rotational band, in spite of these states belong to a continuum spectrum.

^{9}Be is known [see, e.g.[37]) to have two rotational bands, $K = 3/2^-$ based on the ground state and $K = 1/2^+$ based on the 1.68-MeV first excited state. Though the ground state is only 1.6 MeV below the neutron emission threshold, it does not have a halo contrary to the ^{11}Be case.

The $K = 1/2^+$ band is very similar to the analogous band in ^{11}Be (Fig. 4): both bands have the inversed sequence of levels and similar moments of inertia, which are enhanced in comparison with the ^{9}Be g.s. band. However, the head band state ($E_x = 1.68$ MeV, $1/2^+$) is situated in continuum, only 15 keV above the ^{8}Be + n threshold. Our preliminary MDM analysis of the inelastic α-scattering data at 30 and 35 MeV[38,39] showed that all states belonging to the $K = 1/2^+$ band have the enlarged diffraction radii, which are larger than those for the $1/2^+$ g.s. of ^{11}Be and $3/2^-$ g.s. of ^{9}Be.

However, recent results,[40] if confirmed, might change traditional opinion about the structure of ^{9}Be. According to them, the state 4.70 MeV has negative parity and consequently cannot belong to the positive parity rotational band. In this case its radius might be different from the denoted value.

Recently, new measurements of elastic and inelastic scattering of α particles on ^{9}Be with an excitation of 1.68-MeV $1/2^+$ and 2.43-MeV $5/2^-$ states at $E_\alpha = 40$ and 90 MeV were carried out and a new MDM analysis

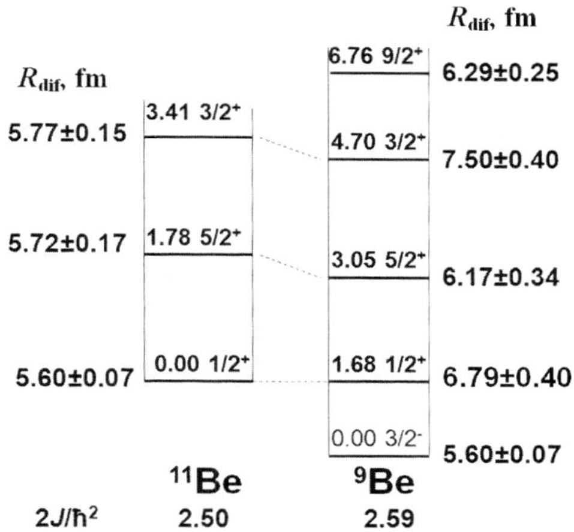

Fig. 4. Plot of the energy levels of ^{11}Be and ^{9}Be belonging to the positive parity rotational bands. The moments of inertia are indicated in the bottom line. The diffraction radii from ^{11}Be + ^{12}C scattering at $E(^{11}\text{Be}) = 737$ MeV and $\alpha +$ ^{9}Be at $E(\alpha) = 30$ MeV are shown in the left and right columns. See text for possible changes in the ^{9}Be level scheme.

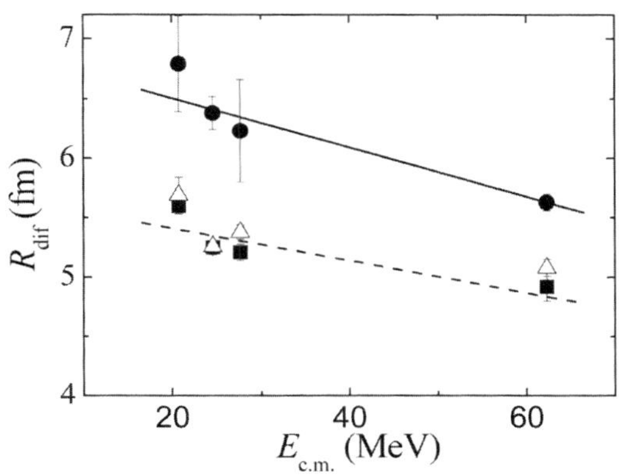

Fig. 5. Energy dependence of the diffraction radii for the $\alpha +$ ^{9}Be system determined from elastic scattering (filled squares) and inelastic scattering to the 2.43-MeV $5/2^-$ state (open triangles) and the 1.68-MeV $1/2^+$ state (filled circles).

was performed.[32] Figure 5 shows the energy dependences of the diffraction radii obtained from the data at $E_\alpha = 30, 35, 40$, and 90 MeV.[32,38,39] The main conclusion is that the diffraction radius of ^{9}Be for the 1.68-MeV state obtained by averaging the data at all four energies is approximately 1 fm larger than that for the g.s. and the 2.43-MeV $5/2^-$ state. The difference

$[R_{\rm dif}\ ({\rm in}) - R_{\rm dif}\ ({\rm el})]$ is energy-independent (within the error bars), which confirms the validity of Eq. (3). According to Eq. (3), the rms radius of the 1.68-MeV state is $R_{\rm rms}(1/2^+) = 3.42 \pm 0.20$ fm, which is close to the value of $R_{\rm rms}(1/2^+) = 3.50 \pm 0.15$ fm, obtained earlier in Ref. 38 and substantially exceeds the radius of ^{9}Be in the ground state, $R_{\rm rms}({\rm g.s.}) = 2.38$ fm.[41]

An estimate for the halo radius in the 1.68-MeV $1/2^+$ state (under the assumption that the radii of ^{8}Be and ^{9}Be are equal) is $R_h \approx 8.2$ fm. This value is close to the halo radius of ^{11}Be in the $1/2^+$ g.s., $R_h = 8.0 \pm 0.2$ fm,[18] which is located only 0.5 MeV below the neutron emission threshold. It is also significantly larger than the halo radius of ^{13}C in the 3.09-MeV $1/2_1^+$ state, $R_h = 5.72 \pm 0.16$ fm,[18] which is 1.86 MeV below the threshold.[9]

Thus, we can state that the 1.68-MeV $1/2_1^+$ state of ^{9}Be really has a neutron halo, in just the same way as the $1/2^+$ g.s. of ^{11}Be and the $1/2_1^+$ state of ^{13}C, but, in contrast to the latter's, it belongs to the continuum.

Observation of halos in nuclei located not only in discreet spectra, but also in continuum, considerably widen the existing conceptions on nuclear structure and require more theoretical studies.

4. The Hoyle state of ^{12}C as a key object for testing the modern cluster theories

A hypothesis of existence of α-particle Bose-Einstein condensation (αBEC) in finite nuclei,[42] which predicted appearance of nuclear states with unusual dilute α-cluster structure resembling a gas of almost non-interacting α particles with considerably enhanced radii, has attracted plenty of attention in the two last decades. This idea gave impetus to further development of the theoretical approaches concerning α-particle clustering initiated more than 50 years ago, and the experimental search for the diluted nuclear excited states. Testing the αBEC model predictions became a challenge to experimental nuclear physics.

Considerable attention has been drawn to the studies of α-cluster states of ^{12}C, especially the famous Hoyle 0_2^+ state, located at $E_x = 7.65$ MeV, which is 0.38 MeV above the 3α threshold. It is well known that this state plays an extremely important role in nucleosynthesis. Calculations by the 3α generator coordinate method (GCM)[43,44] and 3α resonating group method (RGM)[2,45] predicted that ^{12}C in the 0_2^+ state has a loosely coupled 3α structure and an enlarged radius. Modern microscopic calculations in the framework of cluster models such as the antisymmetrized molecular

Table 2. Summary of rms matter radii for the 0_2^+ 7.65-MeV Hoyle state of ^{12}C.

$R_{\rm rms}$ (fm)	4.31	3.83	3.53	3.47	3.38	3.22	3.27	2.93	2.90	2.4	2.89 ± 0.04
Ref.	48	49	50	2	3	51	46	52	53	54	Exp. 5

dynamics[46] and the fermionic molecular dynamics (FMD)[3] also predicted an increased radius of this state. Thus, the Hoyle state is considered to be the best example of the diluted system playing the role of a test for the whole problem of measuring the radii of short-lived nuclear states.[47]

Application of the MDM to the inelastic scattering of ^{2}H, ^{3}He,^{4}He, ^{6}Li, and ^{12}C on ^{12}C in a wide range of energies[4,5,7,9,11] allowed us to determine the consistent values of the rms radii of ^{12}C in the excited states up to $E_x = 11$ MeV. In particular, we showed that in line with the expectations, the Hoyle state has an abnormally large radius, approximately 1.25 times larger than that for the ground state. The results were confirmed by the INRS method, which explored the inelastic rainbow scattering of ^{3}He and ^{4}He on ^{12}C.[4,9,11] The values of the Hoyle state radius obtained by the MDM analysis in comparison with the results of a number of theoretical calculations are shown in Table 2.

The scatter in the predictions shown in Table 2 is evident. Estimates of the αBEC model ($R_{\rm rms}(0_2^+) = 4.31$ fm)[48] is nearly twice as large as the radius of the ground state (2.34 fm). Nevertheless, the recent calculations in the framework of the *ab-initio* Lattice Effective Field Theory (L-EFT)[54] predicted the rms radius (2.4 fm) almost equal to the rms radius of ^{12}C in the g.s. The best agreement with the experimental result showed the AMD calculations,[53] $R_{\rm rms}(0_2^+) = 2.90$ fm, and the no-core symplectic model (NCSpM) calculations,[52] 2.93 fm, while the αBEC model result overestimates the radius of the Hoyle state.

Another important characteristic predicted by the αBEC model is a high probability $W(\alpha)$ (about 70–80%) of the occupation of the lowest s orbit with zero angular momentum for all α.

One of the possible methods to extract information about the α-cluster occupation probability is a comparison of the relative contributions of different relative angular momenta of α and ^{8}Be to the cross section of a direct ^{8}Be transfer in the elastic and inelastic (to the 0_2^+ Hoyle state) $\alpha + {}^{12}$C scattering. In Ref. 55, the angular distributions of $\alpha + {}^{12}$C elastic and inelastic scattering measured at 110 MeV[56] in the full angular range were analyzed.

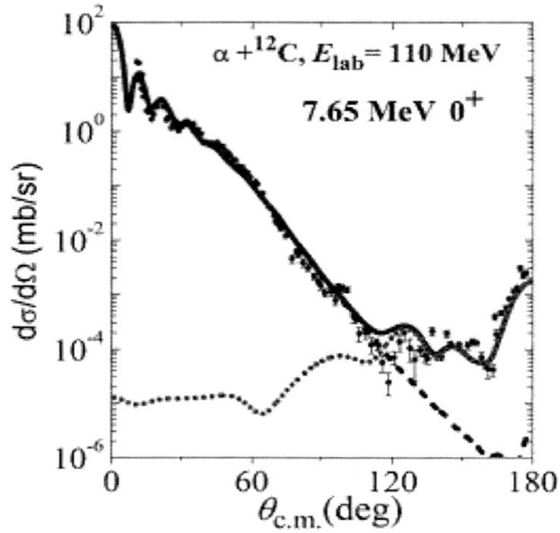

Fig. 6. Differential cross sections of α+ ^{12}C inelastic scattering corresponding to the excitation of the 7.65-MeV 0^+ Hoyle state of ^{12}C. The data are from Ref. 56. Notations are explained in the text.

The data (Fig. 6) showed a clear evidence of the significant back-angle scattering enhancement and strong oscillations at large angles, which are not described by a potential scattering model (dashed curve). These features were attributed to the ^{8}Be direct transfer mechanism (dotted curve). The calculations of the differential cross sections by the coupled reaction channels method involved the coherent sum (interference) over a total angular momentum of the transferred ^{8}Be nucleus (its g.s. and the first excited 2^+ and 4^+ states) and over the relative angular momenta $L = 0, 2,$ and 4 of the $\alpha + {}^8$Be configurations in ^{12}C. The percent ratio of the extracted spectroscopic factors corresponding to the angular momenta 0:2:4 was found 62:36:2 for the 0_2^+ Hoyle state and 19:49:32 for the 0_1^+ g.s. This result means that in the Hoyle state, there exists a dominated (62% of the occupation probability) configuration with all three α clusters on the s orbit, which corresponds to ^{8}Be transfer in its 0^+ g.s.

Calculations of the single-α orbital behavior in 0^+ states as a function of the nuclear radius R_N carried out by Yamada and Schuck[57] demonstrated a correlation between the values of the radius and $W(\alpha)$ shown in Fig. 7. $W(\alpha) = 70\%$ corresponds to a radius as large as 4.3 fm, whereas $W(\alpha) = 60\%$ suits to $R_N = 3$ fm, the value obtained experimentally. These findings resolve apparent inconsistence between both the most important characteristics (the rms radius and $W(\alpha)$). Taking into account all these

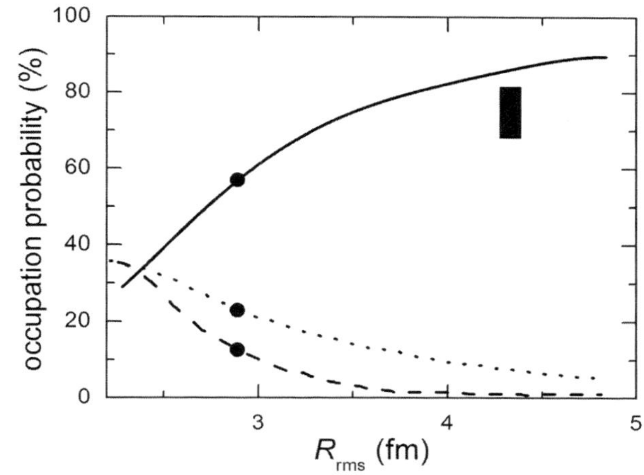

Fig. 7. Dependence of the occupation probabilities of the single-α orbits in the Hoyle state of ^{12}C on its radius. Contribution of the relative angular momenta $L = 0, 2$, and 4 are shown by the solid, dotted, and dashed lines. The experimental data corresponding to the Hoyle state radius $R_{\rm rms} \approx 2.9$ fm are noted by points. Vertical rectangle shows a prediction of the condensate model.

results, we can conclude that the 0_2^+ Hoyle state possesses only some rudimentary features of α-condensation if any.

The Hoyle state is not a unique size isomer in ^{12}C. A considerable increase of the size of ^{12}C was found for other states located above the $^{12}\mathrm{C} \to 3\alpha$ threshold.[5,58] First, the radius of ^{12}C in the 9.64-MeV 3^- state was also determined abnormally large. The MDM analysis provided the rms radius of this state to be $R_{\rm rms}(3^-) = 2.88 \pm 0.11$ fm, which is as large as for the Hoyle state. This result agrees well with the microscopic calculations of Kamimura,[45] the α BEC model predictions,[48] as well as with the molecular dynamics calculations in Ref. 3, all of which predicted enlargement of the radius of the 9.64-MeV 3^- state.

The size isomers, which form the rotational band based on the Hoyle state, will be discussed further in Sec. 5.

5. Rotating size isomers of ^{12}C

As early as 1956, almost after a prediction by Hoyle of the 0_2^+, $E_x = 7.65$ MeV state of ^{12}C, Morinaga[59] had suggested that the Hoyle state could be highly deformed and proposed the exotic linear chain configuration of three

α particles for its structure possible to rotate. According to the 3α RGM calculation by Kamimura[2] and Uegaki *et al.*,[43,44] the 2_2^+ state should locate around $E_x \sim 10$ MeV, and both the Hoyle state and the 2_2^+ state should have a molecule-like structure consisting of three α particles. On the other hand, according to the α BEC model,[57,60] the 2_2^+ state might be an excited state of the Hoyle state and have a structure similar to the 0_2^+ state, with one of the a clusters occupies a D-wave orbit. Consequently, it cannot be a member of the Hoyle rotational band. Just the same, the 2_2^+ state should have an extremely large rms radius ~6 fm.[57]

There was strong evidence for the 2_2^+ state at the excitation energies $E_x = 9.6$–9.8 MeV, which was found by measurements of the $^{12}C(p,p')$[61] and the $^{12}C(\alpha,\alpha')$[62] reactions. Consistent analysis of the proton and α-particle inelastic scattering data[63] have confirmed the existence of the 2_2^+ state of ^{12}C at 9.75 ± 0.15 MeV with a width of 750 ± 150 keV, whereas the measurement and analysis of the $^{12}C(\gamma,\alpha)^8Be$ reaction[64] allowed the authors to unambiguously identify a 2_2^+ state of ^{12}C at 10.03 ± 11 MeV, with a total width of 800 ± 130 keV.

The rms radius of the 2_2^+ state determined by the MDM,[58] $R_{\rm rms}(2_2^+) = 3.07 \pm 0.13$ fm, was found practically the same as that for the Hoyle state $R_{\rm rms}$ (7.65 MeV, 0_2^+) $= 2.89 \pm 0.04$ fm.[5] The rms radius for this state found by Dreyfuss *et al.* applying the NCSpM calculations,[52] $R_{\rm rms}(2_2^+) = 2.93 \pm 0.05$ fm, reasonable agreed with our result. The estimate using the rigid rotator model gives a value 2.7 fm, which is consistent with the radius obtained from the scattering data.[58]

These findings provide arguments in favor of the suggestion that the 0_2^+ and 2_2^+ states really are the members of the second rotational band in ^{12}C. Of course, the decisive conclusion could be done only after identification of the corresponding 4^+ state.

Possible observation of a 4^+ broad state at $E_x = 13.3$ MeV was claimed by Freer *et al.* in Ref. 65. Recently, we reported[66] new measurements of the inelastic $\alpha + {}^{12}C$ scattering at $E_\alpha = 65$ MeV leading to the excitation of the ^{12}C levels in vicinity of 14 MeV. Also a reconstruction of the inelastic scattering data measured previously at 110 MeV[56] was carried out. Besides the well-known 14.08-MeV 4^+ state,[64] a new broad ($\Gamma = 1.4$ MeV) state was observed at $E_x = 13.75$ MeV. The positions of the minima (maxima) of the angular distributions were consistent with the angular momentum transfer $L = 4$, that is $J^\pi = 4^+$ assignment was accepted for both these levels (Fig. 8).

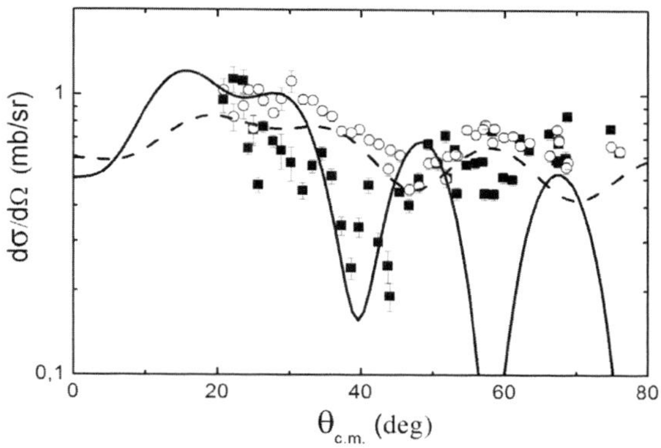

Fig. 8. Differential cross sections of the $\alpha + {}^{12}C$ inelastic scattering at $E(\alpha) = 65$ MeV with the excitation of the 14.08-MeV 4^+ (open circles) and 13.75-MeV (filled squares) states.[66] Solid and dashed lines correspond to the diffraction model calculations.

The MDM analysis used for determining the radii of the 14.08 and 13.75-MeV states has showed that the diffraction radii for these states are R_{dif} (14.08) = 4.2 fm and R_{dif} (13.75) = 5.0 fm, correspondingly. The difference between these radii agrees well with the difference between diffraction radii of the ground and the excited 2_2^+ states (0.8 fm and 0.6 fm correspondingly[58]). All these findings provide evidence that the 13.75-MeV 4^+ state, as well as the 2_2^+ state belong to the rotation band based on the Hoyle state (Fig. 9).

Recent observation Ref. 67 of a 5^- state at 22.4 MeV impelled the authors of cited paper to make a conjecture that the rotation band based on the ^{12}C ground state has a negative parity branch beginning from the 9.64-MeV 3^- state (Fig. 9). Not going into discussion we only note that this idea is hardly compatible with the fact that the 3^- state has a radius significantly larger than that of the positive parity members of the mentioned band.

6. The analogs of the Hoyle state in the neighboring ^{11}B and ^{13}C nuclei

Attention focused on the exotic structure of the Hoyle state was soon extended to the neighboring ^{11}B and ^{13}C nuclei,[46,68,69] which differ from ^{12}C by a proton hole and an extra neutron, correspondingly.

The inelastic-scattering angular distributions of deuterons, ^3He and α particle on ^{11}B and ^{13}C at incident energies of 10–100 MeV/A were analyzed by the MDM[70] with the aim to reveal dilute α-cluster states in

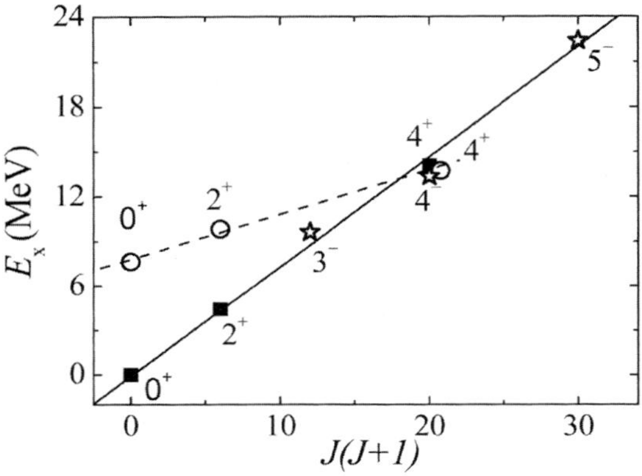

Fig. 9. Rotational bands in ^{12}C.

these nuclei, which can be analogues of the Hoyle state of ^{12}C. The results provided a strong evidence of an enhancement of the diffraction radii and, correspondingly, the rms radii of ^{11}B in the 8.56-MeV $3/2_3^-$ state and ^{13}C in the 8.86-MeV $1/2_2^-$ excited state, located close to the $2\alpha + \mathrm{t}$ and $3\alpha + \mathrm{n}$ thresholds respectively.

The rms radius of ^{11}B in the 8.56-MeV state was found $R_{\mathrm{rms}}(3/2_3^-) = 2.99 \pm 0.18$ fm, i.e., almost equal to the rms radius of ^{12}C in the Hoyle state, R_{rms} (7.65 MeV) $= 2.89 \pm 0.04$ fm, and well agreed with the AMD calculations.[46] A comparison of the angular distributions of the elastic and inelastic scattering of α particles and deuterons on ^{11}B and ^{12}C populated the 8.56-MeV state in ^{11}B and the Hoyle 7.65-MeV in ^{12}C measured at the same incident energies shows remarkable similarity between each others.[70] Thus the 8.56-MeV $3/2^-$ state of ^{11}B can be considered as a size isomer and analogues of the Hoyle state of ^{12}C. Moreover, the analysis fulfilled in Ref. 71 by the MDM got some evidence that the 8.56-MeV $3/2^-$ state really is a head of the predicted in Refs. 72 and 73 rotational band $K_1 = 3/2^-$ whose parameters and the radii of the member states are close to those of the Hoyle state band (Figs. 10 and 11).

The 8.56 MeV state is not a unique size isomer in ^{11}B (Fig. 9). In a very recent experiment,[71,74] the differential cross sections of elastic and inelastic $\alpha + {}^{11}$B scattering were measured at $E_\alpha = 65$ MeV. The radii of about fifteen known states of ^{11}B up to the excitation energy 14 MeV were measured by the MDM analysis.

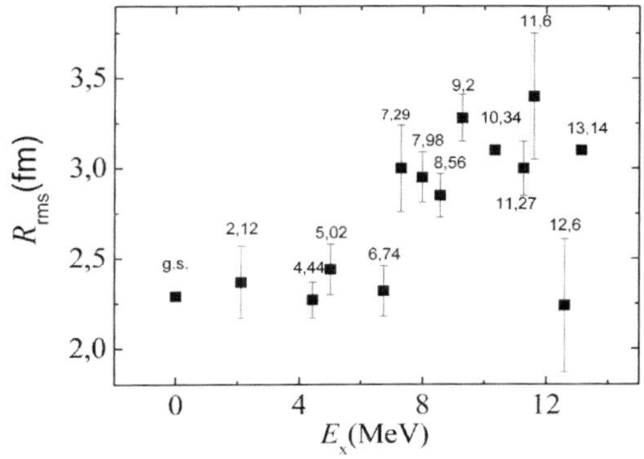

Fig. 10. Rms radii of ^{11}B in different states from the $\alpha + {}^{11}B$ scattering at $E(\alpha) = 65$ MeV.

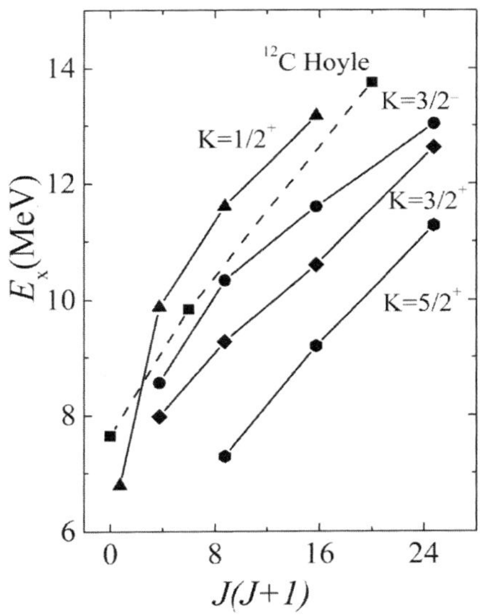

Fig. 11. Predicted rotational bands in ^{11}B at the excitation energies $E_x > 7$ MeV.

Besides the 8.56 MeV state band, the members of three positive-parity rotational bands located at the excitation energies ≥ 7 MeV (Fig. 11) were found to have enhanced radii as well. The latter are larger than the radius of ^{11}B in the g.s. by 0.7 to 1.0 fm and are close to the radius of ^{12}C in the Hoyle state. This result agrees with the predictions[72,73] that the members of the bands with $K = 3/2^+$ and $K = 5/2^+$ should have more complicated $2\alpha + 3N$ structure.

All these identified ^{11}B bands are characterized by high moments of inertia ($2I/\hbar^2 = 3$–4) that are comparable or even larger that the corresponding value for the Hoyle state ($2I/\hbar^2 = 2.7$). In the same way, the low-lying states of ^{11}B ($E_x < 7$ MeV) have both "normal" radii and "reduced" moments of inertia of about $2I/\hbar^2 \sim 1.1$.

To finish the discussion about size isomers in ^{11}B we note that an ambitious hypothesis[75] about an existence in this nucleus a "giant" excited state at $E_x \approx 12.5$ MeV with the radius close to that of uranium nucleus was not confirmed by our measurements.[71,74]

The ^{13}C nucleus also deals with a number of size isomers. The first one was mentioned in section 3 and is associated with the 3.09-MeV $1/2_1^+$ state, which has one-neutron halo. The enhanced rms radius of the 8.86-MeV $1/2_2^-$ excited state of ^{13}C was measured by MDM in Ref. 70, 2.91 ± 0.12 fm, and in Ref. 24, 2.68 ± 0.10. These results satisfactorily agree with the AMD predictions $R_{\rm rms}(1/2_2^-) = 2.9$ fm.[76] The differential cross sections of inelastic α-scattering with excitation of 8.86 MeV state is similar to that leading to formation of the Hoyle state (Fig. 12). So the 8.86-MeV state may be considered as an analog of the Hoyle state. The MDM analysis[70] indicates that the $1/2_3^-$ excited state ($E_x = 11.08$ MeV) of ^{13}C also possibly has the rms radii enhanced by a factor of $1.2 - 1.35$ in comparison with that of the ground state.

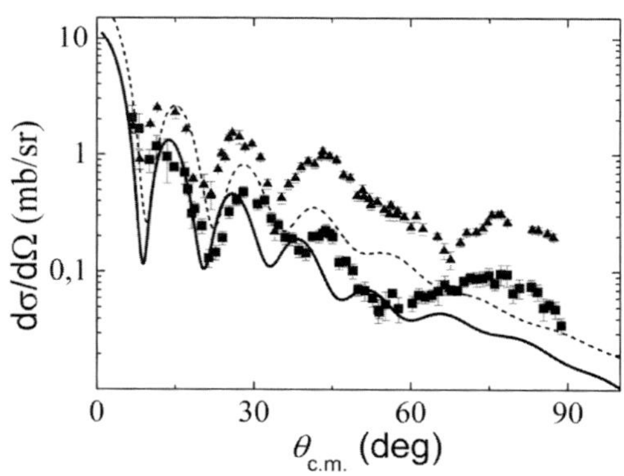

Fig. 12. Differential cross sections of the inelastic $\alpha + {}^{13}$C scattering with excitation of the 8.86 MeV state (squares) and the inelastic $\alpha + {}^{12}$C scattering with excitation of the Hoyle state (triangles) at $E_\alpha = 65$ MeV. Solid and dashed lines correspond to the DWBA calculations. Data is taken from Refs. 12, 24.

Thus, we conclude that there is in ^{11}B and ^{13}C a considerable set of size isomers that have exotic cluster structure. Adding an extra nucleon to the 3α cluster configuration (^{13}C) or making a hole in it (^{11}B) do not destroy its original dilute structure and α-cluster features. This result is in accordance with modern cluster theories. However, we got some evidence that ^{13}C possibly has a state with completely unexpected structure (see next section).

7. "Supercompact" size isomers: Excited states with anomalously small radii

The ^{13}C nucleus is an example of a nucleus in which of a variety of different types of nuclear structure successfully coexist. Besides the regular shell-model levels, there exist dilute states with different structure: the 3.09-MeV $1/2_1^+$ state which contains a neutron halo and the 8.86-MeV $1/2_2^-$ which is considered to be an analogue of the Hoyle state. The $1/2_3^-$ ($E_x = 11.08$ MeV) state possibly also has an enhanced radius.

Recently we have revealed that even more exotic structure can exist in ^{13}C.[77] The rms radius of ^{13}C in the 9.90-MeV $3/2^-$ state obtained by the MDM analysis of the inelastic α-scattering was found 1.89 ± 0.14 fm, i.e. noticeably smaller than the radius of ^{13}C in the ground state (2.33 fm).

An anomalously small radius of the 9.90-MeV state also follows from the comparison of the differential cross sections for the inelastic $\alpha + {}^{13}$C scattering with a population of three states of ^{13}C with the same transfered angular momentum $L = 2$ and close excitation energies (Fig. 13). If positions of the extrema had been in line in all angular distributions then the diffraction radii had to be identical. The shift toward larger momentum transfer, which is observed for the 9.90-MeV level, in fact indicates to a decrease of its radius. It is interesting to note that Milin and von Oertzen[68] suggested that this level is a head of the rotational band $3/2^-(9.90\,\text{MeV}) - 5/2^-(12.13\,\text{MeV}) - 7/2^-(14.98\,\text{MeV})$ and has enhanced radius.[78]

These properties of the 9.90-MeV $3/2^-$ state of ^{13}C provide reason to consider this state as an example of a supercompact size isomer.

8. Radii of dripline nuclei in the excited states

Most of information about size isomers discussed above concerned the nuclei at the stability valley or close to it. One may expect that spreading these

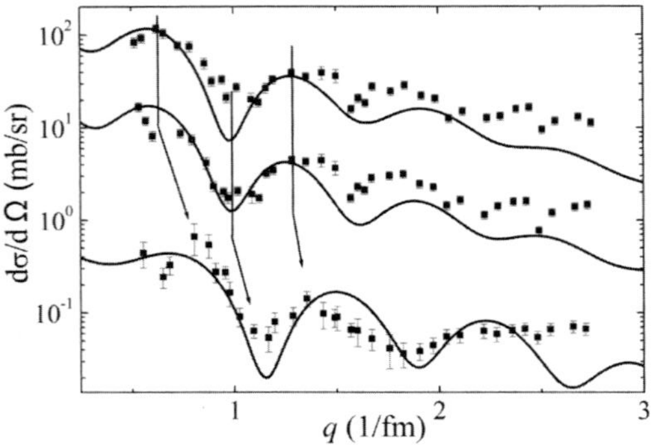

Fig. 13. Differential cross sections of the inelastic $\alpha + {}^{13}C$ scattering populated the states with $E_X = 3.68$ MeV (cross section is multiplied by a factor of 10), 7.55 MeV, and 9.90-MeV $3/2^-$ at $E(\alpha) = 90$ MeV.[77] The DWBA calculations with the angular momentum transfer $L = 2$ are shown by solid lines. The vertical lines are drawn through the diffraction minima and maxima of the cross sections leading to the excitation of the states with $E_X = 3.68$ and 7.55 MeV. The arrows denote the positions of the extremes of angular distributions relating to the formation of the 9.90-MeV $3/2^-$state.

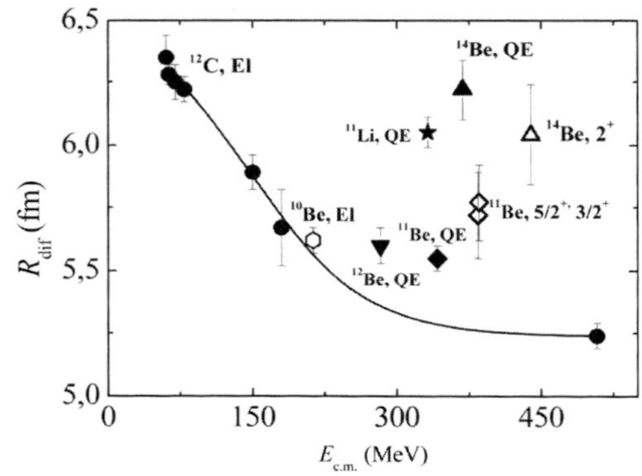

Fig. 14. Diffraction radii for some states of nuclei located close to the neutron drip-line obtained from the elastic (El), quasielastic (QE) and inelastic scattering. The curve is drawn through the data of ${}^{12}C + {}^{12}C$ elastic scattering.

studies to the drip-line nuclei will considerably widen our knowledge on nuclear structure.

Just for getting a general idea about the radii of nuclear states in the region of A $\sim$ 12 close to the neutron drip-line, we estimated the diffraction radii for some nuclei (Fig. 14) using very scarce elastic and inelastic scattering data.

9. Conclusion

Measuring the radii of the short-lived nuclear excited states for a long time was considered as unachievable. Now there exist three methods that allow realizing such investigations: the modified diffraction model, inelastic nuclear rainbow scattering, and asymptotic normalization coefficient method. The MDM and INRS exploit differential cross sections of inelastic scattering, whereas the ANC method is based on the analysis of the transfer reaction cross sections. Though all the methods are model-dependent, and some their details and application areas require further refinement, unique information on nuclear structure was obtained.

The main result consists in discovery of nuclear states with abnormal radii, which we have named *nuclear size isomers.* Among them one may single out two groups of excited states: those with neutron halos and alpha-cluster states. Some of them were predicted by modern nuclear models, some not, providing a challenge to theory. An intriguing feature of these results is that most of these exotic structures were observed in stable nuclei ^{9}Be, ^{11}B, ^{12}C, ^{13}C in experiments with stable beams quite far from the modern mainstream.

The work was supported by Grant No 14-12-00079 of Russian Scientific Foundation.

After the manuscript has been sent to the editor's board a paper (A.A. Ogloblin *et al.*, Phys. Rev.C 94, 051602 ® 2016) was published. No significant radius enhancement in any state of ^{16}O in comparison with its ground state was observed. This result does not confirm the predictions that ^{16}O in its 0_6^+ state has a radius comparable with those of heavy nuclei and can be considered as an analog of the Hoyle state in ^{12}C.

References

1. B. Hahn, D. G. Ravenhall and R. Hofstadter, High-Energy Electron Scattering and the Charge Distributions of Selected Nuclei, *Phys. Rev.* **101**, 1131–1141 (1956).
2. M. Kamimura, Transition densities between the 0_1^+, 2_1^+, 4_1^+, 0_2^+, 2_2^+, 1_1^- and 3_1^- states in ^{12}C derived from the three-alpha resonating-group wave functions, *Nucl. Phys. A* **351**, 456–480 (1981).
3. M. Chernykh, H. Feldmeier, T. Neff, P. von Neumann-Cosel and A. Richter, Structure of the Hoyle State in ^{12}C, *Phys. Rev. Lett.* **98**, 032401 (2007).
4. A. S. Demyanova, A. A. Ogloblin, S. A. Goncharov and T. L. Belyaeva, Study of "condensate" states in ^{12}C and ^{16}O by inelastic scattering, *Int. J. Mod. Phys. E* **17**, 2118–2122 (2008).

5. A. N. Danilov, T. L. Belyaeva, A. S. Demyanova and S. A. Goncharov and A. A. Ogloblin, Determination of nuclear radii for unstable states in ^{12}C with diffraction inelastic scattering *Phys. Rev. C* **80**, 054603 (2009).
6. A. I. Baz, Threshold effects in nuclear reactions, *Advances in Physics* **8** (32), 349–374 (1959); Resonance effects in the scattering of particles near a reaction threshold, *JETP* **36**, 1762–1766 (1959).
7. A. A. Ogloblin, A. N. Danilov, A. S. Demyanova, S. A. Goncharov, T. L. Belyaeva and W. Trzaska, Nuclear size isomers, *EPJ Web of Conferences*, **107**, 09002 (2016).
8. D. M. Brink, in *Many-Body Description of Nuclear Structure and Reactions*, Proceedings of the International School of Physics "Enrico Fermi", Course 36, edited by C. Bloch, Academic Press, New York (1966).
9. A. A. Ogloblin, T. L. Belyaeva, A. N. Danilov, A. S. Demyanova and S. A. Goncharov, Nuclear Threshold States: Yesterday, Today, Tomorrow, *AIP Conference Proceedings*, **1224**, 100–111 (2010).
10. S. Ohkubo and Y. Hirabayashi, Bose-Einstein condensation of α particles and Airy structure in nuclear rainbow scattering *Phys. Rev. C* **70**, 041602(R) (2004); S. Ohkubo and Y. Hirabayashi, Evidence for strong refraction of ^{3}He in an α-particle condensate, *Phys. Rev. C* **75**, 044609 (2007).
11. A. A. Ogloblin, T. L. Belyaeva, A. N. Danilov, A. S. Demyanova and S. A. Goncharov, Measuring the radii of particle unstable nuclear states and search for the signatures of alpha condensation in light nuclei, *Nucl. Phys. A* **834**, 143c–146c (2010).
12. S. A. Goncharov, A. S. Demyanova, Yu. A. Gloukhov *et al.*, Study of the structure of the Hoyle state by refractive α-scattering, *EPJ Web of Conferences* **66**, 03034 (2014).
13. L. D. Blokhintsev, I. Borbely and E. I. Dolinskii, Fiz. Elementar. Chast. Atom. Yad. **8**, 1189–1240 (1977) [English translation Sov. J. Part. Nucl. **8**, 485–535 (1977)].
14. G. R. Satchler, *Direct Nuclear Reactions*, Clarendon Press, Oxford (1983).
15. Z. H. Liu, C. J. Lin, H. Q. Zhang, Z. C. Li, J. S. Zhang, Y. W. Wu, F. Yang, M. Ruan, J. C. Liu, S. Y. Li and Z. H. Peng, Asymptotic normalization coefficients and neutron halo of the excited states in ^{12}B and ^{13}C, *Phys. Rev. C* **64**, 034312 (2001).
16. D. Y. Pang and A. M. Mukhamedzhanov, Asymptotic normalization coefficients and spectroscopic factors from deuteron stripping reactions, *Phys. Rev. C* **90**, 044611 (2014).
17. M. McCleskey, A. M. Mukhamedzhanov, L. Trache *et al.*, Determination of the asymptotic normalization coefficients for $^{14}C + n \leftrightarrow {}^{15}C$, the $^{14}C(n,\gamma)^{15}C$ reaction rate and evaluation of a new method to determine spectroscopic factors, *Phys. Rev. C* **89**, 044605 (2014).
18. T. L. Belyaeva, R. Perez-Torres, A. A. Ogloblin, A. S. Demyanova, S. N. Ershov and S. A. Goncharov, Determination of neutron halo radii in the first excited states of ^{13}C and ^{11}Be with the asymptotic normalization coefficients method, *Phys. Rev. C* **90**, 064610 (2014).

19. T. Otsuka, N. Fukunishi and H. Sagawa, Structure of Exotic Neutron-Rich Nuclei, *Phys. Rev. Lett.* **70**, 1385–1388 (1993).
20. T. Otsuka, M. Ishihara, N. Fukunishi, T. Nakamura and M. Yokoyama, Neutron halo effect on direct neutron capture and photodisintegration, *Phys. Rev. C* **49**, R2289–2292 (1994).
21. A. A. Ogloblin, A. N. Danilov, T. L. Belyaeva, A. S. Demyanova, S. A. Goncharov and W. Trzaska, Effect of neutron halos on excited states of nuclei, *Phys. Rev. C* **84**, 054601 (2011).
22. A. A. Ogloblin, A. N. Danilov, T. L. Belyaeva, A. S. Demyanova, S. A. Goncharov and W. Trzaska, Observation of abnormally large radii of nuclei in excited states in the vicinity of neutron thresholds, *Yad. Fiz.* **74**, 1581–1590 (2011) [English translation *Phys. Atom. Nucl.* **74**, 1548–1561 (2011)].
23. A. S. Demyanova, A. N. Danilov, S. V. Dmitriev *et al.*, *Spectroscopy of exotic states of* ^{13}C, *EPJ Web Conf.* **66**, 02027 (2014).
24. T. Yamada and Y. Funaki, Cluster states and alpha particle condensation in ^{13}C, *Int. J. Mod. Phys. E* **17**, 2101–2105 (2008).
25. J. A. Tostevin and J. S. Al-Khalili, How large are the halos of light nuclei? *Nuclear Physics A* **616**, 418c–425c (1997).
26. I. Tanihata, H. Hamagaki, O. Hashimoto *et al.*, Measurements of Interaction Cross Sections and Nuclear Radii in the Light p-Shell Region, *Phys. Rev. Lett.* **55**, 2676 (1985).
27. I. Tanihata, Neutron halo nuclei, *J. Phys. G: Nucl. Part. Phys.* **22**, 157–198 (1996).
28. K. Riisager, A. S. Jensen and P. Møller, Two-body halos, *Nucl. Phys. A* **548**, 393–413 (1992).
29. P. G. Hansen, A. S. Jensen and B. Jonson, Nuclear Halos, *Annu. Rev. Nucl. Part. Sci.* **45**, 591–634 (1995).
30. K. Riisager, Halos and related structures, *Phys. Scr.* **152**, 014001(pp. 1–13) (2013).
31. I. Izosimov, Borromean halo, Tango halo and halo isomers in atomic nuclei, *EPJ Web of Conferences*, **107**, 09003 (2016).
32. A. S. Demyanova, A. A. Ogloblin, A. N. Danilov *et al.*, Neutron Halo in the Exotic First Excited State of ^{9}Be, *JETP Letters* **102**(7), 413–416 (2015).
33. Z.-T. Lu, P. Mueller, G. W. F. Drake, W. Nörtershäuser, S. C. Pieper and Z.-C. Yan, Laser probing of neutron-rich nuclei in light atoms, *Rev. Mod. Phys.* **85**, 1383–1400 (2013).
34. I. Tanihata, H. Savajols and R. Kanungod, Recent experimental progress in nuclear halo structure studies, *Prog. Part. Nucl. Phys.* **68**, 215–313 (2013).
35. V. Lapoux, N. Alamanos, F. Auger *et al.*, Virtual coupling potential for elastic scattering of ^{10,11}Be on proton and carbon targets, *Phys. Lett. B* **658**, 198–202 (2008).
36. N. Fukuda, T. Nakamura, N. Aoi *et al.*, Coulomb and nuclear breakup of a halo nucleus, *Phys. Rev. C* **70**, 054606 (2004).
37. W. von Oertzen, M. Freer and Y. Kanada-En'yo, Nuclear clusters and nuclear molecules, *Physics Reports* 432, 43–113 (2006).

38. A. S. Demyanova, A. A. Ogloblin, A. N. Danilov *et al.*, Spectroscopy of ^{9}Be and observation of neutron halo structure in the states of positive parity rotational band, *EPJ Web of Conf.* **66**, 02026 (2014).
39. R. J. Peterson, Alpha-particle scattering from ^{9}Be, *Nucl. Phys. A* **377**, 41–52 (1982).
40. Y. Hirayama, T. Shimoda, H. Miyatake *et al.*, Unexpected spin-parity assignments of the excited states in ^{9}Be, *Phys. Rev. C* 91, 024328 (2015).
41. A. Ozawa, T. Suzuki and I. Tanihata, Nuclear size and related topics, *Nucl. Phys. A* **693**, 32–62 (2001).
42. A. Tohsaki, H. Horiuchi, P. Schuck and G. Ropke, Alpha Cluster Condensation in ^{12}C and ^{16}O, *Phys. Rev. Lett.* **87**, 192501 (2001).
43. E. Uegaki, S. Okabe, Y. Abe and H. Tanaka, Structure of the Excited States in ^{12}C, *Prog. Theor. Phys.* **57**, 1262–1276 (1977).
44. E. Uegaki, Y. Abe, S. Okabe and H. Tanaka, On the Positive-Parity States with Anomalous α-Decay Properties in ^{12}C, *Prog. Theor. Phys.* **59**, 1031–1033 (1978); **62**, 1621–1640 (1979).
45. Y. Fukushima and M. Kamimura, in Proceedings of the International *Conference on Nuclear Structure, Tokyo, September 5–10, 1977*, edited by T. Marumori (Physical Society of Japan, Tokyo, 1978).
46. Y. Kanada-En'yo, Negative parity states of ^{11}B and ^{11}C and the similarity with ^{12}C, *Phys. Rev. C* **75**, 024302 (2007).
47. M. Freer and H. O. U. Fynbo, The Hoyle state in ^{12}C, *Progress in Particle and Nuclear Physics*, 78, 1–23 (2014).
48. T. Yamada and P. Schuck, Single α-particle orbits and Bose-Einstein condensation in ^{12}C, *Eur. Phys. J. A* **26**, 185–199 (2005).
49. Y. Funaki, H. Horiuchi, W. von Oertzen, G. Röpke, P. Schuck, A. Tohsaki and T. Yamada, Concepts of nuclear α-particle condensation, *Phys. Rev. C* **80**, 064326 (2009).
50. N. Furutachi and M. Kimura, Bent three-α linear-chain structure of ^{13}C, *Phys. Rev. C* **83**, 021303 (2011).
51. M. Gai, The structure of the Hoyle state and its 2^{+} partner state in 12C, *EPJ Web of Conf.* **38**, 15001 (2012).
52. A. C. Dreyfuss, K. D. Launey, T. Dytrych and J. P. Draayer, Ch. Bahri, Hoyle state and rotational features in Carbon-12 within a no-core shell model framework, *Phys. Lett. B* 727, 511–515 (2013).
53. T. Suhara and Y. Kanada-En'yo, Quadrupole Deformation β and γ Constraint in a Framework of Antisymmetrized Molecular Dynamics, *PTP*, **123**, 303–325 (2010).
54. E. Epelbaum, H. Krebs, T. Lahde, D. Lee and Ulf-G. Meissner, Ab Initio Calculation of the Hoyle State, *Phys. Rev. Lett.* **106**, 192501 (2011).
55. T. L. Belyaeva *et al.*, Large-angle α-particle scattering on ^{12}C and search for signatures of α-particle Bose condensation, *Phys. Rev. C* **82**, 064618, (2010).
56. A. S. Demyanova, T. L. Belyaeva, A. N. Danilov *et al.*, $\alpha+{}^{12}$C scattering at 110 MeV: Has exotic 7.65 MeV Hoyle's state signatures "alpha-condensate" structure? *Phys. At. Nucl.* **72**, 1611– (2009).

57. T. Yamada and P. Schuck, Dilute multi-α cluster states in nuclei, *Phys. Rev. C* **69**, 024309 (2004).
58. A. A. Ogloblin, T. L. Belyaeva, A. N. Danilov, A. S. Demyanova and S. A. Goncharov, Radius of ^{12}C in the excited 2_2^+ Hoyle state, *Eur. Phys. J. A* **49** (46), 1–7 (2013).
59. H. Morinaga, Interpretation of Some of the Excited States of 4n Self-Conjugate Nuclei, *Phys. Rev.*, **101**, 254–258 (1956).
60. Y. Funaki, A. Tohsaki, H. Horiuchi, P. Schuck and G. Ropke, Resonance states in ^{12}C and α-particle condensation, *Eur. Phys. J. A* **24**, 321–342 (2005).
61. M. Freer, H. Fujita, Z. Buthelezi *et al.*, 2^+ excitation of the ^{12}C Hoyle state, *Phys. Rev. C* **80**, 041303 (2009).
62. M. Itoh, H. Akimune, M. Fujiwara *et al.*, Candidate for the 2^+ excited Hoyle state at $E_x \sim 10$ MeV in ^{12}C, *Phys. Rev. C* **84**, 054308 (2011).
63. M. Freer *et al.*, Consistent analysis of the 2^+ excitation of the ^{12}C Hoyle state populated in proton and α-particle inelastic scattering, *Phys. Rev. C* **86**, 034320 (2012).
64. W. R. Zimmerman *et al.* Unambiguous Identification of the Second 2^+ State in ^{12}C and the Structure of the Hoyle State, *Phys. Rev. Lett.* 110 152502 (2013).
65. M. Freer, S. Almaraz-Calderon and A. Aprahamian, Evidence for a new ^{12}C state at 13.3 MeV, *Phys. Rev. C* **83**, 034314 (2011).
66. A. A. Ogloblin, A. S. Demyanova, A. N. Danilov *et al.*, Rotational band in ^{12}C based on the Hoyle state, *EPJ Web of Conf.* **66**, 02074 (2014).
67. D. J. Marín-Lámbarri, R. Bijker and M. Freer, Evidence for Triangular D_{3h} Symmetry in ^{12}C, *Phys. Rev. Lett.* 113, 012502 (2014).
68. M. Milin and W. von Oertzen, Search for molecular bands in ^{13}C, *Eur. Phys. J. A* **14**, 295–307 (2002).
69. T. Kawabata, Y. Sasamoto, M. Fujiwara *et al.*, Cluster states in ^{11}B and ^{13}C, *Journal of Physics: Conference Series*, **111**, 012013 (2008).
70. A. S. Demyanova, A. A. Ogloblin, A. N. Danilov, T. L. Belyaeva and S. A. Goncharov, Radii of cluster states in ^{11}B and ^{13}C, *Int. J. Mod. Phys. E* **20**, 915–918 (2011).
71. A. N. Danilov, A. S. Demyanova, S. V. Dmitriev *et al.*, Study of elastic and inelastic ^{11}B + α scattering and search for cluster states of enlarged radius in ^{11}B, *Yad. Fiz.* **78**(9), 828–844 (2015). [English translation *Phys. At. Nucl.* **78**(9), 777–793 (2015)].
72. T. Suhara and Y. Kanada-En'yo, Cluster structures in ^{11}B, *Phys. Rev. C* **85**, 054320 (2012).
73. H. Yamaguchi, T. Hashimoto, S. Hayakawa *et al.*, α resonance structure in ^{11}B studied via resonant scattering of ^{7}Li + α, *Phys. Rev. C* **83**, 034306 (2011).
74. A. N. Danilov, A. S. Demyanova, A. A. Ogloblin *et al.*, Cluster states in ^{11}B, *EPJ Web of Conf.* **66**, 03007 (2014).

75. T. Yamada and Y. Funaki, $\alpha + \alpha + t$ cluster structures and ^{12}C(0_2^+)-analog states in ^{11}B, *Phys. Rev. C* **82**, 064315 (2010).
76. T. Yoshida, N. Itagaki and T. Otsuka, Appearance of cluster states in ^{13}C, *Phys. Rev. C* **79**, 034308 (2009).
77. A. A. Ogloblin, A. S. Demyanova, A. N. Danilov, S. A. Goncharov, T. L. Belyaeva, W. Trzaska and Yu. G. Sobolev, Possible Observation of an Excited State with an Anomalously Small Radius in the ^{13}C Nucleus, *JETP Lett.* **102**, 199–202 (2015).
78. N. Furutachi and M. Kimura, Bent three-α linear-chain structure of ^{13}C, *Phys. Rev. C* **83**, 021303 (2011).

CHAPTER 12

MANIFESTATIONS OF CLUSTERING IN BINARY AND TERNARY FISSION OF LOW EXCITED HEAVY NUCLEI

Yu. V. Pyatkov

National Nuclear Research University MEPhI (Moscow Engineering Physics Institute), 115409 Moscow, Russia, Joint Institute for Nuclear Research, 141980 Dubna, Russia yvp_nov@mail.ru

D. V. Kamanin

Joint Institute for Nuclear Research, 141980 Dubna, Russia kamanin@jinr.ru

There comes a discussion of the evidences of preformation of a pair of magic nuclei (clusters) such as ^{128,132}Sn and ^{70}Ni, ^{82}Ge in the body of mother system at the early stage of its elongation. It is believed to be a crucial feature of the process of binary fission of low excited heavy nuclei. Strong experimental indication has been obtained that the fragment of conventional binary fission immediately after scission of the fissioning nucleus looks like a di–nuclear system consisting of the magic core and a light cluster to be some part of the neck. Similar clusters manifest themselves in the multibody decays observed for the first time.

1. Cluster nature of fission modes

1.1. *Experimental and theoretical proofs*

First attempts to create a model of the fission process involving shell aspects were taking place in early fifties of the last century. V. V. Vladimirski[1] was perhaps the first to have postulated that the fission probability had a noticeable value only if two cluster structures such as magi cores within the light and heavy fragment corresponding to the $N = 50$ and $Z = 50$, $N = 82$

shells were not destroyed. A dumbbell–like configuration consisting of two magic clusters connected by a flat cylindrical neck was considered as a typical shape of the fissioning system. Formal representation of the fissioning system as a superposition of two interacting clusters (of arbitrary nucleon composition) is under exploration in a cluster model of fission developed by K. Wildermuth.[2] The outlined theory had unfortunately been "too basic" for practical use. There are also well known multiple calculations in the frame of two–center shell model approach.[3,4] One of the most important and reproducible results obtained was to show that specific features of the level scheme of the fission fragments to be borne are already defined at relatively small distance between the centers of the clusters chosen for study. Thus, final fragments can grow from the cores along with elongation of the mother system. Visual proof of clustering of the fissioning system along the descent from the fission barrier was obtained in Refs. 5 and 6. Figure 1 shows potential energy of the fissioning nucleus ^{246}Cm corresponding to the

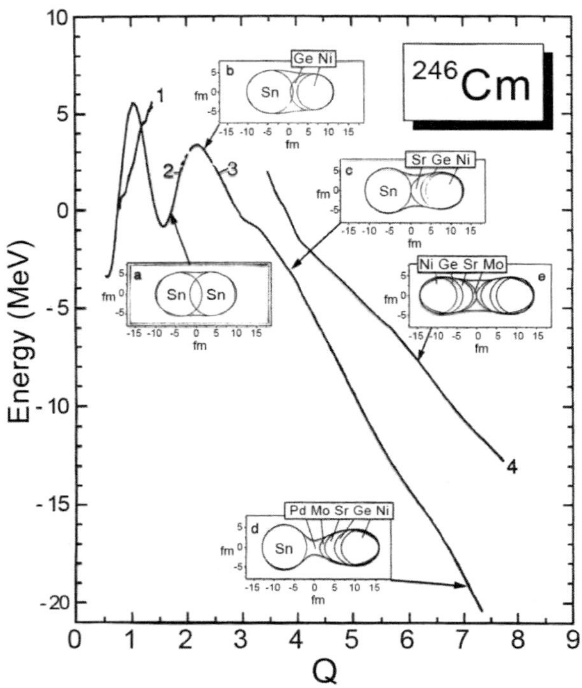

Fig. 1. Potential energy of the fissioning nucleus as a function of its elongation.[6]

bottoms of the potential valleys as the function of parameter Q proportional to the quadrupole moment of the system. The calculations were performed in the frame of Strutinsky approach in ten dimensional deformation space.

Both in valley of mass–asymmetrical shapes (3) and mass–symmetrical shapes (4) the system consists of the pairs of magic clusters (Sn/Ni, Sn/Ge and so on) and the left over nucleons forming a "neck" between the clusters. The map under discussion vividly demonstrates the role of the magic clusters in the body of the fissioning system in the form of fission modes. Actually, the ruptures occurred while the system descents along the distinct valley provide the fission fragments observed experimentally and treated as the appropriate fission mode.[5] Similar "visual" analysis of the nuclear shape in deformation process let authors of work[7] to reveal clustering in quasi–fission reactions leading to super–heavy systems.

Recent studies of the far asymmetric fission (Fig. 2) also let us make inferences concerning the crucial role of just pair of magic clusters in fission process. As it was accentuated in Ref. 8: "...fission is not only determined by the double shell closure in the heavy sphere of the scission point dumbbell configuration around $A = 132\,(Z = 50,\ N \approx 82)$ but also by an effect of the double shell closure of $Z = 28$ and $N \approx 50$ in the corresponding light sphere". As can be inferred from Fig. 2, mass spectra of different actinide nuclei are bounded by mass ~70 amu. It was shown in Ref. 9 that those were really magic isotopes of Ni. The conclusion concerning the light mass boundary of mass spectra of actinides was confirmed as well by the results obtained in the course of studying of the (p, f) reaction.[10]

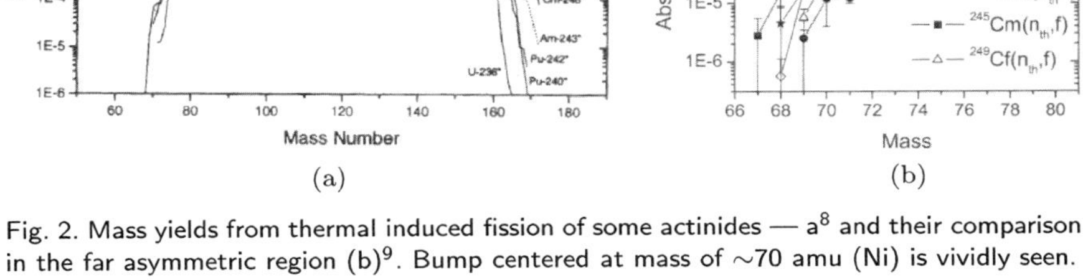

Fig. 2. Mass yields from thermal induced fission of some actinides — a[8] and their comparison in the far asymmetric region (b)[9]. Bump centered at mass of ~70 amu (Ni) is vividly seen.

1.2. *Fine structures in mass–energy distributions of fission fragments*

Specific manifestations of clustering were revealed as a result of processing of the FF mass–energy distributions using methods of image analysis.[11]

It is a new approach to extracting information from known mass–energy distributions of nuclear reaction products by direct processing of two–dimensional data.[11] For example, the typical $M - E$ (mass–energy) distribution of fission fragments in the $^{233}\mathrm{U}(\mathrm{n_{th}}, \mathrm{f})$ reaction seems to looks like a smooth hill (Fig. 4(a)). The more detailed consideration reveals that any E = const section of this distribution (Fig. 4(b)) is not absolutely smooth but on the contrary, exhibits local irregularities (peaks) marked by arrows.

The origin of the peaks becomes clear from the following speculation. The yield $Y(M|E)$ of fission fragments with the mass M and the fixed energy E is

$$Y(M|E) = \sum_{Z} Y(M, Z|E). \tag{1}$$

The summation in expression (1) is performed over all possible values of the nuclear charge Z of the fission fragments. Thus, the spectrum shown in Fig. 4(b) is a superposition of partial mass spectra for fixed charges (the so–called isotopic distributions) known from the experiment.[14] Let us define

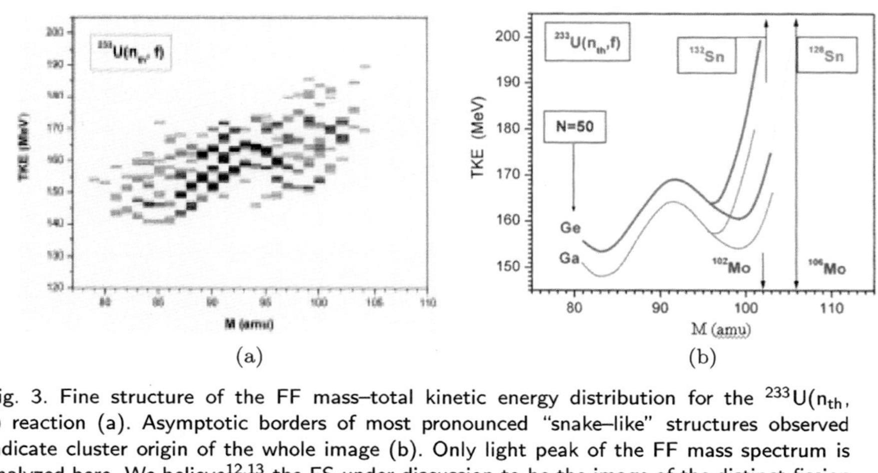

Fig. 3. Fine structure of the FF mass–total kinetic energy distribution for the $^{233}\mathrm{U}(\mathrm{n_{th}}, \mathrm{f})$ reaction (a). Asymptotic borders of most pronounced "snake–like" structures observed indicate cluster origin of the whole image (b). Only light peak of the FF mass spectrum is analyzed here. We believe[12,13] the FS under discussion to be the image of the distinct fission way extending "from the light cluster (Ge, Ga) to the heavy cluster (Sn)".

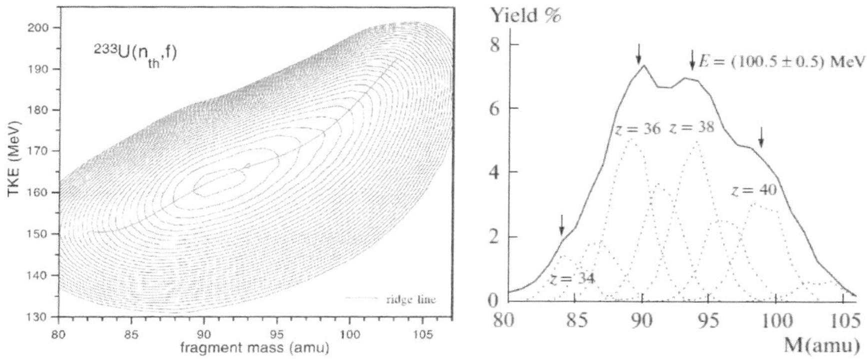

(a) (b)

Fig. 4. The contour map of the experimental mass–energy FF distribution in the $^{233}U(n_{th}, f)$ reaction (a). $E - M$ cross section for the fragment energy $E = (100.5 \pm 0.5)$ MeV.[14] Partial yields for fixed charges are shown by dotted lines.

the term as a "fine structure". By definition, this term means local regions (peaks) of the two–dimensional distribution with a yield higher than that on a smooth substrate.

As it follows from Fig. 4(b), the fine structure in this case is caused by a higher yield of fragments with an even charge (known odd–even modulation of mass yields due to the proton odd–even effect[15]). The peaks in the neighboring E = const sections are correlated and they constitute regular structures in the $E - M$ plane in the form of ridges parallel to the axis of energy E.[14] This structure will further be termed as vertical ridges.

An example of such fine structure in ^{252}Cf (sf) is shown in Fig. 5.[5] The position of the vertical ridges well agrees with the tops of the known isotopic distributions. At lower TKE values, i.e. at higher excitation energies in the scission point $E^* = Q - \mathrm{TKE}$ (where Q is the energy release for corresponding partition of ^{252}Cf and TKE is a total kinetic energy of the detected fragments), there appears an odd–even effect which is presumably associated with the complete clusterization of the fissioning nucleus at large elongations. In fact, the calculation of the potential energy surface (PES) of the fissioning ^{252}Cf nucleus predicts its specific shape at large elongations[5] (Fig. 6(a)), that is the bulk of the nucleons is included into magic clusters of Sn, Mo, and Cd. The predictions well agree with the experimental data. The conditional mass–energy distribution $P(M|E^*)$ (Fig. 6(b)) demonstrates a ridges leading to the partitions predicted in valley 3, 4.

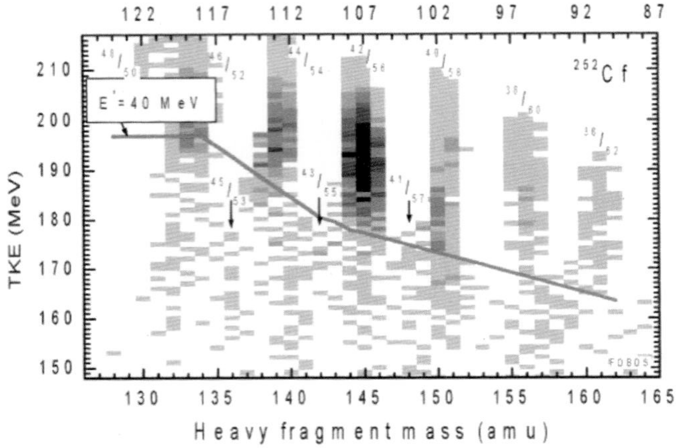

Fig. 5. The fine structure of the TKE–M distribution of ^{252}Cf fission fragments demonstrating the proton odd–even staggering. The FF nuclear charges are indicated near the ridges.[5]

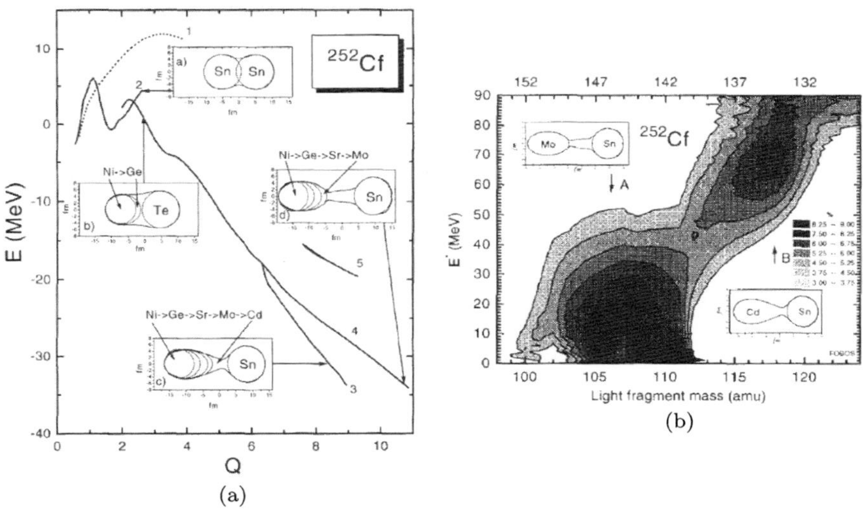

Fig. 6. Potential energy of the fissioning nucleus ^{252}Cf corresponding to the bottoms of the potential valleys, as the function of Q, proportional to its quadrupole moment. The panels depict the shapes of the fissioning system at points marked by arrows (a); (b) — the contour map of the conditional mass–energy distribution $P(M|E^*)$. The panels depict the shapes of the fissioning system following from the PES calculations ascribed to the two dominant structures.

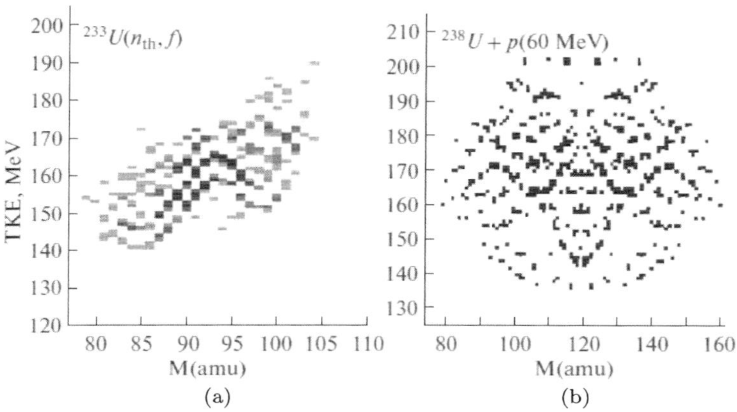

Fig. 7. Snake like fine structures found in the mass–total kinetic energy distribution (M–TKE) of fission fragments in the reactions (a) ^{233}U(n_{th}, f) and (b) ^{238}U + p (60 MeV). Details are presented in the text.

We have posed a question whether the structure in the mass–energy distribution of fission fragments that differs from vertical ridges produced by the odd–even modulation of mass yields, and is therefore caused by other physical reasons. For automatic suppression of vertical ridges in the fine structure search for, we have analyzed M = const sections in which, as it is known,[5,17] the peaks can also be observed. In our studies, we used the algorithms for identifications of peaks known from gamma spectroscopy and from methods for processing noised images.[11] Figure 7 shows examples of a fine structure revealed in the total kinetic energy–mass distributions of fission fragments. The initial data were obtained using the time–of–flight spectrometers of fission fragments described in Refs. 18 and 19. Darker points in the tone diagram (Fig. 7(a)) correspond to higher amplitude of the effect. There is only a light peak shown in the mass distribution of fission fragments. The symmetry of the structure shown in Fig. 7(b) is caused by both the methods of measuring the mass of fission fragments (the method of two velocities[15]) and by the employed filter.[20] The prevailing fine structure is a sequence of snake like curves, sometimes having bifurcation points.[11,20]

What is the purpose of studying the debated fine structures? According to the existing concepts, in fission, for instance, the evolution of a decaying nuclear system is mainly determined by the potential energy surface (PES). Separate potential valleys in PES[21,22] cause the existence of selected trajectories of the system in the deformation space. As it is shown in Ref. 23 in the frame of the time–dependent Hartree–Fock approach, the scission can take place at any point of descent of the system down the fission valley, which is detected as an event of fission in the space of experimentally observed variables.

In other words, a trajectory in the deformation space as a continuous sequence of nuclear states in the fission valley is mapped to continuous trajectories (smooth curves) in the plane of experimentally observed variables,[5] such as coordinates in which Fig. 7(a) has been constructed, namely the total kinetic energy and the mass of fission fragments. Thus, we assume that the fine structure under discussion is a unique image of separate fission trajectories passed by a system for a time as $\sim 10^{-20}$.[24]

Special attention was paid for a quantitative estimate of the reliability of the revealed fine structures.[25]

Analyzing specific features of FS we have put forward a hypothesis concerning the shape of the decaying system along the descent from the fission barrier.[24] Presumably it looks like multibody nuclear molecule based on a pair of magic clusters (Ge/Sn in the most populated valley) while residual nucleons form torus–like "neck" in between gross clusters (Fig. 8). The neck could be also clusterized i.e. consists of He isotopes and neutrons. The evident difference of the proposed compared to those seen in Fig. 6 could be traced back to the fact that only simply connected shapes are adopted in the known calculations of PES.[22,23] Thus, in the frame of the hypothesis under discussion the most compact configuration of the system looks like the one shown in Fig. 8(a). With further elongation the toroidal neck exports nucleons to the space between the tips of the gross clusters. Following[5] we may suppose that the scission of the system can come into being at any stage of the descent (actually according to Ref. 5 it could be tunneling of the system into the valley of separated fragments). As a result, the scission results in forming of two fragments. Heavy fragment unites Sn

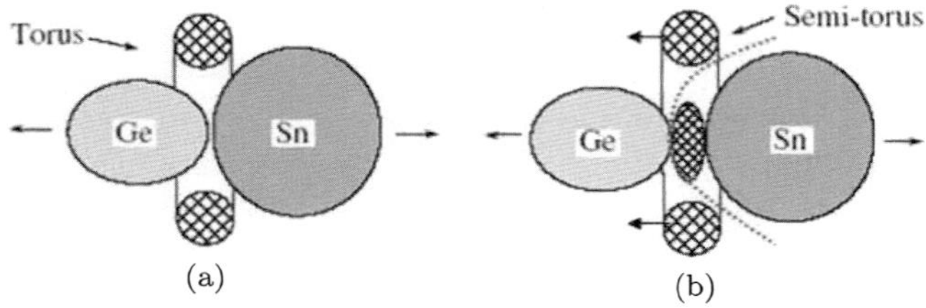

Fig. 8. Presumable shape of the fissioning nucleus along the descent from the fission barrier.[24] See text for details.

cluster and central part of the semi–torus (this part is marked in Fig. 8(b) by the dot line), while the light one consists of Ge cluster and residual part of the torus (marked by two arrows). The scenario is supposed to be a symmetric reference to gross clusters i.e. heavy fragment can be composed of the Sn cluster and the residual part of the semi–torus, while the Ge cluster joins its central part.

In our earlier work[26] a toroidal neck was supposed to be clusterised forming a circle consisted of α–particles and other isotopes of He.

The shape of the fissioning nucleus depicted in Fig. 8 seems to be extremely exotic compared to "classic" dumbbell–like shapes discussed elsewhere.[22] Nevertheless, recent theoretical calculations confirm an actuality of such consideration. A mechanism of true ternary fission was proposed in Ref. 27. This mechanism is driven by a hexadecapole deformation of the fissioning nucleus. Typical shapes of the fissioning ^{252}Cf nucleus are shown in Fig. 9(a). Torus like "neck" unites two other constituents in a single nucleus. The similar shape obtained in the framework of nuclear energy density functionals for ^{28}Si nucleus is shown in Fig. 9(b).[28]

1.3. *Clustering of the fissioning system in third minimum of the fission barrier*

The pronounced third minima have been predicted in theoretical studies of thorium and uranium isotopes, especially in those carried out with the

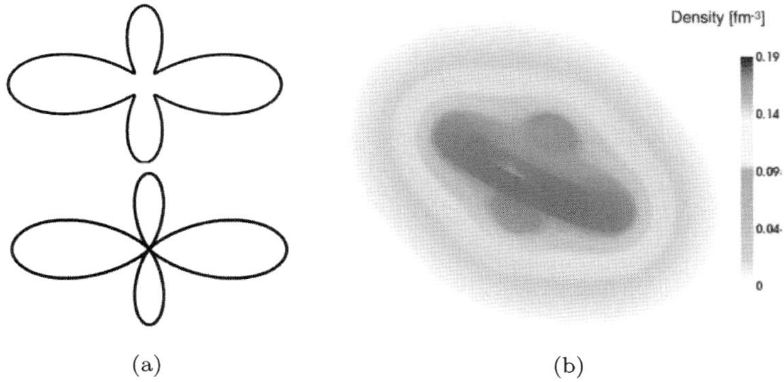

Fig. 9. (Color online) Shapes of the fissioning ^{252}Cf nucleus for the parameter of the hexadecapole deformation $\beta_{40} = 1.7$ and 1.902.[27] Mass ratio for the middle and one of the extreme fragments is equal to 2.13 and 1.87, respectively (a). (b) Self–consistent ground–state density of ^{28}Si calculated in the framework of nuclear energy density functionals[28] vividly demonstrates cluster structure.

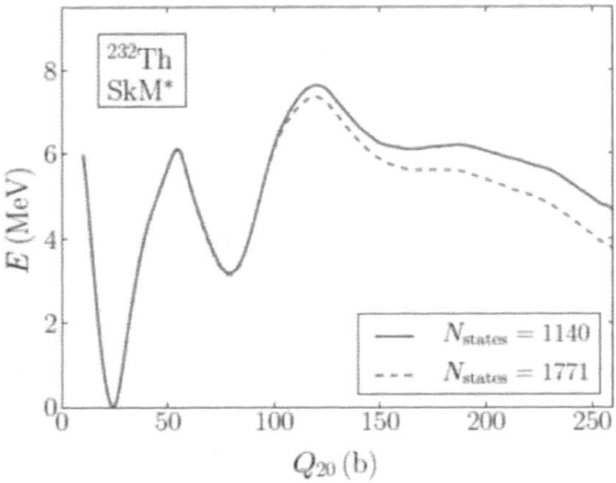

Fig. 10. Potential–energy curve for ^{232}Th with SkM* obtained with a basis of 1140 (solid line) and 1771 (dashed line) basis states.

macroscopic–microscopic (MM) approach.[21,29,30] These predictions were used for the interpretation of resonances observed experimentally. On the other hand, self–consistent studies based on the nuclear density functional theory, as well as recent MM works, typically find a third minimum that is much shallower. To clarify the situation, the self–consistent calculations were carried out in Ref. 31. The authors used the finite–temperature superfluid nuclear density functional theory. They consider two Skyrme energy density functionals: a traditional functional SkM* and a recent functional UNEDF1 optimized for fission studies. The shape of the fission barrier obtained for ^{232}Th is shown in Fig. 10.

The impressive evidences of preformation of magic clusters in the body of the mother fissioning system were demonstrated. The density profiles appeared to correspond to the third minimum of ^{232}Th isotope shown in Fig. 11. The paper demonstrates "that the third minimum can be associated with a dimolecular configuration involving the spherical doubly magic ^{132}Sn and a lighter Zr or Mo fragment in a deformed configuration". The nuclear compositions of the lighter fragments are also close to the magic ones.[32] The effect is reproduced for the series of Th and U isotopes (Fig. 12).

The calculations for the super–heavy nuclei demonstrate similar tendency. For instance, Fig. 13 shows three–humped fission barrier of ^{296}Lv.[33]

The authors conclude: "These intermediate minima correspond to the shape isomer states. From the analysis of the driving potential we may

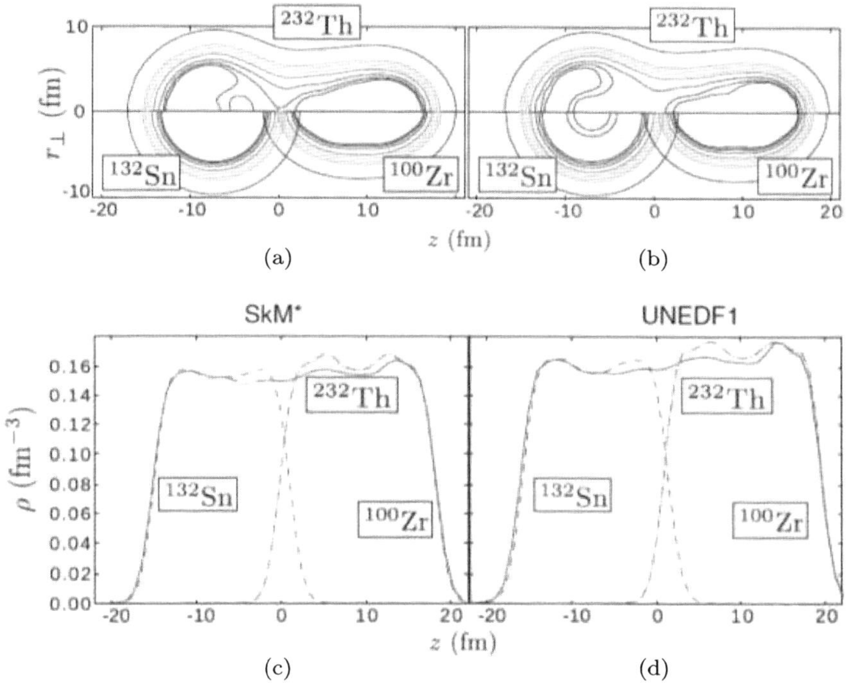

Fig. 11. (Color online) (top) Cross section of total density of ^{232}Th in *yz* plane calculated with (a) SkM* and (b) UNEDF1 at third minimum ($Q20$ = 165 b), compared to cross sections of ^{132}Sn and ^{100}Zr densities. (bottom) Density profiles for ^{232}Th and the ^{132}Sn and ^{100}Zr fragments along the *z* axis obtained in (c) SkM* and (d) UNEDF1.[31]

definitely conclude that *these isomeric states are nothing else but the two-cluster configurations with magic or semi–magic cores surrounded by certain amount of shared nucleons*".

Summing up, one can state an exclusive role of the magic clusters in the mechanism of binary fission. It is substantially stronger assertion than traditional "influence of shell effects" because clustering presupposes space localization of the corresponding nuclear object. The question now arises of whether conventional dumbbell–like shape of the system is really adequate to the fission process or a more complicated multibody nuclear molecule (Fig. 8) should be included into our speculation. It is believed that a new kind of ternary decay discussed below give the arguments in favor of the latter hypothesis.

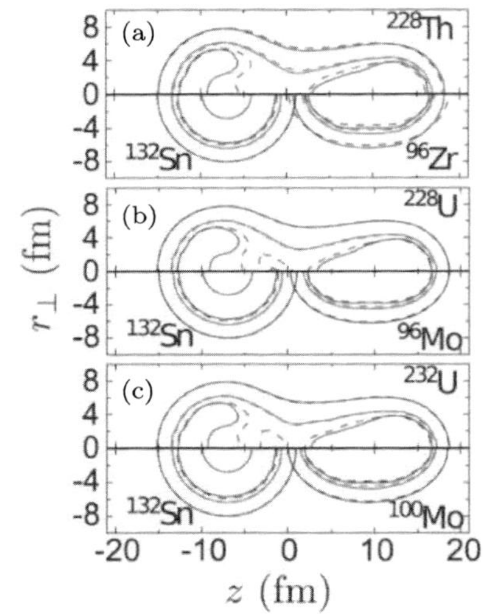

Fig. 12. (Color online) Contour plots for total densities of (a) ^{228}Th, (b) ^{228}U, and (c) ^{232}U calculated with UNEDF1 (solid lines) and SkM* (dashed lines) compared with fragment densities: spherical ^{132}Sn and a lighter deformed nucleus around ^{100}Zr. The contour levels shown are at 50%, 90%, and 95% of the saturation density ($\rho 0 = 0.16$ fm -1).[31]

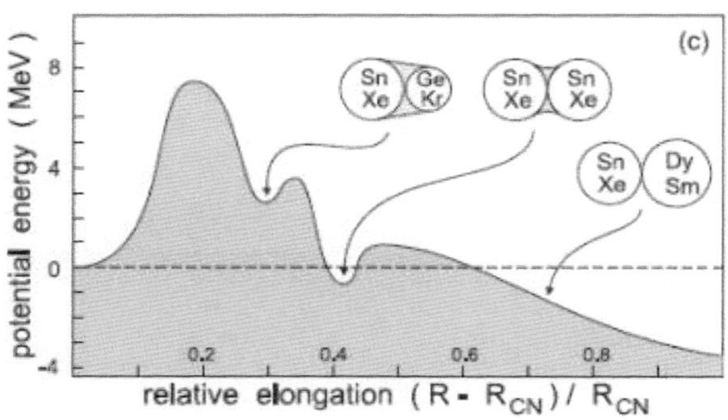

Fig. 13. Three–humped barrier calculated along the fission path of $^{296}_{116}$Lv.

2. Collinear cluster tri–partition

This section is devoted to the observation of a new kind of ternary decay of low–excited heavy nuclei. We have called this decay mode as a "collinear cluster tri–partition" (CCT) in view of the observed features of the effect showing that the decay partners fly apart almost collinearly, and at least one of them definitely has magic nucleon composition. CCT is observed together with conventional binary and ternary fission. It could be one of the rare fission modes, but for the time being it is not a matter–of–fact yet. There have actually been many years between the experimental discovery of the heavy ion radioactivity and the development of the recognized theory of the process.

At the early stage of our work a process of the "true ternary fission", the fission of the nucleus into three fragments of comparable masses was considered to be undiscovered regarding low excited heavy nuclei. Another possible prototype — three body cluster radioactivity — was also unknown. The so called "polar emission"[34] stands the closest, at least cinematically, to the CCT phenomenon, but only very light ions (up to isotopes of Be) have been observed so far.

From the theoretical point of view there are numerous indications on possible ternary decays of the low excited heavy nuclei with comparable masses of the decay products. Within the framework of the liquid drop model (LDM) Swiatecki[35] has shown that the fission into three heavy fragments in energetically more favorable than binary fission for all nuclei with fission parameters $30.5 < Z^2/A < 43.3$. In 1963 Strutinsky[36] has calculated the equilibrium shapes of the fissioning nucleus and showed that along with the ordinary configuration with one neck, there was the possibility of more complicated elongated configurations with two and even three necks, at the same time it was emphasized, that such configurations were much less probable. Later Diehl and Greiner[37,38] showed a preference for prolate over oblate saddle–point shapes for the fission of a nucleus into three fragments of similar size. Such pre–scission configurations could lead to almost collinear separation of the decay partners, at least in a sequential fission process. The results demonstrating a crucial role of shell effects in the formation of the multi–body chain–like nuclear molecules were obtained by Poenaru *et al.*[39] We also would like to refer to the recent theoretical articles devoted to unusual ternary decays of heavy nuclei including CCT.[40–46] The authors analyze the potential energy of different pre–scission configurations leading to ternary decays, and the kinetic energies of the CCT partners[47] are calculated for a sequential decay process. These results can be considered as the first step in the description of the CCT process.

Having in mind both theoretical and experimental background mentioned above we have come to the conclusion, that the collinear tri–partition of low–excited heavy nuclear systems would be a promising field of research. The basic results obtained so far are represented here.

2.1. *The most pronounced manifestation of the CCT*

The bulk of our results has been obtained within the framework of the "missing–mass" approach. With the use two–arm spectrometers there have been measured binary coincidences. This means that only two fragments were actually detected in each fission event, and their total mass, the sum M_s will serve as a sign of multi–body decay in case it appears to be significantly smaller than the mass of the initial system. The registration of a third fragment was blocked, as explained below and shown in Fig. 14.

Figure 15 shows in a logarithmic scale the two–dimensional distribution $(M_2 - M_1)$ of the two registered masses of the coincident fragments.[49] In the experiment performed at the FOBOS setup M_1 is defined as a fragment mass derived from the arm pointing towards the detector arm with the additional dispersive (scattering) materials. There were selected only collinear fission events with a relative angle of 180 ± 2^0, which corresponds to the typical angular spread for conventional binary fission fragments. The "tails" in the mass distributions marked $3-6$ in Fig. 15(a) extending from the regions (1) and (2) which are used to mark the conventional binary

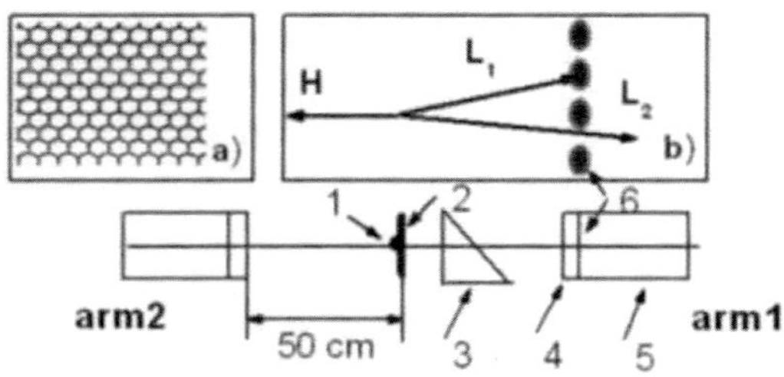

Fig. 14. Scheme for coincidence measurements of two fission fragments of ^{252}Cf (sf). The experiment has been performed at the FOBOS setup.[48] Here: 1 — Cf source, 2 — source backing, 3 — micro–channel plate (MCP) based timing "start" detector, 4 — position sensitive avalanche counter (PSAC) as "stop" detector, 5 — ionization chamber (BIC) with the supporting mesh, 6 — the mesh of the entrance window. The front view of the mesh is shown in the insert (a), an enlarged mesh section is presented in the insert (b). After passage of the two fragments through the source backing, two light fragments *L*1 and *L*2, are obtained with a small angle divergence due to multiple scattering. In (b) we show that one of the fragments (*L*1) can be lost hitting the metal structure of the mesh, while the fragment *L*2 reaches the detectors of the arm1. The source backing (2) exists only on one side and causes the mentioned angular dispersion in the direction towards the right arm1.

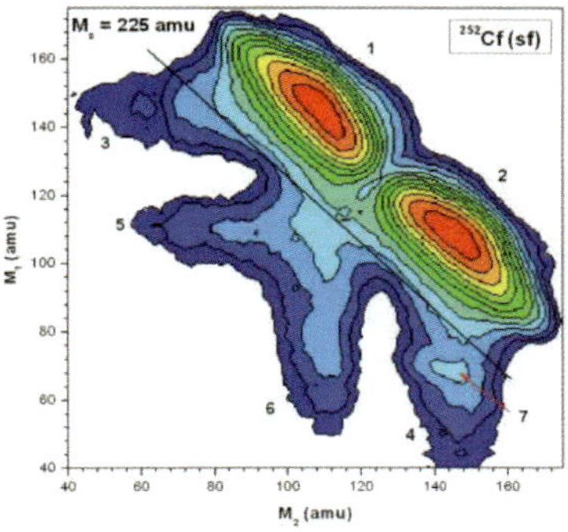

(a)

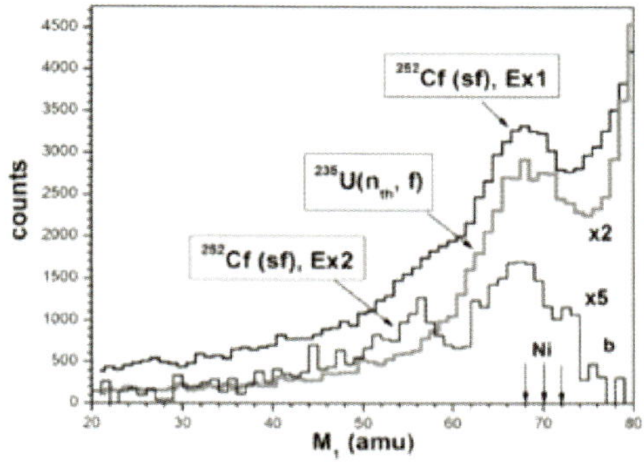

(b)

Fig. 15. (a) Contour map (in logarithmic scale, the steps between the lines are approximately factor 2.5) of the mass–mass distribution of the collinear fragments of ^{252}Cf (sf), detected in coincidence in the two opposite arms of the FOBOS spectrometer. The specific bump in arm1 is marked by an arrow. (b) Projection of the "Ni"–bump onto M_1–axis obtained in three different experiments performed at the FOBOS spectrometer modules.

fission, are mainly caused by the scattering of fragments on both the foils and on the grid edges of the "stop" avalanche counters and in the ionization chambers. We emphasize the small but important asymmetry in the experimental arrangement for the two arms, which consists of the thin source backing (50 μg/cm^2 of Al_2O_3) of the target and the "start" detector foil located only in arm1 (Fig. 15(a)). It is an astonishing difference in the counting rate and in the shapes of the "tails" (3) and (4) that draws our attention. In the case shown in Fig. 15(a) there is a distinct bump, marked (7), on top of the latter "tails" (4). The bump is located in a region corresponding to a large "missing" mass. The statistical significance of the events in the structure (7) can be deduced from Fig. 15(b), where the spectra of the total (summed) masses Ms for the "tails" (4) and (3) are being compared. The yield of the events in the difference spectrum c, is $(4.7 \pm 0.2)^{*}10^{-3}$ relative to the total number of events in the distribution shown in Fig. 15(a). It is only a lower limit of the yield because the following reason: if both fragments ($L1$ and $L2$ in Fig. 14) pass on and enter into the BIC, we register a signal corresponding to the sum of the energies of the two fragments. In this case the event will be treated as binary fission with the unusual parameters. The existence of the bump with the similar yield was confirmed in two other experiments carried out at the modified FOBOS spectrometer.[50]

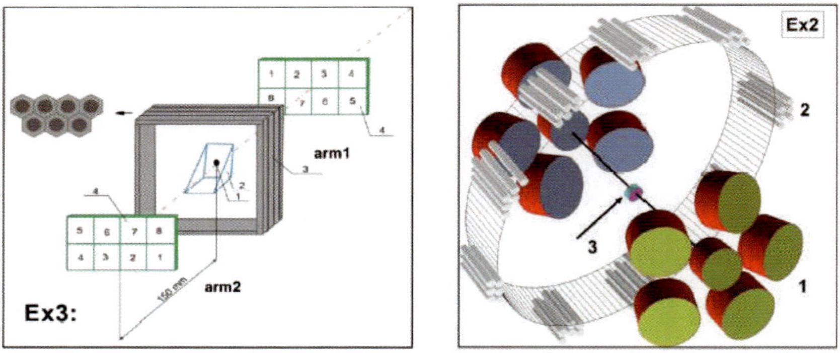

(a) (b)

Fig. 16. Scheme of the COMETA setup, which consists of two mosaics of eight PIN diodes each (4), MCP–based start detector (2) with the ^{252}Cf source inside (1), and a "neutron belt" (3) consisting of 28 ^{3}He–filled neutron counters in a moderator. The cross–section of the belt is marked by an arrow (a). (b) — layout of the experiment for the neutron coincidences with the modified spectrometer based on FOBOS detector modules (1), a "neutron belt" consisting of 140 ^{3}He–filled neutron counters (2), and a "start" avalanche counter with the Cf source inside (3).

As it was emphasized above a specific supporting grid in the FOBOS detecting modules provides the selection of the ternary events. At the same time it serves as a source of the scattered fragments forming the "tails" (Fig. 15(a)) simulating the missing mass events. This effect is essentially suppressed in the mosaic COMETA spectrometer (Fig. 16(a)) which basically allows detecting all the partners of the ternary decay.

This methodically quite different experiment confirms our previous observations concerning the structures in the missing mass distributions. In this case there is no tail caused by the scattering from material in front of the E–detectors. Figure 17(a) shows the region of the mass distribution for the FFs from ^{252}Cf (sf) around the "Ni"–bump ($M_1 = 68 - 80$ amu, $M_2 = 128 - 150$ amu). The structures are seen in the spectrometer arm facing the source backing only. No additional selection of the fission events has been applied in this case which resulted in the experiment having almost no background. A rectangular–like structure below the locus of binary fission is bounded by magic nuclei (their masses are marked by the numbered arrows) namely ^{128}Sn (1), ^{68}Ni (2) and ^{72}Ni (3). In Fig. 17(b) we show the projection of the linear structure seen at the masses of 68 and 72 amu. Two tilted diagonal lines with $Ms = 196$ amu and $Ms = 202$ amu (marked by

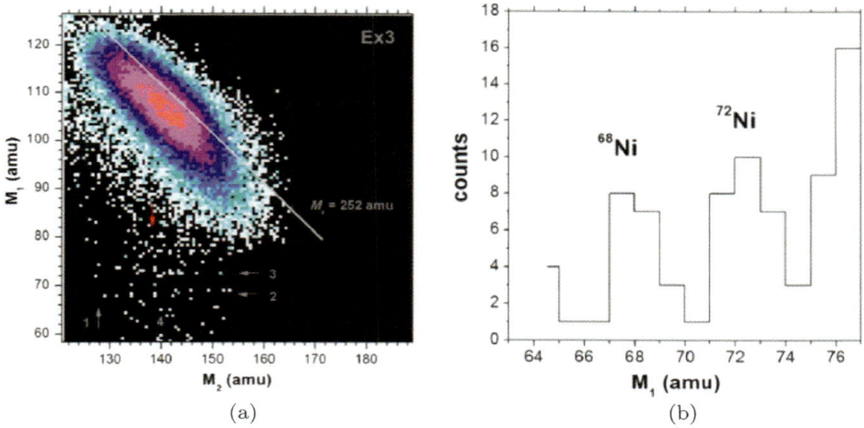

Fig. 17. Region of the mass–mass distribution for the FFs from ^{252}Cf (sf) around the CCT–bump (similar to that marked by an arrow in Fig. 15(a)). The result was obtained at the COMETA setup. No additional gates were applied. The background of the scattered fragments is very low due to the use of PIN–diodes (absence of a grid before the surface). An internal structure of the "bump" seen as horizontal lines from points (marked by arrows, 2 and 3) is shown in Fig. 17(b) as a projection.

number 4) start from the partitions 68/128 and 68/134 (all the nuclei are magic), respectively.

The result for the measured charges (Fig. 18) confirms the previous findings with the mass distributions namely the existence of an additional bump linked with Ni isotopes ("Ni"–bump) in the arm with the scattering medium.

One of the decay modes which contribute to the bump and manifests itself by the tilted ridges $M_1 + M_2 = \text{const}$ can be treated as a new type of cluster decay compared to a well–known heavy ion or lead radioactivity. The key features of both of them are summed up in Fig. 19. The relatively high CCT yield can be well understood once you visualize the collective motion through the hyper–deformed pre–scission shapes of the mother systems, which is supported by the fact that the linear arrangement uses the lowest Coulomb potential energies of three clusters. We also emphasize, that the Q values for ternary fission are more positive by 25–30 MeV, which again is due to the formation of magic fragments like in binary fission. The ternary fission process must be considered to proceed sequentially, with two neck ruptures in a short time sequence.

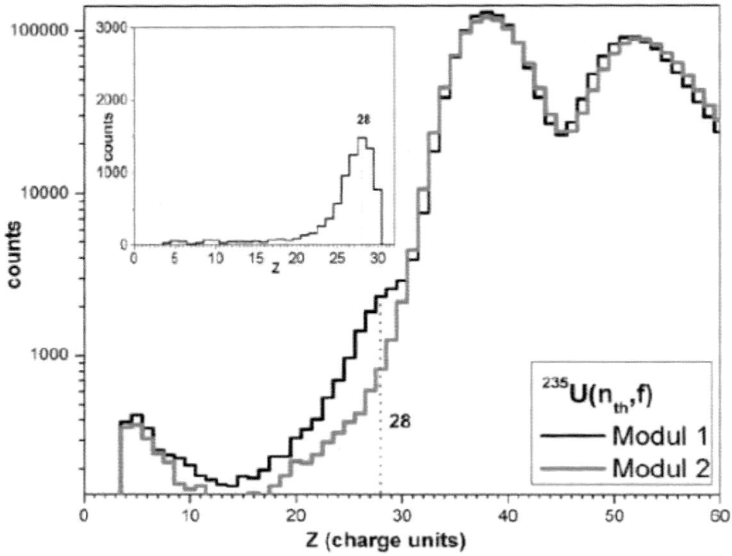

Fig. 18. Nuclear charge spectra for the FF from the reaction $^{235}U(n_{th}, f)$, the FF are detected in the two opposite spectrometer arms. A difference in the yields (bump) presented in the upper panel in a linear scale is visible for the charges around $Z = 28$ (isotopes of N).

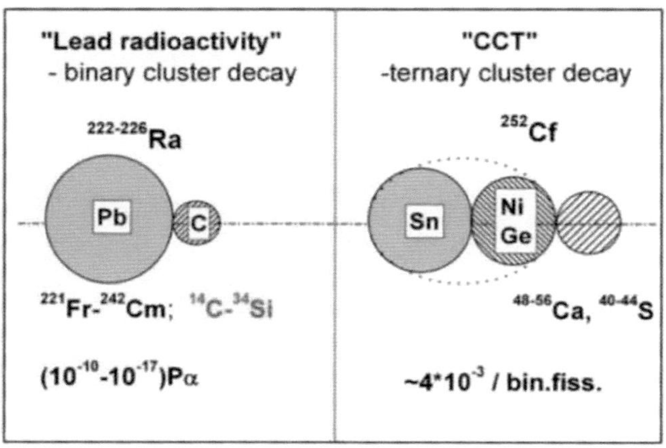

Fig. 19. Cluster scheme for the comparison of the lead radioactivity with collinear cluster tri–partition.

2.2. *Conclusions from the neutron gated data*

The results of our previous experiments allow us to assume that there are several CCT modes[51] with the middle fragment of the three–body pre–scission chain with very low velocity after the scission takes place. Such fragments are expected to emit neutrons almost isotropically. The neutrons emitted from the moving binary fission fragments are focused predominantly along the fission axis. In order to exploit this difference for revealing the CCT events, there was assembled a "neutron belt" in a plane perpendicular to the symmetry axis of the spectrometer (Fig. 16). According to modeling and the previous experiments, the detection efficiency is estimated to be ~5% and ~12% for a neutron emitted in binary fission and from an isotropic source, respectively. The array of neutron counters of the similar geometry was exploited earlier at the modified FOBOS spectrometer.[50] One of the results obtained at the COMETA setup is shown in Fig. 20.

As can be inferred from the figure, the rectangular structure visible in its upper right hand corner is bounded by the nuclei with the masses in the vicinity of known magic nuclei (shown in brackets). These masses (except for double magic ^{132}Sn) were calculated based on the unchanged charge density hypothesis for the fission of ^{252}Cf nucleus. We actually known that

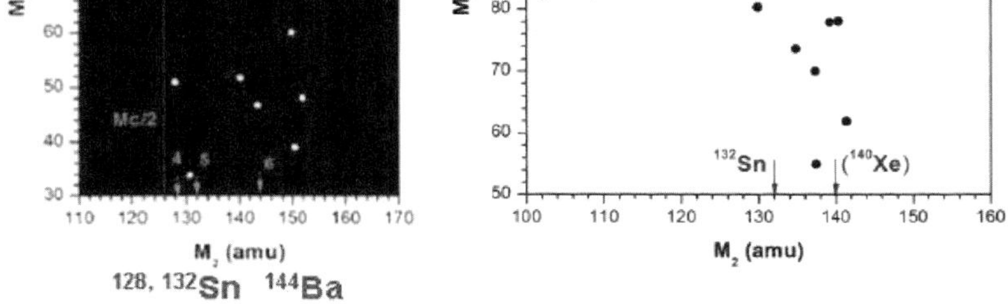

(a) (b)

Fig. 20. Rectangular structures bounded by the magic clusters in the FFs mass correlation distributions obtained at the modified FOBOS spectrometer (a) and COMETA setup (b). See Fig. 16. Selection by the experimental neutron multiplicity was used.[50]

at least three neutrons were emitted in each fission event presented in the figure. A change in the nuclear composition of the mother system can lead to a shift of the masses of the magic nuclei if neutrons were emitted from the decaying system (pre–scission neutrons). This is likely what we observe here. As to the upper right hand corner of the rectangle both mass and charge conservation laws are found only if the upper side of the rectangle corresponds to ${}^{109}{}_{43}$Tc nucleus while the mass of 140 amu corresponds to the nucleon combination of ${}^{140}{}_{55}$Cs.

The structure manifests itself exclusively owing to the difference of the neutron sources for the fragments appearing in both binary fission and CCT, respectively. These two decay modes must differ in the neutron multiplicity or/and in their angular distributions of the emitted neutrons in order to provide higher registration efficiency for neutrons linked with the CCT channel. At the same time the excitation energy of the system at the scission point defined as $Eex = Q - \text{TKE}$ (where Q is the reaction energy, and TKE is the total kinetic energy of all the decay fragments) is known from our experimental data. It does not exceed $Eex = 30$ MeV. This value of the excitation energy is high enough to allow the emission of three or four neutrons, which corresponds to almost the mean neutron multiplicity of binary fission. Thus, the neutron source linked with the new CCT channels must have a much smaller velocity compared to conventional binary fragments, or it can be almost at rest in the extreme case.[47] The latter agrees with the foregoing hypothesis stating that we deal with the pre–scission neutrons, at least, with those in very light missing masses.

Thus, the neutron gated data confirm an existence of the rectangular structures bounded by the magic nuclei in the FF correlation mass distributions. The analysis of the structures allows us to suppose that at least some of the CCT modes are linked with the emission of four pre–scission neutrons on the average.

2.3. *Ternary decays with comparable masses of the fragments*

The effective cleaning of the FF mass–mass distribution from the background linked with the scattered fragments was achieved by way of applying different selection gates. There was a specific rectangular structure revealed (Fig. 21) when the events characterized by the approximately equal momenta and velocities were selected in the mass–mass distribution of the FFs from ^{252}Cf (sf). The structure is bounded by magic isotopes of ^{68}Ni and

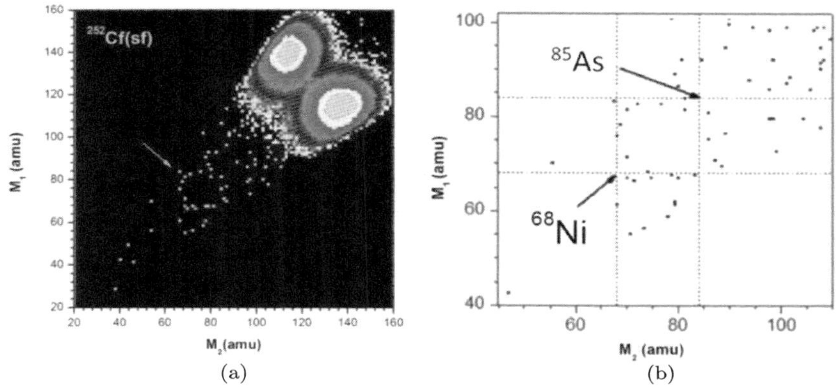

Fig. 21. Correlation mass distribution of the FFs with approximately equal momenta and velocities measured at FOBOS setup (a). Rectangular structure marked by an arrow is presented in section (b) in a larger scale.

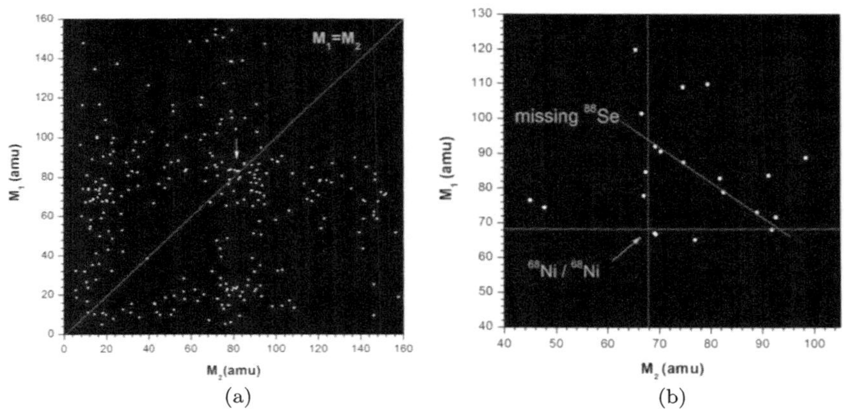

Fig. 22. Results of another gating of the same initial data as those used for obtaining of Fig. 21. Selection by both the FF momenta out of the tails linked with the scattered FFs and experimental neutron multiplicity n ≥ 1 (a). The same distribution under additional condition of approximately equal FFs velocities (b). The condition $M_1 + M_2 = \text{const}$ is met along the titled line. It means that the missing mass is also fixed and is equal to 88 amu (presumably magic ^{88}Se cluster) in this case. Lines in grey are draw to guide the eye.

^{85}As.[52] The results of different processing of the same are shown in Fig. 22. Thus, the rectangular structure that is well seen in Fig. 21 is mainly reproduced within different selection conditions providing additional information on the features of the events involved.

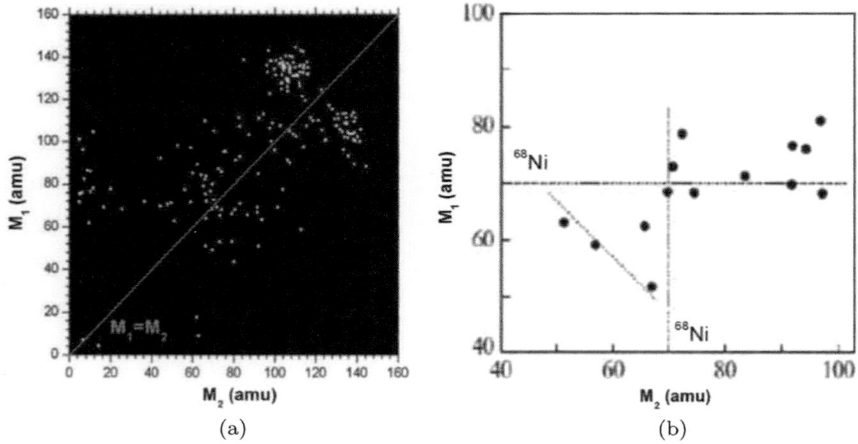

Fig. 23. Mass distribution of coincident FFs from the $^{235}U(n_{th}, f)$ reaction with approximately equal momenta and velocities (a) and distribution of selected events (b) with not only approximately equal momenta and velocities but also with similar nuclear charges; the events at the vertex of the right angle have charges of $Z \approx 28$; for the events lying on the slantwise straight line $M_s = \text{const}$ the "lost" mass corresponds to the deformed magic nucleus ^{118}Pd.

The events characterized by approximately equal momenta and velocities were selected as well in the mass–mass distribution of the FFs from $^{235}U(n_{th}, f)$ reaction. The revealed structure looks like the right angle whose vertex lies on the diagonal of the plot in the vicinity of the points (68, 68) amu (Fig. 23(a)). The assumption that the sides of the right angle in Fig. 23(a) are connected with magic Ni isotopes is confirmed by Fig. 23(b). It depicts the distribution of FFs with approximately equal momenta, velocities, and nuclear charges.

Note, that although the distribution in Fig. 23(b) includes just a few events, the reliability of their analysis is quite high. Indeed, there are no other points in the vicinity of the concentration of points under consideration. Each point results from the measurement of six independent experimental parameters (time–of–flight, energy and drift time at each spectrometer arm).

A less stringent selection made it possible to reveal the whole family of events with comparable masses of the fragments. Figure 24(a) shows the distribution of FFs upon momentum and drift–time selections.[53]

The most intense concentrations of points in Fig. 24(a) remained from the loci of conventional binary fission. The number of points indicated by

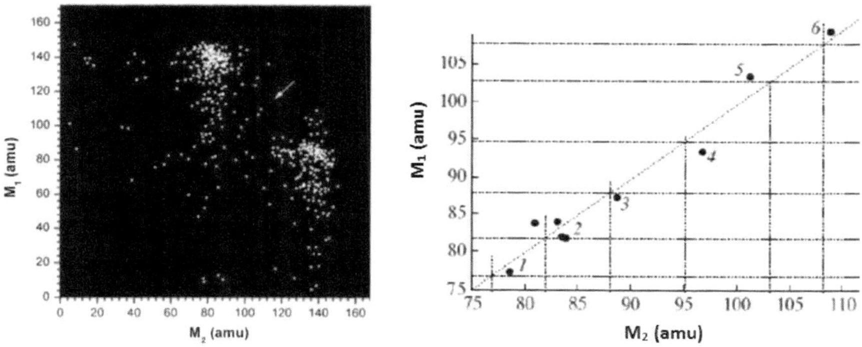

(a) (b)

Fig. 24. (a) Mass–mass distribution of FF selected by the momentum and drift time. The arrow points to the group of events with approximately equal masses. (b) Those events in the larger scale. The numbers label presumably the following pre–scission configurations based on magic clusters: (1) $^{78}Zn_{30}$—$^{80}Ge_{50}$—^{77}Zn, (2) $^{82}Ge_{50}$—$^{72}Ni_{28}$—^{82}Ge, (3) $^{88}Se_{54}$—$^{60(2)}Cr_{24}$—^{88}Se, (4) $^{95}Rb_{58}$—$^{46}Ar_{28}$—^{95}Rb, (5) $^{103}Zr_{40}$—$^{30(2)}Mg$—^{103}Zr, and (6) $^{108}Mo_{66}$—^{20}O—^{108}Mo.

the arrow and situated on the straight line $M_1 = M_2$ increases in line with statistics, and the total yield of these events was $8{*}10^{-6}$ per binary fission.

2.4. *Results from the triple coincidences*

Evidently, for almost collinear decay partners a probability of their independent detection even by the spectrometer of high granularity decreases rapidly along with multiplicity. Nevertheless, ternary events caused by multi–body decays were detected in our experiments at the mosaic spectrometers. For instance, ternary events observed in the reaction $^{238}U + {}^{4}He$ (40 MeV)[54] are given below. In the following three events magic or double magic Sn nuclei were detected as the heaviest fragments (Table 1). The corresponding light fragment (deformed magic nucleus) was clusterized in the scission point forming a di–nuclear system. Presumably, its brake–up comes into beings due to inelastic scattering on the material of start–detector. Such hypothesis is based on the fact that a momentum conservation law is not met in either of the three events.

In the next set of events the decaying system was also fully clusterised i.e. its mass was exhausted by two magic constituents. The "initial" clusters undergo fragmentation leading to formation of two di–nuclear systems. It contrasts with the previous case where both "initial" clusters are relatively

Table 1. Mass conservation law is met in the events presented.

Point number	Decay scheme
1	$^{128}\mathrm{Sn} + {}^{32}\mathrm{Mg} + {}^{80}\mathrm{Ge}$ + **2n**
	$\overline{{}^{112}\mathrm{Ru}}$
2	$^{132}\mathrm{Sn} + {}^{68}\mathrm{Ni} + {}^{42}\mathrm{S}$
	$\overline{{}^{110}\mathrm{Ru}}$
3	$^{130}\mathrm{Sn} + {}^{72}\mathrm{Ni} + {}^{40}\mathrm{S}$
	$\overline{{}^{112}\mathrm{Ru}}$

soft deformed nuclei. Some examples of the events under discussion are given in Table 2. The de facto undetected fragments are highlighted in bold.

3. Observation of shape isomer states in fission fragments

In previous Section 2 we discussed the role of scattering medium in registration of the CCT products, where we stated that even if two CCT partners initially flew in the same direction in perfectly collinear line, they got some angular divergence after having passed the scattering medium on their flight pass due to multiple scattering. Thanks to such effect they can be registered independently in the "stop" mosaic detector, and even thin backing of the radioactive source provides the observable effect.

In order to increase the effect, the additional absorbers (metal foils) were introduced just behind the source at the distance of approximately 1 mm[55,56] (Fig. 25).

The geometry of the source unit is shown in detail in Fig. 25(b). Additional absorber foil was placed at 1 mm distance from the Cf source. Typical fission fragment overcomes such distance in approximately 0.1 ns. There were ternary events analyzed. It means that three fragments were really detected in coincidence in three different silicon detectors. For the sake of convenience, the FFs from such events were labeled as M_1, M_2 and M_3 in an order of decreasing masses in the ternary event. Mass distribution for ternary events is shown in Fig. 26(a). The total mass of the two heavier fragments $M_1 + M_2$ is plotted vs. the mass of the lighter fragments M_3. As

Table 2. Di–nuclear systems based on deformed magic constituents.

Point number	Decay scheme
1	$^{121}\mathrm{Ag} + {}^{23}\mathrm{F} + {}^{65}\mathbf{Mn} + {}^{33}\mathrm{Al}$
	$^{144}\mathrm{Ba}$ $^{98}\mathrm{Sr}$
2	$^{113}\mathrm{Ru} + {}^{31}\mathbf{Mg} + {}^{78}\mathrm{Ni} + {}^{20}\mathrm{Ne}$
	$^{144}\mathrm{Ba}$ $^{98}\mathrm{Sr}$
3	$^{130}\mathrm{Sn} + {}^{14}\mathrm{C} + {}^{62}\mathrm{Cr} + {}^{33}\mathrm{Mg} + \mathbf{3n}$
	$^{144}\mathrm{Ba}$ $^{95}\mathrm{Rb}$
4	$^{140}\mathrm{Xe} + {}^{25}\mathrm{Ne} + {}^{62}\mathbf{Cr} + {}^{15}\mathrm{C}$
	$^{165}\mathrm{Gd}$ $^{77}\mathrm{Zn}$
5	$^{134}\mathrm{Te} + {}^{30}\mathrm{Ne} + {}^{50}\mathrm{Ca} + {}^{27}\mathrm{Ne} + \mathbf{n}$
	$^{164}\mathrm{Gd}$ $^{77}\mathrm{Zn}$
6	$^{110}\mathrm{Te} + {}^{11}\mathrm{Be} + {}^{62}\mathrm{Cr} + {}^{58}\mathbf{V}$
	$^{121}\mathrm{Ag}$ $^{121}\mathrm{Ag}$

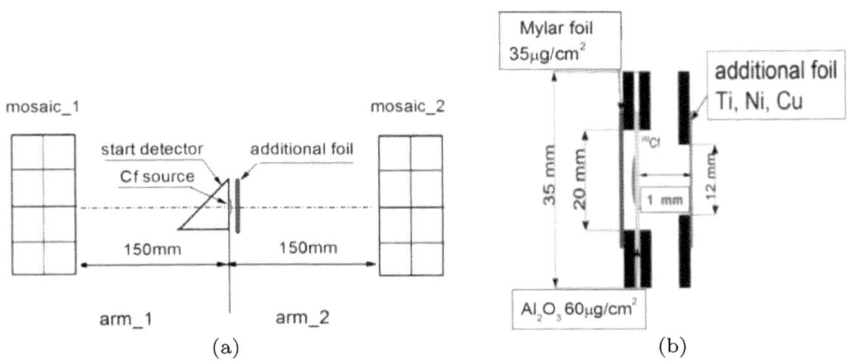

Fig. 25. Layout of the setup used in the experiment aimed at the study of the influence of the additional absorber on the characteristics of the CCT products. The shown double–armed spectrometer consists of the start detector with the Cf source inside and two mosaics eight PIN diodes each (a), (b) ^{252}Cf source and additional foils placed around.

can be inferred from the figure, in all the events where the lightest fragment corresponds to the knocked out Ti ions (M_3 is around 48 amu) we observe a missing mass of fission fragments. It is a very astonishing fact. True, if the elastic Rutherford scattering of one of the fission fragments (FF) took place on the Ti foil, the total mass $M_1 + M_2$ would lie in the vicinity of the mass of the mother system namely 252 amu (taking into account emitted neutrons, approximately four on the average, and the experimental mass resolution). These values are marked in the figure by horizontal lines, while the events resulted from Rutherford scattering are expected to lie in the region marked by the dashed oval.

The mass spectrum of the fragments scattered in the foil and detected in coincidence with the knocked out ions is shown in Fig. 26(b). As can be inferred from the figure this spectrum differs from the FF mass distribution in conventional binary fission. The most impressive difference is observed at the yield of the magic ^{128}Sn nucleus.

The spectrum under discussion can be basically understood through the known manifestations of clustering in binary fission (see Section 1). In accord with these views we suppose that the bulk of the fragments directly after scission of the mother system looks like a di–nuclear system consisting of the magic core and the light cluster which is to be some part of the neck. Actually, this system can be treated as a shape isomer state of the whole fragment.

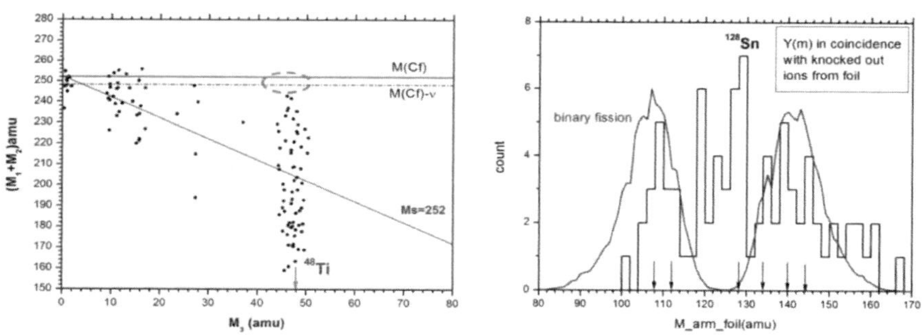

(a) (b)

Fig. 26. Mass distribution for ternary events. Ti foil 2.2 μm thick was used as the additional absorber (a). Mass spectrum of the fragments detected in coincidence with the ions knocked out from the foil and in the same arm with them. The masses of known magic nuclei are marked in the figure by arrows. (b) See text for details.

Typical life–time of the shape isomers observed in our experiments exceeds 1 ns. The indication on the true life–time of these states can be inferred from Ref. 57. The search for isomeric γ decays among fission fragments from 345 MeV/nucleon ^{238}U has been performed at the RIKEN Nishina Center RI Beam Factory. In flight fission of a uranium beam has been used as the production reactions to populate isomers. After having been selected and identified by the superconducting in–flight separator BigRIPS the fission fragments were implanted in an aluminum stopper. The fast isotopic separation and identification of reaction products, which take place in several hundred nanoseconds, allow the event–by–event detection of isomeric γ rays at the focal plane of the separator with small decay losses in flight. A total of 54 microsecond isomers with half–lives of $\sim$0.1–10 μs were identified. At least some of the isomers observed were attributed to shape isomers. It is reasonable to suppose these microsecond isomers' half–lives to be the upper limit for half–lives of the FF shape isomers' states observed in our experiments.

Strong theoretical support for existence of the shape isomer states in fission fragments is also provided by recent calculations carried out by P. Möller and his coworkers.[58] Potential energy surfaces as functions of spheroidal, hexadecapole, and axial asymmetry shape coordinates for several thousands of nuclei from $A = 31$ to $A = 290$ were calculated. The deformations and energies of all minima deeper than 0.2 MeV were tabulated. It gives a good possibility to identify nuclei for which a necessary condition for shape isomers occurs, namely multiple minima in the potential–energy surface. It is known that the life–time of the shape isomer depends on the overlap between the nuclear wave functions of the shape isomer and the ground state, on the excitation energy of the shape isomer, and on the height of the saddle separating the shape isomer and the ground state.[58] Unfortunately, the estimation of the shape isomer life–times is beyond the scope of the outlined work.

Bearing in mind the results of this section two different mechanisms leading to observation of missing mass treated as a label of multi–body decay are likely realized. The first one discussed in Sec. 2.1 is connected with blocking of one of the fragments from ternary decay at the entrance of the "stop" detector. All three fragments are supposed to be born actually in ternary decay of the mother system. A brake–up of the fragment from the conventional binary fission outlined in this section can also be a reason of observable "ternary" decay. Both mechanisms can stay behind the structures seen exclusively in the spectrometer arm faced to the source backing, but only "true" ternary decay can give rise the structures symmetric relative to the spectrometer arms (Sec. 2.3).

4. Summary

Here we mainly present the recent experimental results demonstrating different manifestations of clustering in binary and ternary decays of low excited heavy nuclei. In contrast with the light alpha–cluster systems we deal with very heavy clusters to be magic nuclei with masses of up to 160 amu. The following conclusions can be drawn from the findings outlined above.

(1) There are reliable evidences for the di–cluster mechanism for forming the fission modes in binary fission of actinides. Preformation of a pair of magic nuclei (clusters) such as ^{128,132}Sn and ^{70}Ni, ^{82}Ge in the body of mother system at the early stage of its elongation gives rise to this effect. The feature is also reproduced for the super–heavy fissionning systems.

(2) In the mass correlation distribution of fission fragments (FFs) from ^{252}Cf (sf) and ^{235}U(n_{th}, f) reaction in the region of the big missing masses of about ~50 amu we observe a specific domen of increased yields, what we call as "Ni–bump" by virtue of their presence in the bump of two ridges centered at mass numbers $A = 68, 72$ corresponded to magic isotopes of ^{68,72}Ni. The experimental yield of the events constituting the bump is more then $4^{*}10^{-3}$ per binary fission.

(3) Along with the linear structures $A =$ const the tilted ridges $A_1 + A_2 =$ const (the missing mass is constant) were revealed in the Ni–bump. The ridges starts from the partitions corresponded to the pairs of known magic nuclei ^{68}Ni/^{128}Sn, ^{68}Ni/^{134}Te et cetera. We treat these tilted ridges conditioned by collinear cluster tri–partition as a manifestation of the decay similar to heavy ion or lead radioactivity. In the case the chain of two magic nuclei plays a role of double magic ^{208}Pb.

(4) True ternary fission (the fission with comparable masses of the fragments) events from both ^{252}Cf (sf) and ^{235}U(n_{th}, f) reaction were observed with the yield of $\sim 10^{-5}$ per binary fission. At least two of the detected partners in each ternary event show magic nucleon composition.

(5) The unique rectangular structures bounded by magic nuclei were revealed in the FF mass correlation plots gated by neutrons detected in the "neutron belt" sensitive to the isotropic component of the neutron emission.

(6) It was the first time the binary break–up of the fission fragment due to inelastic scattering in the metal foil had been observed. Based on the obtained results we may suppose the bulk of the fragments directly after scission of the mother system to look like a di–nuclear system consisting of the magic core, and the light cluster to be some part of the neck. This configuration can likely be treated as a shape isomer state of the whole fragment. The effect in hand brightly demonstrates an intimate link between conventional binary fission and ternary decays.

The summing up allows us to state that the evolution of the multicomponent nuclear molecule is strongly believed to be crucial for the low energy fission process both in binary and ternary channels.

Acknowledgements

This text presents research results of the work partially supported by the Russian Science Foundation and fulfilled in the framework of MEPhI Academic Excellence Project (contract 02.a03.21.0005, 27.08.2013), by grant of the Department of Science and Technology of South Africa and by grant of the Federal Ministry of Education and Research (BMBF) of Germany. We are grateful to E. Kuznetsova for her assistance in preparing the text.

References

1. V. V. Vladimirski, On the mechanism of fission of heavy nuclei, *JETP*, **32**(4), 821–825 (1957) (USSR).
2. K. Wildermuth and W. McClure, *Cluster representation of nuclei*. Springer–Verlag, Berlin-Heidelberg-New-York, 1966.
3. U. Mosel and H. W. Schmitt, Potential energy surfaces for heavy nuclei in the two–center model, *Nucl. Phys. A*. **165**, 73–96 (1971).
4. M. G. Mustafa, Transition from mass asymmetry to symmetry in the spontaneous fission of Fm isotopes, *Phys. Rev. C*. **11**, 1059–1063 (1975).
5. Yu. V. Pyatkov *et al.*, Manifestation of clustering in the ^{252}Cf (sf) and ^{249}Cf(n_{th}, f) reactions, *Nucl. Phys. A*. **624**, 140–156 (1997).
6. V. V. Pashkevich, Yu. V. Pyatkov and A. V. Unzhakova, Fission potential energy surface in ten–dimensional deformation space, *Int. J. Mod. Phys. E* **18**, 907–913 (2009).

7. V. Zagrebaev and W. Greiner, Fusion and fission dynamics of heavy nuclear system. In *Proc. of the* 6th *Int. Conf. "Dynamical Aspects of Nuclear Fission"*, (*DANF2006*), pp. 94–111, Smolenice Castle, Slovak Republic (Slovak Republic, 2006).
8. I. Tsekhanovich *et al.*, Mass and charge distributions in the very asymmetric mass region of the neutron induced fission of ^{238}Np, *Nucl. Phys. A* **688**, 633–658 (2001).
9. D. Rochman *et al.*, Super–asymmetric fission in the ^{245}Cm(n_{th}, f) reaction at the Lohengrin fission–fragment mass separator, *Nucl. Phys. A* **735**, 3–20 (2004).
10. D. M. Gorodisskiy *et al.*, Isotopic and isotonic effects in fission–fragment mass yields of actinide nuclei, *Phys. Letters B* **548**, 45–51 (2002).
11. Yu. V. Pyatkov *et al.*, Manifestation of the Fine Structures in the fission fragment mass–energy distribution in the ^{233}U(n_{th}, f) reaction, *Nucl. Instrum. Methods A* **488**, 381–399 (2002).
12. Yu. V. Pyatkov *et al.*, Nontrivial manifestation of clustering in fission of heavy nuclei at low and middle excitations, *Phys. of Atomic Nuclei.* **67**(9), 1726–1730 (2004).
13. Yu. V. Pyatkov *et al.*, Collinear cluster tri–partition of low excited heavy nuclei — status and prospects of experimental investigations. In *Proc. of* 17th *Int. Seminar on Interaction of Neutrons with Nuclei* (*ISINN-17*), pp. 150–157, Dubna, Russia (Russia, 2009).
14. U. Quade *et al.*, Nuclide yields of light fission products from thermal–neutron induced fission of ^{235}U at different kinetic energies, *Nucl. Phys. A* **487**(1), 1–36 (1988).
15. C. Wagemans, ed., *The Nuclear Fission Process.* CRC Press (1991).
16. M. N. Rao *et al.*, Determination of fission fragment angle by the methods of grid pulse height and the time difference between grid and collector pulses in a back–to–back gridded ionisation chamber, *Nucl. Instrum. Methods A* **313**, 227–232 (1992).
17. T. Ohtsuki *et al.*, Binary structure in time distributions of fission fragments in 13–MeV proton–induced fission of ^{232}Th, *Phys. Rev. Lett.* **66**, 17–20 (1991).
18. A. A. Alexandrov *et al.*, A time–of–flight spectrometer for unslowed fission fragments, *Nucl. Instrum. Methods A* **303**, 323–331 (1991).
19. W. H. Trzaska *et al.*, HENDES — high efficiency neutron detection system for correlation measurements with HI beams. In *Proc. of the XIV Int. Conf. on the Application of Accelerators in Research and Industry*, pp. 1059–1062, Denton, Texas (New York, 1997).
20. W. H. Trzaska *et al.*, Manifestations of clustering in fission of excited actinide. In *Proc. Symp. on Nuclear Clusters: from Light Exotic to Superheavy Nuclei*, pp. 237–244, Rauischholzhausen, Germany (Germany, 2002).
21. V. V. Pashkevich, On the asymmetric deformation of fissioning nuclei, *Nucl. Phys. A* **169**, 275–293 (1971).
22. U. Brosa, S. Grossmann and A. Müller, Nuclear scission, *Phys. Rep.* **197**, 167–262 (1990).

23. J. F. Berger, M. Girod and D. Gogny, Microscopic analysis of collective dynamics in low energy fission, *Nucl. Phys. A* **428**, 23–36 (1984).
24. D. J. Hinde *et al.*, Neutron multiplicities in heavy–ion–induced fission: Timescale of fusion–fission, *Nucl. Phys. A* **452**, 550–572 (1986).
25. Yu. V. Pyatkov *et al.*, Estimating the reliability of revealing the fine structure in mass–kinetic energy distributions of nuclear reaction products, *Pattern Recognition and Image Analysis*, **21**(1), 82–87 (2011).
26. Yu. V. Pyatkov *et al.*, Cluster approach to description of fission modes, *Nucl. Phys. A* **611**, 355–369 (1996).
27. F. F. Karpeshin, Ternary fission of nuclei into comparable fragments, *Phys. Atomic Nuclei.* **78**(5), 548–550 (2015).
28. J.-P. Ebran *et al.*, Localization and clustering in the nuclear Fermi liquid, *Phys. Rev. C* **87**, 044307 (2013).
29. P. Möller, Odd–multipole shape distortions and the fission barriers of elements in the region $84 \leq z \leq 120$, *Nucl. Phys. A* **192**, 529–580 (1972).
30. S. Ćwiok, W. Nazarewicz, J. Saladin, W. Płóciennik and A. Johnson, Hyperdeformations and clustering in the actinide nuclei, *Phys. Lett. B* **322**, 304–310 (1994).
31. J. D. McDonnell, W. Nazarewicz and J. A. Sheikh, Third minima in thorium and uranium isotopes in a self–consistent theory, *Phys. Rev. C* **87**, 054327 (2013).
32. H. Märton, private communication.
33. V. Zagrebaev and W. Greiner, Clustering phenomena in fission and fusion processes of heavy nuclei. In *Proc. Int. Symp. on Atomic Cluster Collisions (ISACC07)*, pp. 23–33, GSI Darmstadt (London, 2008), p. 23.
34. E. Piasecki *et al.*, Multiparameter studies of polar emission in ^{236}U fission, *Nucl. Phys. A* **255**, 387–404 (1975).
35. W. J. Swiatecki, Deformation energy of a charged drop. III. Further developments. In *Proc. of the Second UN Conference on the Peaceful Uses of Atomic Energy,* pp. 248–272, Geneva (Geneva, 1958).
36. V. M. Strutinsky, N. Ya. Lyashchenko and N. A. Popov, Symmetrical shapes of equilibrium for a liquid drop model, *Nucl. Phys.* **46**, 639–659 (1963).
37. H. Diehl and W. Greiner, Ternary fission in the liquid drop model, *Phys. Lett. B* **45**, 35–37 (1973).
38. H. Diehl and W. Greiner, Theory of ternary fission in the liquid drop model, *Nucl. Phys. A* **229**, 29–46 (1974).
39. D. N. Poenaru *et al.*, Multicluster accompanied fission, *Phys. Rev. C* **59**, 3457–3460 (1999).
40. R. B. Tashkhodjaev, A. K. Nasirov and W. Scheid, Collinear cluster tripartition as sequential binary fission in the ^{235}U(n$_{\rm th}$, f) reaction, *Eur. Phys. J. A* **47**, art. 136 (2011).
41. K. Manimaran and M. Balasubramaniam, All possible ternary fragmentations of ^{252}Cf in collinear configuration, *Phys. Rev. C* **83**, 034609 (2011).
42. K. Manimaran and M. Balasubramaniam, Ternary fission fragmentation of ^{252}Cf for all possible third fragments, *Eur. Phys. J. A* **45**, 293–300 (2010).

43. V. I. Zagrebaev, A. V. Karpov and W. Greiner, True ternary fission of superheavy nuclei, *Phys. Rev. C.* **81**, 044608 (2010).
44. W. von Oertzen and A. K. Nasirov, True ternary fission, the collinear decay into fragments of similar size in the ^{252}Cf (sf) and ^{235}U(n_{th}, f) reactions, *Physics Letters B* **734**, 234–238 (2014).
45. R. B. Tashkhodjaev *et al.*, Theoretical study of the almost sequential mechanism of true ternary fission, *Phys. Rev. C* **91**, 054612 (2015).
46. W. von Oertzen, A. K. Nasirov and R. B. Tashkhodjaev, Multi-modal fission in collinear ternary cluster decay of ^{252}Cf(sf, fff), *Physics Letters B* 746, 223–229 (2015).
47. K. R. Vijayaraghavan, W. von Oertzen and M. Balasubramaniam, Kinetic energies of cluster fragments in ternary fission of ^{252}Cf, *Eur. Phys. J. A* **48**, art. 27 (2012).
48. H.-G. Ortlepp *et al.*, The 4pi-fragment-spectrometer FOBOS, *Nucl. Instrum. Methods A* **403**, 65–97 (1998).
49. Yu. V. Pyatkov *et al.*, Collinear cluster tri-partition of ^{252}Cf (sf) and in the ^{235}U(n_{th}, f) reaction, *Eur. Phys. J. A* **45**, 29–37 (2010).
50. Yu. V. Pyatkov *et al.*, The collinear cluster tri-partition (CCT) of ^{252}Cf (sf): New aspects from neutron gated data, *Eur. Phys. J. A* **48**, 94–109 (2012).
51. Yu. V. Pyatkov *et al.*, Collinear tripartition of ^{248}Cm and ^{252}Cf nuclei as a probe of clustering. In *Proc. of the Intern. Conf. of Nuclear Physics "Nuclear Shells — 50 Years"*, pp. 144–150, Russia, Dubna (World Scientific, 2000).
52. D. V. Kamanin and Yu. V. Pyatkov, Clusterization in Ternary Fission, *Lecture Notes in Physics 875, "Clusters in Nuclei"* **3**, 183–246 (2013).
53. Yu. V. Pyatkov *et al.*, Collinear Cluster Tripartition Channel in the Reaction ^{235}U(n_{th}, f), *Physics of Atomic Nuclei*, **73**(8), 1309–1316 (2010).
54. Yu. V. Pyatkov *et al.*, Observation of the collinear multi-body decays in the reaction ^{238}U + ^{4}He (40 MeV), *Int. Journal of Modern Physics E* **17**, 2226–2230 (2008).
55. Yu.V. Pyatkov *et al.*, First Results on Observation of New Shape Isomers, *Physics Procedia*, **74**, 67–73 (2015).
56. Yu. V. Pyatkov *et al.*, *Preliminary Results on Observation of New Shape Isomers*, FIAS Interdisciplinary Science Series, Nuclear Physics: Present and Future, pp. 79–88 (2014).
57. D. Kameda *et al.*, Observation of new microsecond isomers among fission products from in-flight fission of 345 MeV/nucleon ^{238}U, *Phys. Rev. C* **86**, 054319 (2012).
58. P. Möller *et al.*, Nuclear shape isomers, *Atomic Data and Nuclear Data Tables*, **98**, 149–300 (2012).

CHAPTER 13

SPONTANEOUS FISSION, CLUSTER RADIOACTIVITY AND ALPHA DECAY OF SUPERHEAVY NUCLEI ^{282,284}Cn AND ^{286}Lv

D. N. Poenaru* and R. A. Gherghescu

Horia Hulubei National Institute of Physics and Nuclear Engineering, Bucharest-Magurele, Romania
**poenaru@nipne.ro*

Experiments on spontaneous fission of superheavy nuclei ^{282,284}Cn and ^{286}Lv are available. We calculate the half-lives of these nuclides against three decay modes: α decay (αD), cluster radioactivity (CR) and spontaneous fission (SF). ASAF (analytical superAsymmetric) and UNIV (universal curve) models are used to investigate αD and CR. The best model to calculate the half-life against αD is semFIS (semiempirical formula based on fission theory). The Strutinsky's macroscopic-microscopic method based on the best two-center shell model allows us to find the deformation energy during the fission. The nuclear inertia tensor is determined with the cranking model. It is employed to evaluate the action integral determining the SF half-life. The strongest decay mode of each of the three nuclides is SF followed by αD.

1. Introduction

The most important decay modes of superheavy (SH) nuclei are α decay (αD) and spontaneous fission (SF).[1–5] SHs are produced in fusion reactions which are almost cold (only one neutron evaporation) when the target is a double magic ^{208}Pb or its neighbour ^{209}Bi, or in fusion reactions induced by ^{48}Ca leading to an excited compound nucleus releasing energy by few (3-4) neutron evaporations and γ ray emission.[6–11] In 2012 the International Union of Pure and Applied Chemistry (IUPAC) approved the names flerovium (symbol Fl) and livermorium (Lv), for the elements with atomic numbers $Z = 114, 116$. On 30 November 2016 IUPAC Announces the Names of the Elements 113 — nihonium (Nh), 115 — moscovium (Mc), 117 — tennessine (Ts), and 118 — oganesson (Og). The credit for 113 is given to Kosuke Morita *et al.* from RIKEN (the first one discovered in Asia). A Russian-American team of scientists at the Joint Institute for Nuclear Research in Dubna and Lawrence Livermore National Laboratory in California had produced sufficient evidence to claim the discovery of elements 115, 117 and 118.

For heavier SHs, with $Z > 121$, there is a good chance for cluster decay modes or cluster radioactivity (CR)[12,13] to compete.[14,15]

Experimental data[16] on SF are leading us to $\log_{10} T_f^{exp}(s) = -3.086, -0.980$ for ^{282,284}Cn, respectively and $T = 140$ ms for ^{286}Fl with SF 60% ($T_f^{exp} = 233$ ms) and αD 40% ($T_\alpha^{exp} = 350$ ms), meaning $\log_{10} T_f^{exp}(s) = -0.632$ and $\log_{10} T_\alpha^{exp}(s) = -0.456$. We use the known relationships

$$\frac{1}{T} = \frac{1}{T_f} + \frac{1}{T_\alpha} \; ; \; T_f = \frac{100}{i_f} T \; ; \; T_\alpha = \frac{100}{i_\alpha} T \tag{1}$$

where $i_f = 60$ is the percentage of SF intensity and $i_\alpha = 40$ the percentage of αD.

For αD we succeeded to find a large number (581) of known nuclei, after a previous 559 emitters[17] and 578 (see Ref. 18) from which it is easy to get in case of three nuclides mentioned above $\log_{10} T_\alpha^{exp}(s) = -0.95, 0.99$ and -0.46, respectively.

A systematic study[19] published in December 2015 shows how well are reproduced the αD experimental data by 18 different relationships. It was found that our "SemFIS2 formula is the best one to predict the alpha-decay half-lives ... In addition, the UNIV2 formula with fewest parameters and the VSS, SP and NRDX formulas with fewer parameters work well in prediction on the SHN alpha-decay half-lives." The semFIS2 and UNIV2 are those variants used in Ref. 20 for the region of alpha emitters with atomic numbers $Z = 92 - 118$.

In the present work we report results obtained by using semFIS (semi-empirical based on fission theory), ASAF (analytical superasymmetric) and UNIV (universal) formulae[13] to calculate αD half-lives and ASAF for CR. A dynamical study based on cranking inertia and the best asymmetric two-center shell model[21] will allow us to find the theoretical value of the SF half-life. The released energy, or Q-value for αD and CD is calculated based on the latest atomic mass evaluation.[22]

2. The models

The measurable decay constant $\lambda = \ln 2/T$, characterizing the well known exponential law $e^{-\lambda t}$ (variation in time of the number of parent nuclei) can be expressed in both alpha-like[23] or fission-like theories, as a product of three (model dependent) quantities

$$\lambda = \nu S P_s \tag{2}$$

where ν is the frequency of assaults on the barrier per second, S is the preformation probability of the cluster at the nuclear surface, strongly dependent on the nuclear structure, and P_s is the quantum penetrability of the external potential barrier.

The half-life of a parent nucleus AZ against the split into a cluster $A_e Z_e$ and a daughter $A_d Z_d$

$$T = [(h \ln 2)/(2E_v)] exp(K_{ov} + K_s) \tag{3}$$

is calculated[13] by using the WKB quasiclassical approximation, according to which the action integral is expressed as

$$K = \frac{2}{\hbar} \int_{R_a}^{R_b} \sqrt{2B(R)E(R)} dR \tag{4}$$

in which $K = K_{ov} + K_s$ for overlapping and separated fragments, respectively. In the simplest case one may assume $B = \mu$ — the reduced mass, which is rigorously true when the fragments are at the touching point or they are separated. Generally speaking one should use the inertia tensor, as shown below.

2.1. *ASAF model — analytical superasymmetric fission model*

In the above eq. $E(R)$ is replaced by $[E(R) - E_{corr}] - Q$. E_{corr} is a correction energy similar to the Strutinsky[24] shell correction, also taking into account the fact that Myers-Swiatecki's liquid drop model (LDM)[25] overestimates fission barrier heights, and the effective inertia in the overlapping region is different from the reduced mass.

The two terms of the action integral K, corresponding to the overlapping (K_{ov}) and separated (K_s) fragments, are calculated by analytical formulas (approximated for K_{ov} and exact for K_s in case of separated spherical shapes within the LDM):

$$K_{ov} = 0.2196(E_b^0 A_e A_d/A)^{1/2}(R_t - R_i)\left[\sqrt{1-b^2} - b^2 \ln \frac{1+\sqrt{1-b^2}}{b}\right] \tag{5}$$

$$K_s = 0.4392[(Q + E_v + E^*)A_e A_d/A]^{1/2} R_b J_{rc}; \quad b^2 = (E_v + E^*)/E_b^0 \tag{6}$$

$$\begin{aligned} J_{rc} &= (c)\arccos\sqrt{(1-c+r)/(2-c)} - [(1-r)(1-c+r)]^{1/2} \\ &\quad + \sqrt{1-c}\ln\left[\frac{2\sqrt{(1-c)(1-r)(1-c+r)} + 2 - 2c + cr}{r(2-c)}\right] \end{aligned} \tag{7}$$

where $r = R_t/R_b$ and $c = rE_c/(Q + E_v + E^*)$. In the absence of the centrifugal contribution ($l = 0$), one has $c = 1$.

We took $E_v = E_{corr}$ in order to get a smaller number of parameters. It is evident that, owing to the exponential dependence, any small variation of E_{corr} induces a large change of T, and thus plays a more important role compared to the preexponential factor variation due to E_v. Shell and pairing effects are included in $E_{corr} = a_i(A_e)Q$ ($i = 1, 2, 3, 4$ for even-even, odd-even, even-odd, and odd-odd parent nuclei). For a given cluster radioactivity we have four values of the coefficients a_i, the largest for even-even parent and the smallest for the odd-odd one (see Fig. 1 of Ref. 26 and the updated Fig. 1 of Ref. 18).

2.2. *UNIV — Universal curve*

A simple universal law may be obtained by making some approximations. We assume the following linear dependence for the preformation probability in the range of emitted clusters already measured:

$$\log S = \frac{(A_e - 1)}{3} \log S_\alpha \tag{8}$$

where the preformation probability of the α particle S_α may be determined by fitting with experimental data. The second approximation: the frequency ν is independent of the emitted cluster and the daughter nucleus $\nu(A_e, Z_e, A_d, Z_d) =$ constant.

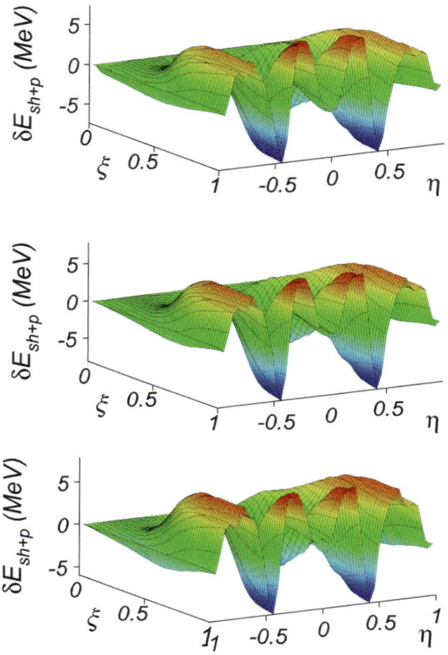

Fig. 1. (Color online) Shell plus pairing corrections versus $\xi = (R - R_i)/(R_t - R_i)$ and $\eta = (A_1 - A_2)/A$ for spontaneous fission of the three nuclei: ^{282}Cn (top); ^{284}Cn (center) and ^{286}Fl (bottom).

From a fit of experimental data we got[27] $S_\alpha = 0.0160694$ and $\nu = 10^{22.01}$ s^{-1}, leading to

$$\log T = -\log P_s - 22.169 + 0.598(A_e - 1) \tag{9}$$

This equation represents a straight line for a given A_e, with a slope equal to unity. The vertical distance between two universal curves corresponding to A_{e1} and A_{e2} is $0.598(A_{e2} - A_{e1})$. For any combination of fragments A_eZ_e, A_dZ_d one can calculate easily

$$-\log P_s = 0.22873(\mu_A Z_d Z_e R_b)^{1/2}\left[\arccos\sqrt{r} - \sqrt{r(1-r)}\right] \tag{10}$$

where $r = R_t/R_b$, $R_t = 1.2249(A_d^{1/3} + A_e^{1/3})$, $R_b = 1.43998 Z_d Z_e/Q$, and $\mu_A = A_d A_e/A$ is the reduced mass number.

2.3. *Semiempirical model based on fission theory — semFIS*

In order to improve the behaviour around a magic number, a numerical coefficient χ was introduced in the equation

$$\log T = 0.43429\chi K_s - 20.446 \tag{11}$$

where

$$K_s = 2.52956 Z_{da}[A_{da}/(AQ_\alpha)]^{1/2}[\arccos\sqrt{x} - \sqrt{x(1-x)}] \tag{12}$$

$$x = 0.423 Q_\alpha (1.5874 + A_{da}^{1/3})/Z_{da} \tag{13}$$

where A_{da}, Z_{da} are the mass and atomic numbers of the daughter nucleus and χ is a second order polynomial

$$\chi = B_1 + B_2 y + B_3 z + B_4 y^2 + B_5 yz + B_6 z^2 \tag{14}$$

with y and z expressing the distance from the closest magic-plus-one neutron and proton numbers N_i and Z_i

$$y \equiv (N - N_i)/(N_{i+1} - N_i) \; ; \; N_i < N \leq N_{i+1} \tag{15}$$

$$z \equiv (Z - Z_i)/(Z_{i+1} - Z_i) \; ; \; Z_i < Z \leq Z_{i+1} \tag{16}$$

$N_i = \ldots, 51, 83, 127, 185, 229, \ldots$, $Z_i = \ldots, 29, 51, 83, 127, 165, \ldots$, and $Z_{da} = Z - 2$, $A_{da} = A - 4$ Example for even-even α emitters $B_1 = 0.993119$, $B_2 = -0.004670$, $B_3 = 0.017009$, $B_4 = 0.045030$, $B_5 = 0.018101$, $B_6 = -0.025097$. A FORTRAN programme with self-improving parameters is available.[28]

2.4. *Dynamics based on Cranking inertia, ATCSM and Y+EM*

We are using the macroscopic-microscopic method[24] to estimate the deformation energy, expressed as a sum of Yukawa-plus-exponential[29] (Y+EM) phenomenological energy, E_{Y+E}, and the shell plus pairing corrections, $\delta E = \delta U + \delta P$ based on the asymmetric two center shell model (ATCSM):[21]

$$E_{def} = E_{Y+E} + \delta E \tag{17}$$

An example of shell plus BCS pairing correction energies for the three SHs studied in the present paper is shown in figure 1. We derived a computer program[30] allowing to calculate E_{Y+E} for binary fragmentation with different charge densities.

The shell- and pairing corrections δU and δP are calculated for proton and neutron level schemes of the ATCSM and then added, e.g. $\delta P = \delta P_p + \delta P_n$. When the parent nucleus and the two fragments are spherical we have only two deformation coordinates (R, R_2) where R is the separation distance of the fragments, R_2 is the radius of the light fragment and $\eta = (A_1 - A_2)/A$. The nuclear inertia may be written, correspondingly, as a function of the three components of the mass tensor

$$B(R) = B_{R_2R_2}\left(\frac{dR_2}{dR}\right)^2 + 2B_{R_2R}\frac{dR_2}{dR} + B_{RR} = B_{22} + B_{21} + B_{11} \quad (18)$$

$R_{20} = R_0 = r_0A^{1/3}$ is equal to the radius of the parent, and the initial and touching point separation distances are $R_i = R_0 - R_{2f}$ and $R_t = R_{1f} + R_{2f}$. The radius constant in Y+EM is $r_0 = 1.16$ fm.

According to the cranking model, after including the BCS pairing correlations,[31] the inertia tensor[32] is given by

$$B_{ij} = 2\hbar^2 \sum_{\nu\mu} \frac{\langle\nu|\partial H/\partial\beta_i|\mu\rangle\langle\mu|\partial H/\partial\beta_j|\nu\rangle}{(E_\nu + E_\mu)^3}(u_\nu v_\mu + u_\mu v_\nu)^2 \quad (19)$$

where H is the single-particle Hamiltonian allowing to determine the energy levels and the wave functions $|\nu\rangle$, u_ν^2, v_ν^2 are the BCS occupation probabilities, E_ν is the quasiparticle energy and β_i, β_j are the independent shape coordinates.

For minimization of the least action trajectory in the plane (R, R_2) we need not only B_{RR} but also the values of $B_{R_2R_2}, B_{R_2R}$ in every point of a grid of 66×24 for 66 values of $(R - R_i)/(R_t - R_i)$ and 24 values of η or R_{2f}. If we assume $R_2 = R_{2f}$ = constant, we are left with only one deformation parameter, R.

3. Results

The half-lives against αD within ASAF, UNIV and semFIS are given in Table 1. There are also the differences between the theoretical and experimental decimal logarithms of half-lives, from which it is clear that the best agreement with experimental data is obtained by using semFIS, followed by UNIV, as was also observed in Ref. 19. Branching ratio relative to spontaneous fission is defined as the ratio $b_f = T_f/T_\alpha$. For all the considered by us three nuclides, SF is the main decay mode.

For CR within ASAF the results of T_c are presented in Table 2, where the branching ratio relative to αD, $b = T_\alpha/T_c$ are showing a too small

Table 1. Alpha decay of ^{282,284}Cn and ^{286}Lv. Differences between theoretical and experimental decimal logarithms of half-lives expressed in seconds within ASAF, UNIV and semFIS. b_f is the branching ratio relative to spontaneous fission.

Parent nucleus	Q (MeV)	$\lg_{10} T_\alpha^{exp}$	$\lg_{10} T_f^{exp}$	ASAF	UNIV	semFIS	$\lg_{10} b_f^{exp}$
^{282}Cn	10.109	−0.95	−3.086	−0.40	−0.26	0.11	−2.136
^{284}Cn	9.540	0.99	−0.980	−0.79	−0.63	−0.18	−1.970
^{286}Lv	10.372	−0.46	−0.632	−1.01	−0.85	−0.44	−0.172

Table 2. Cluster decay of ^{282,284}Cn and ^{286}Lv. Half-lives and branching ratios within ASAF.

Parent nucleus	Emitted cluster	Q (MeV)	$\lg_{10} T_c^{ASAF}$	$\lg_{10} b^{ASAF}$
^{282}Cn	^{74}Zn	246.32	8.41	−6.274
^{284}Cn	^{76}Zn	246.28	7.82	−6.840
^{286}Lv	^{78}Ge	265.22	5.30	−4.668

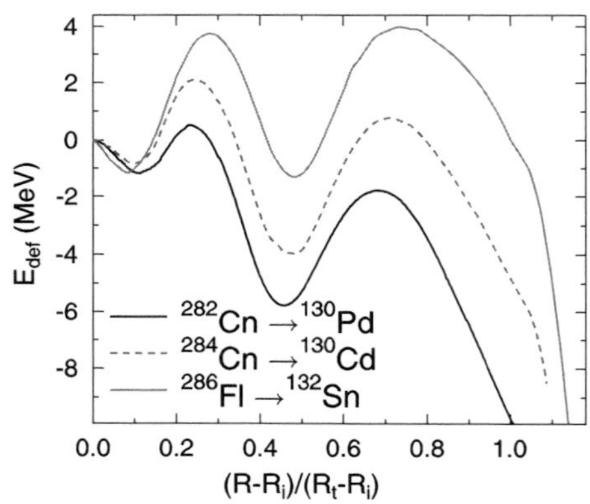

Fig. 2. (Color online) Fission barriers for three processes: $^{282}Cn \rightarrow {}^{130}Pd + {}^{152}$Dy; $^{284}Cn \rightarrow {}^{130}Cd + {}^{154}$Gd and $^{286}Fl \rightarrow {}^{132}Sn + {}^{154}$Gd.

effect, impossible to measure in the region of SHs, despite the much smaller values, from −9 to −17 already measured for parent nuclei with $Z = 87-96$ and $A = 221 - 242$.

Spontaneous fission is the most important decay mode of these three nuclides. We can reproduce the experimental value of SF life-time by using

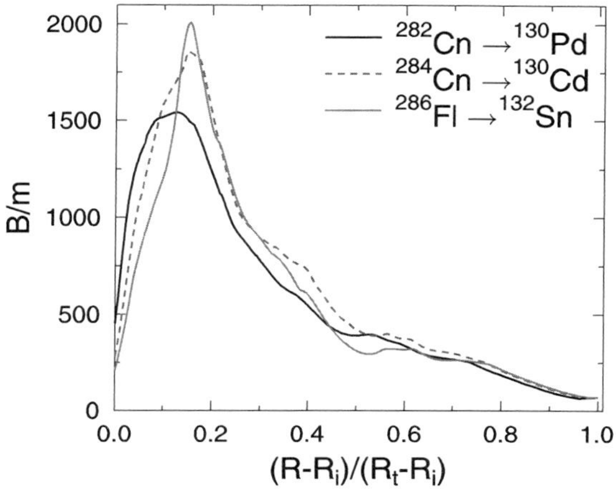

Fig. 3. (Color online) Nuclear inertia for three processes: $^{282}Cn \rightarrow {}^{130}Pd{+}^{152}$Dy; $^{284}Cn \rightarrow {}^{130}Cd{+}^{154}$Gd and $^{286}Fl \rightarrow {}^{132}Sn{+}^{154}$Gd.

Table 3. Spontaneous fission of ^{282,284}Cn and ^{286}Lv. Half-lives and branching ratios relative to cluster decay, b_c. Zero point vibration energy, E_v necessary to reproduce the experiment.

Parent nucleus	Light fragment	η	$\lg_{10} T_f^{exp}$	E_v (MeV)	$\lg_{10} b_c$
^{282}Cn	^{130}Pd	0.0780	−3.086	0.437	−9.36
^{284}Cn	^{130}Cd	0.0845	−0.980	1.921	−6.83
^{286}Lv	^{132}Sn	0.0769	−0.632	1.361	−6.16

one of the most efficient channel for each parent nucleus, as shown in Figs. 2 and 3, where the barriers and nuclear inertia are shown for the following processes:

$$^{282}Cn \rightarrow^{130} Pd +^{152} Dy \tag{20}$$

$$^{284}Cn \rightarrow^{130} Cd +^{154} Gd \tag{21}$$

$$^{286}Fl \rightarrow^{132} Sn +^{154} Gd \tag{22}$$

The corresponding mass asymmetry takes the following three values $\eta = 0.0780; 0.0845; 0.0769$ as shown in the Table 3. These values are not far from $\eta = 0.087$ at the minima around symmetry of shell plus pairing correction energies in Fig. 1, produced due to the shell effects of the doubly magic fragment ^{132}Sn. The most important shell effect is seen around

$\eta = 0.475; 0.465; 0.455$ produced by the doubly magic daughter ^{208}Pb. On the other hand the valley of αD is rather shallow. One has to choose[33] a lighter α emitter nucleus, like ^{106}Te with a daughter in the neighborhood of ^{100}Sn in order to observe a deeper valley.

In conclusion, as may be seen from the Tables 1–3, the main decay mode of each of the three SHs ^{282,284}Cn and ^{286}Fl is SF and the rarest decay mode is CR. Only for ^{286}Fl αD is comparable (40%) with SF (60%).

References

1. D. Poenaru and W. Greiner, eds., *Handbook of Nuclear Properties.* Clarendon Press, Oxford, (1996).
2. D. N. Poenaru, ed., *Nuclear Decay Modes.* Institute of Physics Publishing, Bristol, (1996).
3. M. G. Itkis, E. Vardaci, I. Itkis, G. Knyazheva and E. Kozulin, *Nucl. Phys.* **A 944**, 204–237 (2015).
4. P.-H. Heenen, J. Skalski, A. Staszczak and D. Vretenar, *Nucl. Phys.* **A 944**, 415–441 (2015).
5. X. J. Bao, S. Q. Guo, H. F. Zhang, Y. Z. Xing, J. M. Dong and J. Q. Li, *J. Phys. G: Nucl. Part. Phys.* **42**(8), 085101 (2015).
6. J. Khuyagbaatar *et al.*, *Phys. Rev. Lett.* **112**, 172501 (2014).
7. S. Hofmann, *Radiochim. Acta.* **99**, 405–428 (2011).
8. J. H. Hamilton, S. Hofmann and Y. Oganessian, *Ann. Rev. Nucl. Part. Sci.* **63**, 383–405 (2013).
9. Y. T. Oganessian and V. K. Utyonkov, *Nucl. Phys. A* **944**, 62–98 (2015).
10. K. Morita *et al.*, *J. Phys. Soc. Jpn.* **76**, 045001 (2007).
11. A. Sobiczewski, *Radiochim. Acta.* **99**, 395–404 (2011).
12. Encyclopaedia Britannica Online. Web site http://www.britannica.com/EBchecked/topic/465998/.
13. D. N. Poenaru and W. Greiner. Cluster radioactivity. In ed. C. Beck, *Clusters in Nuclei Vol. 1. Lecture Notes in Physics vol. 818*, pp. 1–56. Springer, Berlin (2010).
14. D. N. Poenaru, R. A. Gherghescu and W. Greiner, *Phys. Rev. C* **85**, 034615 (2012).
15. D. N. Poenaru, R. A. Gherghescu and W. Greiner, *Phys. Rev. Lett.* **107**(6), 062503 (2011).
16. G. Audi, F. Kondev, M. Wang, B. Pfeiffer, X. Sun, J. Blachot and M. MacCormick, *Chinese Physics, C* **36**(12), 11571286 (2012).
17. D. N. Poenaru, R. A. Gherghescu and W. Greiner, *J. Phys. G: Nucl. Part. Phys.* **40**(10), 105105 (2013).
18. D. N. Poenaru, R. A. Gherghescu, W. Greiner and N. S. Shakib. In ed. W. Greiner, *Nuclear Physics: Present and Future (FIAS Interdisciplinary*

Science Series), pp. 131–140 (Springer International Publishing Switzerland, 2015).
19. Y. Wang, S. Wang, Z. Hou and J. Gu, *Phys. Rev. C* **92**, 064301 (June, 2015).
20. D. N. Poenaru, R. A. Gherghescu and N. Carjan, *Europhysics Letters*, **77**, 62001/1–4 (2007).
21. R. A. Gherghescu, *Phys. Rev.* **C 67**, 014309/1–20 (2003).
22. M. Wang, G. Audi, A. H. Wapstra, F. Kondev, M. MacCormick, X. Xu and B. Pfeiffer, *Chinese Physics, C* **36**(12), 16032014 (2012).
23. R. Blendowske, T. Fliessbach and H. Walliser. In ed. D. N. Poenaru, *Nuclear Decay Modes*, chapter 7, pp. 337–349. IOP Publishing, Bristol (1996).
24. V. M. Strutinsky, *Nucl. Phys. A* **95**, 420–442 (1967).
25. W. D. Myers and W. J. Swiatecki, *Nucl. Phys. A.* **81**, 1–60 (1966).
26. D. N. Poenaru, D. Schnabel, W. Greiner, D. Mazilu and R. Gherghescu, *Atomic Data Nucl. Data Tables*, **48**, 231–327 (1991).
27. D. N. Poenaru and W. Greiner, *Physica Scripta* **44**, 427–429 (1991).
28. D. N. Poenaru, M. Ivaşcu and D. Mazilu, *Comp. Phys. Communications* **25**, 297–309 (1982).
29. H. J. Krappe, J. R. Nix and A. J. Sierk, *Phys. Rev. C.* **20**, 992–1013 (1979).
30. D. N. Poenaru, M. Ivaşcu and D. Mazilu, *Comp. Phys. Communications.* **19**, 205–214 (1980).
31. J. Bardeen, L. Cooper and J. Schrieffer, *Phys. Rev. C* **108**, 1175 (1957).
32. M. Brack, J. Damgaard, A. Jensen, H. C. Pauli, V. M. Strutinsky and G. Y. Wong, *Rev. Mod. Phys.* **44**, 320–405 (1972).
33. D. N. Poenaru, R. A. Gherghescu and W. Greiner, *Il Nuovo Cimento* **111 A**, 887–893 (1998).

CHAPTER 14

PREDICTIONS ON THE FEASIBLE ALPHA AND CLUSTER DECAYS FROM $^{298-336}126$ SUPERHEAVY NUCLEI

K. P. Santhosh and B. Priyanka

School of Pure and Applied Physics,
Kannur University, Swami Anandatheertha Campus
Payyanur 670327, Kerala, India
drkpsanthosh@gmail.com

The Coulomb and proximity potential model (CPPM), which comes under the class of fission models, has been used for the extensive study on the cluster decay process in $^{298-336}126$ superheavy nuclei, and thus the probable cluster decays from the various isotopes of $Z = 126$ have been investigated and predicted. The Universal formula for cluster decay (UNIV) of Poenaru *et al.*, the Universal Decay Law (UDL) of Qi *et al.*, and the Scaling Law of Horoi *et al.*, have also been used for evaluating the cluster decay half lives of these nuclei, and a comparison of the predicted decay half lives with the values evaluated using CPPM reveals that our predicted values matches well with these theoretical values. The behaviour of the cluster half lives with the neutron number of the daughter nuclei can be clearly seen from the plots for $\log_{10}(T_{1/2})$ against the neutron number of the daughter nuclei in the corresponding decay, and these plots depicts that for most of the decays, the half life are the minimum for the decay leading to a daughter with $N = 184$. The odd-even staggering (OES) could be seen to be more prominent in the emission of heavy odd-mass clusters. The Geiger-Nuttall plots of $\log_{10}(T_{1/2})$ versus $Q^{-1/2}$ and the Universal curve between -lnP and $\log_{10}(T_{1/2})$ for various clusters ranging from ^{14}C to ^{30}Mg from $^{288-339}126$ isotopes have also been studied, and are found to be linear. Our study also reveals fact that, in cluster decays of SHN, the role of neutron shell closure is crucial than proton shell closure. Also, most of the predicted half lives are well within the present experimental upper limit (10^{30}s) and lower limit (10^{-6}s) for measurements and hence these predictions may be of great use for further experimental investigation on alpha and cluster decay in the superheavy region.

Keywords: Cluster decay, Alpha decay, Superheavy Nuclei.

PACS numbers: 23.70.+j; 23.60.+e; 27.90.+b

1. Introduction

Since the first observation of the α-cluster radioactivity by Rutherford in 1899,[1] it became evident that an atomic nucleus may have different types of decays. Further, different types of radioactivities were discovered subsequently, of which the phenomenon of cluster decay may be considered to the rarest. The process of cluster decay was theoretically predicted by Sandulescu *et al.*,[2] in 1980, and 4 years later, the first experimental observation of this novel process was reported by Rose and Jones[3] for a spontaneous ^{14}C decay of ^{223}Ra. Though the early literatures convey the fact that there existed a fission data, as early as 1951, for ^{24}Ne decay of ^{232}U,[4] but the authors could not identify the process to be a cluster decay. The pioneering experiment performed in 1984 was further followed by several experiments,[5–8] in the later years, to identify the major clusters that could be emitted from various parents, and the works of Bonetti *et al.*,[5,7,8] is highly remarkable. Further, different theoretical groups have performed a vast number of theoretical studies[9–13] on the cluster decays from various parent nuclei, based on the fission models and cluster models, and these studies have provided valuable information on this rare and cold process which can be identified as an intermediate process between alpha decay and fission. Here, the pioneering works of Gupta *et al.*,[9] and Poenaru *et al.*,[10] has to be mentioned with great importance. Poenaru *et al.*, have also extended[14,15] the cluster studies on the SHN, within the Analytical Superasymmetric Fission Model (ASAFM), being used by the authors since 1984. Through these works, the Poenaru *et al.*, revealed the possibility of observing regions in the superheavy nuclei where heavy particle radioactivity (HPR) is stronger than alpha radioactivity.

In the recent years, the theoretical studies on the nuclear properties, decay mechanisms and the decay half lives of extremely neutron rich nuclei, and the shell evolution have gained much regard in the field of nuclear physics. In order to understand the shell closures beyond $Z = 82$ and $N = 126$, both experimental and theoretical studies on the superheavy nuclei (SHN) should be carried out in a tremendous manner. On the basis of the cold fusion[16] and hot fusion[17] reactions, Oganessian *et al.*, and Hoffman *et al.*, respectively, have been successful in synthesising the SHN upto $Z = 118$,[18–22] and in 2009, Oganessian *et al.*, have also performed an attempt[23] to synthesise the SHN with $Z = 120$. Different theoretical models suggests, predicts and points towards different proton and neutron numbers as the next magic numbers beyond $Z = 82$ and $N = 126$. In the early 1990's, based on the Strutinsky's approach, Sobiczewski,[24] Möller *et al.*,[25] and Smolanczuk *et al.*,[26] have predicted $Z = 114$ to be the most stable, and later, based on the Hartree-Fock calculations, $Z = 120$, 124, and 126 were proposed[27,28] to have the highest stability. Further studies of Gupta *et al.*,[29] and Saldanha *et al.*,[30] put forward a prediction that $^{292}120$ ($Z = 120$, $N = 172$) to be doubly magic, and very recently Kumar[31] have also identified $N = 172$ to be a magic number. Recent studies of Adamian *et al.*[32] on the evaporation-residue cross sections have concluded the next magic nucleus to have $Z \geq 120$, and several studies performed in the recent years have given some interesting predictions on the next sub-magic, and spherical doubly magic nucleus[33] and other proton and neutron magicities.[34–36]

Santhosh *et al.*, within the Coulomb and proximity potential model (CPPM)[13] which comes under the broad class of fission models have performed extensive studies on the cluster decay from both heavy and superheavy nuclei, since 2000. The studies on the exotic decay of various proton-rich parents in the trans-tin region emitting clusters ranging from ^{4}He to ^{34}Si,[37–39] the cluster decay study on the superdeformed Sr isotopes,[40] study on the transition point from cluster mode to fission mode,[41] a new semi-empirical model for cluster radioactivity,[42] the cluster decay studies from SHN,[43,44] and the fine structure in cluster decay[45] are the some of the major works thoroughly studied within CPPM. As the search for the next doubly magic nuclei beyond ^{208}Pb has gained great importance in the past few years, several theoretical and experimental efforts have been made on the investigation of long lived SHN. In this context, the studies on the alpha decay, cluster decay and spontaneous fission of SHN become relevant, and the works done by Santhosh *et al.*,[46–50] in these areas are highly mentionable. Some amusing predictions on the proton and neutron magic numbers have also been proposed by Santhosh *et al.*, through the cluster decay studies on both heavy and superheavy nuclei. The studies on the cluster decay from heavy nuclei[48,49] have implied on $N = 126$ to be a magic number, and the studies on the cluster decay from SHN have always stressed on the fact that $N = 184$ to be the magic number beyond $N = 126$, and have also proposed[44,51] $^{298}114$ ($Z = 114$, $N = 184$) to be the next spherical doubly magic nucleus beyond ^{208}Pb. It is highly commendable that these predictions go hand-in-hand with the earlier predictions given by various theoretical groups.

While the studies on the alpha decay and spontaneous fission of SHN can provide clarity on the nuclear species at the beginning of the decay chains and gives information on the mode of decay, the cluster decay studies may be used to identify the magicities of the nuclides in the reactions. Eventhough several theoretical studies may be quoted to have been performed on the SHN within $Z = 125$,[52] and leading to daughter nuclei with $Z = 126$ or $N = 126$, the behaviour of the isotopes of SHN with $Z \geq 126$ against alpha decay[53] and cluster decay remains unknown. Being the next predicted magic number, in the present manuscript, the cluster decay properties of all the available isotopes of $Z = 126$, within $298 \leq A \leq 336$, have been studied with an aim on predicting the neutron magicity beyond $N = 126$. In addition, the mien of the cluster decay half lives against the neutron number of the daughter nuclei, the behaviour of the Geiger-Nuttall plots for $\log_{10}(T_{1/2})$ versus $Q^{-1/2}$ for the emission of various clusters from the isotopes of $Z = 126$, and the manner in which the negative logarithm of penetrability ($-\ln P$) varies with $\log_{10}(T_{1/2})$ of all the decays, have been studied in this manuscript.

The details on the model used for the present study, the Coulomb and Proximity Potential Model (CPPM) have been described in Section 2. The results obtained, and the related discussion on the cluster decay of the nuclei under study has been presented in section 3. The section 4 gives a conclusion on the entire work.

2. The coulomb and proximity potential model (CPPM)

The potential energy barrier in CPPM is taken as the sum of Coulomb potential, proximity potential and centrifugal potential, for the touching configuration and for the separated fragments, and for the pre-scission (overlap) region, the simple power law interpolation done by Shi *et al.*,[11] is used. Shi and Swiatecki[11] were the first to use the proximity potential in an empirical manner and later on, Malik *et al.*,[12] have quite extensively used it in the preformed cluster model (PCM), based on pocket formula of Blocki *et al.*,[54] given as,

$$\Phi(\epsilon) = -\left(\frac{1}{2}\right)(\epsilon - 2.54)^2 - 0.0852(\epsilon - 2.54)^3, \quad for\ \epsilon \leq 1.2511 \tag{1}$$

$$\Phi(\epsilon) = -3.437 \exp\left(\frac{-\epsilon}{0.75}\right), \quad for\ \epsilon \geq 1.2511 \tag{2}$$

where Φ is the universal proximity potential. For studying the fusion cross section of different target-projectile combinations, Dutt *et al.*,[55,56] have used different versions of proximity potentials. Another formulation of proximity potential[57] has been used in the present model, as given by eqs. 6 and 7, and the assault frequency ν is calculated for each parent-cluster combination which is associated with vibration energy. But, for even-A parents and for odd-A parents, Shi and Swiatecki[58] got ν empirically, unrealistic values as 10^{22} and 10^{20} respectively.

The interacting potential barrier for a parent nucleus exhibiting cluster decay is given as,

$$V = \frac{Z_1 Z_2 e^2}{r} + V_P(z) + \frac{\hbar^2 \ell(\ell+1)}{2\mu r^2}, \quad for\ z > 0 \tag{3}$$

Here Z_1 and Z_2 are the atomic numbers of the daughter and emitted cluster, z is the distance between the near surfaces of the fragments, r is the distance between fragment centers and is given as $r = z + C_1 + C_2$, where, C_1 and C_2 are the Süsmann central radii of fragments. The term ℓ represents the angular momentum, μ the reduced mass and V_P is the proximity potential. The proximity potential V_P is given by Blocki *et al.*[54] as,

$$V_P(z) = 4\pi\gamma b \left[\frac{C_1 C_2}{C_1 + C_2}\right] \Phi\left(\frac{z}{b}\right) \tag{4}$$

with the nuclear surface tension coefficient,

$$\gamma = 0.9517[1 - 1.7826(N-Z)^2/A^2]\ \ MeV/fm^2 \tag{5}$$

where N, Z and A represent neutron, proton and mass number of parent, Φ represents the universal proximity potential[57] given as,

$$\Phi(\epsilon) = -4.41 e^{-\epsilon/0.7176}, \quad for\ \epsilon > 1.9475 \tag{6}$$

$$\Phi(\epsilon) = -1.7817 + 0.9270\epsilon + 0.0169\epsilon^2 - 0.05148\epsilon^3, \quad for\ 0 \leq \epsilon \leq 1.9475 \tag{7}$$

with $\epsilon = z/b$, where the width (diffuseness) of the nuclear surface $b \approx 1$ fm and Süssmann central radii C_i of the fragments related to sharp radii R_i is,

$$C_i = R_i - \left(\frac{b^2}{R_i}\right) \tag{8}$$

For R_i we use semi empirical formula in terms of mass number A_i as,[54]

$$R_i = 1.28 A_i^{1/3} - 0.76 + 0.8 A_i^{-1/3} \tag{9}$$

The potential for the internal part (overlap region) of the barrier is given as,

$$V = a_0 (L - L_0)^n\ for\ z < 0 \tag{10}$$

where $L = z + 2C_1 + 2C_2$ and $L_0 = 2C$, the diameter of the parent nuclei. The constants a_0 and n are determined by the smooth matching of the two potentials at the touching point.

Using one dimensional WKB approximation, the barrier penetrability P is given as,

$$P = exp\left\{-\frac{2}{\hbar}\int_a^b \sqrt{2\mu(V-Q)}dz\right\} \tag{11}$$

Here the mass parameter is replaced by $\mu = mA_1A_2/A$, where m is the nucleon mass and A_1, A_2 are the mass numbers of daughter and emitted cluster respectively. The turning points a and b are determined from the equation, $V(a)=V(b)=Q$. The above integral can be evaluated numerically or analytically, and the half life time is given by,

$$T_{1/2} = \left(\frac{\ln 2}{\lambda}\right) = \left(\frac{\ln 2}{\nu P}\right) \tag{12}$$

where, $\nu = (\frac{\omega}{2\pi}) = (\frac{2E_v}{h})$ represents the number of assaults on the barrier per second and λ the decay constant. E_v is the empirical vibration energy and is given as,[59]

$$E_v = Q\left\{0.056 + 0.039\exp\left[\frac{(4-A_2)}{2.5}\right]\right\}, \quad for\ A_2 \geq 4 \tag{13}$$

Classically, the α particle is assumed to move back and forth in the nucleus and the usual way of determining the assault frequency is through the expression given by $\nu = velocity/(2R)$, where R is the radius of the parent nuclei. But, as the alpha particle has wave properties, a quantum mechanical treatment is more accurate. Thus, assuming that the alpha particle vibrates in a harmonic oscillator potential with a frequency ω, which depends on the vibration energy E_v, we can identify this frequency as the assault frequency ν given in Eqs. (12) and (13).

3. Results and discussions

In the decay transitions between the ground state energy levels of the parent nuclei and the ground state energy levels of the daughter nuclei, the energy released is given as eqn. (14), and the possibility to have a cluster decay process is related to its exotermicity, $Q > 0$.

$$Q_{gs\to gs} = \Delta M_p - (\Delta M_c + \Delta M_d) \tag{14}$$

Here ΔM_p, ΔM_d, ΔM_c are the mass excess of the parent, daughter and cluster respectively. The mass excess values to be used for evaluating the Q values for cluster decays have been extracted from the experimental mass tables of Wang *et al.*,[60] and for those isotopes of which the experimental mass excess were unavailable in Ref. 60, the mass excess values have been extracted from the theoretical mass tables of Möller *et al.*,[61] and Koura *et al.*[62] The present work aims at predicting the probable clusters that could be emitted from the isotopes of superheavy nuclei with $Z = 126$ within the range $298 \leq A \leq 336$, concentrating mainly on the light and medium, even and odd clusters like ^{4}He, ^{8,10}Be, ^{14}C, ^{15}N, ^{18,20,22}O, ^{23}F, ^{22,24,25,26}Ne and ^{28,29,30}Mg. The study has been performed within the CPPM, where the external drifting potential barrier is taken as the sum of the Coulomb potential, proximity potential and centrifugal potential for the touching configuration and for the separated fragments.

In addition to the calculations on the cluster decay half lives performed within CPPM, the Universal (UNIV) curve of Poenaru *et al.*,,[63,64] the Universal Decay Law (UDL) of Qi *et al.*,[65,66] and the Scaling Law of Horoi *et al.*,[67] have also been used for studying the cluster decay half lives for the isotopes of $Z = 126$. A short description on these formalisms has been given below.

In UNIV, based on the quantum mechanical tunnelling process,[68] the disintegration constant λ and the partial decay half life T of the parent nucleus is related as,

$$\lambda = ln2/T = \nu S P_s \tag{15}$$

where the three model-dependent quantities ν, S and P_s are the frequency of assaults on the barrier per second, the pre-formation probability of the cluster at the nuclear surface, and the quantum penetrability of the external potential barrier respectively.
By using the decimal logarithm on Eq. (15), the equation trasforms as,

$$\log_{10} T(s) = -\log_{10} P - \log_{10} S + [\log_{10}(\ln 2) - \log_{10} \nu] \tag{16}$$

The UNIV was derived by assuming the quantities ν as a constant, and S to be depending only on the mass number of the emitted particle A_e..[63,68] Using a fit with experimental data for α decay, the corresponding numerical values[63] were obtained as, $S_\alpha = 0.0143153$ and $\nu = 10^{22.01} s^{-1}$. Now, in Eq. (16), the decimal logarithm of the pre-formation factor is given as,

$$\log_{10} S = -0.598(A_e - 1) \tag{17}$$

and the additive constant for an even-even nucleus is,

$$c_{ee} = [-\log_{10}\nu + \log_{10}(\ln 2)] = -22.16917 \tag{18}$$

The penetrability of an external Coulomb barrier, having separation distance at the touching configuration $R_a = R_t = R_d + R_e$ as the first turning point, and the second turning point defined by $e^2 Z_d Z_e/R_b = Q$, may be found analytically as,

$$-\log_{10} P_s = 0.22873(\mu_A Z_e Z_d R_b)^{1/2} \times [\arccos\sqrt{r} - \sqrt{r(1-r)}] \tag{19}$$

Here, μ_A is the reduced mass, $r = R_t/R_b$, $R_t = 1.2249(A_d^{1/3} + A_e^{1/3})$ and $R_b = 1.43998 Z_d Z_e/Q$, where Q is the released energy, and the liquid-drop-model radius constant $r_0 = 1.2249 fm$.

The Universal Decay Law (UDL) for α-decay and cluster decay modes was introduced by Qi *et al.*,[65,66] where it was made possible to represent, on the same plot with a single straight line, the logarithm of the half lives minus some quantity versus one of the two parameters (χ' and ρ') that depend on the atomic and mass numbers of the daughter and emitted particles as well as the Q value. In UDL, the half lives of monopole radioactive decays, the Q values of the outgoing particles, and the masses and charges of the nuclei involved in the decay are related as,

$$\log_{10}(T_{1/2}) = aZ_cZ_d\sqrt{\frac{A}{Q_c}} + b\sqrt{AZ_cZ_d(A_d^{1/3} + A_c^{1/3})} + c \tag{20}$$

$$= a\chi' + b\rho' + c, \tag{21}$$

where the quantity $A = \frac{A_d A_c}{A_d + A_c}$ and the constants $a = 0.4314$, $b = -0.4087$ and $c = -25.7725$ are the coefficient sets of Eq. (21), determined by fitting to experiments of both α and cluster decays.[65] The effects that induce the clusterization in the parent nucleus are included in the term $b\rho' + c$, and as this relation holds for the monopole radioactive decays of all clusters, it is called the Universal Decay Law (UDL).[65]

An empirical formula for cluster decay was introduced by Horoi *et al.*,[67] in order to determine the half lives of both the alpha and cluster decays, and is given as,

$$\log_{10}(T_{1/2}) = (a_1\mu^x + b_1)\left[(Z_1Z_2)^y/\sqrt{Q} - 7\right] + (a_2\mu^x + b_2), \tag{22}$$

where μ is the reduced mass, and the six parameters are $a_1 = 9.1, b_1 = -10.2, a_2 = 7.39, b_2 = -23.2, x = 0.416$ and $y = 0.613$.

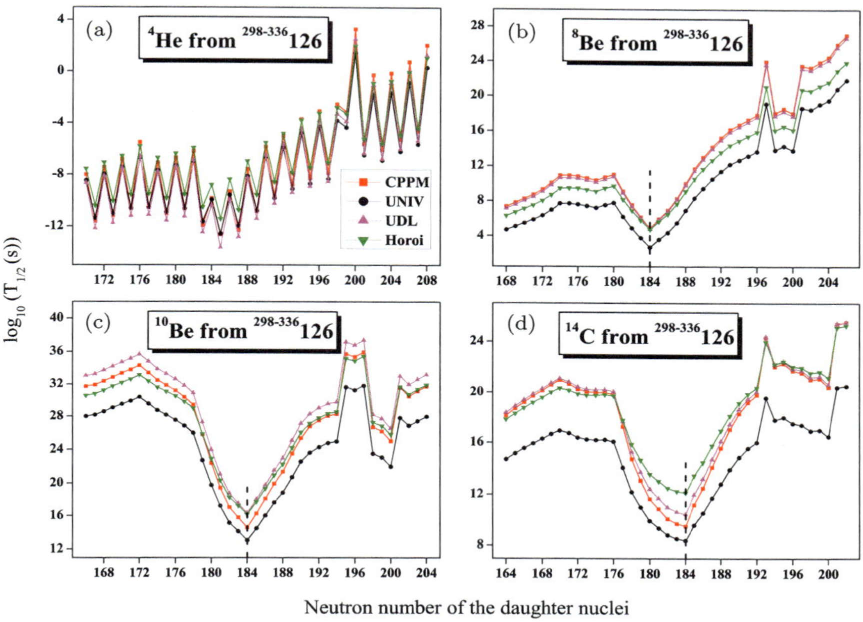

Fig. 1. Plot for the computed $\log_{10}(T_{1/2})$ values vs. neutron number of daughter nuclei for the emission of ^{4}He, ^{8,10}Be and ^{14}C from $^{298-336}$126 SHN.

The entire calculations have been represented as Figs. 1–4, which give the plot for $\log_{10}(T_{1/2})$ vs. neutron number of the daughter nuclei for the emission of various clusters from $^{298-336}$126 superheavy nuclei. The behaviour of the cluster half lives computed within CPPM towards the neutron number of the daughter nuclei could be clearly seen from these figures. The plot for the cluster emission of ^{4}He, ^{8,10}Be, and ^{14}C from $^{298-336}$126 has been given in Fig. 1, and Fig. 2 represents the plot for the cluster emission of ^{18,20,22}O and ^{22}Ne from $^{298-336}$126 superheavy nuclei. From these figures, it can be clearly seen that the minima of the logarithmic half lives for the emission of ^{8,10}Be, ^{14}C and ^{22}O, as shown in Figs. 1(b), (c), (d) and 2(c) respectively, are found for the decay leading to the daughter nuclei ^{306}122 (N = 184), ^{304}120 (N = 184) and ^{302}118 (N = 184) respectively, while for the emission of ^{4}He, the minimum is for the decay leading to the near neutron magic daughter nuclei ^{309}124 (N = 185). In Figs. 3 and 4, the plots for the cluster emission of even clusters ^{24,26}Ne, ^{28,30}Mg, and odd clusters ^{15}N, ^{23}F, ^{25}Ne, ^{29}Mg are given respectively. From

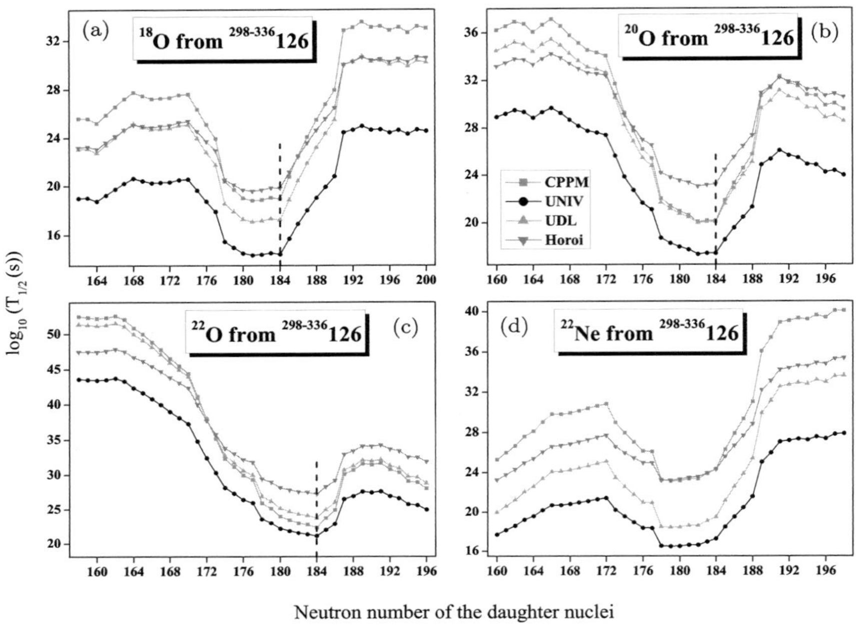

Fig. 2. Plot for the computed $\log_{10}(T_{1/2})$ values vs. neutron number of daughter nuclei for the emission of ^{18,20,22}O and ^{22}Ne from $^{298-336}$126 SHN.

Fig. 3(b), it can be seen that the minima of the logarithmic half lives for the emission of ^{26}Ne is found for the decay leading to the daughter nucleus ^{300}116 (N = 184). In the case of emission of odd clusters ^{15}N and ^{23}F, the minima of the logarithmic half lives are found for the decay leading to the near neutron magic daughter nuclei ^{302}119 (N = 183) and ^{300}117 (N = 183) respectively, while in the case of emission of odd-cluster ^{25}Ne, the minima of the logarithmic half lives is found for the decay leading to ^{300}116 (N= 184). Thus, the Figures 1 and 2 clearly evince the fact that, the minima of the logarithmic half lives for most of these cluster emission are found for the decay leading to a daughter with N= 184 or its neighbouring daughter nuclei. As a minimum in the decay half lives corresponds to the greater barrier penetrability, and in turn adverts the neutron/proton shell closure of the daughter nuclei, the role of neutron magicity N = 184 in cluster radioactivity is evident in these figures. It is well known and has been unveiled through the earlier studies on the cluster decay studies on heavy nuclei that, the half life to be the minimum for those decays leading

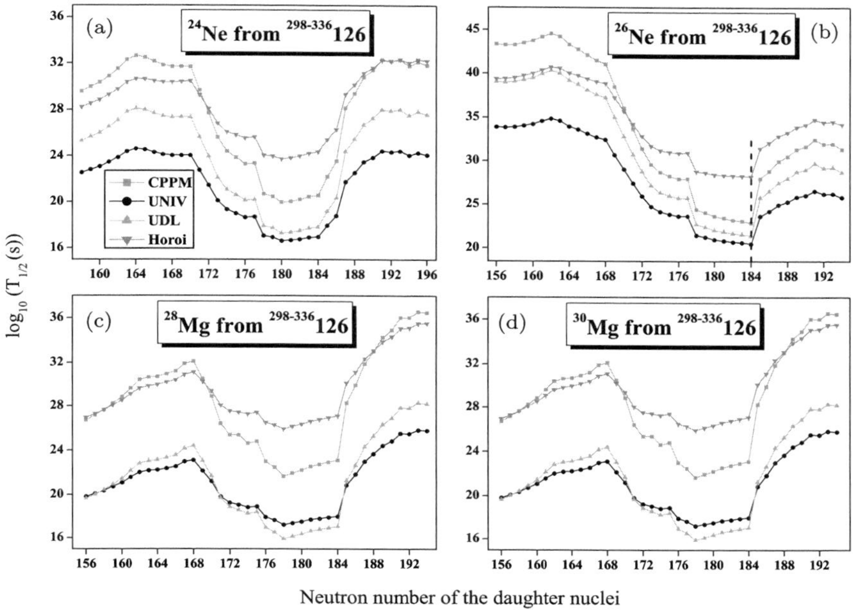

Fig. 3. Plot for the computed $\log_{10}(T_{1/2})$ values vs. neutron number of daughter nuclei for the emission of ^{24,26}Ne and ^{28,30}Mg from $^{298-336}126$ SHN.

to the doubly magic daughter ^{208}Pb ($Z = 82$, $N = 126$) or its neighboring nuclei. The present study on the cluster decay half lives of the superheavy nuclei gives a pronounced minima for the daughter with $N = 184$ and its neighbouring nuclei. The fact that the neutron shell closure plays a decisive role in the cluster decays of superheavy nuclei can be thus highlighted, and the minima at $N = 184$ may be interpreted as a result of the strong shell effect of the assumed magic number of the neutrons.

It is well known that, usually, for the cluster emission of odd clusters from parent nuclei, the odd-even staggering (OES) effects are found to be predominant. The abrupt changes in binding energy as one goes from a nucleus with an even number of neutrons (or protons) to its neighbour with an odd number of nucleons are known as odd-even-staggering (OES) effects, which can be attributed to the existence of nucleonic pairing correlations.[69,70] From the Fig. 4(c) and 4(d), for the emission of the odd clusters ^{25}Ne and ^{29}Mg respectively from $^{298-336}126$, the OES can be seen predominantly, than for the emission of other odd clusters ^{15}N and ^{23}F in

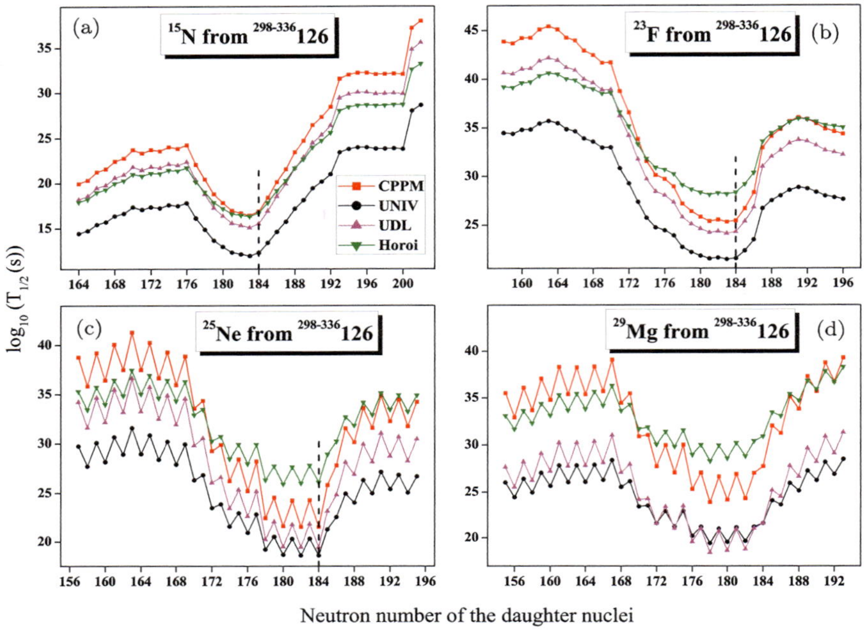

Fig. 4. Plot for the computed $\log_{10}(T_{1/2})$ values vs. neutron number of daughter nuclei for the emission of odd clusters ^{15}N, ^{23}F, ^{25}Ne and ^{29}Mg from $^{298-336}$126 SHN.

Fig. 4(a) and (b). From these observations, it is evident that OES becomes more and more dominant when the emitted clusters are heavy.

The cluster decay half lives evaluated using the UNIV, UDL and scaling law of Horoi have also been depicted in these figures, and each of these show a similar trend as that of our evaluated values. It is also highly mentionable that most of the decays in Figs. 1 and 2 are well within the present experimental upper and lower limits (10^{30}s and 10^{-6}s respectively) of half lives favourable for the cluster decay measurements, and hence could be detected via experiments.

The Geiger-Nuttall plots for $\log_{10}(T_{1/2})$ versus $Q^{-1/2}$ for the emission of various clusters from the parents $^{298-336}$126 superheavy nuclei have been represented in Fig. 3, and the plots are found to be linear with different slopes and intercepts. In the case of the emission of ^{4}He from the isotopes of $Z = 126$, two different GN plots could be observed which corresponds to the emission of ^{4}He from the even-even and even-odd parent isotopes of $Z = 126$. Here we would like to point towards the fact that the Geiger-Nuttall

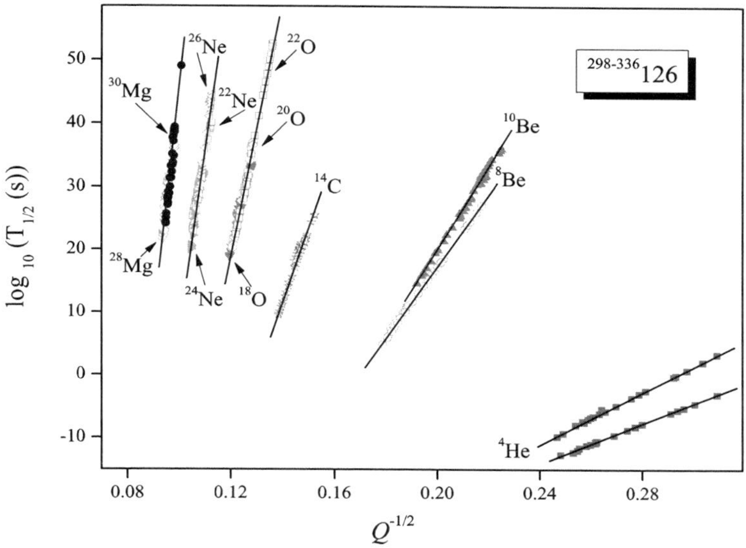

Fig. 5. Geiger-Nuttall plot of $\log_{10}(T_{1/2})$ values versus $Q^{1/2}$ for various clusters from $^{298-336}126$ superheavy nuclei.

law is for pure Coulomb potential. But, in the present study, the interaction potential barrier has been taken as the sum of Coulomb potential, proximity potential, and the centrifugal potential. Still, the Geiger-Nuttall plots open up that the inclusion of proximity potential will not produce much deviation to the linear nature of these plots, which also agree with our earlier observations. It has also to be mentioned that presence of proximity potential (nuclear structure effect) and shell effect (through Q value) can be clearly evidenced from the observed variation in the slope and intercept of Geiger-Nuttall plots for different clusters from various parents. Figure 6 shows the plot (Universal curve) between negative logarithm of penetrability (-ln P) and the logarithmic half-lives ($\log_{10}(T_{1/2})$ of all the decays. A linear graph with slope B $=$ 0.43567 and the intercept A $=$ 21.55704, shows that the assault frequency is nearly a constant for all the decays.

A more close observation of the cluster decay half lives reveals that most of the clusters are probable for emission, and a comparison of the cluster decay half lives evaluated within CPPM and the other formalisms gives the same information. Also, the half lives evaluated using various theoretical models shows that, for most of the decays, the CPPM values matches well with the UDL values and the values obtained using the Scaling Law of Horoi

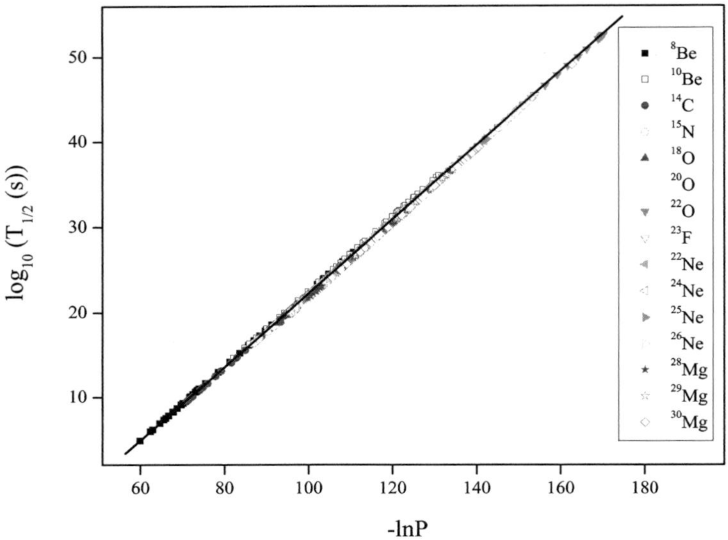

Fig. 6. The universal curve for calculated logarithmic half-lives versus negative logarithm of penetrability for various clusters from $^{298-336}126$ superheavy nuclei.

than that of the UNIV. As the alpha and cluster decay half lives predicted through our study are much below the experimental limit ($T_{1/2} < 10^{30}$s), we hope these observations to provide a ray of hope to the experimentalists, and to pave a new path for the future experiments.

4. Conclusions

The feasibility for the emission of the even clusters ^{4}He, ^{8,10}Be, ^{14}C, ^{18,20,22}O, ^{22,24,26}Ne and ^{28,30}Mg, and the emission of odd clusters ^{15}N, ^{23}F, ^{25}Ne and ^{29}Mg from the superheavy nuclei with $Z = 126$ within the range $298 \leq A \leq 336$ has been investigated within the CPPM. The Universal formula for cluster decay (UNIV) of Poenaru *et al.*, the Universal Decay Law (UDL) and the Scaling Law of Horoi *et al.*, has also been used for evaluating the cluster decay half lives. A comparison of the calculated alpha and cluster half lives, within CPPM, with the values evaluated using other theoretical models show a similar trend. The plots for $\log_{10}(T_{1/2})$ against the neutron number of the daughter in the corresponding decay reveals that, for most of the decays, the half life is minimum for the decay leading to a daughter with $N = 184$, or its neighbouring daughter nuclei, and the plots for the odd-clusters provide the information on the odd-even staggering (OES) to be seen as more prominent in the emission of heavy odd-mass clusters. The predictions on the alpha and cluster decay half lives of $Z = 126$, performed within CPPM, may be of great use for further experimental investigation on cluster decay in the superheavy region, as most of the predicted half lives are well within the present upper limit for measurements.

References

1. E. Rutherford, *Phil. Mag. Series* 5, **47**, 100 (1899).
2. A. Sandulescu, D. N. Poenaru and W. Greiner, *Sov. J. Part. Nucl.* **11**, 528 (1980).
3. H. J. Rose and G. A. Jones, *Nature* **307**, 245 (1984).
4. A. H. Jaffey and A. Hirsch, *unpublished data, quoted in: R. Vandenbosch and J. R. Huizenga, Nuclear Fission, Academic Press, New York* (1973).
5. R. Bonetti, L. Milazzo Colli and G. D. Sassi, *J. Phys. G: Nucl. Phys.* **3**, 1111 (1977).
6. Yu. Ts. Oganessian, Yu. A. Lazarev, V. L. Mikheev, Yu. A. Muzychka, I. V. Shirokovsky, S. P. Tretyakova and V. K. Utyonkov, *Z. Phys. A* **349**, 341 (1994).
7. R. Bonetti, E. Pioretto, H. C. Migliorino, A. Pasinetti, F. Barranco, E. Vigezzi and R. A. Broglia, *Phys. Lett. B* **241**, 179 (1990).
8. R. Bonetti and A. Guglielmetti, *Rom. Rep. Phys.* **59**, 301 (2007).
9. R. K. Gupta and W. Greiner, *Int. J. Mod. Phys. E* **3**, 335 (1984).
10. D. N. Poenaru, M. Ivascu, A. Sandulescu and W. Greiner, *Phys. Rev. C* **32**, 572 (1985).
11. Y. J. Shi, W. J. Swiatecki, *Nucl. Phys. A* **438**, 450 (1985).
12. S. S. Malik, R. K. Gupta, *Phys. Rev. C* **39**, 1992 (1989).
13. K. P. Santhosh and Antony Joseph, *Pramana- J. Phys.* **55**, 375 (2000).
14. D. N. Poenaru, R. A. Gherghescu and W. Greiner, *Phys. Rev. Lett.* **107**, 062503 (2011).
15. D. N. Poenaru, R. A. Gherghescu, W. Greiner, *J. Phys. G: Nucl. Part. Phys.* **39**, 015105 (2012).
16. S. Hofmann and G. Münzenberg, *Rev. Mod. Phys.* **72**, 733 (2000).
17. Yu. Ts. Oganessian, *J. Phys. G: Nucl. Part. Phys.* **34**, R165 (2007).
18. V. Ninov, K. E. Gregorich, W. Loveland, A. Ghiorso, D. C. Hoffman, D. M. Lee, H. Nitsche, W. J. Swiatecki, U. W. Kirbach, C. A. Laue, J. L. Adams, J. B. Patin, D. A. Shaughnessy, D. A. Strellis and P. A. Wilk, *Phys. Rev. Lett.* **83**, 1104 (1999).
19. K. E. Gregorich, T. N. Ginter, W. Loveland, D. Peterson, J. B. Patin, C. M. Folden III, D. C. Hoffman, D.M. Lee, H. Nitsche, J. P. Omtvedt, L. A. Omtvedt, L. Stavsetra, R. Sudowe, P. A. Wilk, P. M. Zielinski and K. Aleklett, *Eur. Phys. J. A* **18**, 633 (2003).

20. Yu. Ts. Oganessian, V. K. Utyonkov, Yu. V. Lobanov, F. Sh. Abdullin, A. N. Polyakov, I. V. Shirokovsky, Yu. S. Tsyganov, G. G. Gulbekian, S. L. Bogomolov, B. N. Gikal, A. N. Mezentsev, S. Iliev, V. G. Subbotin, A. M. Sukhov, O. V. Ivanov, G. V. Buklanov, K. Subotic, A.A. Voinov, M. G.Itkis, K. J. Moody, J. F. Wild, N. J. Stoyer, M. A. Stoyer, R. W. Lougheed and C. A. Laue, *Eur. Phys. J. A* **15**, 201 (2002).
21. Yu. Ts. Oganessian, V. K. Utyonkov, Yu. V. Lobanov, F. Sh. Abdullin, A. N. Polyakov, R. N. Sagaidak, I. V. Shirokovsky, Yu. S. Tsyganov, A. A. Voinov, G. G. Gulbekian, S. L. Bogomolov, B. N. Gikal, A. N. Mezentsev, S. Iliev, V. G. Subbotin, A. M. Sukhov, K. Subotic, V. I. Zagrebaev, G. K. Vostokin, M. G. Itkis, K. J. Moody, J. B. Patin, D. A. Shaughnessy, M. A. Stoyer, N. J. Stoyer, P. A. Wilk, J. M. Kenneally, J. H. Landrum, J. F. Wild and R. W. Lougheed, *Phys. Rev. C* **74**, 044602 (2006).
22. Yu. Ts. Oganessian, F. Sh. Abdullin, C. Alexander, J. Binder, R. A. Boll, S. N. Dmitriev, J. Ezold, K. Felker, J. M. Gostic, R. K. Grzywacz, J. H. Hamilton, R. A. Henderson, M. G. Itkis, K. Miernik, D. Miller, K. J. Moody, A. N. Polyakov, A. V. Ramayya, J. B. Roberto, M. A. Ryabinin, K. P. Rykaczewski, R. N. Sagaidak, D. A. Shaughnessy, I. V. Shirokovsky, M. V. Shumeiko, M. A. Stoyer, N. J. Stoyer, V. G. Subbotin, A. M. Sukhov, Yu. S. Tsyganov, V. K. Utyonkov, A. A. Voinov and G. K. Vostokin, *Phys. Rev. Lett.* **109**, 162501 (2012).
23. Yu. Ts. Oganessian, V. K. Utyonkov, Yu. V. Lobanov, F. Sh. Abdullin, A. N. Polyakov, R. N. Sagaidak, I. V. Shirokovsky, Yu. S. Tsyganov, A. A. Voinov, A. N. Mezentsev, V. G. Subbotin, A. M. Sukhov, K. Subotic, V. I. Zagrebaev, S. N. Dmitriev, R. A. Henderson, K. J. Moody, J. M. Kenneally, J. H. Landrum, D. A. Shaughnessy, M. A. Stoyer, N. J. Stoyer and P. A. Wilk, *Phys. Rev. C* **79**, 024603 (2009).
24. A. Sobiczewski, *Phys. Part. Nucl.* **25**, 119 (1994).
25. P. Möller and J. R. Nix, *J. Phys. G: Nucl. Part. Phys.* **20**, 1681 (1994).
26. R. Smolanczuk, J. Skalski and A. Sobiczewski, *Phys. Rev. C* **52**, 1871 (1995).
27. S. Cwiok, J. Dobaczewski, P. H. Heenen, P. Magierski and W. Nazarewicz, *Nucl. Phys. A* **611** (1996) 211
28. K. Rutz, M. Bender, T. Brvenich, T. Schilling, P. G. Reinhard, J. A. Maruhn and W. Greiner, *Phys. Rev. C* **56**, 238 (1997).
29. R. K. Gupta, G. Mnzenberg and W. Greiner, *J. Phys. G: Nucl. Part. Phys.* **23**, L13 (1997).
30. A. A. Saldanha, A. R. Farhan and M. M. Sharma, *J. Phys. G: Nucl. Part. Phys.* **36**, 115103 (2009).
31. S. Kumar, *Phys. Rev. C* **85**, 024320 (2012).
32. G. G. Adamian, N. V. Antonenko and W. Scheid, *Acta Phys. Polo. B* **40**, 737 (2009).
33. H. F. Zhang, Y. Gao, N. Wang, J. Q. Li, E. G. Zhao and G. Royer, *Phys. Rev. C* **85**, 014325 (2012).
34. S. Kumar, S. Thakur and R. Kumar, *J. Phys. G.* **36**, 105104 (2009).
35. S. Thakur, S. Kumar and R. Kumar, *Braz. J. Phys.* **43**, 152 (2013).

36. Q. Mo, M. Liu and N. Wang, *Phys. Rev. C* **90**, 024320 (2014).
37. K. P. Santhosh and Antony Joseph, *Pramana- J. Phys.* **58**, 611 (2002).
38. K. P. Santhosh and Antony Joseph, *Pramana- J. Phys.* **62**, 957 (2004).
39. K. P. Santhosh and Antony Joseph, *Ind. J. Pure. App. Phys.* **42**, 806 (2004).
40. K. P. Santhosh and Antony Joseph, *Pramana- J. Phys.* **64**, 39 (2005).
41. K. P. Santhosh and Antony Joseph, *Pramana- J. Phys.* **59**, 599 (2002).
42. K. P. Santhosh, R. K. Biju and Antony Joseph, *J. Phys. G: Nucl. Part. Phys.* **35**, 085102 (2008).
43. K. P. Santhosh and B. Priyanka, *Int. J. Mod. Phys. E* **23**, 1450059 (2014).
44. K. P. Santhosh and B. Priyanka, *Nucl. Phys. A* **929**, 20 (2014).
45. K P Santhosh, *Phys. Scr.* **81**, 015203 (2010).
46. K. P. Santhosh, B. Priyanka, G. J. Jayesh and S. Sabina, *Phys. Rev. C* **84**, 024609 (2011).
47. K. P. Santhosh, Indu Sukumaran and B. Priyanka, *Nucl. Phys. A* **935**, 28 (2015).
48. K. P. Santhosh, B. Priyanka and M. S. Unnikrishnan, *Nucl. Phys. A* **889**, 29 (2012).
49. K. P. Santhosh and B. Priyanka, *Eur. Phys. J. A* **49**, 66 (2013).
50. K.P. Santhosh and B. Priyanka, *Nucl. Phys. A* **940**, 21 (2015).
51. K. P. Santhosh and R. K. Biju, *J. Phys. G: Nucl. Part. Phys.* **36**, 015107 (2009).
52. M. M. Sharma, A. R. Farhan and G. Münzenberg, *Phys. Rev. C* **71**, 054310 (2005).
53. H. C. Manjunatha, *Nucl. Phys. A* **945**, 42 (2016).
54. J. Blocki, J. Randrup, W. J. Swiatecki and C. F. Tsang, *Ann. Phys. NY* **105**, 427 (1977).
55. I. Dutt and R. K. Puri, *Phys. Rev. C* **81**, 064608 (2010).
56. I. Dutt and R. K. Puri, *Phys. Rev. C* **81**, 064609 (2010).
57. J. Blocki and W. J. Swiatecki, *Ann. Phys. NY* **132**, 53 (1981).
58. Y. J. Shi and W. J. Swiatecki, *Nucl. Phys. A* **464**, 205 (1987).
59. D. N. Poenaru, M. Ivascu, A. Sandulescu and W. Greiner, *Phys. Rev. C* **32**, 572 (1985).
60. M. Wang, G. Audi, A. H. Wapstra, F. G. Kondev, M. MacCormick, X. Xu and B. Pfeiffer, *Chin. Phys. C* **36**, 1603 (2012).
61. P. Moller, J. R. Nix, W. D. Myers and W. J. Swiatecki, *At. Data Nucl. Data Tables* **59**, 185 (1995).
62. H. Koura, T. Tachibana, M. Uno and M. Yamada, *Prog. Theor. Phys.* **113**, 305 (2005).
63. D. N. Poenaru and W. Greiner, *Phys. Scr.* **44**, 427 (1991).
64. D. N. Poenaru, R. A. Gherghescu, W. Greiner, *Phys. Rev. C* **83**, 014601 (2011).
65. C. Qi, F. R. Xu, R. J. Liotta and R. Wyss, *Phys. Rev. Lett.* **103**, 072501 (2009).

66. C. Qi, F. R. Xu, R. J. Liotta, R. Wyss, M. Y. Zhang, C. Asawatangtrakuldee and D. Hu, *Phys. Rev. C* **80**, 044326 (2009).
67. M. Horoi, A. Brown and A. Sandulescu, *arXiv:nucl-th/9403008* (1994).
68. R. Blendowske and H. Walliser, *Phys. Rev. Lett.* **61**, 1930 (1988).
69. A. Bohr, B. R. Mottelson and D. Pines, *Phys. Rev.* **110**, 936 (1958).
70. W. A. Friedman, arXiv:0911.0634v1 [nucl-th], 2009.

Part 5

Cluster Effects in Nuclear Reactions

CHAPTER 15

MEAN-FIELD INSTABILITIES AND CLUSTER FORMATION IN NUCLEAR REACTIONS

M. Colonna*, P. Napolitani† and V. Baran‡

INFN-Laboratori Nazionali del Sud, I-95123, Catania, Italy
†*Institut de Physique Nucleaire, CNRS/IN2P3, Université Paris-Sud, Université Paris-Saclay, 91406 Orsay cedex, France*
‡*Faculty of Physics, University of Bucharest, Romania*

We review recent results on intermediate mass cluster production in heavy ion collisions at Fermi energy and in spallation reactions. Our studies are based on modern transport theories, employing effective interactions for the nuclear mean- field and incorporating two-body correlations and fluctuations. Namely we will consider the Stochastic Mean Field (SMF) approach and the recently developed Boltzmann-Langevin One Body (BLOB) model. We focus on cluster production emerging from the possible occurrence of low-density mean-field instabilities in heavy ion reactions. Within such a framework, the respective role of one and two-body effects, in the two models considered, will be carefully analysed. We will discuss, in particular, fragment production in central and semi-peripheral heavy ion collisions, which is the object of many recent experimental investigations. Moreover, in the context of spallation reactions, we will show how thermal expansion may trigger the development of mean-field instabilities, leading to a cluster formation process which competes with important re-aggregation effects.

1. Introduction

Nuclei are quantum many-body systems consisting of protons and neutrons strongly bound together. Understanding the properties of such complex systems in terms of their constituent particles and the interaction among them is a true challenge. The original quantal many-body problem, is often approached adopting the mean-field approximation, yielding a so-called effective interaction.[1–4] This scheme is quite successful in describing the features of the nuclear ground state, which exhibits a shell structure, where nucleons move almost independently in an averaged field (i.e. the mean field), analogously to an atomic system. Other relevant aspects of the nuclear structure, such as the excitation modes linked to single-particle motion and/or to the emergence of collective phenomena, are well understood within such a framework. However, nuclear systems also manifest unique features, different from other many-body systems. Indeed, a nucleus is a self-bound system, where spatial correlations among nucleons can be rather strong. As a consequence, assembling and disassembling of nucleons may occur in several ways. Moreover, owing to the weak dependence of binding energy and saturation density on the mass number, disassembling of nucleons and re-aggregation in new configurations may happen at relatively low excitation energy. As well known, these characteristics lead to the appearing of cluster structures, in which a nucleus is divided into several subunits (clusters) and nucleons are confined within each cluster. This already happens in the ground state configuration of selected nuclei.[5,6] However, this fragmentation process becomes quite important in excited systems, as those formed in intermediate-energy heavy- ion collisions, where several clusters and nuclear fragments are produced from a hot source whose excitation energy is typically comparable to the binding energy (per nucleon) of a nucleus. Nuclear clusters are predicted to appear also in equilibrated nuclear matter below saturation density, i.e. in conditions encountered in the inner crust of neutron stars and/or along supernova explosion processes.[7]

Suitable extensions of mean-field models have to be introduced to take explicitly into account the effects of relevant interparticle correlations. Focusing on nuclear dynamics, an intense theoretical work on correlations and density fluctuations has started in the past years, also stimulated by the availability of large amounts of experimental data on fragment formation in intermediate energy heavy ion collisions and the possibility to observe volume (spinodal) instabilities, thus assimilating the nuclear disassembling to the occurrence of liquid-gas phase transitions.[8–13]

The dynamics of nuclear collisions at intermediate energy is often investigated within the framework of semi-classical transport theories, such as the Nordheim approach, in which the Vlasov equation for the one-body phase space density, $f(\mathbf{r},\mathbf{p},t)$, is supplemented with a Pauli-blocked Boltzmann collision term,[14,15] which accounts for the average effect of the two-body residual interaction (quantal correlations are not included). The basic ingredients that enter the resulting transport equation, often called Boltzmann-Uehling-Uhlenbeck (BUU) or Boltzmann-Nordheim-Vlasov

(BNV) equation, are the self-consistent mean-field potential and the two-body scattering cross sections. In order to introduce fluctuations and further (many-body) correlations in transport theories, a number of different avenues have been taken, that can be essentially reconducted to two different classes of models. One is the class of molecular dynamics (MD) models[13,16–20] while the other kind is represented by stochastic mean-field approaches.[12,21–23] In molecular dynamics models the many-body state is represented by a simple product wave function, with or without antisymmetrization. The single particle wave functions are assumed to have a fixed Gaussian shape. In this way, though nucleon wave functions are supposed to be independent (mean-field approximation), the use of localised wave packets induces many-body correlations both in mean-field propagation and hard two body scattering (collision integral), which is treated stochastically. Hence this way to introduce many-body correlations and produce a trajectory branching, leading to a variety of clustered configurations, is essentially based on the use of empirical Gaussian wave packets.

This localisation of the nucleon wave packet has shown to be quite successful in describing the clustered structures characterising the ground state of several light nuclei.[6] However, as far as nuclear dynamics is concerned, while the wave function localisation appears appropriate to describe final fragmentation channels, where each single particle wave function should be localised within a fragment, the use of fixed shape localised wave packets in the full dynamics could affect the correct description of one-body effects, such as spinodal instabilities and zero sound propagation.[20,24]

On the other side, in the so-called stochastic mean-field approaches, which we will adopt in the following, the stochastic extension of the transport treatment for the one-particle density is obtained by introducing a stochastic term representing the fluctuating part of the collision integral,[21–23] in close analogy with the Langevin equation for a Brownian motion. This can be derived as the next-order correction, in the equation describing the time evolution of f, with respect to the standard average collision integral, leading to the Boltzmann-Langevin (BL) equation. Thus, the system is still described solely in terms of the reduced one-body density f, but this function experiences a stochastic time evolution in response to the random effect of the fluctuating collision term. In this way density fluctuations, resulting from many-body correlations, are introduced, that are amplified when instabilities or bifurcations occur in the dynamics. This procedure is suitable for addressing multifragmentation phenomena, where clusters emerge from the growth of density inhomogeneities, driven by the unstable mean-field. However, the fluctuations introduced in this way are not strong enough to fully account for the production of light clusters, which are loosely bound by the mean-field and would require a strong localisation of the nucleon wave packet (as in MD approaches). In the following we will concentrate on the production mechanism of medium size ($Z > 2$) clusters, emerging in nuclear reactions from the disassemby of excited (warm and/or diluted) sources. In particular, we will illustrate how the clustering process is ruled by the interplay between the presence of many-body correlations and the role of mean-field instabilities, which amplify the cluster aggregation process in low-density matter. Moreover, we will discuss the sensitivity of the cluster features to relevant properties of the nuclear effective interaction. The paper is organised as follows: in Sec. 2 we review the theoretical framework. Section 3 is devoted to a survey of results for central heavy ion reactions at Fermi energies. In Sec. 4 we discuss semiperipheral reactions and neck dynamics, whereas Sec. 5 presents some results for fragmentation in spallation reactions. Finally conclusions and perspectives are drawn in Sec. 6.

2. Theoretical description of nuclear reactions

2.1. *The Stochastic Mean Field (SMF) model*

Nuclear reactions can be modelled by solving transport equations based on mean field theories, with short range (2p-2h) correlations included via hard nucleon- nucleon elastic collisions and via stochastic forces, selfconsistently evaluated from the mean phase-space trajectory.

The SMF model[25] can be considered as an approximate tool to solve the BL equation:[21]

$$\frac{df}{dt} = \frac{\partial f}{\partial t}|\{f, H\} = I_{\rm coll}[f] + \delta I[f], \tag{1}$$

where $f(\mathbf{r}, \mathbf{p}, t)$ is the one-body distribution function, the semi-classical analog of the Wigner transform of the one-body density matrix, $H(\mathbf{r}, \mathbf{p}, t)$ the mean field Hamiltonian, $I_{\rm coll}$ the two-body collision term incorporating the Fermi statistics of the particles, and $\delta I[f]$ its fluctuating part. The coordinates of isospin are not shown for brevity. Eq. (1) is solved adopting the test particle method, i.e. each nucleon is associated with a given number, $N_{\rm test}$, of test particles.

It should be noticed that in the BUU/BNV models, the fluctuating term $\delta I[f]$ is neglected.[14,15] In the present SMF treatment we project the fluctuations of the distribution function, generated by the stochastic collision integral in Eq. (1), on the coordinate space and consider local density fluctuations, which could be implemented as such in a numerical calculation. We make the further assumption of local thermal equilibrium, thus being able to derive analytic expressions for the density fluctuations.[25]

When instabilities are encountered along the reaction path, the evolution of the fluctuation "seeds" introduced by the SMF method is then determined by the dissipative dynamics of the BNV evolution, allowing the system to choose its trajectory through the fragmentation configuration. In this way we create a series of "events" in a nuclear collision, which can then be analysed and sampled in various ways.

2.2. *Boltzmann-Langevin dynamics: the BLOB model*

The implementation of the full structure in phase space of the original BL term can still be considered as an important goal. In fact, this allows one to treat a more general class of phenomena, where the correct description of fluctuations and correlations in $\mathbf{p}$ space is essential (such as fragment velocity correlations for instance). We also stress the general interest of this effort. Indeed transport phenomena occur in many physical systems, for which a more precise description of the time evolution of the one-body distribution function, including the effect of many-body correlations, would be important.

The recently introduced BLOB[26] model aims at solving the full BL equation in phase space. While the BLOB model inherits the mean-field description from the SMF model, a different approach in treating the collision integral $\bar{I}[f]$ and the fluctuation term $\delta I[f]$ (i.e. the right hand side of Eq. 1) is employed. As it will be discussed in the following, the numerical implementation of the BLOB approach imposes that the residual term $\bar{I}[f] + \delta I[f]$ agitates extended portions of the phase space in each single scattering event. However, the model differs substantially from an earlier similar strategy, used in the Bauer-and-Bertsch approach,[27] because it constrains the fluctuating term $\delta I[f]$ to act on phase-space volumes with the correct occupation variance. Such constraint avoids that the Pauli blocking could be violated, and it imposes to pay special attention to the metrics of the phase space (see discussion in Ref. 28). Nuclear-matter calculations in a periodic box, in one dimension[29] and in three dimensions[30] have shown that the BLOB approach describes, for low-density nuclear matter, the growth rate of unstable modes in correct connection with the form of the mean-field potential, as ruled by the dispersion relation.[31] The BLOB model for heavy-ion collision is constructed as based on this efficient description of the dispersion relation.

The solution of the BL equation in full phase space is obtained by replacing the conventional Uehling-Uhlenbeck average collision integral by a similar form where one binary collision does not act on two test particles a, b but it rather involves extended phase-space agglomerates of test particles of equal isospin $\mathrm{A} = a_1, a_2, \ldots, \mathrm{B} = b_1, b_2, \ldots$ to simulate wave packets:

$$\bar{I}[f] + \delta I[f] = g \int \frac{\mathrm{d}\mathbf{p}_b}{h^3} \int \mathrm{d}\Omega \; W(\mathrm{AB} \leftrightarrow \mathrm{CD}) F(\mathrm{AB} \rightarrow \mathrm{CD}), \tag{2}$$

where W is the transition rate, in terms of relative velocity between the two colliding agglomerates and differential nucleon-nucleon cross section

$$W(\mathrm{AB} \leftrightarrow \mathrm{CD}) |v_\mathrm{A} - v_\mathrm{B}| \frac{\mathrm{d}\sigma}{\mathrm{d}\Omega}, \tag{3}$$

and F contains the products of occupancies and vacancies of initial and final states calculated for the test-particle agglomerates

$$F(\mathrm{AB} \rightarrow \mathrm{CD}) = [(1 - f_\mathrm{A})(1 - f_\mathrm{B}) f_\mathrm{C} f_\mathrm{D} - f_\mathrm{A} f_\mathrm{B} (1 - f_\mathrm{C})(1 - f_\mathrm{D})]. \tag{4}$$

At each interval of time, by scanning all phase space in search of collisions, and by redefining all test-particle agglomerates accordingly in phase-space cells of volume h^3, nucleon-nucleon correlations are introduced. Since N_{test} test particles are involved in one collision, and since those test particles could be sorted again in new agglomerates to attempt new collisions in the same interval of time as far as the collision is not successful, the nucleon-nucleon cross section contained in the transition rate W should be divided by N_{test}: $\sigma = \sigma_{\mathrm{NN}} / N_{\mathrm{test}}$. Special attention should be payed to the metrics when defining the test-particle agglomeration: the agglomerates are searched requiring that they are the most compact configuration in the phase space metrics which does neither violate Pauli blocking in the initial and in the final states, nor energy conservation in the scattering. The localisation in momentum space makes the collisions more effective in agitating the phase space, and the localisation in coordinate space is needed to keep hydrodynamic effects like the flow dynamics.

The correlations produced through this approach are then exploited within a stochastic procedure, which consists in confronting the effective collision probability $W \times F$ with a random number. As a consequence, fluctuations develop spontaneously in the phase-space cells of volume h^3 with the correct fluctuation amplitude. A precise shape-modulation technique[32] is applied to ensure that the occupancy distribution does not exceed unity in any phase-space location in the final states. This leads to a correct Fermi statistics for the distribution function f, in terms of mean value and variance.

2.3. *Ingredients of the calculations*

Within our framework, the total energy of the system can be written as:

$$E_{\text{tot}} = \frac{1}{N_{\text{test}}} \sum_i p_i^2/(2m) + \int d\text{r}\ \rho(\text{r}) E_{\text{pot}}(\rho_n, \rho_p) + \int d\text{r} \rho_p(\text{r}) E_{\text{pot}}^{\text{Coul}}(\rho_p)/2, \quad (5)$$

where ρ_n, ρ_p denote neutron and proton densities, $\rho = \rho_n + \rho_p$, E_{pot} is the potential energy per nucleon, connected to the mean-field interaction, and $E_{\text{pot}}^{\text{Coul}}$ denotes the Coulomb potential.

Effective interactions, associated with a given Equation of State (EOS) can be considered as an input of all transport codes. We adopt a soft isoscalar EOS (compressibility $K = 200$ MeV). We notice that the considered compressibility value is favoured e.g. from flow, monopole oscillation and multifragmentation studies.[12,33] The choice considered corresponds to a Skyrme-like effective interaction, namely SKM^*, for which we take the effective mass as being equal to the nucleon bare mass. Then E_{pot} can be written as:

$$E_{\text{pot}}(\rho) = \frac{A}{2}\tilde{\rho} + \frac{B}{\sigma+1}\tilde{\rho}^{\sigma} + \frac{C_{\text{surf}}}{2\rho}(\nabla\rho)^2 + \frac{1}{2}C_{\text{sym}}(\rho)\tilde{\rho}\beta^2, \quad (6)$$

where $\tilde{\rho} = \rho/\rho_0$ (ρ_0 denotes the saturation density), $A = -356$ Mev, $B = 303$ Mev, $\sigma = 7/6$. We notice that surface effects are automatically introduced in the dynamics when considering finite width wave packets for the test particles employed in the numerical resolution. An explicit surface term is also added (third term of Eq. (6)) and tuned in such a way that the total surface energy reproduces the surface energy of nuclei in the ground state.[34] This procedure yields $C_{\text{surf}} = -6/\rho_0^{5/3}$ MeV fm^5. The last term of Eq. (6) represents the potential part of the symmetry energy per nucleon, with $\beta = (\rho_n - \rho_p)/\rho$.

For some of the reaction mechanisms analyzed with SMF and BLOB, the sensitivity of the simulation results, and of the corresponding cluster features, is tested against different choices of the density dependence of $C_{\rm sym}$: the asysoft, $C(\rho) = \rho_0(482 - 1638\rho)\,\mathrm{MeV}$, the asystiff, $C(\rho) = 32\,\mathrm{MeV}$, and the asysuperstiff, $C(\rho) = 32\frac{2\rho}{\rho+\rho_0}\,\mathrm{MeV}$.[35] The value of the symmetry energy, $E_{\rm sym}/A = \frac{\epsilon_F}{3} + \frac{1}{2}C_{\rm sym}(\rho)\tilde{\rho}$, at

asy-EoS	$E_{\rm Sym}/A$	L(MeV)
asysoft	30.	14.
asystiff	28.	73.
asysupstiff	28.	97.

saturation, as well as the slope parameter, $L = 3\rho_0\frac{dE_{\rm sym}/A}{d\rho}|_{\rho=\rho_0}$, are reported in Table 1 for each of these asy-EOS (ϵ_F denotes the Fermi energy). Just below the saturation density the asysoft parameterisation exhibits a weak variation with density, while the asysuperstiff shows a rapid decrease. Momentum-dependent effective interactions may also be implemented into Eq. (1).[36,37]

Two-body correlations are taken into account through the collision integral, r.h.s of Eq. (1). The main ingredient entering this process is the nucleon-nucleon cross section $\sigma_{\rm NN}$, for which we use the free isospin, angle and energy dependent values or in-medium modified parameterisations.[38]

It should be noticed that the procedure adopted to solve the collision integral, which employs random numbers, is stochastic. In SMF, owing to the fact that collisions are treated for pairs of test particles, fluctuations are reduced by $1/N_{\rm test}$. Thus an explicit fluctuation term is needed, as indicated in Eq. (1) and explained above, to account for the stochastic nature of the nucleon-nucleon collision process. On the other hand, in BLOB collisions are treated directly for nucleons, i.e. for test particle agglomerates, independently of the value of $N_{\rm test}$.

3. Survey of results for central collisions

Due to compression and/or thermal effects, the composite systems formed in central heavy ion collisions at Fermi energies (30–60 AMeV) may reach low density values, attaining the co-existence zone of the nuclear matter phase diagram. In this situation, the system may undergo a spontaneous phase separation, breaking up into several fragments,[12,31] as a consequence of the development of mean-field spinodal instabilities. However, already in the high density phase, nucleon correlations are expected to be rather large, due to the huge amount of two-body nucleon-nucleon collisions. Hence some memory of these high density correlations could be kept along the clusterisation process. According to the theoretical description adopted here, clusters emerge essentially from the occurrence of mean-field instabilities. However, many-body correlations play an essential role in any case because they provide the seeds for the nucleon assembly into clusters. The interplay between mean-field and correlations effects can be investigated comparing the results obtained with SMF and with BLOB. Indeed the latter has been conceived with the purpose of including fluctuations in full phase space, thus improving the treatment of fluctuations and correlations, but preserving, at the same time, mean-field features such as the dispersion relation for unstable modes. Fluctuations act on both isoscalar and isovector degrees of freedom. It is interesting to underline that, in neutron-rich systems, the cluster formation mechanism also keeps the fingerprints of the isovector channel of the nuclear effective interaction, which is related to the symmetry energy term of the nuclear EoS.[35,39] Indeed one observes that the clusters (liquid drops) which emerge from the low-density nuclear matter have a lower N/Z ratio, with respect to the surrounding nucleons and light particles. This effect, the so-called isospin distillation, is connected to the density derivative of the symmetry energy and leads to the minimisation of the system potential energy.[36]

The mechanism of cluster formation by mean-field instabilities is explored in Fig. 1(a), in the system ^{136}Xe+^{124}Sn for central collisions at 25 and 32 AMeV; a probability map shows how ripples in the potential landscape evolve in size as a function of time [from Ref. 26]. At around 100fm/c large sizes, corresponding to the whole composite system, coexists with small sizes (especially at 25 AMeV), which are consistent with the leading wavelength of the dispersion relation, i.e. about the size of nuclei in the region of oxygen and neon.[31] If in this latter situation all potential concavities could come apart into fragments, a pure signature of the spinodal decomposition would stand out as a set of equal-size fragments.[40] However, when the radial expansion of the system is not sufficient to outweigh the mean- field resilience, the density landscape continues to evolve

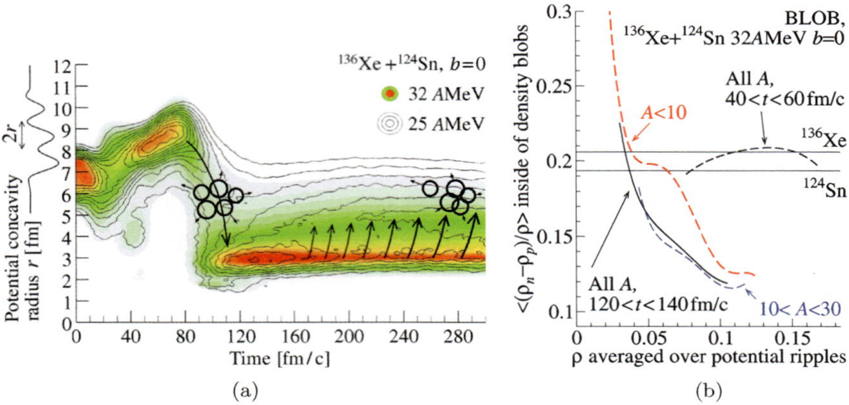

Fig. 1. (a) Evolution of the size of potential ripples in ^{136}Xe+^{124}Sn at 25 and 32 AMeV in central collisions (adapted from Ref. 26). Spinodal-like fragmentation occurring at around 80–100fm/c is followed by a process of recombination at later times. (b) Average isovector density in potential ripples as a function of the local density averaged on potential ripples at different intervals of time. Different sets of potential-ripple sizes are selected and indicated by the corresponding mass A.

towards a more compact shape so that, if the system still succeeds in disassembling, potential concavities can merge together, giving rise to larger fragments and producing asymmetric fragment configurations. This process of recombination prevails at 25 AMeV and weakens at larger incident energies. Figure 1(b) shows the action of the isovector terms on the same system in a process of isospin distillation. An asystiff EOS is used in Fig. 1(a) and 1(b). As a function of the local density averaged on potential ripples, and for different sizes of those latter, the average isospin distribution for the corresponding sites is studied at different intervals of time. Different sets of potential-concavity sizes are indicated by extracting a corresponding mass A. From an initial situation where the system is close to saturation density and the average isospin is defined by the target and projectile nuclei, densities drop to smaller values and the isospin distribution extends over a large range. In particular, the smaller is the local density, the larger is the isospin measured in corresponding sites, as neutrons favour the most volatile phase. Such ordering has been intensively studied within the SMF approach.[36]

The improvement introduced with the BLOB approach is primarily providing a correct sampling of the isoscalar amplitudes and yielding a consistent description of the threshold toward multifragmentation. Figure 2 shows how in the system ^{136}Xe+^{124}Sn, analysed for central impact parameters

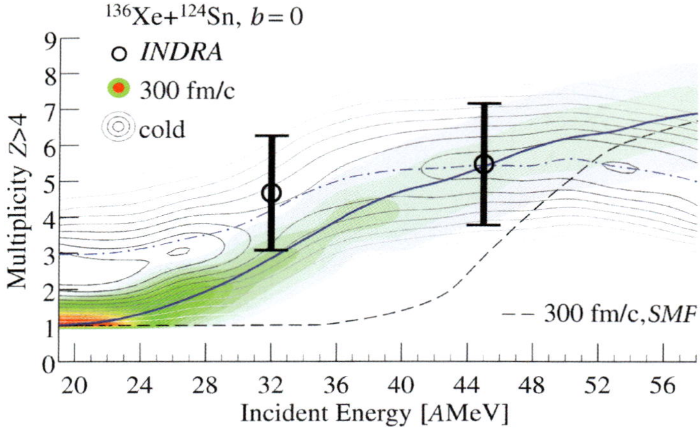

Fig. 2. BLOB simulation: evolution of IMF ($Z > 4$) multiplicity, as a function of incident energy for the system ^{136}Xe+^{124}Sn and a selection of central collisions at 300 fm/c (colour shades) and for the cold system (grey contours). Corresponding mean values are indicated for the BLOB simulation and for a corresponding SMF calculation (adapted from Ref. 26). Corresponding experimental data from Indra experiments[41,42] are added for comparison, giving average (symbols) and variance of a gaussian fitted to the multiplicity distributions (bars).

at 300 fm/c, fragmentation events start competing with the predominant fusion mechanism already beyond 20 *A*MeV per nucleon of incident energy and, while the multiplicity of intermediate mass fragments (IMF) with $Z > 4$, registered at 300 fm/c, continues to grow, it exhibits a maximum at around 45 *A*MeV when considering cold fragments. A calculation undertaken with SMF for the same system results not well adapted when exploring too low incident energies, due to the smaller amplitude of the isoscalar fluctuations. In the figure, two experimental points from INDRA[41,42] indicate the contribution from a compact source; they are compared to the calculated cold distribution and indicate that the BLOB simulation is quantitatively consistent. The cooling of the hot system (300 fm/c) is undertaken by the use of the decay model Simon.[43] Additional details on the comparison between theory and experiment[41,42] are provided in Fig. 3, where the difference between SMF and BLOB is tested around the multifragmentation threshold (32 *A*MeV) and in full multifragmentation regime (45 *A*MeV).

4. Survey of results for semi-peripheral collisions

As discussed above, for central collisions, the nuclear multifragmentation can be associated with a liquid-gas phase transition in a composite system.

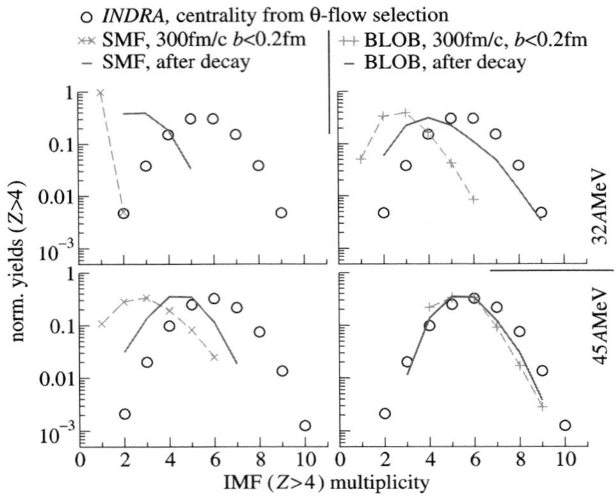

Fig. 3. BLOB and SMF simulations: IMF ($Z > 4$) multiplicity distributions for the systems ^{136}Xe+^{124}Sn at 32 and 45 AMeV, for a selection of central collisions at 300 fm/c and after secondary decay are compared to experimental data from Indra, with average multiplicity and variance given in Refs. 41,42 The experimental distributions correspond to the bars shown in Fig. 2.

The kinetics of this phase transition is related to spinodal decomposition in two-component nuclear matter accompanied by the isospin distillation. At semicentral impact parameters the mechanism changes from fusion to binary channels at low incident energy; in this transition an intermediate mechanism may appear where a neck is produced.[44,45] The neck fragmentation with a peculiar intermediate mass fragment ($2 < Z < 20$) distribution and an entrance channel memory was observed experimentally and predicted by various transport models.[36,46–48] In this case, the low-density neck region triggers an isospin migration from the higher density regions corresponding to the projectilelike fragment (PLF) and targetlike fragment (TLF). Therefore, the isospin content of the IMFs is expected to reflect the isospin enrichment of the midvelocity region. Many investigations have been carried out with the SMF approach.[36,49] For even more peripheral collisions, an essentially binary reaction in the exit channel can by accompanied by a dynamically induced fission of the participants, and for N/Z-asymmetric entrance channel combinations, isospin diffusion drives the system toward charge equilibration.[49,50] Consequently, the isospin degree of freedom can be seen as a precious tracer providing additional information about the physical processes taking place during the evolution of the colliding systems. Moreover, from a comparison between the experimental data and the theoretical model predictions, isospin dynamics allows one to investigate the density and/or temperature dependence of the symmetry energy.[35] More exclusive analyses from the new experimental facilities will certainly impose severe restrictions on various models and parameterisations concerning this quantity. An interplay between statistical and dynamical mechanisms is expected and we clearly evidence the development of hierarchy effects in the transverse velocity of IMFs, which can be a signal of the cluster formation time scale and of the importance of many-body correlations. Moreover, new interesting correlations between kinematic features of the fragments and isospin dynamics, which can provide clues in searching for the most sensitive observables to the symmetry energy, are noted.

These effects are illustrated below, in the context of SMF and BLOB models, for symmetric reactions.

Figure 4 shows results obtained for the system ^{124}Sn+^{124}Sn, at 50 *A*MeV, b = 4 fm, with the asysoft EoS. Transverse velocity distributions are presented for events with three IMF's, ordering the fragments, in each event, according to their size. One can see that the smallest fragment may acquire a large transverse velocity, especially on the reaction plane, indicating that it is emitted on shorter time scales.

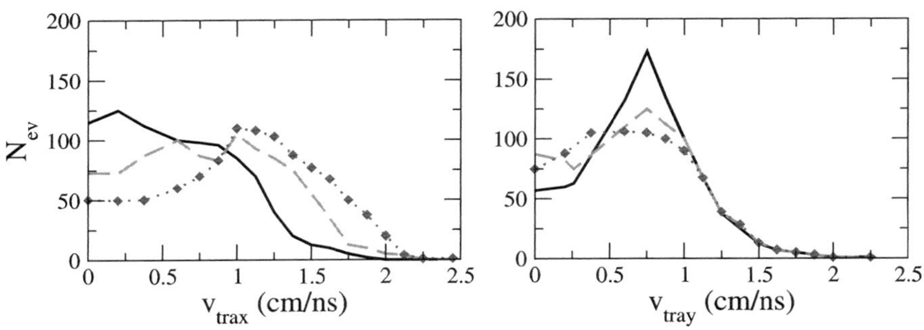

Fig. 4. Left panel: transverse velocity in reaction plane distribution for fragmentation events with three IMFs, as obtained in the reaction ^{124}Sn+^{124}Sn, at 50 *A*MeV, b = 4 fm. Right panel: transverse velocity out of reaction plane distribution for the same set of events. Heaviest (black, continuous line), second heaviest (red, long-dashed line), and third heaviest (blue, dotted line) fragments in the hierarchy. The asysoft EoS is adopted in the calculations. Adapted from Ref. 44.

The emission time has important consequences on the fragment features and, in particular, on the isospin content. Indeed the isospin migration mechanism will be more effective for fragments which stay longer in contact with PLF and TLF, as illustrated in Figs. 5 and 6. We observe in fact that smaller fragments are generally more neutron-rich. In addition, the isospin content decreases with the transverse velocity. These effects are more pronounced in the asystiff case, which leads to the largest migration effects.[36]

This is a clear example of the impact of the reaction dynamics and of the cluster emission time scale, which is strongly influenced by correlation effects, on the cluster properties.

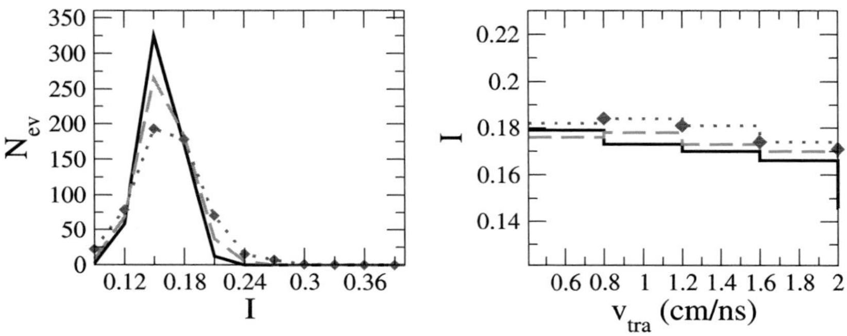

Fig. 5. Asysoft EOS choice. Events with three IMFs. Left panel: Distribution of the isospin parameter $I = (N - Z)/A$. Right panel: Fragment isospin content as a function of transverse velocity. All lines are like in the caption of Fig. 4. Adapted from Ref. 44.

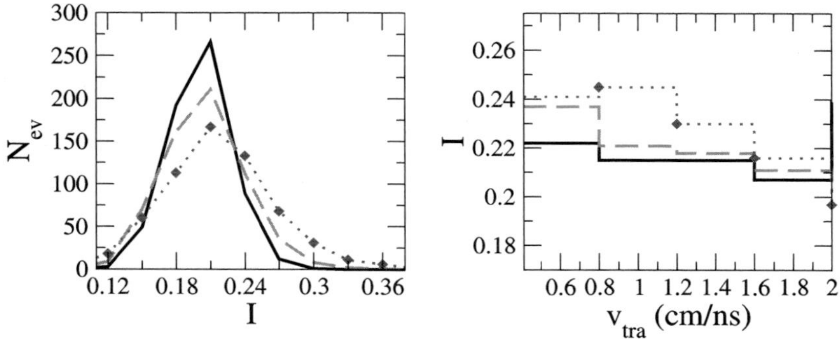

Fig. 6. Like in Fig. 5, for the asysuperstiff EOS. Adapted from Ref. 44.

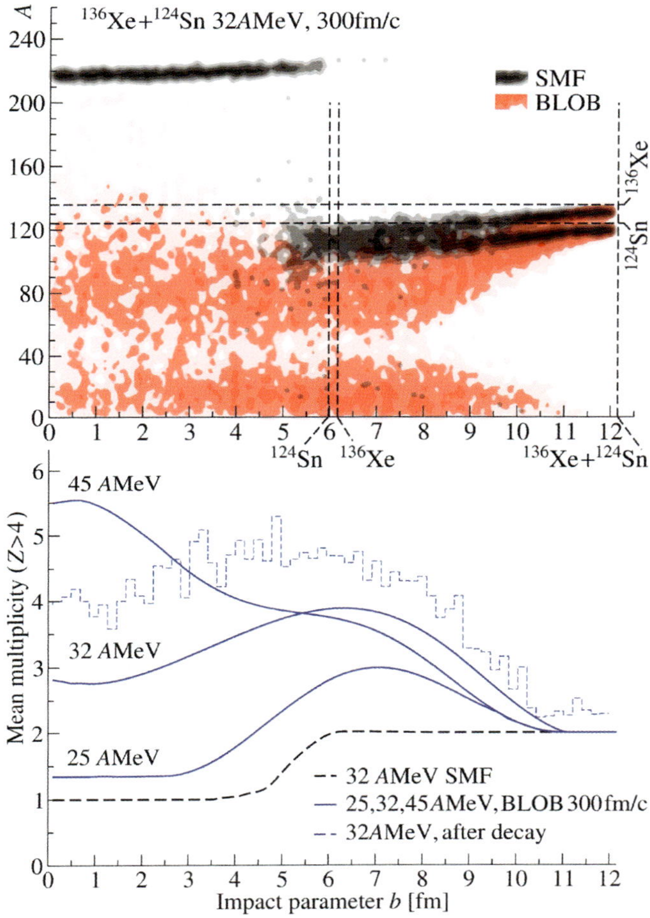

Fig. 7. Bottom. Survey of average fragment multiplicity in peripheral collisions for BLOB and SMF simulations for the system ^{136}Xe+^{124}Sn at 25, 32 and 45 *A*MeV at 300 fm/c. The 32 *A*MeV system is also shown after secondary decay. Top. corresponding mass distributions at 32 *A*MeV for the hot system.

Figure 7 extends the survey of Fig. 2 to all impact parameters for the same system ^{136}Xe+^{124}Sn. The improvement that the BLOB approach introduces is not directly in the description of the isovector modes, but rather in the more complete description of the isoscalar modes. Those latter allow for extending the mechanism of neck fragmentation down to low incident energies (below 15 *A*MeV) and accessing isospin migration when the process of ternary splits is the most relevant for semiperipheral impact parameters. This is illustrated in the bottom panel of Fig. 7, where ternary and quaternary splits (related to neck formation) at 300 fm/c are shown to dominate the semiperipheral impact parameters below 32 *A*MeV. A maximum of IMF multiplicity (quaternary splits) is reached at 32 *A*MeV; on the contrary, at this same energy, the rate of ternary breakups in SMF is still negligible.

The top panel of Fig. 7, adds information on the mass of the IMFs registered at 300 fm/c for the 32 *A*MeV incident energy. While SMF jumps from fusion to binary splits, BLOB produces IMF at all impact parameters b, reaching a maximum of production above $b = 6$ fm for fragments with $A < 40$. This region is further studied in Fig. 8, where the isotopic content N/Z of the IMFs is shown as a function of the transverse velocity component with respect to the quasi-projectile-quasi-target axis for a stiff and a soft form of the symmetry-energy potential component. The dramatically reduced production of IMF in SMF is accompanied by a much smaller mean transverse velocity component; this points to a more explosive dynamics in BLOB, resulting from the improved treatment of many-body correlations. The isovector features are expected from the properties of the employed

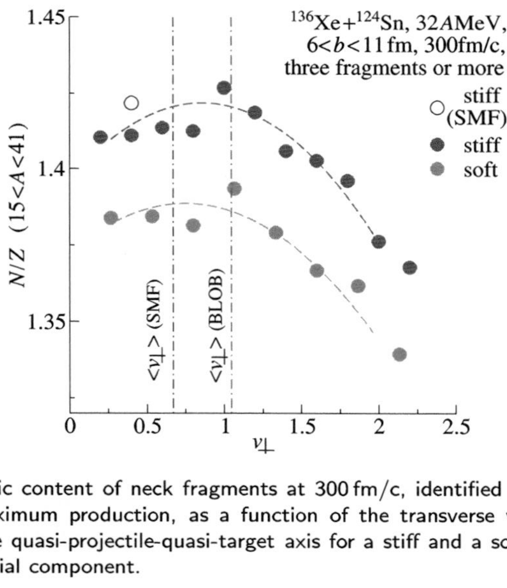

Fig. 8. Isotopic content of neck fragments at 300 fm/c, identified in the impact parameter region of maximum production, as a function of the transverse velocity component with respect to the quasi-projectile-quasi-target axis for a stiff and a soft form of the symmetry energy potential component.

interaction. A stiff form of the symmetry-energy potential component is associated with a larger variation with density, thus enhancing the isospin migration towards the diluted neck region and resulting into a larger isospin content of the IMF. Smaller transverse velocity components are related to fragment configurations more aligned along the quasi-projectile-quasi-target axis, exploiting a longer interval of time available for the process of isospin migration, and finally favour more neutron rich fragments. The neutron enrichment reduces for less aligned configurations and larger transverse velocities. This study is consistent with the experimental investigation reported in ref.[51]

5. Spallation reactions induced by relativistic protons and deuterons

In nuclear spallation reactions induced by relativistic light projectiles, out-of-equilibrium processes are mainly associated to the very initial instants of the collision, when the light projectile traversing the heavy target induces a process of cavitation and the prompt emission of light ejectiles.[52–54] Since this initial stage does not produce mayor dynamical deformations, the successive evolution of the system may be efficiently assimilated to the decay of a fully thermalised source.[55] Even though this two-stage picture yields a successful description of both the heavy nuclide and the light particle spectra, it is not fully compatible with the production of clusters and fragments of intermediate mass (around oxygen, neon and above). Those latter indeed manifest a kinematics which is more consistent with multifragmentation than to asymmetric fission.[56] Statistical multifragmentation approaches could handle these features[57,58] in a macroscopic description, but a microscopic time-dependent picture of the process requires a transport formalisation where a source of dynamical fluctuations is included. So, also in this case, clusters may emerge from the interplay between low-density instabilities, which can be encountered because of the thermal expansion, and dynamical correlations. For this purpose, stochastic transport approaches have been applied to heated systems where the initial conditions were prepared as the outcome of a schematic intranuclear cascade treatment.[59,60] In this framework, it was found that after the light projectile leaves the target, this latter undergoes a dynamical process which may progress for a rather long time. If the target nucleus is sufficiently excited and brought to rather low density, mechanical instabilities can set in and even amplify, like for the scenario of heavy-ion collisions at Fermi energies, discussed in

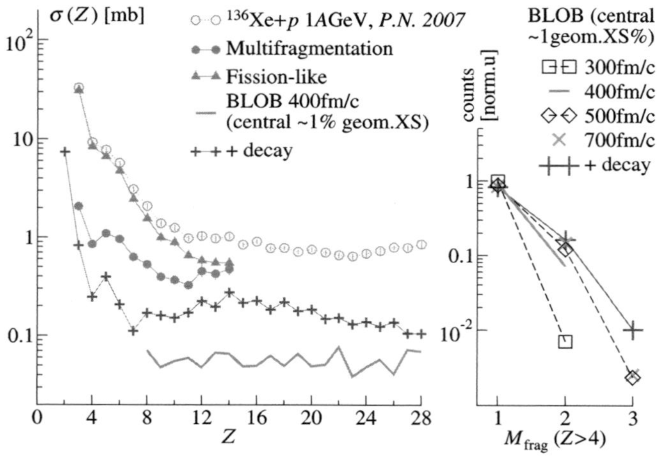

Fig. 9. Left panel: Production cross section of light elements emitted in $p+^{136}$Xe at 1 AGeV. Data were measured at the FRS (Darmstadt);[61] contributions related to multifragmentation or to fission-like events could be separated.[56] The full production of light elements is compared to a BLOB simulation restricted to central impact parameters covering 1% of the total geometric cross section. Right panel: corresponding multiplicity distribution, as obtained in BLOB calculations at different times and for the cold system.

the previous sections. Some of those inhomogeneities seldom prevail, separate into fragments, and leave the target nucleus. Alternatively, they may combine in larger blobs of matter so that, even when the system undergoes a multifragmentation event, the multiplicity of fragments is small or even reduced to two; this last situation is particularly interesting because it gives rise to a binary split from a process which is fundamentally different from ordinary asymmetric fission. Figure 9 illustrates the light elements produced in $p+^{136}$Xe at 1AGeV, measured at the FRS (Darmstadt)[61] and indicates an experimental selection of contributions which are consistent either to multifragmentation events or to binary splits;[56] the data are compared qualitatively to a BLOB calculation restricted to central impact parameters covering 1% of the total geometric cross section (left panel). It is also shown (right panel) that the calculated multiplicity of fragments is predominantly reduced to two and, more rarely, to three fragments.

This scenario of frustrated multifragmentation does not only recall heavy-ion collisions approaching Fermi energies, but it can also extend to the spectator region of peripheral heavy-ion collisions at relativistic energies.

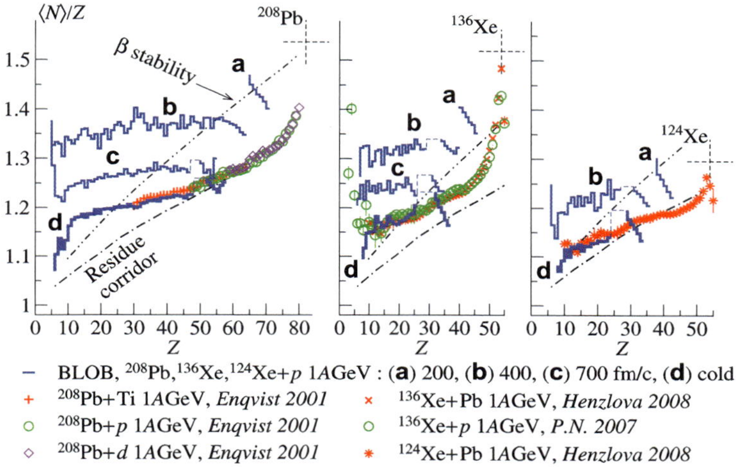

Fig. 10. Study of the evolution in time of the isotopic content of fragments produced in three spallation systems at relativistic energy. The simulation, relying on the BLOB model,[60] is compared to several experimental data measured at FRS (Darmstadt) for fragments produced in proton/deuteron-induced spallation and for fragments emitted by spectator sectors in relativistic peripheral ion-ion collisions.[61–65]

A study where several experimental data measured at FRS (Darmstadt) both from spallation reactions and from peripheral heavy-ion collisions are compared to a transport simulation relying on the BLOB model is presented in Fig. 10 for the isotopic component of the nuclide production in $p+ {}^{124}$Xe, ^{136}Xe, ^{208}Pb at 1 AGeV.[61–65] In particular, it is shown that the isotopic content of the lightest IMFs evolves in time from approximately the value of the target to a value which approaches but does not reach the residue corridor.[66] This effect[67] and the collapsing of all datasets on the same ridge for a given isotopic content is a well known signature of the presence of a limiting temperature in the disassembling of the system, in connection with the liquid-gas phase transition of nuclear matter. This feature also demonstrates the compatibility of the transport approach, which does not impose any prior equilibrium condition, with a statistical picture of a thermalised system.[68]

6. Conclusions

We have discussed a mechanism for cluster production based on the development of mean-field instabilities occurring for nuclear matter at low density and moderate temperature. These conditions can be reached during the expansion phase of a heavy ion reaction at Fermi energy or due to thermal expansion of an excited nuclear source. Though the mechanism is driven by the mean-field, many-body correlations play an essential role because they provide the initial seeds of the cluster assembly. The relative importance of mean-field vs. correlations effects in spinodal decomposition has been investigated employing and comparing the results of two transport models: SMF and BLOB. The latter implements fluctuations in full phase space, thus improving the description of many-body correlations. As a result, one observes an increased clustering probability, shifting the threshold for multifragmentation towards lower excitation energies, in agreement with experimental data. The clusters also acquire a large collective velocity. Fragmentation features are also explored for semi-peripheral reactions where new interesting aspects emerge from the neck dynamics. Also in this case events with larger IMF multiplicity are obtained with BLOB, with also larger transverse velocity. Comparison with experimental data is in progress. Finally the BLOB model well describes also the possible occurrence of clustering, counterbalanced by mean-field resilience effects, in hot (but not initially diluted) sources, such as the nuclear systems formed in spallation reactions. From these results it clearly emerges the important role of many-body correlations even in describing processes which are driven by mean-field instabilities. Further extensions of the model, towards a better description of the production of very light clusters, are envisaged.

References

1. J. W. Negele and D. Vautherin, *Phys. Rev. C* **5**, 1472 (1972).
2. J. Dechargé and D. Gogny, *Phys. Rev. C* **21**, 1568 (1980).
3. J. R. Stone and P.-G. Reinhard, *Prog. Part. Nucl. Phys.* **58**, 587 (2007), and refs. therein.
4. M. Bender, P.-H. Heenen and P.-G. Reinhard, *Rev. Mod. Phys.* **75**, 121 (2003), and refs. therein.
5. J.-P. Ebran, E. Khan, T. Nikšić and D. Vretenar *Phys. Rev. C* **90**, 054329 (2014).
6. Y. Kanada-En'yo, Masaaki Kimura and Akira Ono *Prog. Theor. Exp. Phys.*, 01A202 (2012).
7. J. M. Lattimer and M. Prakash, *Phys. Rep.* **333**, 121 (2000).
8. R. Botet *et al.*, *Phys. Rev. Lett.* **86**, 3514 (2001).
9. J. B. Elliott *et al.*, *Phys. Rev. Lett.* **88**, 042701 (2002).

10. M. D'Agostino *et al.*, *Nucl. Phys.* **A699**, 795 (2002).
11. G. Tabacaru *et al.*, *Eur. Phys. J. A* **18**, 103 (2003).
12. Ph. Chomaz, M.Colonna and J.Randrup, *Phys. Rep.* **389**, 263 (2004).
13. A. Ono, *Phys. Rev. C* **59**, 1999 (853).
14. G. F. Bertsch and S. Das Gupta, *Phys. Rep.* **160**, 189 (1988).
15. A. Bonasera, F. Gulminelli and J. Molitoris, *Phys. Rep.* **243**, 1 (1994).
16. J. Aichelin, *Phys. Rep.* **202**, 1991 (233).
17. H. Feldmeier, *Nucl.Phys.* **A515**, 147 (1990).
18. A. Ono, H. Horiuchi, T. Maruyama and A. Ohnishi, *Phys. Rev. Lett.* **68**, 2898 (1992); A. Ono, H. Horiuchi, T. Maruyama and A. Ohnishi, *Prog. Theor. Phys.* **87**, 1185 (1992).
19. M. Papa, T. Maruyama and A. Bonasera, *Phys. Rev. C* **C64**, 024612 (2001).
20. M. Colonna and Ph.Chomaz, *Phys. Lett. B* **436**, 1 (1998).
21. S. Ayik, C. Gregoire, *Phys. Lett. B* **212**, 269 (1988); S.Ayik and C.Gregoire, *Nucl. Phys.* **A513**, 187 (1990).
22. Y. Abe, S. Ayik, P.-G. Reinhard and E. Suraud, *Phys. Rep.* **275**, 49 (1996).
23. J. Randrup and B. Remaud, *Nucl. Phys.* **A514**, 339 (1990).
24. J. Rizzo, M. Colonna and A. Ono, *Phys. Rev. C* **76**, 024611 (2007).
25. M. Colonna *et al.*, *Nucl. Phys.* **A642**, 449 (1998).
26. P. Napolitani and M. Colonna, *Phys. Lett. B* **726**, 382 (2013).
27. W. Bauer, G. F. Bertsch and S. Das Gupta, *Phys. Rev. Lett.* **58**, 863 (1987).
28. F. Chapelle, G. F. Burgio, Ph.Chomaz and J. Randrup, *Nucl. Phys.* **A540**, 227 (1992).
29. J. Rizzo, Ph.Chomaz and M. Colonna, *Nucl. Phys.* **A806**, 40 (2008).
30. P. Napolitani, M. Colonna and V. de la Mota, Conf. proc. IWM2014-EC, Catania 2014, EPJ Web of Conf. 88, 00003 (2015).
31. M. Colonna and Ph. Chomaz, *Phys. Rev. C* **49**, 1908 (1994).
32. P. Napolitani and M. Colonna, Conf. proc. IWM2011-EC, Caen 2011, EPJ Web of Conf. **31**, 00027 (2012).
33. B. Borderie and M. F. Rivet, *Progr. in Part. and Nucl. Phys.* **61**, (Book Series), 551 (2008), and refs. therein.
34. A. Guarnera, M. Colonna and Ph. Chomaz, *Phys. Lett. B* **373**, 297 (1996).
35. M. Colonna, V. Baran and M. Di Toro, *Eur. Phys. J. A* **50**, No.2, 30 (2014).
36. V. Baran, M. Colonna, V. Greco and M. Di Toro, *Phys. Rep.* **410**, 335 (2005).
37. J. Rizzo, M. Colonna and M. Di Toro, *Phys. Rev. C* **72**, 064609 (2005).
38. P. Danielewicz, *Acta Phys.Polon.* **B33**, 45 (2002).
39. B. A. Li, L. W. Chen and C. M. Ko, *Phys. Rep.* **465**, 113 (2008).
40. B. Borderie *et al.* (INDRA Collaboration), *Phys. Rev. Lett.* **86**, 3252 (2001).
41. F. Gagnon-Moisan *et al.* (INDRA Collaboration), *Phys. Rev. C* **86**, 044617 (2012).
42. G. Ademard *et al.* (INDRA Collaboration), *Eur. Phys. J. A* **50**, 33 (2014).
43. D. Durand, *Nucl. Phys.* **A541**, 226 (1992).
44. V. Baran, M. Colonna, M. Di Toro and R. Zus, *Phys. Rev. C* **85**, 054611 (2012).

45. C. Rizzo, M. Colonna, V. Baran and M. Di Toro *Phys. Rev. C* **90**, 054618 (2014).
46. R. Lionti, V. Baran, M. Colonna and M. Di Toro, *Phys. Rev. C* **71**, 044602 (2005).
47. M. Di Toro, A. Olmi and R. Roy, *Eur. Phys. J. A* **30**, 65 (2006) and refs. therein.
48. B. Djerroud *et al.*, *Phys. Rev. C* **64**, 034603 (2001); J. Colin *et al.* (INDRA Coll.), *Phys. Rev. C* **67**, 064603 (2003); A. A. Stefanini *et al. Z. Phys. A* **351**, 167 (1995).
49. J. Rizzo, M. Colonna, V. Baran, M. Di Toro, H. H. Wolter and M. Zielinska-Pfabe *Nucl. Phys.* **A806**, 79 (2008).
50. M. B. Tsang *et al.*, *Phys. Rev. Lett.* 102, 122701 (2009).
51. De Filippo E. *et al.*, *Phys. Rev. C* **71**, 044602 (2005).
52. J. Cugnon, T. Mizutani and J. Vandermeulen, *Nucl. Phys.* **A352**, 505 (1981).
53. J. Cugnon, *Nucl. Phys.* **A462**, 751 (1987).
54. J. Cugnon, C. Volant and S. Vuillier, *Nucl. Phys.* **A620**, 475 (1997).
55. V. E. Viola *et al.*, *Phys. Rep.* **434**, 1 (2006).
56. P. Napolitani, K.-H. Schmidt and L. Tassan-Got, *J. Phys. G: Nucl. Part. Phys.* **38**, 115006 (2011).
57. P. Napolitani, K.-H. Schmidt, A. S. Botvina, F. Rejmund, L. Tassan-Got and C. Villagrasa, *Phys. Rev. C* **70**, 054607 (2004).
58. S. R. Souza, B. V. Carlson, R. Donangelo, W. G. Lynch, A. W. Steiner and M. B. Tsang, *Phys. Rev. C* **79**, 054602 (2009).
59. M. Colonna, J. Cugnon and E. C. Pollacco, *Phys. Rev. C* **55**, 1404 (1997).
60. P. Napolitani and M. Colonna, *Phys. Rev. C* **92**, 034 (2015).
61. P. Napolitani *et al.*, *Phys. Rev. C* **76**, 064609 (2007).
62. T. Enqvist *et al.*, *Nucl. Phys.* **A686**, 481 (2001).
63. P. Napolitani *et al.*, Conf. proc. VII Int. School Seminar on Heavy-Ion Phyics, Dubna, Phys. Atom. Nucl. **66**, 1471 (2003).
64. T. Enqvist *et al.*, *Nucl. Phys.* **A703**, 435 (2002).
65. D. Henzlova *et al.*, *Phys. Rev. C* **78**, 044616 (2008).
66. R. J. Charity, *Phys. Rev.* **58**, 1073 (1998).
67. K.-H. Schmidt, M. V. Ricciardi, A. S. Botvina and T. Enqvist, *Nucl. Phys.* **A710**, 157 (2002).
68. A. H. Raduta, M. Colonna, V. Baran and M. Di Toro, *Phys. Rev. C* **74**, 034604 (2006).

CHAPTER 16

CLUSTERS IN HEAVY-ION COLLISION DYNAMICS

Akira Ono

Department of Physics, Tohoku University, Sendai 980-8578, Japan

Clusters are copiously formed in heavy-ion collisions. In typical cases of central collisions at several ten to several hundred MeV/nucleon, it has been known for a long time in experiments that about 80–90% of protons in the system are bound in light clusters and heavier fragments in the final state. This has been a challenging problem for transport theories which are normally based on single-nucleon motions. Even in molecular dynamics approaches, the clusters should be explicitly treated as quantum correlations. I will review how clusters are introduced in antisymmetrized molecular dynamics, and discuss the strong impacts of cluster correlations on reaction dynamics and observables.

1. Clusters and fragments in collisions

1.1. *Overview and experimental facts*

In heavy-ion collisions in the medium energy region of the incident energy from several ten to several hundred MeV/nucleon, various kinds of excited nuclear many-body systems are realized at intermediate stages depending on the incident energy, the impact parameter and the sizes of the projectile and target nuclei. In head-on collisions, a compressed system may be formed which expands and disintegrates into fragment nuclei at low densities. In peripheral collisions, moderately excited projectile-like and target-like nuclei may be formed which are dexcited by emitting light particles or disintegrates into several fragment nuclei. The projectile and target parts are often connected by a neck region which can also break up into fragments. Fragmentation or emission of nuclei with various sizes is one of the typical features in these reactions.

It has been quite essential to try to understand such interesting phenomena from two different points of view.[1] One is the understanding or modeling by thermodynamics or statistical mechanics, assuming thermal equilibrium of a finite nuclear system or nuclear matter. Statistical fragmentation models[2,3] have rather successfully described important features of fragmentation assuming some input parameters of the system at freeze out. It has also been quite attractive to understand fragmentation in terms of nuclear liquid-gas phase transition and/or the spinodal decomposition of low-density unstable matter.[4] A natural question to this way of understanding is that the reactions often seem quite dynamical without complete thermal equilibrium with a very long time scale. A lot of efforts have been made in order to develop dynamical transport models which are categorized into two classes: mean-field models and molecular-dynamics models. See Ref. 5 for a recent international activity to compare the results of various transport codes. It is most probably important to confirm that the dynamical models are consistent with thermal properties of equilibrium if the actual situations of reactions are not very far from equilibrium. Some molecular dynamics calculations, however, claim that the fragments which will be emitted at later times have already been formed and can be recognized at early times when the density is still high.[6]

In this article, the two words 'cluster' and 'fragment' will be used to distinguish light nuclei with mass numbers $A \leq 4$ and heavier fragment nuclei with $A \geq 5$, respectively. We don't have any prejudice on how they are actually formed when these two words are used separately, even though the original meanings of these words might imply different mechanisms of formation. This distinction is not quite arbitrary in that a cluster has a single or very few bound states while a fragment has more bound states and resonances. When the character of isospin fractionation/distillation is analyzed by changing the density dependence of the symmetry energy, clusters ($A \leq 3$) and fragments ($Z \geq 3$) actually behave as gas and liquid phases, respectively, in the antisymmetrized molecular dynamics (AMD) calculations,[7] for example.

For several decades, it has been known well in experiments that fragments and clusters are copiously produced in heavy-ion collisions and in other reactions.[8] For example, in Xe + Sn central collisions at 50 MeV/nucleon, the INDRA data[9] show that only 10% of total protons in the system are emitted as free protons and all the other protons are bound in fragments and clusters in the final state. About 20% of protons are bound in α particles, about 10% are bound in clusters of $A = 2$ and 3, and the other 60% are bound in fragments. Even at higher incident energies, 250 MeV/nucleon for example, the ratio of free protons is still about

20% as shown by the FOPI data[10,11] for the central Au + Au collisions. In these collisions, the system is first compressed and then expands relatively rapidly. The importance of clusters can also be found in the data of Ref. 12 which shows the cluster mass fractions as functions of the excitation energy of the *vaporized* ^{36}Ar projectile nucleus. The formation of fragments and clusters should be closely linked to the properties of expanding nuclear matter. The cluster composition in excited low-density nuclear matter can also be an important factor to describe core-collapse supernovae.[13]

1.2. *Theoretical approaches for dynamics*

It has been a longstanding problem to describe fragment formation from microscopic points of view. Transport theories, which solve the time evolution of the many-nucleon system or its one-body distribution, have generally been a powerful approach to understand collision dynamics in the Fermi energy region and in higher energies. Among transport models, the molecular dynamics models were supposed to be suitable for fragmentation reactions because many-nucleon correlations are included at least in the sense of classical dynamics. However, it turned out that some quantum features are required in order to properly describe the whole dynamics of fragment formation. The AMD approach[14] takes into account the full antisymmetrization of many-nucleon wave function, which is an important quantum feature. Emergence of many fragmentation channels was also a crucial model extension for a consist understanding of the fragment formation and the single-particle dynamics.[15,16] However, it must be noted here that this extension has been introduced mainly considering single-nucleon motions in the mean field and thus it may not be very clear how it is related to many-body correlations, in particular, to the cluster correlations of forming light nuclei such as deuterons and α particles.

When fragmentation is studied in transport models, not much has been argued on the roles of cluster correlations unfortunately. It was difficult for most models to describe clusters because the models are based on single-nucleon motions as mentioned above. To predict cluster observables, coalescence prescriptions have been often applied to the result of transport calculations,[17] though these kinds of approaches are not treating cluster correlations and collision dynamics consistently, i.e., it is assumed that correlations do not influence the time evolution of the single-nucleon distribution. The important exception is a version of BUU by Danielewicz *et al.*,[18] in which clusters up to ^{3}H and ^{3}He have been introduced as new particle species and the equations for the distribution functions are coupled by the collision terms representing various reactions of clusters and nucleons. A recent calculation by this model demonstrates that collision dynamics is influenced by the clusters.[19] However, α clusters have not been introduced in this specific model, and the description of formation of fragments ($A \geq 5$) might not be straightforward in mean-field-based models.

When clusters are produced in reactions, they should be still surrounded by other particles. The properties of a cluster in idealistic nuclear medium are calculated with an inmedium Schrödinger equation by Röpke.[20,21] The effect of Pauli blocking is taken into account as a medium effect. A cluster can exist as a bound state, depending on the density of the medium and the momentum of the cluster relative to the medium. Then, in actual systems, many clusters may coexist and change the nuclear matter properties and the collision dynamics. During the time evolution, cluster creation and breakup will be repeated by reactions such as $p + n + X \leftrightarrow d + X'$, $p + n + d + X \leftrightarrow \alpha + X'$ and many others.

Recently the heavy-ion data from Texas A&M have been analyzed to extract the cluster composition and other properties of nuclear matter along a path of the evolution in the plane of the temperature and the density.[22] Comparison of the extracted equilibrium constants for the cluster composition with various equation-of-state (EOS) models suggests necessity of a cluster suppression mechanism such as medium effects.[23] Nevertheless, in general, the link between heavy-ion collisions and EOS should be carefully checked because of the dynamical nature of collisions.

It is desired that the transport models solve the time evolution of collisions by correctly handling cluster correlations. In addition to the extended version of BUU by Danielewicz *et al.*[18] mentioned above, we have extended the AMD approach to introduce cluster correlations by improving the two-nucleon collision procedure.[24,25]

In Sec. 2, we will briefly review approaches without explicit treatment of clusters for the description of fragmentation. We will see that, even tough these approaches may be suitable for the description of fragment formation, it is difficult to describe the light-cluster production consistently. In Sec. 3, the formulation and results of AMD with explicit treatment of cluster correlations will be reviewed. We will see that the treatment of clusters improves the formation of clusters and fragments as well as that the cluster correlations have significant impacts on the reaction dynamics in some aspects.

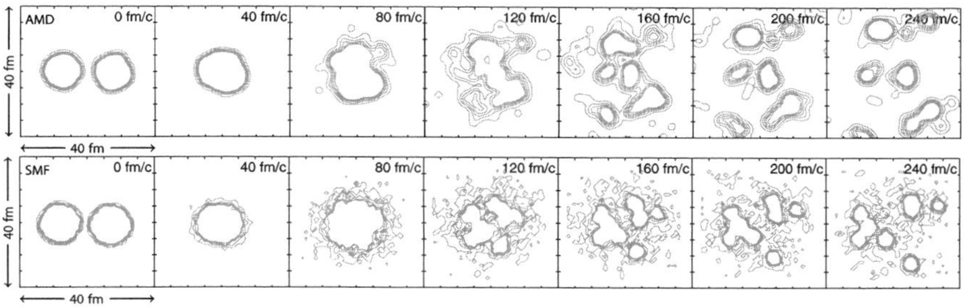

Fig. 1. Time evolution of density profiles obtained by the AMD (upper) and SMF (lower) models, for the head-on collision of ^{112}Sn + ^{112}Sn at 50 MeV/nucleon. From Ref. 27.

2. Single-particle-based approaches for fragmentation

Fragmentation is an interesting problem in excited nuclear systems, in which quantum many-body correlations play essential roles as well as the existence of nuclear liquid-gas phase transition. Figure 1 shows a typical example of the time evolution of ^{112}Sn + ^{112}Sn head-on collisions at 50 MeV/nucleon.[26,27] The calculation by the stochastic mean field (SMF) model[29] in the lower row is compared with the AMD calculation[15,16] in the upper row. In both cases, the system compressed at an early stage starts to expand almost spherically. As the expansion proceeds, the density fluctuation develops to form many fragments. Quantitative differences are found in this figure between the two models. The origin of these differences are to be reviewed later.

Both AMD and SMF describe a reaction by the stochastic time evolution of single-particle distribution function $f(\boldsymbol{r}, \boldsymbol{p}, t)$ or almost equivalently by a single Slater determinant. In SMF, there is no restriction on $f(\boldsymbol{r}, \boldsymbol{p})$ except that it is represented by test particles. In AMD, the single-particle wave functions are restricted to the form of Gaussian wave packets with the centroids $\boldsymbol{Z}_1, \boldsymbol{Z}_2, \ldots, \boldsymbol{Z}_A$ and with a width parameter $\nu = (2.5 \text{ fm})^{-2}$, so that $f(\boldsymbol{r}, \boldsymbol{p})$ is limited to the form

$$f(\boldsymbol{r}, \boldsymbol{p}) = 8 \sum_j \sum_k e^{-2\nu(\boldsymbol{r}-\boldsymbol{R}_{jk})^2 - (\boldsymbol{p}-\boldsymbol{P}_{jk})^2/2\hbar^2\nu} B_{jk} B_{kj}^{-1} \tag{1}$$

with $\boldsymbol{R}_{jk} = (\boldsymbol{Z}_j^* + \boldsymbol{Z}_k)/\sqrt{\nu}$, $\boldsymbol{P}_{jk} = 2i\hbar\sqrt{\nu}(\boldsymbol{Z}_j^* - \boldsymbol{Z}_k)$ and B_{jk} being the overlap matrix of two wave packets. In both models, the time evolution is governed by the single-particle motions in the mean field and the two-nucleon collisions, and the latter are considered to be a source of stochastic branching or bifurcation into different final channels.

In spite of these conceptual similarity, the actual implementations are different. In AMD, a part of the single-particle motion is treated by the deterministic equation of motion for the centroids while the effect of the change of the shape of each wave packet is treated by giving fluctuation to the centroid (wave packet splitting). The two-nucleon collision process has stochastic nature because the scattering angle is chosen randomly. In SMF, the mean-field term and the collision term are both deterministic terms in the BUU-type equation, but the fluctuation term is also taken into account.[28] The fluctuations have been introduced based on the principle that the variance of the occupation of each phase-space cell should follow $\sigma_{\bar{f}}^2 = \bar{f}(1-\bar{f})$, which is equivalent to require that each cell should be fully occupied ($f=1$) or empty ($f=0$). If a phase-space cell is identified with a Gaussian wave packet, the SMF and AMD models are conceptually similar, but the results can be different due to the different approximate treatments of fluctuations. In fact, the density fluctuation (among different events) is already developing in AMD at the relatively early stage of $50 \lesssim t \lesssim 100$ fm/c, while the fluctuation develops in SMF only at a later stage $t \sim 100$ fm/c suggesting a fragmentation mechanism by spinodal decomposition.[4] This difference can be interpreted as the origin of the differences in the expansion velocity, the nucleon emission and so on predicted by these models.[26,27] More recently the fluctuation term in SMF was improved in the BLOB approach by considering the complete fluctuation in the phase space at every two nucleon collision.[30] The new BLOB calculation shows the onset of fragmentation at relatively low energies compared to the SMF case.

The final results of the AMD and SMF models for the fragment charge distribution have been compared with the experimental data with reasonable successes.[15,31] Figure 2 shows the AMD results compared with data for central Xe + Sn collisions at 50 MeV/nucleon. We find that the result depends very much on the choice of the coherence time τ, namely, how long the coherence of the single-particle wave function is kept before it loses coherence due to many-body correlations. When the coherence is assumed to be lost when the nucleon collide with another nucleon ($\tau = \tau_{\mathrm{NN}}$), the fragment yields for $Z \gtrsim 3$ are well reproduced. However, a problem is found in the yields of light particles. The α-particle multiplicity $M_\alpha \approx 7$ is too small and the proton multiplicity $M_p \approx 20$ is too large compared to the experimental data $M_\alpha \approx M_p \approx 10$.

Compared to the very early version of AMD without wave packet splitting,[14] the reproduction of fragment yield was improved very much in many

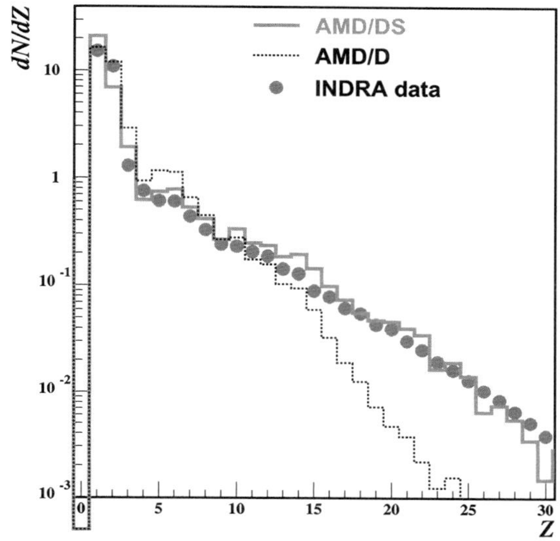

Fig. 2. The charge distribution of the produced clusters in ^{129}Xe + Sn collisions at 50 MeV/nucleon with the impact parameter $0 < b < 4$ fm, after calculating the secondary decay of excited clusters and applying the experimental filter for the detector setup. Solid histogram (labeled AMD/DS) shows the result of AMD with the coherence time $\tau = \tau_{\mathrm{NN}}$, while the dotted histogram (labeled AMD/D) shows the result with the strongest decoherence $\tau \to 0$. The INDRA experimental data are shown by solid points. From Ref. 15.

cases by introducing wave-packet splitting as a quantum branching process based on the single-particle motions with decoherence.[15,16,32] However, a consistent reproduction of the yields of light clusters and light intermediate mass fragments has not been achieved by a single set of model parameters for all the reaction systems. This may suggest that the correlations to produce light clusters should be treated more explicitly.

It should be noted that the AMD wave function or Eq. (1) is suitable for describing cluster correlations in the ground state and low-lying excited states of nuclei.[33] When a cluster is placed in a nucleus within AMD, a reasonable value of the cluster-binding energy is obtained depending on the location and the center-of-mass momentum of the cluster (see Fig. 3 taken from Ref. 34). An important question is, however, whether such clusterized states are realized with correct probabilities during the evolution of reactions. In fact, after a two-nucleon collision, one (or both) of the scattered nucleons may accidentally form a cluster with surrounding nucleon(s) if the wave packet centroids are close in phase space. The probability is governed

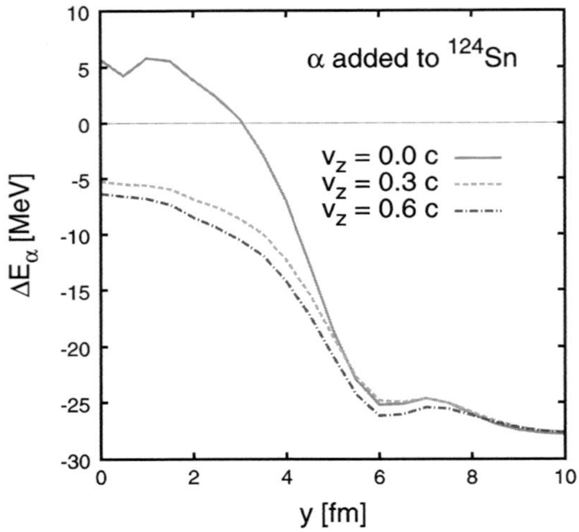

Fig. 3. Internal binding energy of an α cluster placed in the ground state ^{124}Sn nucleus as a function of the position $(0, y, 0)$ and the velocity $(0, 0, v_z)$ of the cluster relative to the center of the nucleus. Calculated with the SLy4 effective interaction. From Ref. 34.

by the classical phase space or the classical density of states $\tilde{D}(E)$ of the internal degrees of freedom of a cluster to be formed. The density of states starts at $E = E_{\rm gs}$ with $\tilde{D}(E) = 0$ for $E \le E_{\rm gs}$ below the ground state energy of the cluster. Since the cluster-binding energy $|E_{\rm gs}|$ is small due to the cancellation of the potential and kinetic energies, the bound phase-space volume $\int_{E_{\rm gs}}^{0} \tilde{D}(E)dE$ cannot be large enough to correspond to a single quantum bound state. Therefore cluster correlations do not emerge dynamically with reasonable probabilities if the time evolution of wave packets is determined by the equation of motion with stochastic two-nucleon collisions. It should be noted that irreversible stochastic processes, such as the wave packet splitting, can change the phase-space weights and improve the cluster multiplicities as in the case of the strongest decoherence ($\tau \to 0$) in Fig. 2, though a fully consistent reproduction of fragmentation is difficult by adjusting only the coherence time.

3. AMD approach with clusters

3.1. *Formulation*

In (quantum) molecular dynamics approaches for heavy-ion collisions, many two-nucleon collisions occur during the time evolution of a reaction. Each two-nucleon collision is treated as a random process from the initial state to a final state, which should be interpreted as simulating quantum transitions with transition rates

$$v d\sigma = \frac{2\pi}{\hbar} |\langle \Phi_f | V | \Phi_i \rangle|^2 \delta(E_f - E_i) \frac{p_{\rm rel}^2 dp_{\rm rel} d\Omega}{(2\pi\hbar)^3}. \tag{2}$$

In the case of AMD, both the initial state $|\Phi_i\rangle$ and the final state $|\Phi_f\rangle$ of this collision are AMD wave functions, and $(p_{\rm rel}, \Omega)$ specifies the relative momentum between the two nucleons in the final state. In general, medium modification is considered for the scattering matrix elements. The Pauli blocking for the scattered nucleons is taken into account.[14]

When two-nucleon collisions are introduced in a usual way, the wave-packet momentum centroids of only the colliding two nucleons are changed by a collision (in the 'physical coordinate' representation).[14] This method was generalized[24,25] to allow the possibility that each colliding nucleon may form a cluster of mass numbers $A = 2, 3$ or 4 with some other wave packets in the final state $|\Phi_f\rangle$. Namely, in the calculation with cluster correlations, when two nucleons N_1 and N_2 collide, we consider the process

$$N_1 + N_2 + B_1 + B_2 \to C_1 + C_2 \tag{3}$$

in which each of the scattered nucleons N_j $(j = 1, 2)$ may form a cluster C_j with a spectator particle B_j. This process includes the collisions without cluster formation as the special case of $C_j = N_j$ with empty B_j. The formulation was given in Refs. 24 and 25. The essential ideas are briefly repeated below.

The transition rate of the cluster-forming process is given by Eq. (2) with the suitable choice of the set of final states $|\Phi_f\rangle$. When a cluster is formed, the corresponding wave packets are placed at the same phase-space point, i.e., the cluster internal state is represented by the harmonic-oscillator $(0s)^n$ configuration. Denoting the initial and final states of the $N_j + B_j$ system by $|\varphi_j\rangle$ and $|\varphi'_j\rangle$, respectively, we have the transition rate

$$v d\sigma = \frac{2\pi}{\hbar} |\langle \varphi'_1 | \varphi_1^{\boldsymbol{q}} \rangle|^2 |\langle \varphi'_2 | \varphi_2^{-\boldsymbol{q}} \rangle|^2 |M|^2 \delta(E_f - E_i) \frac{p_{\rm rel}^2 dp_{\rm rel} d\Omega}{(2\pi\hbar)^3}, \tag{4}$$

where $|\varphi_j^{\pm \boldsymbol{q}}\rangle = e^{\pm i\boldsymbol{q}\cdot\boldsymbol{r}_j} |\varphi_j\rangle$ are the states after the momentum transfer $\pm\boldsymbol{q}$ to the nucleons N_j $(j = 1, 2)$, and $(p_{\rm rel}, \Omega)$ is the relative momentum between N_1 and N_2 in these states. The matrix element $|M|^2$ is essentially the same as for the usual two-nucleon collisions.

The actual situation of a two-nucleon collision requires more considerations because there are many possible ways of forming a cluster for each N of the scattered nucleons N_1 and N_2. For a scattered nucleon N, we first consider the possibility that N may form a cluster with one of the nucleons $\{B_k;\ k = 1, 2, \dots\}$ which have the same spin-isospin state. This spin-isospin state that is studied first is randomly decided. The cluster-formed state is denoted by $|\Phi'_k\rangle$ which is obtained, by first changing the state to $|\Phi^{\boldsymbol{q}}\rangle$ by the momentum transfer $\boldsymbol{q}$ to N, and then moving the two wave packets of N and B_k to the same phase-space point without changing their center of mass. Since the different final states are not orthogonal $\mathcal{N}_{kl} = \langle\Phi'_k|\Phi'_l\rangle \neq \delta_{kl}$, the probability that N forms a cluster with one of $\{B_k\}$ should be calculated as

$$P = \sum_{kl} \langle\Phi^{\boldsymbol{q}}|\Phi'_k\rangle \mathcal{N}^{-1}_{kl} \langle\Phi'_l|\Phi^{\boldsymbol{q}}\rangle. \tag{5}$$

This probability is calculated with an approximation that the many-body state is a direct product of wave packets centered at the 'physical coordinates'.[14] With the calculated probability P, a cluster will be formed with one of $\{B_k\}$ and, with the rest of the probability $(1-P)$, N does not form a cluster with a nucleon of this spin-isospin state. The procedure is repeated for other spin-isospin states for $\{B_k\}$. Thus the formation of light clusters is considered up to an α particle. The factor $|\langle\varphi'_1|\varphi_1^{\boldsymbol{q}}\rangle|^2|\langle\varphi'_2|\varphi_2^{-\boldsymbol{q}}\rangle|^2$ in Eq. (4) should be replaced by the probability determined by this prescription.

Even when the cluster formation is introduced, the many-body state is always represented by an AMD wave function which is a Slater determinant of nucleon wave packets. The time evolution of the many-body state is solved just as usual without depending on whether some of the wave packets form clusters due to collisions in the past (except for the inter-cluster binding process in the next paragraph). This is in contrast to the version of BUU by Danielewicz *et al.*[18] where clusters are treated as new particle species. In our approach, a nucleon in a formed cluster may collide with some other nucleon so that the cluster is broken. It may be the case that the scattered nucleon forms the same cluster as before, so that an elastic scattering of the cluster is possible. All of these kinds of processes are based on the nucleon-nucleon scattering matrix elements, without introducing parameters to control individual channels of cluster formation. It is, of course, possible to modify the cluster formation probabilities. For example, in the calculation of Ref. 25, we suppressed the overall cluster production probability when the momentum transfer is extremely small, by a factor

$1 - \exp[-\boldsymbol{q}^2/(50\ \mathrm{MeV}/c)^2]$. In the next subsection, however, we will review the results without such suppression.

It has been found that the correlations to bind light clusters to form heavier fragments should be taken into account on top of the usual time evolution of AMD.[24] Many of light nuclei (Li, Be *etc.*) have only one or a few bound states which may be regarded as bound states of internal clusters. By the same reason for the issue of light clusters, the quantum-mechanical probability of forming such a nucleus is not consistent with the semiclassical phase space with which it can be formed in the standard treatment of AMD. Therefore, for a better description, inter-cluster correlation is introduced as a stochastic process of binding clusters. The basic idea is to replace the relative momentum between clusters by zero if moderately separated clusters are moving away from each other with a small relative kinetic energy. For specific implementations and choices of parameters, see Refs. 25, 35 and 36. It is important how to ensure the total energy conservation.

3.2. *Yields of clusters and fragments*

Recent results of AMD with clusters for heavy-ion fragmentation reactions were reported in Refs. 35 and 36. The main difference of the models in these two references is that, in Ref. 36, the momentum fluctuation, originating from the width of Gaussian wave packets, are taken into account for emitted particles. See Ref. 36 for more details. In this subsection, we will mainly review the results of Ref. 36 for fragmentation. The results shown here in figures are essentially the same as in Ref. 36 but the statistics has been improved.

Figure 4 shows the comparison for Xe + Sn collisions at 50 and 32 MeV/nucleon. The light particle multiplicities (multiplied by the particle mass number) are also shown in the inset. The overall distribution of the fragments ($Z \geq 3$) and the individual multiplicities of light charged particles are all reproduced very well by this version of AMD with cluster correlations, except for the deviations around $Z = 4$ which have not been well understood. As well as the cluster correlations (to form clusters up to α particles), the inter-cluster correlations are quite important to reproduce the data at both incident energies simultaneously.

The fragment multiplicities were also calculated for violent collisions of ^{40}Ca+^{40}Ca at 35 MeV/nucleon (Fig. 5) and central collisions of Au+Au at 250 MeV/nucleon (Fig. 6). The reproduction of data is almost satisfactory if

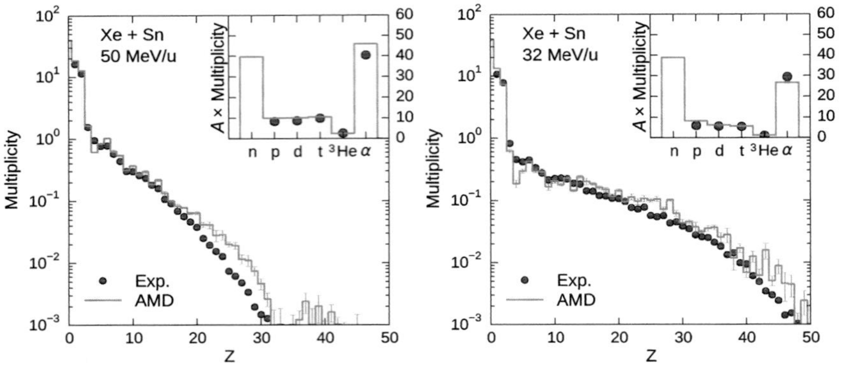

Fig. 4. Fragment charge distribution in central Xe + Sn collisions ($0 < b < 2$ fm) at the incident energies of 50 (left) and 35 (right) MeV/nucleon, calculated by AMD with cluster correlations. The inset shows the multiplicities of light particles multiplied by the mass number. The INDRA experimental data are from Ref. 9.

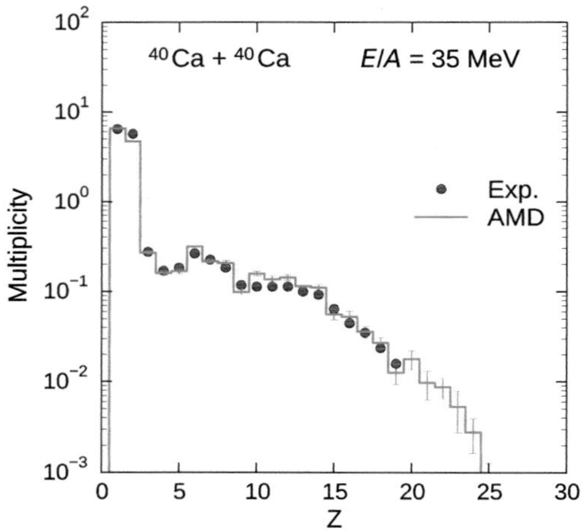

Fig. 5. Fragment charge distribution in violent events in $^{40}Ca + ^{40}Ca$ collisions at 35 MeV/nucleon. A filter for the experimental setup has been applied to the calculated events, and the violent events are selected by the same condition as for the experimental data.[37]

we consider the wide difference of the system sizes and the incident energies. The reproductions of these systems are much more consistent than with any other previous versions of AMD.

The yields of clusters and fragments represent the essential information on how many-body spatial correlations develop in expanding systems with

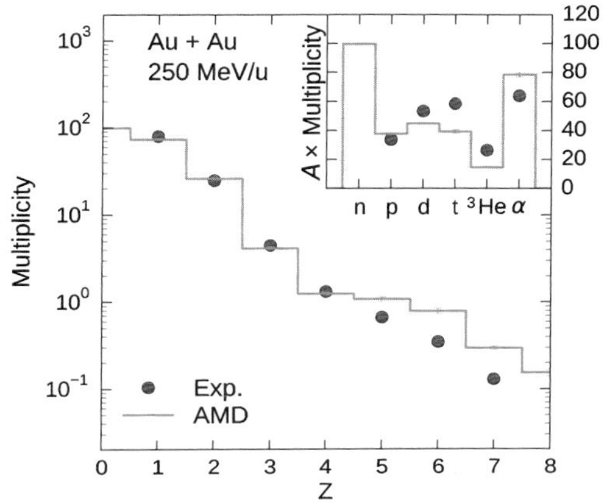

Fig. 6. Fragment charge distribution in central Au + Au collisions at 250 MeV/nucleon. The calculated result by AMD with cluster correlations is compared to the FOPI experimental data.[11]

various expansion speeds and/or thermal excitations. A consistent description of this feature is an important milestone to enable precise studies of other features, as in the following subsections.

3.3. *Neutron-to-proton ratio*

We discuss here a typical system of central ^{124}Sn + ^{124}Sn collisions at 50 MeV/nucleon based on the same calculation as in Ref. 36 but with improved statistics. Such a neutron-rich system is suitable to see the effects of the density dependence of the symmetry energy $S(\rho)$ or the parameter $L = 3\rho_0(dS/d\rho)_{\rho_0}$. In the late stage of the reaction at this energy, the calculation clearly shows that the symmetry energy influences the degree of fractionation, i.e., how much the gas part is neutron rich compared with the liquid part of the system.[7] Fragment nuclei with $Z \gtrsim 3$ are usually regarded as belonging to the liquid part, while the gas part consists of emitted nucleons and light clusters. Direct information of the liquid part may be obtained from the isotope distribution of fragment nuclei. However, in order to obtain a robust constraint on the L parameter from the comparison with experimental data, it is necessary to better understand the decay of primary fragments. On the other hand, direct information of the gas part should be observed in the yields and spectra of nucleons and light clusters (see Ref. 38 and references therein). For this purpose, it is evidently important to have a good theoretical description of cluster emission.

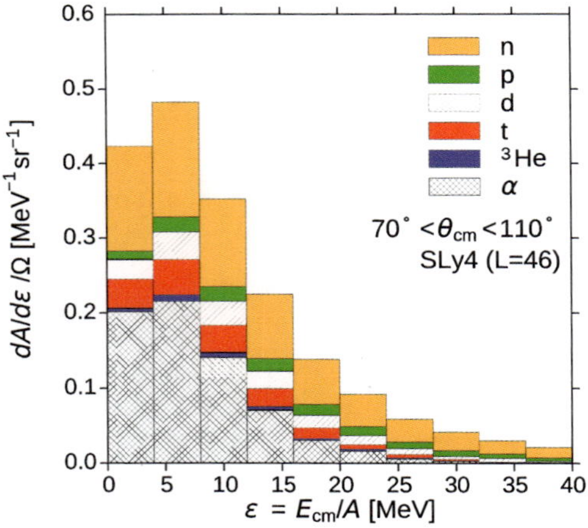

Fig. 7. The calculated mass-weighted distributions of the energies $\varepsilon = E/A$ of light particles emitted to the angles $70^\circ < \theta < 110^\circ$ in the center-of-mass system of $^{124}\mathrm{Sn} + {}^{124}\mathrm{Sn}$ central collisions at 50 MeV/nucleon.

In order to better understand the composition of the gas part, Fig. 7 shows the distributions of the kinetic energy per nucleon $\varepsilon = E_{\mathrm{cm}}/A$ of the nucleons and clusters emitted in transverse directions $70^\circ < \theta_{\mathrm{cm}} < 110^\circ$. Each colored area shows the number of the clusters multiplied by the cluster mass number A. Therefore the figure shows the decomposition of gas nucleons into clusters at each ε. The α cluster formation is quite important since about half of gas nucleons are bound in α particles, except for large ε. The neutron-to-proton ratio is $(N/Z)_{\mathrm{gas}} \approx 2$ for the sum of these particles ($A \leq 4$). This is significantly larger than the ratio $(N/Z)_{\mathrm{system}} = 1.48$ for the total system, which is understood as the effect of neutron fractionation/distillation. Since the same numbers of neutrons and protons are bound in many α particles (and deuterons), only a small number of protons and ^{3}He are left. Consequently the free n/p ratio and the $t/{}^3$He ratio can become very large, because of the simple relation

$$\frac{N_{\mathrm{gas}} - 2n_\alpha}{Z_{\mathrm{gas}} - 2n_\alpha} > \frac{N_{\mathrm{gas}}}{Z_{\mathrm{gas}}}, \tag{6}$$

with n_α being the number of α particles. Figure 8 shows that the calculated free n/p ratio is actually very large ($\sim$10 at low energies). This large ratio

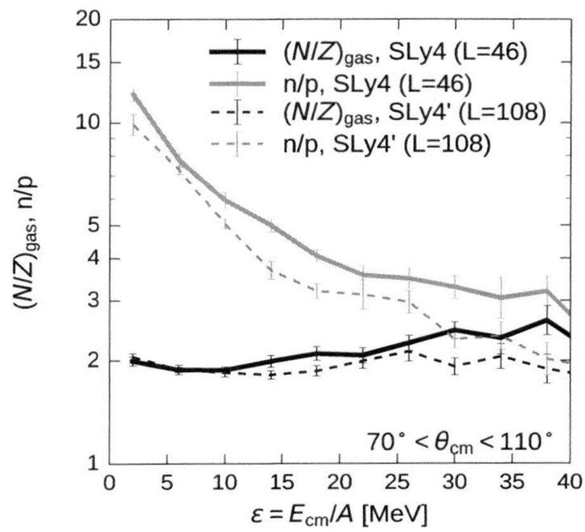

Fig. 8. The spectral n/p ratio of free neutrons and protons (red) and the $(N/Z)_{\rm gas}$ ratio for the sum of light particles (black). The lines show the calculated results by AMD with cluster correlations. The adopted effective interaction for the solid lines is the Skyrme SLy4 interaction corresponding to a relatively soft symmetry energy $L = 46$ MeV, and that for the dashed lines is a modified version of SLy4 which corresponds to a stiffer symmetry energy $L = 108$ MeV.

is consistent with what is implied by the experimental data of Ref. 38. The exact value of the n/p ratio is probably quite sensitive to the details of the formation of α clusters. On the other hand, the $(N/Z)_{\rm gas}$ ratio for the sum of these light particles takes a moderate value ($\sim$2) without a strong dependence on the energy ε.

In order to study the effect of the density dependence of the symmetry energy, Fig. 8 shows the results of two calculations with different density dependence of the symmetry energy. One is the Skyrme SLy4 interaction with $L = 46$ MeV and the other is a force corresponding to $L = 108$ MeV obtained by modifying the SLy4 force. Dependence on the L parameter is seen in the n/p and $(N/Z)_{\rm gas}$ ratios by comparing the solid and dashed lines. This effect is consistent with the stronger fractionation for the lower value of L which corresponds to the larger value of the symmetry energy $S(\rho)$ at $\rho \leq \rho_0$. It might be possible to obtain a constraint on L or $S(\rho < \rho_0)$ by the comparison with experimental data.

However, a comparison of the results of two different versions (Refs. 35 and 36) suggests that the neutron-to-proton ratio is also sensitive to the details of the cluster formation process and/or the treatment of the momentum fluctuation. Even though we are close to the realistic situation owing to the extension of AMD for cluster correlations, it is still important to check the agreement of model predictions and experimental data for various observables.

3.4. *Impacts of clusters on collision dynamics*

It is usually considered that clusters and fragments are formed in low-density matter produced as a consequence of the collective dynamics of heavy-ion collisions. Therefore, it is a natural and reasonable idea that clusters and fragments should be influenced by collision dynamics. However, we here consider the opposite relation. Namely, the question is whether the cluster correlations have impacts on collision dynamics.

Figure 9 shows the time evolution of number of nucleons in clusters with mass numbers 2, 3 and 4 in an AMD calculation with cluster correlations for Xe + Sn central collisions at 50 MeV/nucleon. The number of clusters can be counted by following the history of two-nucleon collisions which create and break clusters. The mean-field part of the equation motion can also break clusters due to the different forces to the nucleons in a cluster. In this approach, the time evolution does not depend on whether a group of nucleons are regarded as a cluster or just as nucleons that happen to be close in phase space, except for the procedure for the inter-cluster correlations. However, these counted numbers of clusters show that the cluster correlation is very strong at $t \gtrsim 50$ fm/c in such a way that more than 2/3

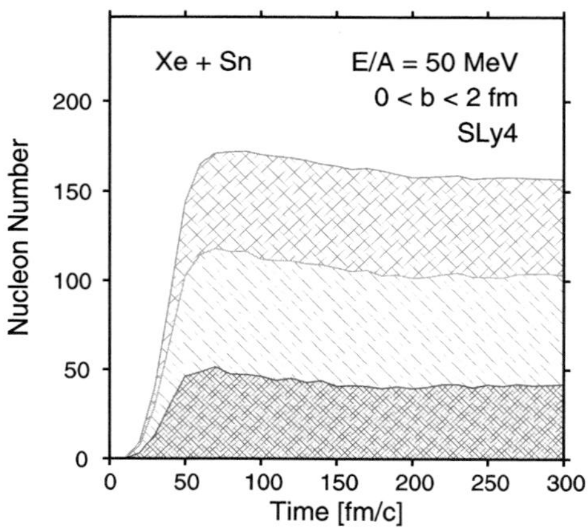

Fig. 9. Time evolution of the numbers of nucleons in clusters in Xe + Sn central collisions at 50 MeV/nucleon. The parts, from bottom to top, represent the number of nucleons in 4-, 3- and 2-nucleon clusters, respectively. Taken from Ref. 24.

of nucleons constitute clusters. The numbers are counted by assuming that there are no clusters at $t = 0$ in this case, though the numbers at $t \gtrsim 50$ fm/c do not depend very much on this assumption.

Such a strong cluster correlation can change the thermal and other properties of excited nuclear many-body systems. When correlated and uncorrelated systems are compared at the same energy, the temperature of the correlated system is significantly higher owing to the different numbers of degrees of freedom and the effect of cluster binding energies. In fact, the calculations of AMD with and without clusters show quite different evolution of global collision dynamics as presented in Ref. 24. The simple AMD without clusters cannot explain sufficient expansion and fragmentation of the system in most cases, while the AMD with clusters shows strong tendency of vaporization into the gas of light clusters. To obtain good reproductions of experimental data as in the previous subsections, the treatment of inter-cluster correlation was very important in addition to the cluster correlations.

Let us now turn to the central collisions of neutron-rich nuclei ^{132}Sn + ^{124}Sn at 300 MeV/nucleon. This system was recently studied by AMD in Ref. 25 to relate emitted pions and the dynamics of neutrons and protons which carries precious information of symmetry energy at high densities $\rho \sim 2\rho_0$.[39] The essential idea behind is that the π^-/π^+ ratio should be approximately related to the $(N/Z)^2$ ratio of nucleons in high density region. The effect of cluster correlations was studied as well as that of symmetry energy by comparing the results of AMD calculations with different options. A hadronic cascade model called JAM was used for the treatment of Δ resonances and pions on top of the nucleon dynamics calculated by AMD.

Figure 10 summarizes various isospin-related ratios for different AMD+JAM calculations, with soft ($L = 46$ MeV) or stiff ($L = 108$ MeV) symmetry energies and with or without cluster correlations in AMD. The simple JAM calculation without any mean field term is also shown for comparison. See Ref. 25 for the exact definitions of these ratios. The left-most column represents the neutron-to-proton squared ratio $(N/Z)^2_\rho$ in the high density region with $\rho > \rho_0$. The value of $(N/Z)^2_\rho$ is lower than the $(N/Z)^2_{\rm sys}$ ratio of the total system because of the symmetry energy effect that favors isospin symmetry ($N \approx Z$) of the high density part. The difference between soft and stiff symmetry energies, for each calculation with and without clusters, shows the degree of the sensitivity to the density dependence of symmetry energy. An important question is how this kind of symmetry

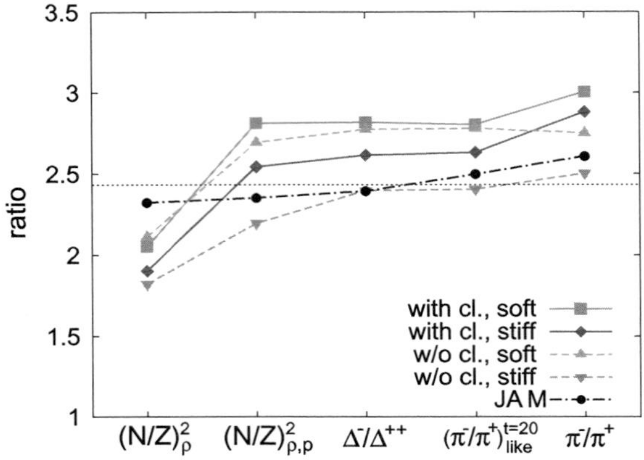

Fig. 10. The nucleon ratios $(N/Z)^2_\rho$ and $(N/Z)^2_{\rho,p}$ in high-density region without and with high-momentum condition, respectively, the Δ^-/Δ^{++} production ratio, the pion-like ratio at $t = 20$ fm/c, and the final π^-/π^+ ratio. Each line connects the ratios for each of the five cases of calculation for central collisions of ^{132}Sn + ^{124}Sn at 300 MeV/nucleon. The horizontal line represents the $(N/Z)^2_{\text{sys}}$ ratio of the total system. From Ref. 25.

energy effects in the nucleon dynamics may be reflected in the final observables such as the π^-/π^+ ratio that is shown in the right-most column of Fig. 10. The final pion ratio is larger than $(N/Z)^2_{\text{sys}}$ at this incident energy, as also observed in the Au + Au experiment[11] and in other transport calculations.[40–42] The value of the final pion ratio does not actually agree with the $(N/Z)^2_\rho$ ratio in high density region.

The understanding of the relation between the nucleon dynamics and the pion ratio was improved in Ref. 25 by looking at the phase space region of high density *and* high momentum. The high-momentum condition is natural because the Δ-resonance production is possible only if enough energy is available. The second column of Fig. 10 shows the $(N/Z)^2_{\rho,p}$ ratio calculated for the nucleons that exist in the high density region $\rho > \rho_0$ and have high momenta $|\boldsymbol{p} - \boldsymbol{p}_{\text{rad}}| > 480$ MeV/c, where the collective radial momentum $\boldsymbol{p}_{\text{rad}}$ has been subtracted. We can see that the high-momentum region is more neutron-rich in all the cases of calculations. Furthermore, this $(N/Z)^2_{\rho,p}$ ratio well agrees with the Δ^-/Δ^{++} ratio of the number of Δ production reactions, $nn \to p\Delta^-$ and $pp \to n\Delta^{++}$, as shown in the third column of the figure. This agreement is confirmed by the comparison of Fig. 11 and Fig. 12 which show the time evolution of these two kinds of ratios, respectively. The Δ^-/Δ^{++} production ratio also agrees with the

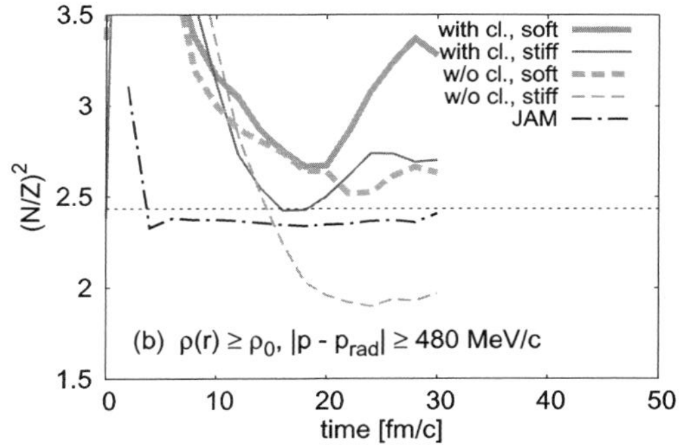

Fig. 11. The time evolution of the squared ratio of neutrons and protons $(N/Z)^2$ calculated for the selected nucleons that exist in the high density region $\rho(r) \geq \rho_0$ and have high momenta $|p - p_{\mathrm{rad}}| \geq 480$ MeV$/c$. From Ref. 25.

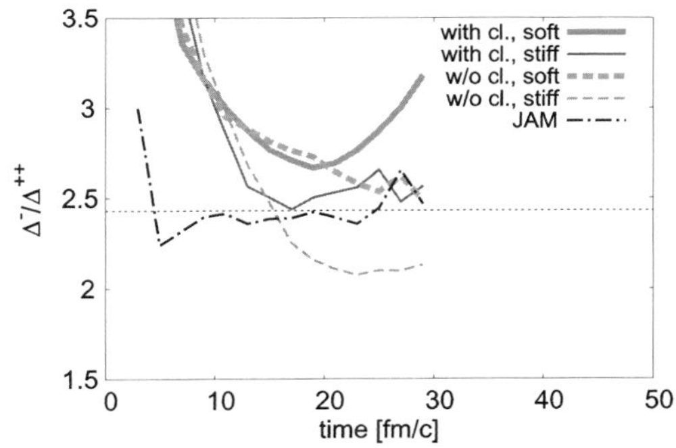

Fig. 12. The time evolution of the Δ^-/Δ^{++} ratio of the Δ production rates. The five different lines show the calculations. The horizontal line represents the $(N/Z)^2_{\mathrm{sys}}$ ratio of the total system. From Ref. 25.

$(\pi^-/\pi^+)_{\mathrm{like}}$ ratio at $t = 20$ fm$/c$ which is shown at the fourth column of Fig. 10. The pion-like numbers are counted taking into account the number of existing Δ resonances with the weights of the decay branching ratios. The time $t = 20$ fm$/c$ approximately corresponds to the maximum compression of the system. Due to the dynamics after $t = 20$ fm$/c$, the pion ratio is modified as shown in Fig. 10.

In these results of Ref. 25, effects of cluster correlations can be observed in some aspects. First of all, the sensitivity to the symmetry energy is weaker with clusters than without them in the neutron-to-proton ratios, $(N/Z)^2_\rho$

and $(N/Z)^2_{\rho,p}$ in Fig. 10. This is understandable because the cluster correlation forces some neutrons and protons to move together and therefore the symmetry forces acting differently on the neutrons and the protons are averaged out. By comparing the change from $(N/Z)^2_\rho$ to $(N/Z)^2_{\rho,p}$, we also notice that the high-momentum part is more neutron-rich when the cluster correlations are turned on. A plausible reason for this is as follows. If clusters are formed, they tend to have smaller momenta per nucleon on average than non-clustered nucleons, as in thermally equilibrated systems. In addition, clusters tend to contain similar number of neutrons and protons like α clusters. Therefore, the low-momentum part of the phase space becomes more symmetric ($N \approx Z$) and thus the high-momentum part becomes more neutron-rich owing to the cluster correlations if the numbers of neutrons and protons are conserved. We also notice that, when the pion ratio is modified from $(\pi^-/\pi^+)_{\text{like}}$ at $t = 20$ fm/c to the final π^-/π^+ ratio, the ratio is increased in the calculations with cluster correlations. This implies that the exterior region which the created pions have to pass through is more neutron-rich in the calculations with cluster correlations.

The information of the symmetry energy effects on nucleon dynamics should be reflected not only in the pion ratio but also in the nucleon observables. In the literature, however, the n/p spectral ratio has been found to be insensitive to the density dependence of the symmetry energy, while it is sensitive to the neutron-proton effective mass difference[43] based on the SMF calculation without cluster correlations. On the other hand, the present AMD calculation with cluster correlations shows that the symmetry-energy effect in the neutron-proton dynamics at early times is rather directly linked to the final N/Z spectral ratio of nucleons and light clusters.[35] The cluster correlations are essential in this respect. In fact, in the calculation of AMD without cluster correlations, we find the sensitivity to the symmetry energy is lost in the N/Z spectral ratio. This suggests that the expansion dynamics is largely influenced by cluster correlations, which is understandable because the number of particles (namely clusters and nucleons) is small if the cluster correlation is strong. The interaction between particles ceases at an earlier time with cluster correlations, while the interaction between nucleons continues longer in the calculations without clusters so that the final neutron and proton spectra are modified from those at early times. This effect is similar to what we saw in Sec. 2 in a different context in the comparison of AMD and SMF.[27]

4. Summary

Clusters are important in heavy-ion collisions, not only simply because they are emitted, but also because formation and existence of light clusters influence the global reaction dynamics and the nuclear matter properties. AMD has been extended to include cluster correlations in the final states of two-nucleon collisions. The inter-cluster correlations to bind several clusters should also be considered. A consistent reproduction of basic characters of fragmentation by recent calculations[36] is much more satisfactory than before. This suggests that the excited nuclear many-body systems realized in heavy-ion collisions have stronger cluster correlations than usually expected from single-particle motions in the mean field picture. Some observables, such as the n/p spectral ratio, that are supposed to be sensitive to the symmetry energy, have to be carefully analyzed by well understanding the influence of cluster correlations. A recent calculation with the AMD+JAM approach[25] showed that Δ resonances and pions are actually good probes for the neutron-proton dynamics at high densities in collisions of neutron-rich nuclei at 300 MeV/nucleon. We emphasize again that the cluster correlations have considerable impacts on the neutron-proton dynamics as well as the high-density symmetry energy does.

Acknowledgments

This work was supported by JSPS KAKENHI Grant Number 24105008. Numerical computation in this work was partly carried out at the Yukawa Institute Computer Facility.

References

1. "Dynamics and Thermodynamics with Nuclear Degrees of Freedom", P. Chomaz *et al.* (Eds.), *Eur. Phys. J. A* **30** (2006).
2. J.P. Bondorf, A.S. Botvina, A.S. Iljinov, I.N, Mishustin and K. Sneppen, *Phys. Rep.* **257**, 133 (1995).
3. D.H.E. Gross, *Phys. Rep.* **279**, 119 (1997).
4. P. Chomaz, M. Colonna and J. Randrup, *Phys. Rep.* **389**. 263 (2004).
5. J. Xu *et al.*, *Phys. Rev. C* **93**, 044609 (2016).
6. P.B. Gossiaux, R.K. Puri, Ch. Hartnack and J. Aichelin, *Nucl. Phys. A* **619**, 379 (1997).
7. A. Ono, *J. Phys.: Conf. Ser.* **436**, 012068 (2013).
8. B. Borderie and M.F. Rivet, *Prog. Part. Nucl. Phys.* **61**, 551 (2008).
9. S. Hudan *et al.*, *Phys. Rev. C* **67**, 064613 (2003).

10. W. Reisdorf *et al.*, *Nucl. Phys. A* **612**, 493 (1997).
11. W. Reisdorf *et al.*, *Nucl. Phys. A* **848**, 366 (2010).
12. B. Borderie *et al.*, *Eur. Phys. J. A* **6**, 197 (1999).
13. K. Sumiyoshi and G. Röpke, *Phys. Rev. C* **77**, 055804 (2008).
14. A. Ono, H. Horiuchi, T. Maruyama and A. Ohnishi, *Prog. Theor. Phys.* **87**, 1185 (1992).
15. A. Ono, S. Hudan, A. Chbihi and J. D. Frankland, *Phys. Rev. C* **66**, 014603 (2002).
16. A. Ono and H. Horiuchi, *Porg. Part. Nucl. Phys.* **53**, 501 (2004).
17. L.W. Chen, C.M. Ko and B.A. Li, *Nucl. Phys. A* **729**, 809 (2003).
18. P. Danielewicz and G.F. Beartsch, *Nucl. Phys. A* **533**, 712 (1991).
19. D.D.S. Coupland, W.G. Lynch, M.B. Tsang, P. Danielewicz and Y. Zhang, *Phys. Rev. C* **84**, 054603 (2011).
20. G. Röpke, *Nucl. Phys. A* **867**, 66 (2011).
21. G. Röpke, *Phys. Rev. C* **92**, 054001 (2015).
22. L. Qin *et al., Phys. Rev. Lett.* **108**, 172701 (2012).
23. M. Hempel, K. Hagel, J. Natowitz, G. Röpke and S. Typel, *Phys. Rev. C* **91**, 045805 (2015).
24. A. Ono, *J. Phys.: Conf. Ser.* **420**, 012103 (2013).
25. N. Ikeno, A. Ono, Y. Nara and A. Ohnishi, *Phys. Rev. C* **93**, 044612 (2016).
26. J. Rizzo, M. Colonna and A. Ono, *Phys. Rev. C* **76**, 024611 (2007).
27. M. Colonna, A. Ono and J. Rizzo, *Phys. Rev. C* **82**, 054613 (2010).
28. M. Colonna, M. Di Toro, A. Guarnera, S. Maccarone, M. Zielinska-Pfabé and H.H. Wolter, *Nucl. Phys. A* **642**, 449 (1998).
29. J. Rizzo, M. Colonna, M. Di Toro and V. Greco, *Nucl. Phys. A* **732**, 202 (2004).
30. P. Napolitani and M. Colonna, *Phys. Lett. B* **726**, 382 (2013).
31. J.D. Frankland *et al., Nucl. Phys. A* **689**, 940 (2001).
32. A. Ono and H. Horiuchi, *Phys. Rev. C* **53**, 2958 (1996).
33. Y. Kanada-En'yo, M. Kimura and A. Ono, *Prog. Theor. Exp. Phys.* **2012**, 01A202 (2012).
34. A. Ono, *J. Phys.: Conf. Ser.* **569**, 012086 (2014).
35. A. Ono, *EPJ Web of Conferences* **117**, 07003 (2016).
36. A. Ono, *in Proceedings of the 5th International Workshop on Compound-Nuclear Reactions and Related Topics, Tokyo, Japan, October 19–23, 2015,* to be published in EPJ Web of Conferences.
37. K. Hagel *et al.*, *Phys. Rev. C* **50**, 2017 (1994).
38. D.D.S. Coupland *et al.*, *Phys. Rev. C* **94**, 011601(R) (2016).
39. B.A. Li, *Phys. Rev. Lett.* **88**, 192701 (2002).
40. Z. Xiao, B. A. Li, L. W. Chen, G.C. Yong and M. Zhang, *Phys. Rev. Lett.* **102**, 062502 (2009).
41. Z. Q. Feng and G. M. Jin, *Phys. Lett. B* **683**, 140 (2010).
42. J. Hong and P. Danielewicz, *Phys. Rev. C* **90**, 024605 (2014).
43. V. Giordano, M. Colonna, M. Di Toro, V. Greco and J. Rizzo, *Phys. rev. C* **81**, 044611 (2010).

CHAPTER 17

MOLECULAR STRUCTURES IN SLOW NUCLEAR COLLISIONS*

Alexis Diaz-Torres
Department of Physics,
University of Surrey, Guildford, GU2 7XH, UK
a.diaztorres@surrey.ac.uk

A modern two-center shell model and its usefulness for addressing low-energy reaction dynamics of light and heavy nuclei are presented. A perspective for further developments in nuclear reaction theory is given.

1. Introduction

The physics of low-energy nuclear reactions is crucial for understanding energy production and nucleosynthesis in the Universe.[1] I have been interested in this field since my years as a graduate student[2] in Cuba (1994–1997), as it integrates many-body nuclear structure, reaction dynamics and mechanisms as well as quantum mechanics.

The two-center shell model (TCSM) has been a key tool for my studies on reaction theory. This concept, which was first introduced in practice by the Frankfurt school in heavy-ion physics,[3–5] is very useful and has helped me understand, from a microscopical perspective, the low-energy reaction dynamics of complex nuclei and its impact on observables such as mass and charge distributions of reaction products.[6] The TCSM combined with Strutinsky's method provides collective potential-energy landscapes (PES) which a nuclear molecule or dinuclear system (DNS) formed in low-energy collisions may explore.[7] The two-center single-particle (sp) levels and the dynamics at their avoided crossings also determine the collective inertia and, consequently, the kinetic energy of the DNS in the PES. I used both the

*To the memory of Prof. Dr. Walter Greiner

adiabatic and diabatic TCSM during my PhD work[8] in Giessen (1998–2000) for investigating the basic assumptions of the DNS model of fusion.[9] The diabaticity of the sp motion in the entrance phase of a heavy-ion collision produces a short-range repulsive nucleus-nucleus potential,[10] hampering the fusion of the nuclei along the internuclear radius.[11,12] The TCSM based cranking-mass parameter for the neck coordinate of the DNS appears to be much larger than the corresponding hydro-dynamical value, justifying a small radius of the neck between the touching nuclei.[13] These findings support the correctness of the DNS concept, which explains many observations.[14] I have also used the TCSM for addressing the reaction dynamics of light nuclei, such as ^{9}Be, ^{12}C and ^{16}O.[15–17] I have developed the TCSM further,[17,18] using spherical and arbitrarily oriented deformed Woods-Saxon potentials, and have recently applied it to understanding and quantifying the sub-Coulomb fusion of ^{12}C+^{12}C with the time-dependent wave-packet (TDWP) method.[19] The TDWP method also seems to be useful for addressing low-energy heavy-ion collisions forming heavy and super-heavy nuclei. The role of decoherence and dissipation in the reaction dynamics can be studied with the coupled-channels density-matrix (CCDM) method,[20,21] which is based on the Lindblad equation for the density matrix of an open quantum system. The usefulness of the Lindblad equation for solving problems in heavy-ion physics was first pointed out by Sandulescu *et al.* in Refs. 22.

The present contribution provides a survey of my recent works which use the nuclear molecular picture. Various quantum dynamical reaction models based on this picture will be presented, highlighting the crucial role of the dinuclear structure formed by the strongly interacting nuclei (ranging from light to heavy ions) in the fusion process. In my opinion, this picture is useful for guiding and interpreting measurements with low-energy exotic beams at new generation facilities, such as FRIB and FAIR.

2. The two-center shell model

The TCSM is a basic microscopic model to describe the sp motion in a heavy-ion collision close to the Coulomb barrier. It is physically justified, provided the relative motion of the colliding nuclei is much slower than the sp motion in the two mean-field potentials.[5] The sp states are determined solving the Schrödinger equation with a phenomenological two-center potential. Most applications in fission, cluster radioactivity, fusion and heavy-ion collisions have been based on a double oscillator potential.[3–5]

Improved versions of TCSM based on oscillator potentials were developed for treating either very asymmetric fission[23] or asymmetric fission with deformed fragments.[24] The TCSM problem with realistic finite depth potentials has been solved with the wave function expansion method (diagonalization procedure) in Refs. 25–27. This issue can also be solved with the potential separable expansion (PSE) method.[28] Using the rigorous PSE method I have developed a TCSM for fusion,[17,18] in which both spherical and arbitrarily oriented deformed Woods-Saxon (WS) potentials can be employed. This TCSM is realistic regarding bound and continuum sp states, and is computationally demanding for dealing with heavy systems.

2.1. *The TCSM with woods-saxon potentials*

The TCSM potential for neutrons can be constructed by the superposition of two shifted and rotated WS potentials, which for protons includes Coulomb potentials as well.[17,18] For instance, the neutron two-center potential reads as:

$$V = \sum_{s=1}^{2} \exp(-i\,\mathbf{R}_s\,\hat{k})\,\hat{U}(\Omega_s)\;V_s\;\hat{U}^{-1}(\Omega_s)\;\exp(i\,\mathbf{R}_s\,\hat{k}), \qquad (1)$$

where $\hat{k} = \hbar^{-1}\hat{p}$ is the sp wave-number operator. The centers are located at $\mathbf{R}_1$ and $\mathbf{R}_2$, the relative coordinate is $\mathbf{R} = \mathbf{R}_1 - \mathbf{R}_2$, and $\hat{U}(\Omega_s)$ are the corresponding rotation operators with the Euler angles Ω_s. The WS potentials, V_s, which include spin-orbit interactions, are approximately represented within a truncated sp harmonic oscillator basis.[17,18] To describe the fusion process, the WS potential parameters are interpolated between their values for the separated nuclei and the spherical compound nucleus, making use of the condition of volume conservation for an equipotential surface that determines the nuclear shapes.[18] The parameters of the asymptotic WS potentials including the spin-orbit term reproduce the experimental single-particle energy levels around the Fermi surface of the colliding nuclei, whereas for the spherical compound nucleus the parameters of the global WS potential by Soloviev[29] are used, its depth being adjusted to reproduce the experimental single-particle separation energies.[30] The two-center problem is solved in the momentum representation by means of a set of linear algebraic equations,[17,18] the sp states being determined by the zero values of its Fredholm determinant.

As an example, Fig. 1 presents the (adiabatic) neutron energy levels including all bound states with different magnetic sub-states for ^{16}O +

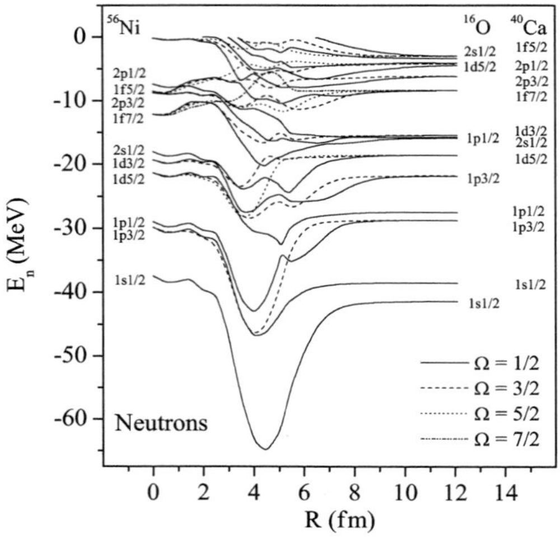

Fig. 1. The adiabatic neutron-levels correlation diagram for $^{16}O + {}^{40}Ca \rightarrow {}^{56}Ni$, which is calculated with the TCSM using two spherical Woods-Saxon potentials. See Ref. 18 for further details.

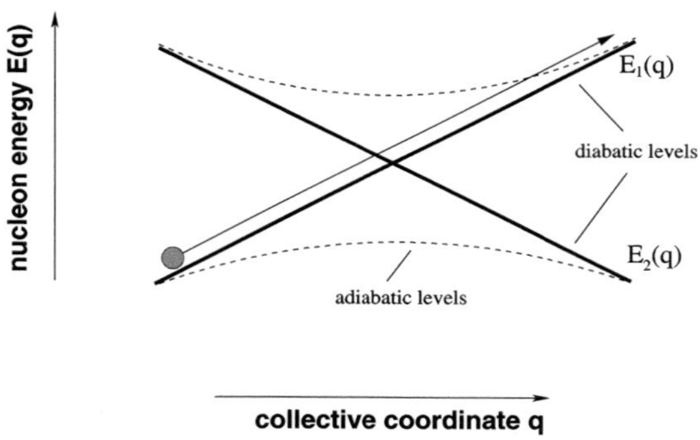

Fig. 2. Illustration of the diabatic single-particle motion (solid curve) at an avoided crossing of two adiabatic single-particle levels (dotted curves) with the same quantum numbers as those shown in Fig. 1.

$^{40}Ca \rightarrow {}^{56}Ni$. Most of the avoided crossings in the adiabatic level diagram turn into real crossings in the diabatic sp motion, as illustrated in Fig. 2. Adiabatic sp states diagonalize the two-center sp Hamiltonian, whereas the diabatic states minimize the strong dynamical non-adiabatic radial coupling[18] at an avoided crossing between two adiabatic sp levels with the same symmetry. Diabatic states with the same quantum numbers can cross each other because they are not solutions of an eigenvalue problem.

3. The dynamical PES in heavy element formation

The solution of the two-center problem provides the *microscopic* ingredients (sp energies and wave-functions) to calculate the *macroscopic* quantities (collective PES, transport coefficients, etc.) that determine the dynamical evolution of the nuclear shapes.

The total (time-dependent) collective PES of the fusing system can be written[7] as the sum of a reference adiabatic potential, which is calculated with Strutinsky's macroscopic-microscopic method, along with a centrifugal part due to collective rotation, and a diabatic contribution due to particle-hole excitations in the entrance phase of the reaction. The macroscopic liquid drop part of the potential energy is calculated with the Yukawa plus exponential method,[31] while the shell corrections to the ground-state energy are calculated using a novel method.[32] The diabatic contribution is determined using the adiabatic and diabatic orbitals resulting from the solution of the two-center problem and the occupation numbers of these states.

The occupation numbers of the diabatic sp orbitals, which determine the diabatic contribution to the collective PES, are obtained solving a linearized relaxation equation,[12] while the adiabatic occupations obey an equilibrated Fermi distribution for a finite temperature as the fusing system heats up during the relaxation of the dynamical collective PES. The diabatic contribution is initially maximal, but gradually decreases when the diabatic sp occupations approach the adiabatic occupations. The dynamical collective PES describes a continuous transition from the initial diabatic collective PES to the asymptotic adiabatic one, and this concept was first suggested in Ref. 33.

Figure 3 shows the initial diabatic and the asymptotic adiabatic PES for $^{48}Ca + ^{208}Pb$ ($\eta_0 = 0.625$) $\rightarrow$ ^{256}No, where all the DNS nuclei forming ^{256}No are considered spherical.[7] The entrance mass asymmetry η_0 determines the initial diabatic PES. Shell effects are not only manifested in the collective PES by means of the static ground-state shell corrections, but also by means of the diabatic contribution that results from the sp motion through the avoided crossings of the shell structure of the different nuclear shapes (see Fig. 2).

Following capture (touching configuration), the relaxation equation describing the occupation numbers of the diabatic sp orbitals is coupled to

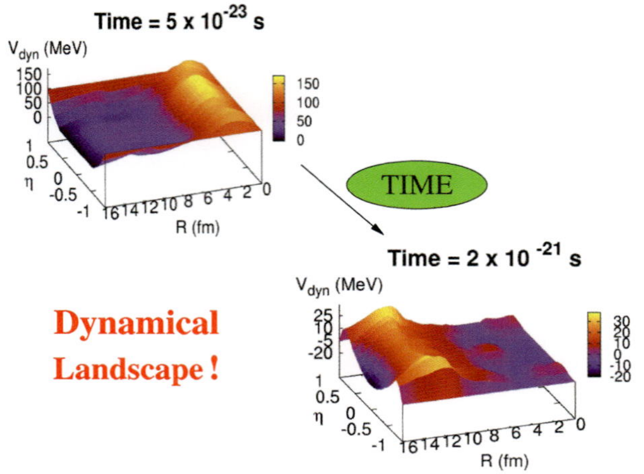

Fig. 3. Dynamical collective PES for the central collision ^{48}Ca + ^{208}Pb ($\eta_0 = 0.625$) $\rightarrow$ ^{256}No, as a function of the mass asymmetry and the internuclear distance. See Ref. 7 for further details.

the Pauli master equation[7] that describes the time-dependent probability of the fusing system to occupy each one of a discrete set of nuclear shapes (for practical reasons the collective space of shape coordinates is discretized), as shown in Fig. 4. Figure 5 illustrates a 2D-model for spherical nuclei, in which a mesh is defined in terms of the radius R between the nuclei and the mass asymmetry coordinate η describing different fragmentations of the united system. A node of the mesh is related to a nuclear shape. The model may be extended to more dimensions (*hyper-mesh*) with the inclusion of other relevant collective coordinates.

The space of nuclear shapes can be divided into three regions, e.g., see Fig. 5: (i) region of compact shapes around the near-spherical shape of the compound nucleus (*fusion region*), (ii) region of separated fragments beyond the Coulomb barrier (*quasi-fission region*), and (iii) region of intermediate shapes which could lead to fusion or quasi-fission (*competition region*).

The line or surface defining the fusion region is determined by a set of collective coordinates $\mathbf{q}$ from which the shell structure of the near-spherical compound nucleus manifests.[7] This can be clearly identified studying the two-center single-particle levels diagram.

Knowing the solution of the master equation,[7] the time-dependent probability for compound nucleus formation P_{CN} (*population of the fusion*

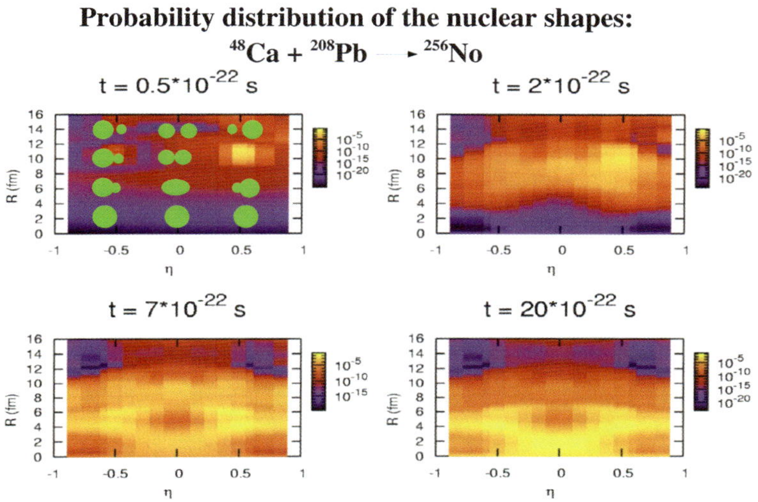

Fig. 4. Time-dependent probability distribution of the nuclear shapes for ^{48}Ca + ^{208}Pb ($\eta_0 = 0.625$) $\rightarrow$ ^{256}No, as a function of the mass asymmetry and the internuclear distance. See Ref. 7 for further details.

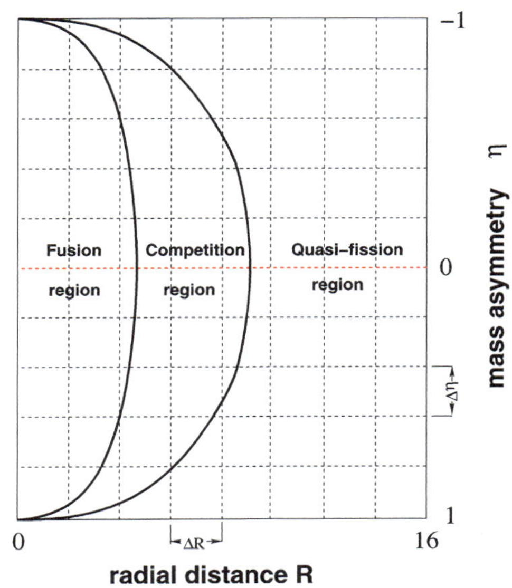

Fig. 5. Illustration of a 2D-mesh defined in terms of the separation R between the nuclei and the mass asymmetry coordinate η describing different fragmentations of the united system. A node of the mesh represents a nuclear shape. The thick solid curves separate the fusion, competition and quasi-fission regions. See Ref. 7 for further details.

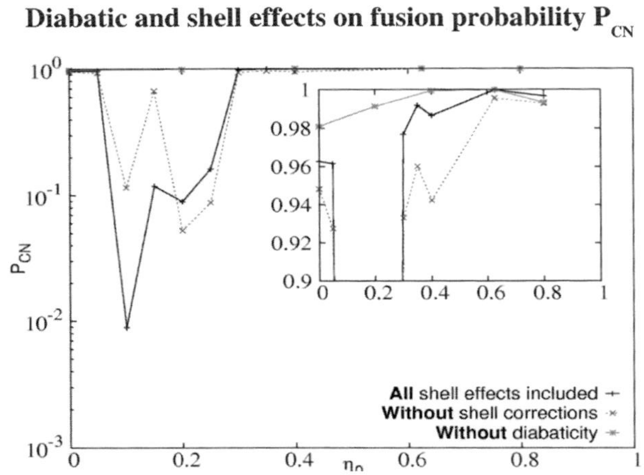

Fig. 6. P_{CN} as a function of the entrance mass asymmetry η_0 for central collisions forming the ^{256}No compound nucleus (CN). See Ref. 7 for further details.

region) and the quasi-fission probability P_{QF} (*population of the quasi-fission region*) can be easily calculated. The time scale for fusion-quasi-fission is obtained from the condition that, following capture, the initial (unit) probability of the colliding nuclei occupying the contact configuration (located in the competition region) becomes zero, as the probability distribution bifurcates into the fusion and the quasi-fission regions.

Figure 6 depicts the P_{CN} value as a function of the initial target-projectile mass asymmetry η_0. It appears that (i) the effect of diabaticity in suppressing P_{CN} can be very strong in near-symmetric collisions, and (ii) the ground-state shell corrections can play a key role in establishing the P_{CN} value.

The mass and charge distributions of quasi-fission (experimentally accessible quantities) can be calculated by projecting[7] the quasi-fission probability P_{QF} on the mass and charge asymmetry coordinates, respectively. Direct measurements of the quasi-fission yields along with P_{CN} would help clarify the correct fusion scenario.

4. The astrophysical S-factor of key heavy-ion systems

The deep sub-Coulomb fusion cross section, σ_{fus}, of reactions involving ^{12}C and ^{16}O is critical for calculating astrophysical reaction rates in dying massive stars, in which ^{16}O + ^{16}O and ^{12}C + ^{12}C are important reactions for the oxygen and carbon burning phases, respectively. This cross section is usually represented by the S-factor ($S = \sigma_{fus} E e^{2\pi\eta}$, where η is the Sommerfeld parameter), as it facilitates the extrapolation of high-energy fusion data because direct experiments at very low energies are very difficult to carry out. Unfortunately, there is a huge uncertainty in the S-factor stemmed from the extrapolation of different phenomenological parametrizations that explain the high-energy data, as shown for instance in Ref. 34 for ^{16}O + ^{16}O. For ^{12}C + ^{12}C, the presence of pronounced resonance structures makes it much more uncertain.[35–39] These extrapolated values lead to reactions rates that differ by many orders of magnitude.[37–39] Therefore, a direct calculation of the S-factor at energies of astrophysical interest (< 3 MeV) is very important.

4.1. $^{16}O + {}^{16}O$

The adiabatic collective PES is obtained with Strutinsky's method, while the radial-dependent collective mass parameter is calculated with the cranking mass formula.[16] For simplicity, the pairing contribution to the collective potential and radial mass has been neglected. The rotational moment of inertia of the DNS is defined as the product of the cranking mass and the square of the internuclear distance. The macroscopic part of the potential results from the finite-range liquid drop model[31] and the nuclear shapes of the TCSM.[18] The microscopic shell corrections to the potential are calculated with the method of Ref. 32. The TCSM is used to calculate the neutron and proton energy levels,[18] E_i, as a function of the separation R between the nuclei along with the radial coupling[18] between these levels that appears in the numerator of the cranking mass expression:

$$M(R) = 2\hbar^2 \sum_{i=1}^{A} \sum_{j>A} \frac{|\langle j|\partial/\partial R|i\rangle|^2}{E_j - E_i}. \tag{2}$$

Figure 7(a) shows the s-wave molecular adiabatic potential (thick solid curve) as a function of the internuclear separation, which is normalized with the experimental Q-value of the reaction ($Q = 16.54$ MeV). The sequence of nuclear shapes related to this potential is also presented. For comparison we show the Krappe-Nix-Sierk (KNS) potential[40] (thin solid curve) and the empirical Broglia-Winther (BW) potential[40] (dotted curve). Effects of neck between the interacting nuclei, before they reach the geometrical contact separation (arrow), are not incorporated into the KNS potential. The

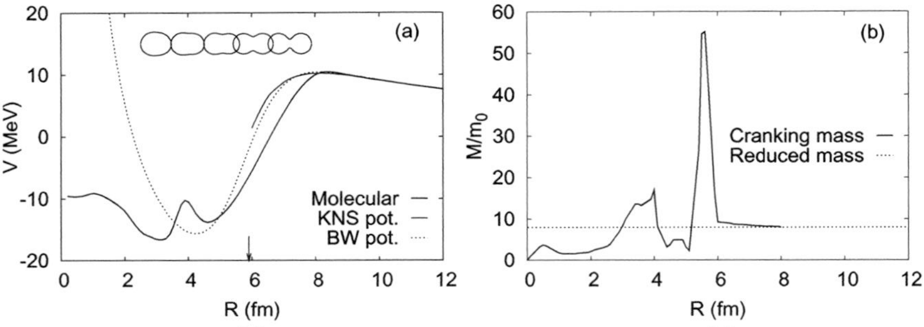

(a) (b)

Fig. 7. (a) The collective PES as a function of the internuclear radius for the central collision $^{16}O + {}^{16}O$. The arrow indicates the geometrical touching radius. (b) The radial collective mass parameter (in units of nucleon mass m_0).

concept of nuclear shapes is not embedded in the BW potential which tends to be similar to the KNS potential. Comparing the KNS potential to the molecular adiabatic potential we note that the neck formation substantially decreases the potential energy after passing the barrier radius ($R_b = 8.4$ fm). The inclusion of neck effects is crucial to successfully explain the existing S-factor data.[41]

Figure 7(b) shows the radial-dependent cranking mass (thick solid curve), while the asymptotic reduced mass is indicated by the dotted line. Just passing the barrier radius, when the neck between the nuclei starts to develop, the cranking mass slightly increases compared to the reduced mass and pronounced peaks appear inside the touching configuration. For $^{16}O + {}^{16}O$, these peaks are mainly caused by the strong change of the sp wave functions during the rearrangement of the shell structure of the asymptotic nuclei into the compound-system shell structure. The peaks could also be due to avoided crossings between the adiabatic molecular sp states,[18] which can make the denominator of the cranking mass expression (4) very small. The amplitude of these peaks may be reduced by (i) the pairing correlation that spreads out the sp occupation numbers around the Fermi surface, and (ii) the diabatic sp motion at avoided crossings, which can change those populations. For the strongest peak in Fig. 7(b), which is located very close to the internal turning point for a wide range of sub-barrier energies, only aspect (i) may be relevant as the radial velocity of the nuclei is rather small there, suppressing the Landau-Zener transitions. For compact shapes, aspect (ii) may lead to intrinsic excitation of the composite system, but this is not important here for the calculation of the fusion cross section which is determined by the barrier-penetration model.

Having the adiabatic potential and the adiabatic mass parameter, the radial Schrödinger equation is exactly solved with the modified Numerov method and the ingoing wave boundary condition imposed inside (about 2 fm) the Coulomb barrier. The fusion cross section σ_{fus} is calculated taken into account the identity of the interacting nuclei and the parity of the wave function for the relative motion (only even partial waves L are included here), i.e., $\sigma_{fus} = \pi\hbar^2/(\mu E)\sum_L(2L+1)\mathcal{T}_L$, where μ is the asymptotic reduced mass, E is the center-of-mass incident energy and $\mathcal{T}_L$ is the partial transmission coefficient.

Figure 8 shows the S-factor excitation curve. For a better presentation, the experimental data of each set[41] are binned into $\Delta E = 0.5$ MeV energy intervals. In this figure the following features can be observed:

(i) the molecular adiabatic potential in Fig. 7(a) correctly (thick and thin solid curves) explains the measured data, in contrast to either the results obtained with the BW potential (dotted curve) or the calculations within the Fermionic Molecular Dynamics (FMD) approach[16] (dashed curve). Since the width of the barrier decreases for the molecular adiabatic potential in Fig. 7(a), it yields larger fusion cross sections than those arising from the shallower KNS and BW potentials.

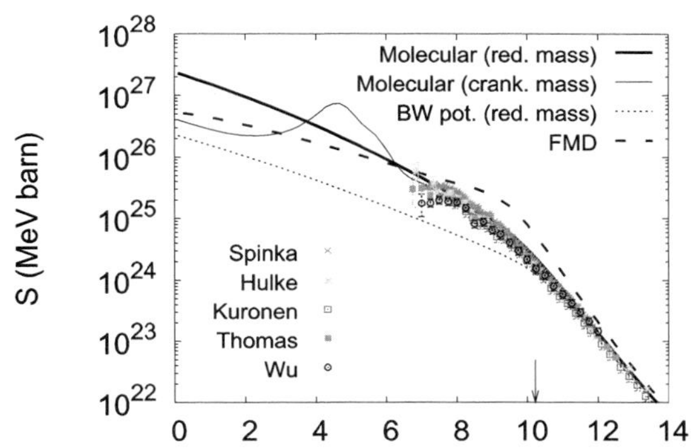

Fig. 8. (Color online) The S-factor excitation function for $^{16}O + ^{16}O$. The theoretical curves are compared with experimental data. The arrow indicates the Coulomb barrier of the molecular potential in Fig. 7(a).

(ii) the use of the cranking mass parameter in Fig. 7(b) substantially impacts on the low energy S-factor, which is revealed by the comparison between the thick and thin solid curves. It starts reducing the S-factor around 7-8 MeV energy region and produces a local maximum around 4.5 MeV. At the lowest incident energies (below 4 MeV) the S-factor is suppressed by a factor of five compared to that arising from a constant reduced mass. The peak in the S-factor is due to an increase of the fusion transmission coefficient, which is caused by the resonant behavior of the collective radial wave function at small internuclear distances inside the Coulomb barrier radius, as shown in Fig. 4 in Ref. 16.

4.2. $^{12}C + {}^{12}C$

The sub-Coulomb fusion of ^{12}C + ^{12}C has been recently addressed with the coupled channels method.[42,43] These calculations indicate important effects of both the low-lying low-density energy spectrum of ^{24}Mg and the ^{12}C Hoyle state on the low-energy fusion cross section. The theoretical fusion excitation curves are smooth, without resonant structures. This feature seems to be due to the use of a strong absorption model that does not include the physics of intermediate structure (nuclear molecule).[5] The key role of intermediate structure in fusion can be addressed with a new quantitative approach based on the TDWP method.[19]

4.2.1. *The TDWP method*

The TDWP method involves three steps[45]:

(1) the definition of the initial wave function $\Psi(t = 0)$,
(2) the propagation $\Psi(0) \to \Psi(t)$, dictated by the time evolution operator, $\exp(-i\hat{H}t/\hbar)$, where $\hat{H}$ is the total Hamiltonian that is time-independent,
(3) after a long propagation time, the calculation of observables (cross sections, spectra, etc) from the time-dependent wave function, $\Psi(t)$.

The wave function and the Hamiltonian are represented in a multi-dimensional numerical grid. For ^{12}C + ^{12}C, these are considered a function of five collective coordinates : the internuclear distance R, and the (θ_1, ϕ_1) and (θ_2, ϕ_2) spherical angles of the ^{12}C nuclei symmetry axis, thus reducing the complexity of the quantum many-body reaction problem. Moreover, the wave function is *not* expanded in any intrinsic basis (e.g., rotational or

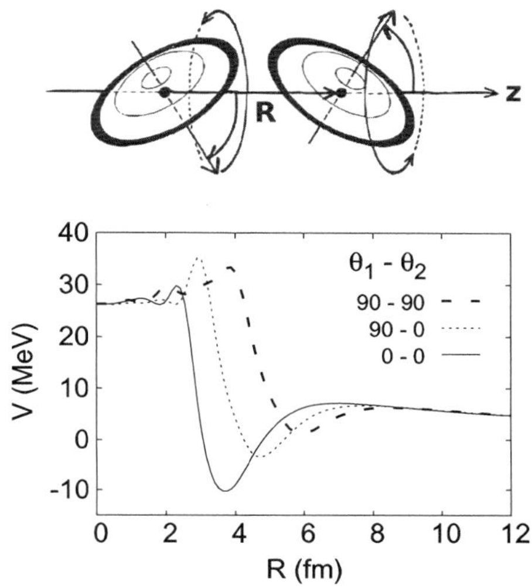

Fig. 9. Cuts in the collective potential-energy landscape for the $^{12}C + ^{12}C$ nuclear molecule as a function of the internuclear distance and alignment, $V(R, \theta_1, \phi_1 = 0, \theta_2, \phi_2 = 0)$. The 90-90 alignment (dashed line) facilitates the access by tunneling to the potential pockets. These are explored by the system, guided by the kinetic-energy operator which drives non-axial symmetric configurations towards the potential pocket of 0-0 alignment (solid line), where fusion happens by a strong absorption.

vibrational states of the individual nuclei), but it is calculated directly. The outgoing-wave-boundary condition at large internuclear distances as well as the *irreversible* process of fusion at small internuclear distances (usually described with an ingoing-wave-boundary condition) are here treated with the absorbing-boundary-condition.[44] The low-energy collision is described in the rotating center-of-mass frame within the nuclear molecular picture.[5]

Figure 9 shows the $^{12}C + ^{12}C$ nuclear molecule (upper panel) along with the collective potential-energy landscape (lower panel) as a function of the internuclear distance and the alignment between the two deformed nuclei. The potential curves are presented for fixed orientation of the two ^{12}C intrinsic symmetry axis relative to the internuclear axis; the three axes are coplanar. The potential of the non-axial symmetric configurations very weakly depends on the azimuthal angle of the ^{12}C intrinsic symmetry axis. The collective potential energy has been calculated with the finite-range liquid-drop model[31] using nuclear shapes of a realistic two-center shell model.[17] The large oblate deformation of ^{12}C ($\beta_2 = -0.5$ and moment of inertia $I = 0.67\,\hbar^2\text{MeV}^{-1}$ which is consistent with the observed 2^+

excitation energy [$E_{2^+} = \frac{\hbar^2}{2I}\,2(2+1) = 4.43$ MeV]) results in a continuous of Coulomb barriers and potential pockets, which are distributed in a range of radii ($R = 4-8$ fm). The lowest barrier ($90-90$ alignment) favors the initial approach of ^{12}C nuclei which must re-orientate in order to get trapped in the deepest pocket of the potential ($0-0$ alignment) where fusion occurs. In transit to fusion, the ^{12}C + ^{12}C nuclear molecule can populate quasi-stationary (doorway) states belonging to the shallow potential pockets of the non-axial symmetric configurations (shape coexistence). These doorway states may also decay into scattering states, instead of feeding fusion, as the ^{12}C nuclei largely keep their individuality within the molecule.[17] The complex motion of the ^{12}C + ^{12}C nuclear molecule through the potential energy landscape is driven by the kinetic energy operator that includes Coriolis interaction between the total angular momentum of the system and the intrinsic angular momentum of the ^{12}C nuclei. We use an exact expression of the kinetic energy operator.[46]

Having determined the total collective Hamiltonian of the ^{12}C + ^{12}C system in terms of the radial coordinate, R, and the spherical angles of the ^{12}C symmetry axis, θ_i and ϕ_i, the time propagation of an *initial* wave function has been carried out using the modified Chebyshev propagator for the evolution operator.[47] The initial wave function is determined when the ^{12}C nuclei are far apart at their ground-states (0^+), the radial and the internal coordinates being decoupled:

$$\Psi_0(R,\theta_1,k_1,\theta_2,k_2) \;=\; \chi_0(R)\,\psi_0(\theta_1,k_1,\theta_2,k_2), \tag{3}$$

where k_i are conjugate momenta of the ϕ_i azimuthal angles, so (3) is in a mixed representation. The radial component, $\chi_0(R)$, is considered a Gaussian wave-packet, while $\psi_0(\theta_1,k_1,\theta_2,k_2)$ is the internal symmetrized wave function due to the exchange symmetry of the system:

$$\begin{aligned}\psi_0(\theta_1,k_1,\theta_2,k_2) = &\big[\zeta_{j_1,m_1}(\theta_1,k_1)\zeta_{j_2,m_2}(\theta_2,k_2)\\ &+(-1)^J\zeta_{j_2,-m_2}(\theta_1,k_1)\zeta_{j_1,-m_1}(\theta_2,k_2)\big]\\ &/\sqrt{2+2\,\delta_{j_1,j_2}\delta_{m_1,-m_2}},\end{aligned} \tag{4}$$

where $\zeta_{j,m}(\theta,k) \;=\; \sqrt{\frac{(2j+1)(j-m)!}{2\,(j+m)!}}\;P_j^m(\cos\theta)\,\delta_{km}$, and P_j^m are associated Legendre functions. At the ^{12}C ground-state, $j_1 = j_2 = 0$ and $m_1 = m_2 = 0$. Please note that the radial and internal coordinates are strongly coupled when the ^{12}C nuclei come together, so a product state like Eq. (3) is only justified asymptotically.

Since the initial radial wave-packet, $\chi_0(R)$, contains different translational energies, an energy projection method is required. The energy-resolved scattering information can be obtained using a window operator.[48] The key idea is to calculate the energy spectrum, $\mathcal{P}(E_k)$, of the initial and final wave functions. E_k is the centroid of a *total energy bin* of width 2ϵ. A vector of reflection coefficients, $\mathcal{R}(E_k)$, is determined by the ratio[45]:

$$\mathcal{R}(E_k) = \frac{\mathcal{P}^{final}(E_k)}{\mathcal{P}^{initial}(E_k)}. \tag{5}$$

The transmission coefficients are:

$$\mathcal{T}(E_k) = 1 - \mathcal{R}(E_k). \tag{6}$$

The energy spectrum $\mathcal{P}(E_k) = \langle\Psi|\hat{\Delta}|\Psi\rangle$, where $\hat{\Delta}$ is the window operator[48]:

$$\hat{\Delta}(E_k, n, \epsilon) \equiv \frac{\epsilon^{2^n}}{(\hat{\mathcal{H}} - E_k)^{2^n} + \epsilon^{2^n}}, \tag{7}$$

$\hat{\mathcal{H}}$ is the system asymptotic Hamiltonian, and n determines the shape of the window function. As n is increased, this shape rapidly becomes rectangular with very little overlap between adjacent energy bins,[48] the bin width remaining constant at 2ϵ. The spectrum is constructed for a set of E_k where $E_{k+1} = E_k + 2\epsilon$. Thus, scattering information over a range of incident energies can be extracted from a time-dependent wave function that has been calculated on a grid. In this work, $n = 2$ and $\epsilon = 50$ keV. Solving two successive linear equations for the vector $|\chi\rangle$:

$$(\hat{\mathcal{H}} - E_k + \sqrt{i}\epsilon)(\hat{\mathcal{H}} - E_k - \sqrt{i}\epsilon)|\chi\rangle = |\Psi\rangle, \tag{8}$$

yields $\mathcal{P}(E_k) = \epsilon^4 \langle\chi|\chi\rangle$.

At deep sub-Coulomb energies, $\mathcal{R}(E_k) \approx 1$, and Eq. (6) becomes numerically unstable. The transmission coefficient is then obtained from:

$$\mathcal{T}(E_k) = \frac{-(8/\hbar v_k)\,\epsilon^4 \langle\chi|\mathrm{Im}(\hat{W})|\chi\rangle}{\mathcal{P}^{initial}(E_k)}, \tag{9}$$

where $\mathrm{Im}(\hat{W}) < 0$ denotes the imaginary potential that operates at the fusion pocket in Fig. 9, while $v_k = \sqrt{2E_k/\mu}$ is the asymptotic relative velocity. Eqs. (6) and (9) provide the same results at energies around the Coulomb barrier, which is a good test of the calculation. As an example and sake of simplicity, Fig. 10 shows the transmission-coefficient excitation function for the $^{16}O + {}^{16}O$ central collision, which is determined by

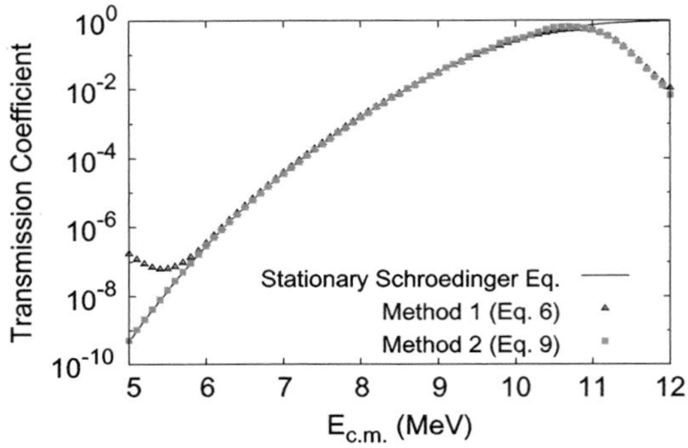

Fig. 10. Transmission-coefficient excitation function for the ^{16}O + ^{16}O central collision through the Coulomb barrier of the BW potential in Fig. 7(a), calculated with various methods indicated. The Coulomb-barrier energy is about 10 MeV.

three methods: solving the stationary Schrödinger equation (solid line) and employing Eqs. (6) and (9) (symbols). The good agreement among these methods demonstrates the reliability of the present TDWP approach.

The model calculations are performed on a five-dimensional grid, i.e., a Fourier radial grid ($R = 0 - 1000$ fm) with 2048 evenly spaced points,[49] and for the angular variables, (θ_1, k_1) and (θ_2, k_2), a grid based on the extended Legendre discrete-variable representation (KLeg-DVR) method.[50] The KLeg-DVR grid-size is determined by the values of j_{max} and k_{max},[50] which are set as 4 and 0, respectively. Total angular momenta up to $J = 10\hbar$ are included. The cross sections are provided by a single wave-packet propagation with the initial, average total energy $E_0 = 3$ MeV. The initial wave-packet was centered at $R_0 = 400$ fm, with width $\sigma = 10$ fm, and was boosted toward the collective PES in Fig. 9 with the appropriate average kinetic energy for the E_0 required. The calculations are considered preliminary, as these do not include Coriolis effects yet and the number of points in the KLeg-DVR grid is limited.

The sub-Coulomb S-factor excitation function for ^{12}C + ^{12}C is presented in Fig. 11 which shows interesting features:

(1) The total angular momenta $J = 0$ and $2\hbar$ clearly determine the observed resonant structures in the 4-5 MeV energy window.
(2) The $J = 0$ component is *not* dominant at deep sub-Coulomb energies, where various partial waves substantially contribute to the total S-curve. At these energies, the total S-curve is smooth and overestimates the

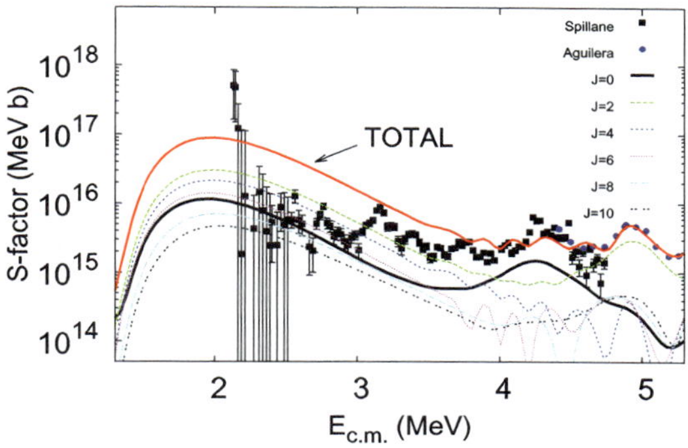

Fig. 11. The sub-Coulomb S-factor excitation function for $^{12}C + {}^{12}C$. Measurements[35,36] are compared to preliminary theoretical calculations within the time-dependent wave-packet method, indicating that molecular structure and fusion are closely connected.

data. For $E < 2$ MeV, the fusion cross section declines strongly. More complete and precise calculations as well as more accurate measurements are required at the astrophysically important energy region ($E < 3$ MeV), which are challenging and are being pursued currently.

Preliminary calculations are very promising, indicating that molecular structure and fusion are closely connected in both the $^{12}C + {}^{12}C$ and $^{16}O + {}^{16}O$ systems.[16] Calculations for other astrophysically important heavy-ion systems (for e.g., $^{16}O + {}^{12}C$) are in progress. The present method might be a more suitable tool for expanding the cross section predictions towards lower energies than the commonly employed potential-model approach. It is worth mentioning that the TDWP method has also been applied to addressing low-energy collisions involving weakly bound nuclei.[51–54]

5. Dissipative quantum dynamics in nuclear collisions

To treat irreversibility in quantum dynamical reaction models is difficult. The stationary-state multichannel scattering theory including complex potentials has been very successful in describing low-energy nuclear reaction dynamics.[55] Nevertheless, this does not seem to account for quantum decoherence which is an important aspect of irreversibility in open dynamical systems,[56] when a limited number of relevant degrees of freedom and reaction channels are included.[57] Decoherence, which always accompanies dissipation in open quantum systems, means dynamical dislocalization of coherent quantum superpositions owing to the interaction of the reduced system with its environment.[58] Coherent quantum superpositions are the basis of the coherent coupled-channels approach to nuclear reaction dynamics near the Coulomb barrier.[59] While this quantum approach is able to explain various reaction observables, important issues are unresolved. It includes the description of dissipation of energy and angular momentum, as revealed in heavy-ion deep-inelastic scattering.[59] The Lindblad equation applied to a few relevant collective coordinates that include mass and charge asymmetries seems to be a good starting point for a practical and rigurous quantum treatment of the problem.[22,60] The CCDM method[20,21] is based on the coupled-channels approach formulated with the Lindblad equation for the time-dependent density matrix, instead of the time-dependent Schrödinger equation for the wave function. In contrast to this approach, most of the dynamical models[20] of dissipative nuclear collisions do *not* treat the relative motion of the nuclei quantum mechanically and/or use *incoherent* (statistically averaged) rather than *decoherent* (partially coherent) reaction channels. The CCDM approach incorporates environmental effects into the coupled-channels model that uses a limited number of important reaction channels. A number of environments can coexist in a nuclear collision, which may be specific to particular degree of freedom, such as isospin asymmetry or weak binding. Among these environments, which can be coupled to specific states or to all states of the reduced system, are (1) the high state-density of multinucleonic excitations in different mass and charge partitions (transfer), (2) the continuum of nonresonant decay states of weakly bound nuclei (breakup), and (3) the innumerable compound-nucleus states (fusion). Simplified, exploratory CCDM calculations have been carried out for the tunneling probability in the ^{16}O + ^{144}Sm sub-Coulomb fusion[20] and for the angular distribution of the ^{154}Sm excitation probabilities in the ^{16}O induced reaction.[21] The calculations are very time-consuming and memory-demanding, but energy-shifting formulae appear to be useful for performing more complete CCDM calculations.[61] Investigations on constructing realistic Lindblad operators are also required. The construction of specific Lindblad operators in reactions involving weakly bound nuclei could be guided by the continuum-discretized coupled-channels (CDCC) method, as the CCDM outcomes can be compared with the CDCC results for elastic-scattering differential cross sections.[62]

6. Outlook

The nuclear molecular picture in low-energy reaction dynamics of light and heavy nuclei is realistic and useful. This picture emphasizes the importance of the dinuclear structure formed by the strongly interacting nuclei for understanding the reaction outcomes such as the fusion cross section. A two-center shell model based on finite depth potentials, which describes the single-particle continuum properly, is useful for planning and interpreting low-energy reaction measurements involving stable and exotic beams. Both the time-dependent wave-packet and the coupled-channels density-matrix methods, using a few important collective coordinates connected with direct observations such as the mass and charge asymmetry coordinates, may improve our undertanding of observables in collisions of heavy-ions and exotic nuclei. These approaches can provide a quantitative, quantum dynamical model for reactions forming the heaviest nuclei, which is essential for further progress in the field.

Acknowledgements

I am very thankful to Werner Scheid, Walter Greiner, Nikolai Antonenko, Gurgen Adamian, Vazgen Sargsyan, David Hinde, Nanda Dasgupta, Jeff Tostevin, Gerald Milburn, Martin Freer, Christian Beck, Leandro Gasques and Michael Wiescher for insightful discussions on the topics addressed in the present contribution. I would like to acknowledge the kind permission of Elsevier Science Publishers to use some figures.

References

1. C. Rolfs and W. S. Rodney. Cauldrons in the Cosmos, University of Chicago Press, Chicago, 1988.
2. A. Diaz-Torres, F. Guzman-Martinez and R. Rodriguez-Guzman, "Level density and collective enhancement factor of a compound-nucleus in a non-adiabatic approach", *Z. Phys. A*, vol. 354, pp. 409–416, December 1996.
3. P. Holzer, U. Mosel and W. Greiner, "Double-centre oscillator and its application to fission", *Nucl. Phys. A*, vol. 138, pp. 241–252, December 1969.
4. J. A. Maruhn and W. Greiner, "The asymmetric two center shell model", *Z. Phys. A*, vol. 251, pp. 431–457, October 1972.
5. W. Greiner, J. Y. Park and W. Scheid. Nuclear Molecules, World Scientific, Singapore, 1994.

6. A. Diaz-Torres, G. G. Adamian, N. V. Antonenko and W. Scheid, "Quasifission process in a transport model for a dinuclear system", *Phys. Rev. C*, vol. 64, pp. 024604-1-9, June 2001; "Potential in mass asymmetry and quasifission in a dinuclear system", *Nucl. Phys. A*, vol. 679, pp. 410–426, January 2001.
7. A. Diaz-Torres, "Modeling of compound nucleus formation in the fusion of heavy nuclei", *Phys. Rev. C*, vol. 69, pp. 021603-1-5(R), February 2004; "Competition between fusion and quasifission in a heavy fusing system: Diffusion of nuclear shapes through a dynamical collective potential landscape", *Phys. Rev. C*, vol. 74, pp. 064601-1-13, December 2006.
8. A. Diaz-Torres, "Effects of diabaticity on fusion of heavy nuclei in the dinuclear model", Ph.D. thesis, Justus-Liebig-Universität Giessen, Giessen, Hessen, Germany, 2000. [Online] Available: http://geb.uni-giessen.de/geb/volltexte/2000/218. [Accessed Jan. 29, 2015].
9. N. V. Antonenko, E. A. Cherepanov, A. K. Nasirov, V. B. Permjakov and V. V. Volkov, "Competition between complete fusion and quasifission in reactions between massive nuclei. The fusion barrier", *Phys. Lett. B*, vol. 319, pp. 425–430, December 1993; G. G. Adamian, N. V. Antonenko, W. Scheid, and V. V. Volkov, "Fusion cross sections for superheavy nuclei in the dinuclear system concept", *Nucl. Phys. A*, vol. 633, pp. 409–420, April 1998.
10. W. Nörenberg, "Memory effects in the energy dissipation for slow collective nuclear motion", *Phys. Lett. B*, vol. 104, pp. 107–111, August 1981; W. Cassing, A. K. Dhar, A. Lukasiak and W. Nörenberg, "Diabatic aspects of the single-particle motion within the time-dependent Hartree-Fock theory", *Z. Phys. A*, vol. 314, pp. 309–316, October 1983; A. Lukasiak, W. Cassing, and W. Nörenberg, "The diabatic two-center shell model", *Nucl. Phys. A*, vol. 426, pp. 181–204, September 1984.
11. A. Diaz-Torres, N. V. Antonenko and W. Scheid, "Dinuclear system in diabatic two-center shell model approach", *Nucl. Phys. A*, vol. 652, pp. 61–70, May 1999.
12. A. Diaz-Torres, G. G. Adamian, N. V. Antonenko and W. Scheid, "Melting or nucleon transfer in fusion of heavy nuclei?", *Phys. Lett. B*, vol. 481, pp. 228–235, May 2000.
13. G. G. Adamian, N. V. Antonenko, A. Diaz-Torres and W. Scheid, "Dynamical restriction for a growing neck due to mass parameters in a dinuclear system", *Nucl. Phys. A*, vol. 671, pp. 233–254, May 2000.
14. G. G. Adamian, N. V. Antonenko and W. Scheid. Lecture Notes in Physics 848: Clusters in Nuclei, vol. 2, Ed. C. Beck, Springer Verlag, Heidelberg, 2012.
15. A. Diaz-Torres, I. J. Thompson and W. Scheid, "Alpha particle production by molecular single-particle effect in reactions of 9Be just above the Coulomb barrier", *Phys. Lett. B*, vol. 533, pp. 265–270, May 2002; "Breakup of 9Be on 209Bi above and near the Coulomb barrier as a molecular single-particle effect: its influence on complete fusion and scattering", *Nucl. Phys. A*, vol. 703, pp. 83–104, May 2002.

16. A. Diaz-Torres, L. R. Gasques and M. Wiescher, "Effects of nuclear molecular configurations on the astrophysical S-factor for 16O+16O", *Phys. Lett. B*, vol. 652, pp. 255–258, September 2007.
17. A. Diaz-Torres, "Solving the two-center nuclear shell-model problem with arbitrarily oriented deformed potentials", *Phys. Rev. Lett.*, vol. 101, pp. 122501-1-4, September 2008.
18. A. Diaz-Torres and W. Scheid, "Two center shell model with Woods-Saxon potentials: adiabatic and diabatic states in fusion", *Nucl. Phys. A*, vol. 757, pp. 373–389, August 2005.
19. A. Diaz-Torres and M. Wiescher, "Quantifying the 12C+12C sub-Coulomb fusion with the time-dependent wave-packet method", *AIP Conf. Proc.*, vol. 1491, pp. 273–276, October 2012; "Quantum partner-dance in the 12C+12C system yields sub-Coulomb fusion resonances", *J. of Phys: Conf. Series*, vol. 492, pp. 012006-1-5, March 2014; "Relating molecular structure and low-energy fusion through time-dependent wave-packet dynamics: the 12C+12C collision", *EPJ Web of Conf.*, vol. 93, pp. 02017-1-4, May 2015.
20. A. Diaz-Torres, D. J. Hinde, M. Dasgupta, G. J. Milburn and J. A. Tostevin, "Dissipative quantum dynamics in low-energy collision of complex nuclei", *Phys. Rev. C*, vol. 78, pp. 064604-1-6, December 2008.
21. A. Diaz-Torres, "Coupled-channels density-matrix approach to low-energy nuclear collision dynamics: A technique for quantifying quantum decoherence effects on reaction observables", *Phys. Rev. C*, vol. 82, pp. 054617-1-6, November 2010.
22. A. Sandulescu, H. Scutaru and W. Scheid, "Open quantum system of two coupled harmonic oscillators for application in deep-inelastic heavy-ion collisions", *J. of Phys. A*, vol. 20, pp. 2121–2132, June 1987; A. Sandulescu and H. Scutaru, "Open quantum systems and the damping of collective modes in deep inelastic collisions", *Ann. of. Phys.*, vol. 173, pp. 277–317, February 1987.
23. M. Mirea, "Superasymmetric two-center shell model for spontaneous heavy-ion emission", *Phys. Rev. C*, vol. 54, pp. 302–314, July 1996.
24. R. A. Gherghescu, "Deformed two-center shell model", *Phys. Rev. C*, vol. 67, pp. 014309-1-20, January 2003.
25. K. Pruess and P. Lichtner, "Calculations of single-particle polarization using a realistic two-center shell model", *Nucl. Phys. A*, vol. 291, pp. 475–509, November 1977.
26. G. Nuhn, W. Scheid and J. Y. Park, "Two-center shell model for deformed and arbitrarily orientated nuclei", *Phys. Rev. C*, vol. 35, pp. 2146–2155, June 1987.
27. M. Mirea, "Two center shell model with Woods-Saxon potentials", *Rom. Rep. in Phys.*, vol. 59, pp. 523–531, February 2007.
28. F. A. Gareev, M. Ch. Gizzatkulov and J. Revai, "A new method for solving the two-center problem with realistic potentials", *Nucl. Phys. A*, vol. 286, pp. 512–522, August 1977.
29. V. G. Soloviev. Theory of Complex Nuclei, Pergamon Press, Oxford, 1976.

30. G. Audi and A. H. Wapstra, "The 1993 atomic mass evaluation: (I) Atomic mass table", *Nucl. Phys. A*, vol. 565, pp. 1–65, December 1993.
31. K. T. R. Davies and J. R. Nix, "Calculation of moments, potentials, and energies for an arbitrarily shaped diffuse-surface nuclear density distribution", *Phys. Rev. C*, vol. 14, pp. 1977–1994, November 1976; P. Möller, and J. R. Nix, "Atomic masses and nuclear ground-state deformations calculated with a new macroscopic-microscopic model", Atomic Data and Nuclear Data Tables, vol. 26, pp. 165–196, March 1981.
32. A. Diaz-Torres, "Shell corrections for finite depth potentials with bound states only", *Phys. Lett. B*, vol. 594, pp. 69–75, June 2004.
33. W. Nörenberg and C. Riedel, "Entrance-channel coherence in dissipative heavy-ion collisions and compound-nucleus formation", *Z. Phys. A*, vol. 290, pp. 335–336, September 1979.
34. C. L. Jiang, K. E. Rehm, B. B. Back and R. V. F. Janssens, "Expectations for 12C and 16O induced fusion cross sections at energies of astrophysical interest", *Phys. Rev. C*, vol. 75, pp. 015803-1-11, January 2007.
35. E. F. Aguilera *et al.*, "New gamma-ray measurements for 12C+12C sub-Coulomb fusion: Toward data unification", *Phys. Rev. C*, vol. 73, pp. 064601-1-12, June 2006.
36. T. Spillane *et al.*, "12C+12C fusion reactions near the Gamow energy", *Phys. Rev. Lett.*, vol. 98, pp. 122501-1-4, March 2007.
37. L. R. Gasques *et al.*, "Implications of low-energy fusion hindrance on stellar burning and nucleosynthesis", *Phys. Rev. C*, vol. 76, pp. 035802-1-10, September 2007.
38. R. L. Cooper, A. W. Steiner and E. F. Brown, "Possible resonances in the 12C+12C fusion rate and superburst ignition", *Astrophys. J.*, vol. 702, pp. 660–671, September 2009.
39. M. E. Bennett *et al.*, "The effect of 12C+12C rate uncertainties on s-process yields", *J. of Phys: Conf. Ser.*, vol. 202, pp. 012023-1-4, February 2010.
40. W. Reisdorf, "Heavy-ion reactions close to the Coulomb barrier", *J. Phys. G*, vol. 20, pp. 1297–1353, May 1994.
41. H. Spinka and W. Winkler, "Experimental investigation of the 16O+16O total reaction cross section at astrophysical energies", *Astrophys. J.*, vol. 174, pp. 455–461, June 1972; G. Hulke, C. Rolfs, and H. P. Trautvetter, "Comparison of the fusion reactions 12C+20Ne and 16O+16O near the Coulomb barrier", *Z. Phys. A*, vol. 297, pp. 161–183, June 1980; A. Kuronen, J. Keinonen, and P. Tikkanen, "Cross section of 16O+16O near the Coulomb barrier", *Phys. Rev. C*, vol. 35, pp. 591–595, February 1987; J. Thomas *et al.*, "Sub-barrier fusion of oxygen isotopes: A more complete picture", *Phys. Rev. C*, vol. 33, pp. 1679–1689, May 1986; S. C. Wu, and C. A. Barnes, "Fusion and elastic scattering cross sections for the 16O+16O reactions near the Coulomb barrier", *Nucl. Phys. A*, vol. 422, pp. 373–396, June 1984.
42. C. L. Jiang *et al.*, "Origin and consequences of 12C+12C fusion resonances at deep sub-barrier energies", *Phys. Rev. Lett.*, vol. 110, pp. 072701-1-5, February 2013.

43. M. Assuncao and P. Descouvemont, "Role of the Hoyle state in 12C+12C fusion", *Phys. Lett. B*, vol. 723, pp. 355–359, June 2013.
44. M. Ueda M, K. Yabana and T. Nakatsukasa, "Absorbing boundary condition approach to breakup reactions of one-neutron halo nuclei", *Nucl. Phys. A*, vol. 738, pp. 288–292, June 2004.
45. D. J. Tannor. Introduction to Quantum Mechanics: A Time-Dependent Perspective, University Science Books, Saulito, 2007.
46. F. Gatti *et al.*, "Rotational excitation cross sections of para-H2 + para-H2 collisions. A full-dimensional wave-packet propagation study using an exact form of the kinetic energy", *J. Chem. Phys.*, vol. 123, pp. 174311-1-13, November 2005.
47. V. A. Mandelshtam and H. S. Taylor, "A simple recursion polynomial expansion of the Green's function with absorbing boundary conditions. Application to the reactive scattering", *J. Chem. Phys.*, vol. 103, pp. 2903–2907, August 1995.
48. K. J. Schafer and K. C. Kulander, "Energy analysis of time-dependent wave functions: Application to above-threshold ionization", *Phys. Rev. A*, vol. 42, pp. 5794–5797, November 1990.
49. R. Kosloff, "Propagation methods for quantum molecular dynamics", *Ann. Rev. Phys. Chem.*, vol. 45, pp. 145–178, October 1994.
50. S. Sukiasyan and H.-D. Meyer, "On the effect of initial rotation on reactivity. A multi-configuration time-dependent Hartree (MCTHD) wave packet propagation study on the H+D2 and D+H2 reactive scattering systems", *J. Phys. Chem. A*, vol. 105, pp. 2604–2611, February 2001.
51. K. Yabana, "Low energy reactions of halo nuclei in a three-body model", *Prog. Theor. Phys.*, vol. 97, pp. 437–450, March 1997; M. Ito, K. Yabana, T. Nakatsukasa and M. Ueda, "Suppressed fusion cross section for neutron halo nuclei", *Phys. Lett. B*, vol. 637, pp. 53–57, June 2006.
52. M. Boselli and A. Diaz-Torres, "Unambiguous separation of low-energy fusion processes of weakly bound nuclei", *J. Phys. G*, vol. 41, pp. 094001-1-11, July 2014.
53. M. Boselli and A. Diaz-Torres, "Quantifying low-energy fusion dynamics of weakly bound nuclei from a time-dependent quantum perspective", *Phys. Rev. C*, vol. 92, pp. 044610-1-7, October 2015.
54. M. Boselli and A. Diaz-Torres, "Reaction dynamics of weakly-bound few-body nuclei at energies around the Coulomb barrier", *Few-Body Syst.*, vol. 57, pp. 177–184, March 2016.
55. J. R. Taylor. Scattering Theory, Wiley and Sons, New York, 1972.
56. H.-P. Breuer and F. Petruccione. The Theory of Open Quantum Systems, Oxford University Press, Oxford, 2002.
57. A. Diaz-Torres, "Absence of decoherence in the complex-potential approach to nuclear scattering", *Phys. Rev. C*, vol. 81, pp. 041603-1-4(R), April 2010.
58. H. D. Zeh, "Roots and fruits of decoherence", Sémin. Poincaré, vol. 1, pp. 115–129, November 2005.

59. D. J. Hinde, M. Dasgupta, A. Diaz-Torres and M. Evers, "Quantum coherence and decoherence in low energy nuclear collisions: from superposition to irreversibility", *Nucl. Phys. A*, vol. 834, pp. 117c–122c, March 2010.
60. M. Genkin and W. Scheid, "A two-dimensional inverse parabolic potential within the Lindblad theory for application in nuclear reactions", *J. Phys. G*, vol. 34, pp. 441–450, March 2007.
61. A. Diaz-Torres, G. G. Adamian, V. V. Sargsyan and N. V. Antonenko, "Energy-shifting formulae yield reliable reaction and capture probabilities", *Phys. Lett. B*, vol. 739, pp. 348–351, November 2014.
62. A. Diaz-Torres and A. M. Moro, "Insights into low-energy elastic scattering of halo nuclei", *Phys. Lett. B*, vol. 733, pp. 89–92, April 2014.

CHAPTER 18

DYNAMICAL COLLECTIVE CLUSTERIZATION IN HOT AND ROTATING OR NON-ROTATING COMPOUND NUCLEI INCLUDING SPONTANEOUS CLUSTER RADIOACTIVITY

Raj K. Gupta*

Physics Department, Panjab University,
Chandigarh-160014, India,
rajkgupta.chd@gmail.com†

In this review, the dynamical cluster-decay model (DCM) is first analyzed for the clustering phenomenon in the decay of a hot and rotating compound nucleus (CN), which means a non-statistical description of the decay of a CN to light particles (n, p, α-particle and γ-rays), the fusion-fission which includes the intermediate mass fragments (both light and heavy), and the non-compound nucleus (nCN) contribution treated as the quasi-fission-like process. The model has been applied to a large number of different reactions belonging to various mass regions from light to superheavy compound systems, with the compact non-coplanar degree of freedom Φ_c included ($\Phi_c \neq 0$) or not included ($\Phi_c = 0$), and is illustrated here for the case of significant non-coplanarity ($\Phi \neq 0$), in particular for the nCN content in fusion cross section. Another important application discussed is that of recently extended $\ell = 0$ version of DCM, the preformed cluster model PCM(T $\neq$ 0), to the observed α-decay chains and predicted heavy cluster radioactivity of superheavy nuclei (SHN) where the recoil of residual SHN occurs due to the x-neutrons emission from the superheavy compound system. A brief presentation of PCM(T=0) applied to spontaneous cluster radioactivity, and presence of clustering in light halo-nuclei in terms of cluster-core model (CCM) is shown to support the cluster + valence-nucleon picture at T=0.

1. Introduction

Heavy-ion fusion reactions is an important and exciting research area of nuclear physics. This has received much attention in recent years since

*Professor Emeritus, Panjab University
†Alternate e-mail address: rkgupta@pu.ac.in

it facilitated the investigation of the production and reaction mechanism of new, heavy and superheavy nuclei with the formation of a compound nucleus (CN) via complete fusion process. At below barrier energies, the CN are highly excited (i.e., hot with temperature T) that carry large angular momentum ℓ, and hence decay by emitting neutrons or multiple light particles (LPs: $A \leq 4$, $Z \leq 2$, like n, p, α), and their heavier counterparts and γ-rays, termed the evaporation residue ER, and fusion-fission (ff) consisting of near-symmetric and symmetric fission fragments (nSF and SF), including also the intermediate mass fragments (IMFs) of masses $5 \leq A \leq 20$ and $2 < Z < 10$. Also, a non-compound nucleus (nCN) decay process could contribute to the (total) fusion cross section σ_{fus}, which makes the study a bit more difficult and hence exciting. Figure 1 (right-hand panel) illustrates schematically the various components of CN fusion/decay cross section, also called the CN production cross section, or simply σ_{fus}, given as

$$\begin{aligned}\sigma_{fus} &= \sigma_{CN} + \sigma_{nCN} \\ &= \sigma_{ER} + \sigma_{ff} + \sigma_{nCN} \qquad (1)\end{aligned}$$

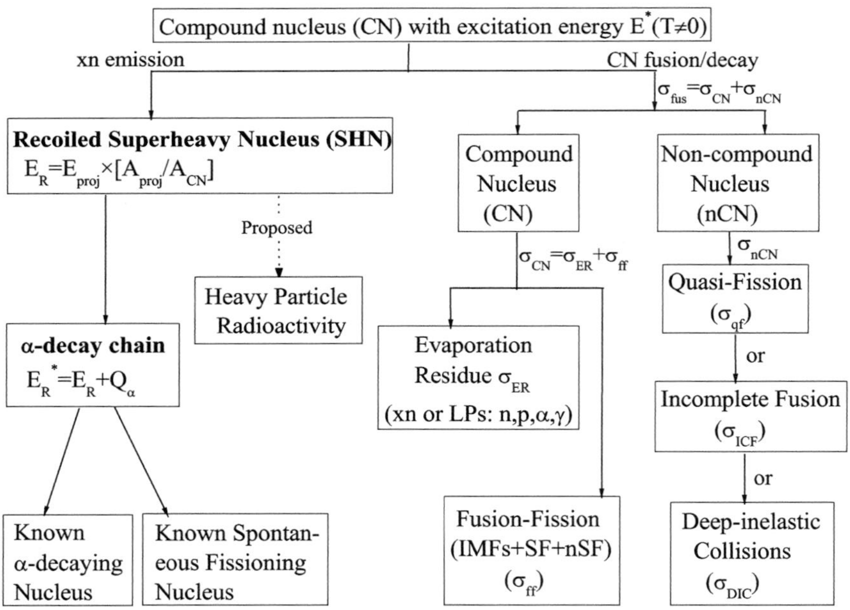

Fig. 1. Schematic of compound nucleus decay process via the observed α-decay chain and/or proposed heavy-particle radioactivity after x-neutrons (xn) emission, and the compound nucleus fusion/decay path via xn or light-particles and fusion-fission to calculate the compound nucleus, and the non-compound nucleus, decay cross sections.

All the components of σ_{fus} are individually measurable. However, if the nCN component σ_{nCN} were not measured, it can be estimated empirically from the calculated and measured quantities, as

$$\sigma_{nCN} = \sigma_{fus}^{Expt.} - \sigma_{fus}^{Cal.}. \tag{2}$$

Different combinations of these processes (ER, IMFs, ff and nCN), or a single one of them as the dominant mode, occur for different mass regions of CN. Then, the CN fusion/formation probability P_{CN} and the CN survival probability P_{surv} against fission are defined as[1,2]

$$P_{CN} = \frac{\sigma_{CN}}{\sigma_{fus}} = 1 - \frac{\sigma_{nCN}}{\sigma_{fus}}, \tag{3}$$

and

$$P_{surv} = \frac{\sigma_{ER}}{\sigma_{CN}} = 1 - \frac{\sigma_{ff}}{\sigma_{CN}}. \tag{4}$$

Thus, P_{CN} and P_{surv} each, give the deviation from unity by the nCN and ff content, respectively. In other words, P_{CN}=1, i.e., a pure CN process, if σ_{nCN}=0, and P_{surv}=1, i.e., de-excitation of CN by ER alone, if σ_{ff}=0.

Alternatively, on a different time scale and so-far observed only for superheavy elements (SHEs), another possible decay mode of CN is the formation of a relatively long lived residual superheavy nucleus (SHN) after the emission of x-neutrons, also illustrated schematically in Fig. 1 (left-hand pannel). In a heavy ion reaction between a projectile of mass A_{proj} with energy E_{proj} and target of mass A_{targ} at rest ($E_{targ} = 0$), $A_{CN} = A_{proj} + A_{targ}$ for the complete fusion process since full momentum transfer occurs from the projectile to CN. Then, if x-neutron emission occurs from the CN, the evaporation residue product, the superheavy nucleus (SHN), has a maximum of cross section corresponding to projectile energy, just close to the Coulomb barrier, and recoils with a recoil energy

$$E_R = E_{proj} \times \frac{A_{proj}}{A_{CN}} \tag{5}$$

such that the residue SHN moves forward with this much kinetic energy. Note that, if the target nucleus is a heavy nucleus of thin layer, the recoiled residue SHN may leave the target and get collected at the focal plane of the separator. The residue SHN, carrying the kinetic energy E_R, decays primarily through consecutive α-emissions, the observed α-decay chain(s), with an excitation energy

$$E_R^* = E_R + Q_\alpha. \tag{6}$$

Eventually, the chain ends at either a known α-decaying or a known spontaneous fissioning nucleus, thereby identifying the SHN uniquely.[3] The other competing decay mode of SHN is the heavy-cluster/heavy-particle radioactivity (HPR), not-yet observed but proposed in many different theoretical works.[4–8]

For the decay of a hot and rotating (with T and ℓ effects included) CN, a dynamical collective clusterization picture, referred to as the Dynamical cluster-decay model (DCM), is developed which contains the spontaneous (T $= 0$, $\ell = 0$) ground-state decay (cluster radioactivity) and decay of hot, non-rotating (T, $\ell = 0$) recoiled SHN as its special cases. In DCM, this is achieved by introducing collective coordinates of mass (and charge) fragmentation $\eta(\eta_Z)$ together with the relative separation coordinate R, deformations $\beta_{\lambda i}$ ($\lambda = 2, 3, 4, i = 1, 2$) and orientations ($\theta_i$, Φ); θ_i for co-planar, and θ_i and Φ (azimuthal angle) for non-coplanar nuclei. The DCM uses the concept of preformed clusters (thereby including nuclear structure effects explicitly) penetrating the confining nucleus-nucleus interaction barrier, first advanced as a Preformed cluster model [PCM(T $= 0$)] for exotic cluster radioactivity observed at T $= 0$, $\ell = 0$, and more recently applied as PCM($T \neq 0$) to the observed α-decay chains and for predicting heavy particle/ cluster radioactivity (HPR) of SHN carrying a recoil energy E_R without ℓ effects ($\ell = 0$). Within the same picture, clustering is also shown to be present in exotic light-halo nuclei, treated within a cluster-core model (CCM) based on cluster + valence-nucleon picture at T $= 0$.

The DCM is applied to a large number of reactions with different entrance channels leading to different compound systems (see, e.g. Refs. 1 and 2). The calculations are made for various center-of-mass energies, and the possible role of deformed, oriented configurations and different nuclear proximity interactions is analyzed. The deformation effects are included up to hexadecapole deformations ($\beta_2, \beta_3, \beta_4$) with 'compact' orientations[9] $\theta_{ci}, i = 1, 2$ or up to only quadrupole deformation (β_{2i}) with optimum orientations[10] $\theta_i^{opt.}$ of "hot" fusion process, for both the cases of co-planar (azimuthal angle Φ=0^0) and non-coplanar nuclei ($\Phi \neq 0^0$); see Fig. 2 and Fig. 3, respectively, in Ref. 1.

Section 2 gives a brief description of the DCM and PCM with effects of temperature, deformations and orientation degrees-of-freedom included. The calculations are discussed in Sec. 3 for the (total) fusion cross section for a hot and rotating CN reaction with non-coplanar ($\Phi \neq 0^0$) compact configurations and noncompound-nucleus effects included [DCM], the α-decay

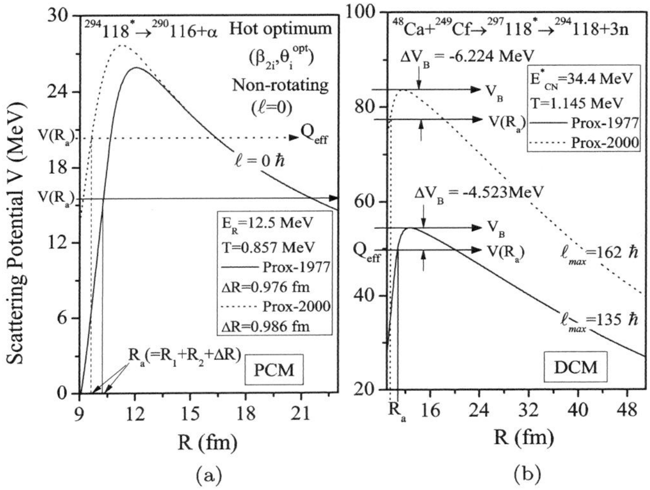

Fig. 2. (a) Scattering potentials $V(R)$ for α-decay of $^{294}118^*$, formed in the ^{48}Ca+^{249}Cf reaction after 3n emission, using the PCM at T = 0.857 MeV (equivalently, E_R = 12.5 MeV) for non-rotating $\ell = 0$ case. (b) The same as for (a) but for $^{297}118^* \to ^{294}118+3n$ using the DCM, at fixed T = 1.145 MeV, corresponding to E^*_{CN} = 34.4 MeV, calculated at $\ell = \ell_{max}$. The concept of barrier lowering $\Delta V_B = V(R_a) - V_B$ is also shown. Solid lines are for the Prox-1977 and dotted lines for Prox-2000 nuclear proximity potentials for deformed nuclei. This figure is based on Fig. 1 in Ref. 18.

chain(s) and competing heavy particle/ cluster radioactivity (HPR) of hot, non-rotating (T, $\ell = 0$) recoiled SHN [PCM(T $\neq$ 0) $\equiv$ DCM(ℓ = 0)], spontaneous cluster radioactivity [PCM(T = 0)], and cluster structure of neutron-, proton-rich halo nuclei. A summary of our results and conclusions are presented in Sec. 4.

2. The dynamical cluster-decay model and its $\ell = 0$ version preformed cluster model

The dynamical cluster-decay model (DCM) for hot and rotating CN is a reformulation of the preformed cluster model (PCM) of Gupta and collaborators[11–13] for ground state decays in exotic cluster radioactivity and related phenomenon. Following the Quantum mechanical fragmentation theory (QMFT),[14,15] as already introduced above, both models are worked out in terms of the coordinates of mass (and charge) asymmetry $\eta = (A_1 - A_2)/A$ (and $\eta_Z = (Z_1 - Z_2)/Z$), the relative separation distance R, the multipole deformations $\beta_{\lambda i}$ ($\lambda = 2, 3, 4; i = 1, 2$), orientations θ_i, and angle Φ between

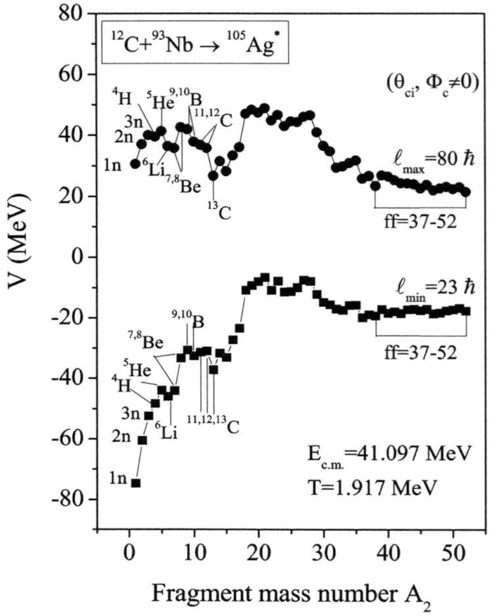

Fig. 3. Mass fragmentation potential minimized in charge-asymmetry coordinate η_Z for the decay of $^{105}Ag^*$ formed in $^{12}C+^{93}Nb$ reaction at $E_{c.m.} = 41.097$ MeV and at ℓ_{min} and ℓ_{max} values. This figure is based on Fig. 3 in Ref. 40.

the principal planes of two nuclei (1 and 2 stand, respectively, for heavy and light fragments). Using decoupled approximation to R- and η-motions, the DCM (see, e.g., the review in Ref. 16) defines the CN decay or fragments production cross section, in terms of ℓ partial waves, as

$$\sigma_{(A_1,A_2)} = \sum_{\ell=0}^{\ell_{max}} \sigma_\ell = \frac{\pi}{k^2} \sum_{\ell=0}^{\ell_{max}} (2\ell+1) P_0 P; \quad k = \sqrt{\frac{2\mu E_{c.m.}}{\hbar^2}} \tag{7}$$

where, P_0, the preformation probability, refers to η-motion and P, the penetrability, to R-motion, both depending on $\ell, T, \beta_{\lambda i}, \theta_i$ and Φ. Here $\mu = mA_1A_2/(A_1 + A_2)$ is the reduced mass with m as the nucleon mass. ℓ_{max} is the maximum angular momentum, defined for light particles evaporation residue cross section $\sigma_{ER} \rightarrow 0$. The same formula (7) is applicable to the nCN decay process, calculated as the quasi-fission-like decay channel, where P_0=1 for each decay channel. Then, it follows from Eq. (7) that

$$\sigma_{ER} = \sum_{A_2=1}^{4 \text{ or } 5} \sigma_{(A_1,A_2)} \quad \text{or} \quad = \sum_{x=1}^{4 \text{ or } 5} \sigma_{xn}, \tag{8}$$

and

$$\sigma_{ff} = 2 \sum_{A/2-x}^{A/2} \sigma_{(A_1,A_2)}, \tag{9}$$

where, e.g., $x = 16$,[17] giving $\sigma_{CN} = \sigma_{ER} + \sigma_{ff}$. Eq. (9) is also applicable to IMFs with sum taken from 5 (or 6) up to the maximum measured value of A_2 and without the multiplying factor 2.

Then, for $\ell = 0$ case, i.e., PCM(T $\neq$ 0), the $\ell = 0$ reduced version of DCM,

$$\sigma_0 = \frac{\pi}{k^2} P_0 P, \tag{10}$$

which is an equivalent of the decay constant

$$\lambda = \nu_0 P_0 P \qquad \left[\text{or decay half-life} \quad T_{1/2} = \frac{\ln 2}{\lambda}\right] \tag{11}$$

with ν_0 as the barrier assault frequency, given simply as

$$\nu_0 = \text{velocity}/R_0 = (2E_2/\mu)^{1/2}/R_0, \tag{12}$$

with total kinetic energy $E_1 + E_2 = Q_{eff}$, with $E_2 = (A_1/A)Q_{eff}$ for the case of decay from hot CN. This is illustrated in Fig. 2(a) for α-decay of recoiled SHN. Thus, PCM(T $\neq$ 0) is equivalent of using DCM($\ell = 0$ case).

The preformation probability P_0 of fragments in both PCM and DCM is the solution of stationary Schrödinger equation controlling η motion at a fixed R = R$_a$, the first turning point of penetration path, taken same for different ℓ-values,

$$\left[-\frac{\hbar^2}{2\sqrt{B_{\eta\eta}}}\frac{\partial}{\partial\eta}\frac{1}{\sqrt{B_{\eta\eta}}}\frac{\partial}{\partial\eta} + V_R(\eta, T)\right]\psi^\nu(\eta) = E_\eta^\nu \psi^\nu(\eta), \tag{13}$$

with $\nu = 0, 1, 2..$ referring to ground state (ν=0) and excited state solutions and the term B$_{\eta\eta}$ represents the smooth hydrodynamical masses.[19]

The solution of Eq. (13), after normalization, gives P_0 as,

$$P_0 = |\psi[\eta(A_i)]|^2 \frac{2}{A_{CN}} \sqrt{B_{\eta\eta}}, \tag{14}$$

where, for a Boltzmann-like function,

$$|\, \psi \,|^2 = \sum_{\nu=0}^{\infty} |\, \psi^\nu \,|^2 \, exp(-E^\nu/T). \tag{15}$$

P_0 contains the structural information of the CN which enters through the fragmentation potential $V_R(\eta, T)$, defined as:

$$\begin{aligned} V_R(\eta,T) &= \sum_{i=1}^{2}[V_{LDM}(A_i,Z_i,T)] + \sum_{i=1}^{2}[\delta U_i]\exp(-T^2/T_0^2) \\ &\quad + V_C(R,Z_i,\beta_{\lambda i},\theta_i,T) + V_P(R,A_i,\beta_{\lambda i},\theta_i,T) \\ &\quad + V_\ell(R,A_i,\beta_{\lambda i},\theta_i,T). \end{aligned} \tag{16}$$

Here, V_{LDM} is T-dependent liquid drop model energy of Davidson *et al.*[20] and δU is the "empirical" shell corrections from Myers and Swiatecki.[21] The other three terms V_C, V_P and V_ℓ correspond to T-dependent Coulomb, nuclear proximity and centrifugal potentials, respectively, for deformed, oriented (co-planar or non-coplanar) nuclei. The moment of inertia used in V_ℓ is in the sticking limit, having the form $I_S(T) = \mu R^2 + \frac{2}{5}A_1 m R_1(\alpha_1,T)^2 + \frac{2}{5}A_2 m R_2(\alpha_2,T)^2$. Here, angle α_i is the angle between the symmetry axis and the radius vector $R_i(\alpha_i,T)$ of the colliding nuclei, measured in the clockwise direction from the symmetry axis, whereas the orientation angle θ_i is the angle between the nuclear symmetry axis and the collision Z-axis, measured in the counter clockwise direction.

The nuclear proximity potential for additional attraction is defined as,[22]

$$V_P(s_0(T)) = 4\pi\bar{R}\gamma b(T)\phi(s_0(T)), \tag{17}$$

where $b(T) = 0.99(1+0.009T^2)$ is the nuclear surface thickness, γ is the surface energy constant and $\bar{R}(T)$ is the mean curvature radius. ϕ is a universal function which is independent of the shapes of nuclei or the geometry of the nuclear system but depends on minimum separation distance $s_0(T)$. Different versions of proximity potential, say, Prox-1977[23] and Prox-2000,[24] having different isospin and asymmetry dependant parameters, have different universal functions. For Prox-1977 of Blocki *et al.*,[23] respectively, for $s_0(T) \leq 1.2511$ and $s_0(T) \geq 1.2511$,

$$\phi(s_0) = \begin{cases} -\frac{1}{2}(s_0-2.54)^2 - 0.0852(s_0-2.54)^3 \\ -3.437exp(-\frac{s_0}{0.75}) \end{cases} \tag{18}$$

Calculations[8,25,26] for heavy and superheavy nuclei, in general, support Prox-1977. The other nuclear potentials used are the various Skyrme forces (SIII and GSkI) in Skyrme energy density formalism (SEDF).[27–29]

The barrier penetration probability P in Eq. (7) is the WKB integral

$$P = \exp\Big[-\frac{2}{\hbar}\int_{R_a}^{R_b} \{2\mu[V(R) - Q_{eff}]\}^{1/2} dR\Big] \tag{19}$$

where R_b is the second turning point satisfying, $V(R_a, T) = V(R_b, T) = Q_{eff}$, with Q_{eff} as the effective Q-value of the decay process. The first turning point R_a, taken to be the same for all ℓ values, is defined as

$$\begin{aligned} R_a &= R_1(\alpha_1, T) + R_2(\alpha_2, T) + \Delta R(T) \\ &= R_t(\alpha, T) + \Delta R(T). \end{aligned} \tag{20}$$

where radius vectors $R_i(\alpha_i, T) = R_{0i}(T)[1 + \sum_\lambda \beta_{\lambda i} Y_\lambda^{(0)}(\alpha_i)]$, with nuclear radii[30] $R_{0i}(T) = \Big[1.28A_i^{1/3} - 0.76 + 0.8A_i^{-1/3}\Big](1 + 0.0007T^2)$. Here the neck-length parameter ΔR is the separation distance between surfaces of two fragments, that assimilates deformation and neck formation effects between two nuclei, introduced within the extended model of Gupta and collaborators.[12,31,32] This method of introducing ΔR is similar to that used in scission-point[33] and saddle-point[34] statistical fission models. The DCM calculations are found to be extremely sensitive to the choice of ΔR, which in turn decides the entry point of barrier penetration as well as of cluster's preformation probability P_0. The choice of parameter R_a (equivalently, ΔR) for the best fit to the data corresponds to effects of "barrier lowering" for each decay channel, defined for each ℓ as the difference between $V_B(\ell)$ and $V(R_a, \ell)$, the barrier height and the actually used barrier, as

$$\Delta V_B = V(R_a, \ell) - V_B(\ell). \tag{21}$$

This is a built-in property of the DCM, illustrated in Fig. 2(b) for $3n$ decay of $^{297}118^*$. It may be worth noting that the use of the neck-length parameter, equivalently, "barrier modification", is the only acceptable explanation to-date to account for the phenomenon of fusion hindrance at sub-barrier energies, since the effective barrier gets significantly modified.[35,36]

The compound nucleus temperature T in DCM is given by $E_{c.m.}$ and the Q-value for incoming channel, as

$$E^*_{CN} = E_{c.m.} + Q_{in} \tag{22}$$

shown in Fig. 2(b) for the illustrative case of $3n$ emission from $^{297}118^*$ nucleus. On the other hand, for PCM(T $\neq$ 0), T is related to the recoil energy E_R of the (recoiled) residue SHN[3] whose excitation energy E^*_R for

α-decay is given by Eq. (6), i.e., $E_R^* = E_R + Q_\alpha$, depicted in Fig. 2(a) for $^{294}118^*$ nucleus, where both E_{CN}^* and E_R^* are related to T (in MeV) as

$$E_{CN}^* = E_R^* = \frac{1}{11}AT^2 - T. \tag{23}$$

For the recoil energy E_R, we have taken a value of 12.5 MeV which is chosen in reference to the average of the measured E_R range (7-18 MeV)[37] for α-decay chain of $^{294}118^*$ nucleus. The Q values are calculated by using binding energies from 2003 experimental compilation of Audi *et al.*[38] and theoretical estimates of Möller Nix,[39] whenever not available in Ref. 38.

3. Calculations and discussion of results

Historically speaking, though PCM(T = 0) was introduced first [for spontaneous (T = 0, $\ell = 0$) cluster radioactivity], then the DCM (for hot and rotating heavy ion reactions, successively including deformation and orientation effects, first for co-planar and then for non-coplanar nuclei), the PCM(T $\neq$ 0) is introduced very recently for the observed α-decay chain(s) and proposed heavy cluster radioactivity, in particular, of superheavy nuclei where recoil of residual SHN occurs due to x-neutrons emission from the superheavy compound system. In the following, we begin with application of the DCM, then of PCM(T $\neq$ 0) and PCM(T = 0), the last one being of the cluster-core model (CCM) for cluster structure of n- and p-halo nuclei.

3.1. *Decay of hot and rotating compound nucleus ^{105}Ag*: Non-compound nucleus effects and non-coplanarity of nuclei*

As an illustrative example of the DCM applied to a hot and rotating complex, we consider the ^{12}C+^{93}Nb reaction at some near- and below-barrier energies, using for nuclear interaction potential the pocket formula of Blocki *et al.*,[23] for $\beta_{2i} - \beta_{4i}$ deformations and 'compact' orientations θ_{ci}, $i = 1, 2$ of co-planar ($\Phi_c = 0^0$) and non-coplanar ($\Phi_c \neq 0^0$) nuclei.[17,40]

The excitation functions (EFs) of the observed decay products, the heavy residues, for the CN ^{105}Ag* formed in ^{12}C+^{93}Nb reaction are measured[41] whose complementary light fragments are the LPs (equivalently, evaporation residue ER) and intermediate mass fragments (IMFs, $4 \leq A \leq 13$). The observed ER consist of 2n, 3n, 4n, and possibly also ^{4}H, and the IMFs are ^{4,5,6}H, ^{4,5}He, ^{8}Li, ^{9,10,11}Be, ^{12}B, and ^{13}C. Note, here one-neutron (1n) emission is found missing (cross section $\sigma_{1n} = 0$). Then, there are the, so-far un-observed, nCN and ff-component in the neighborhood of symmetric fission. Our DCM calculations for co-planar ($\Phi = 0^0$) nuclei,[17] using the pocket formula for nuclear interaction potential, shows that a best fit to the data is obtained only if a large nCN contribution, calculated as a quasi-fission-like process, were allowed in the measured fusion crosssection $\sigma_{fus}(= \sigma_{ER} + \sigma_{IMFs})$. Furthermore, if fits were attempted only for ER crosssection, 1n contribution is large, rather largest, which, in turn reduces to zero or becomes relatively small if the data for (total) σ_{fus} were considered. For use of SEDF nuclear interaction potentials,[42] the DCM fits only some data (mainly IMFs), and for only some Skyrme forces, and hence is once again a case of large nCN component, whose empirically estimated content is fitted nearly exactly for use of DCM with fragment preformation factor taken equal to one, i.e., using DCM($P_0 = 1$). Thus, 'barrier modification' effect is introduced via changing the only parameter of the model, the neck-length parameter ΔR, for a best fit to the empirical nCN data in each (ER and IMFs) decay channel. Also, independent of the choice of Skyrme or proximity force, the fusion-fission (ff) cross section σ_{ff}, is predicted around the symmetric mass $A/2 \pm 16$, and is $\sim 10^1$ to 10^3 mb, depending on the choice of neck-length parameter for fission region.

With the inclusion of Φ degree-of-freedom,[40] the nCN cross section gets further enhanced at the higher two energies, and hence the pure CN cross section content is decreased, which has interesting consequences for CN formation probability P_{CN} since the calculated (total) fusion cross section σ_{fus} is fitted to experimental data.

Following the prescription of Ref. 9, we calculate the compact θ_{ci} and Φ_c for all the possible fragmentations (A_1, A_2) of ^{105}Ag*, and use it to calculate the fragmentation potential $V(A_2)$, minimized in η_Z coordinate, for the best fitted ΔR values to both the measured ER and summed-IMFs cross sections. This is illustrated in Fig. 3 for $E_{c.m.} = 41.097$ MeV (T = 1.917 MeV) and at ℓ_{min} and ℓ_{max} values fixed, respectively, via calculated P_0 and P plotted as a function of ℓ for the $\sigma_{ER} \rightarrow 0$, i.e., the contributions of P_0 and P become negligible ($< 10^{-10}$). We find that ℓ_{max} increases in going from $\Phi_c = 0^0$ to $\Phi_c \neq 0^0$, say from $75\hbar$ to $82\hbar$ at $E_{c.m.} = 54.205$ MeV, and that some maxima and minima in $V(A_2)$ of Fig. 3 change due to the non-zero Φ_c value (see Ref. 17 for $\Phi_c = 0^0$ case). Also, we notice in Fig. 3 that lower ℓ's contribute more to LPs and IMFs and the higher to

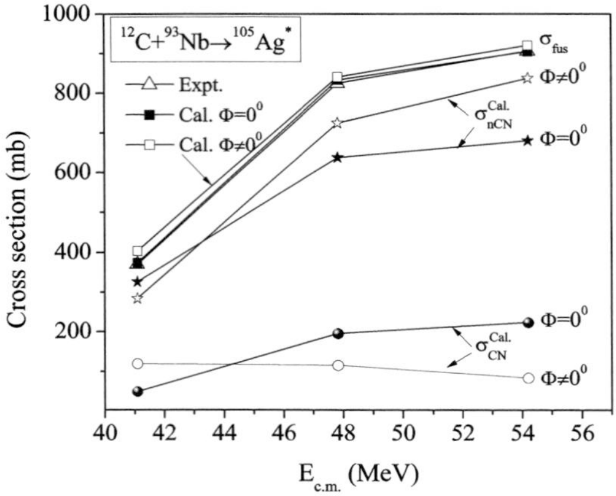

Fig. 4. The DCM-calculated CN cross section $\sigma_{CN}^{Cal.}$, the nCN component $\sigma_{nCN}^{Cal.}$, and their sum, the fusion cross section $\sigma_{fus}^{Cal.}$, for ^{12}C+^{93}Nb→^{105}Ag* reaction, for $\Phi_c \neq 0^0$ case,[40] plotted as a function of $E_{c.m.}$, compared with $\Phi_c=0^0$ case.[17] The experimental data[41] for $\sigma_{fus}^{Expt.}$ is also shown.

near-symmetric fission (nSF) region ff. Interestingly, the nSF mass region of ff (marked in Fig. 3 for $\Phi_c \neq 0^0$ case) remains the same as $A/2 \pm 16$ for $\Phi_c = 0^0$.

Using fragmentation potential $V(A_2)$ of Fig. 3 and scattering potential of one like in Fig. 2(b), we present, in Fig. 4, the DCM calculated CN cross section $\sigma_{CN}^{Cal.}$, the nCN contribution $\sigma_{nCN}^{Cal.}$ (fitted to empirically estimated $\sigma_{nCN}^{emp.}$, given as $\sigma_{fusion}^{Expt.} - \sigma_{CN}^{Cal.}$), and their sum, the fusion cross section $\sigma_{fus}^{Cal.}$ for $\Phi_c \neq 0^0$, compared with $\sigma_{fus}^{Expt.}$ and our earlier calculation[17] for $\Phi_c = 0^0$, at the three chosen $E_{c.m.}$'s. We notice that, in going from $\Phi_c = 0^0$ to $\Phi_c \neq 0^0$, the $\sigma_{CN}^{Cal.}$ decreases considerably, and consequently $\sigma_{nCN}^{Cal.}$ (fitted to $\sigma_{nCN}^{emp.}$) increases, at the higher two energies. An interesting result is the existence of large nCN contribution in both $\Phi_c = 0^0$ and $\Phi_c \neq 0^0$ cases, whose corresponding cross sections $\sigma_{nCN}^{emp.}$ are fitted almost exactly with in the DCM($P_0 = 1$). Also, in Fig. 4 is shown that the calculated $\sigma_{fus}^{Cal.}$ for two cases ($\Phi_c = 0^0$ and $\Phi_c \neq 0^0$) compare nicely with measured $\sigma_{fus}^{Expt.}$; the small difference being due to a relatively larger uncertainty in fitting procedure. We shall see in the following [Fig. 6(a)] that this result of $\sigma_{nCN}^{Cal.}$ becoming larger in the case of $\Phi_c \neq 0^0$ has important consequences for P_{CN}.

Figure 5 presents the details of the fits obtained in the case of $\sigma_{CN}^{Cal.}$: (i) at $E_{c.m.} = 54.205$ MeV, σ_{2n} seems best fitted for both $\Phi_c = 0^0$ and

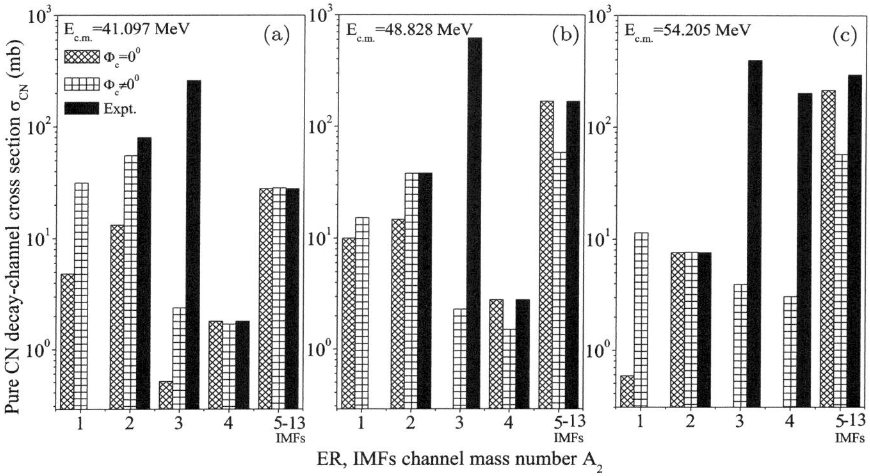

Fig. 5. The DCM-calculated CN decay channel (ER and IMFs) cross sections for the two cases of $\Phi_c = 0^0$ and $\Phi_c \neq 0^0$ degrees-of-freedom,[17,40] compared with experimental data at the three $E_{c.m.}$'s, for ^{105}Ag* formed in ^{12}C+^{93}Nb reaction.

$\Phi_c \neq 0^0$, with a small contribution of σ_{1n} (0.587 and 11.4 mb compared with experimental $\sigma_{1n} = 0$); (ii) at $E_{c.m.} = 47.828$ MeV, σ_{2n} is again best fitted for $\Phi_c \neq 0^0$, and $\sigma_{^4He}$ and σ_{IMFs} in the case of $\Phi_c = 0^0$. Again σ_{1n} contribution is small, (iii) at $E_{c.m.} = 41.097$ MeV, the best fits are obtained for $\sigma_{^4H}$ and σ_{IMFs} in both $\Phi_c = 0^0$ and $\Phi_c \neq 0^0$ cases, though σ_{2n} is under estimated in both cases. The fits for other decay channels, in particular 3n, are rather poor for the pure CN decay process. Also, note that in going from $\Phi_c = 0^0$ to $\Phi_c \neq 0^0$, the ff component $\sigma^{Cal.}_{IMFs}$ in $\sigma^{Cal.}_{CN}$ is decreased at the higher two energies, which will be reflected in the behavior of P_{surv} (Fig. 6(b)).

Figure 6 shows our calculated P_{CN} and P_{surv} for ^{105}Ag* formed in ^{12}C+^{93}Nb reaction, $\Phi_c \neq 0^0$ case[2] compared with $\Phi_c = 0^0$ case.[1] In Fig. 6(a), we notice that P_{CN} of ^{105}Ag* for $\Phi = 0^0$ case show an increasing behavior with increasing E^*_{CN}, which for $\Phi_c \neq 0^0$ becomes a decreasing function of E^*_{CN}, i.e., with Φ degree-of-freedom included the trend of variation of P_{CN} with E^*_{CN} for ^{105}Ag* becomes decreasing, instead of increasing. Note that ^{105}Ag* is a weakly fissioning nucleus, but its P_{CN} for $\Phi_c = 0^0$ case belongs to a group of strongly fissioning superheavy nuclei (illustrated for Z = 112, ^{286}Cn* and Z = 114, ^{292}Fl*), but the same for $\Phi_c \neq 0^0$ to the group of weakly fissioning nuclei, like ^{188}Pt*. This change in behavior of

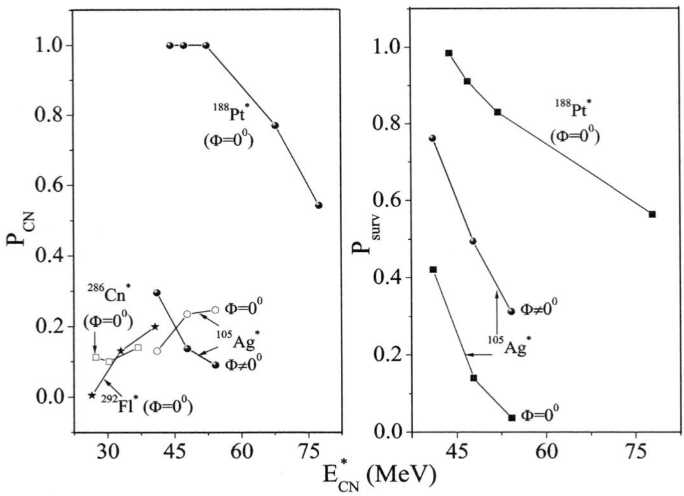

Fig. 6. DCM-calculated (a) P_{CN} and (b) P_{surv} for ^{105}Ag* formed in ^{12}C+^{93}Nb reaction for $\Phi_c \neq 0^0$ and $\Phi_c = 0^0$ cases,[1,2] compared with $\Phi_c = 0^0$ case of strongly fissioning superheavy nuclei ^{286}Cn* and ^{292}Fl*, and a weakly fissioning nucleus ^{188}Pt*, with $\sigma^{Cal.}_{IMFs}$ representing the *ff* component.

the variation of P_{CN} with E^*_{CN} (or $E_{c.m.}$) happens because, in going from $\Phi_c = 0^0$ to $\Phi_c \neq 0^0$, at the higher two energies, the $\sigma^{Cal.}_{CN}$ decreases considerably, and consequently $\sigma^{Cal.}_{nCN}$ increases, as discussed above in Fig. 4. On the other hand, in Fig. 6(b), the trend of variation of P_{surv} with E^*_{CN} for the two cases of Φ remains the same, i.e., decreases with increasing E^*_{CN}, which matches the group of weakly fissioning nuclei, like ^{188}Pt*. Putting the results of P_{CN} and P_{surv} together, an important result is that their variations with E^*_{CN} (or $E_{c.m.}$) in case of $\Phi_c \neq 0^0$ are now both of weakly fissioning nuclei, thereby resolving a kind of anomaly in the case of $\Phi_c = 0^0$ where P_{CN} belonged to strongly fissioning superheavy nuclei and P_{surv} to weakly fissioning nuclei.

Finally, we have also compared the best fitted ΔR values for the two cases of CN and nCN processes, using $\Phi_c \neq 0^0$ and $\Phi_c = 0^0$. We find that $\Delta R_{nCN} < \Delta R_{CN}$ for all the decay channels, and hence the reaction times are larger for the nCN process. Next for the CN process, in going from $\Phi_c = 0^0$ to $\Phi_c \neq 0^0$, the ΔR_{CN} is in general increased, and hence reaction time decreased, as was also observed for ^{164}Yb* system formed in ^{64}Ni+^{100}Mo reaction at both below and above barrier energies.[36] Hence, an important result is that for non-coplanar ($\Phi_c \neq 0^0$) collisions the reaction time for CN process is smaller, and the same for nCN is larger.

Concluding, in this sub-section, we have presented a model (the DCM) for hot and rotating CN, using various nuclear interaction potentials with deformation effects included upto hexadecapole deformation, and for co-planar ($\Phi_c = 0^0$) and non-coplanar ($\Phi_c \neq 0^0$) nuclei having compact orientations of the hot fusion process. For the illustrative example of the decay of $^{105}Ag^*$, formed in ^{12}C+^{93}Nb reaction, we have investigated the role of Φ degree-of-freedom on the predicted large nCN component in the measured CN fusion cross section for co-planar nuclei. The only parameter of the model is neck-length parameter ΔR, which vary smoothly with CN excitation energy or temperature of the system, whose values stay with in the nuclear proximity limits of ~2 fm. The empirically determined nCN cross section is also nearly exactly fitted within the same formalism of using ΔR in DCM with fragment preformation factor $P_0 = 1$, i.e., a quasi-fission-like process. A comparative study of nCN effects in different CN,[1,36,43] presents an important implication that non-coplanarity of nuclei works as an independent degree-of-freedom like the deformations themselves.

3.2. *Decay of hot and non-rotating (recoiled) superheavy nucleus: α-decay chain and heavy particle radioactivity*

Figure 1 (left-hand panel) shows the path of α-decay chain from a hot, recoiled ($\ell = 0$, E_R^*) superheavy nucleus (SHN) formed after an emission of xn from the hot superheavy CN. The chain is considered to end at either a known α-decaying nucleus or a known spontaneous fissioning nucleus, and we consider here the first case[3] of ending at a known α-decaying nucleus (the other case of ending at a known spontaneous fissioning nucleus could also be dealt in a similar way, except that the final nucleus is now a case of formation rather than decay via α-particle). The other possible decay path of SHN is the, not-yet observed, heavy-particle radioactivity (HPR).[4–8] Both decay modes are discussed in the following sub-subsections.

3.2.1. *α-decay chains of recoiled SHN*

In order to identify the SHN uniquely, it is necessary to measure as precisely as possible the α-decay energies and the respective decay half-life times of the entire α-decay chain. Such a data are available for the ^{48}Ca induced "hot fusion" reactions forming SHN with Z=113-118 after 3n−5n emission from the respective superheavy CN. Using PCM(T $=0$), recently,[44] the half-lives of α-decay chains of $^{289,288,287}115^*$ and more recently[18] of $^{294}118^*$, were calculated by us, where fragments are considered to be in the ground state (T $=0$), like in spontaneous α-decay. The calculated half-lives are found to agree with experimental data, but only within a constant empirical factor of 10^4. However, experimently,[37,45] the synthesized SHN, before α-decay chain begins, possess a recoil energy E_R (e.g., 7-18 MeV for $^{294}118^*$), which is used in PCM(T $\neq 0$) to provide the temperature effects for α-decay chains. Therefore, very recently,[3,18] the α-decay chains have been analyzed for the first time by using the PCM(T $\neq 0$) with deformations included up to β_2 and "hot optimum" orientations of nuclei. The interesting result of using the PCM(T $\neq 0$) [$\equiv$DCM($\ell=0$)] is that now no empirical factor is required to fit the data. Our calculations are illustrated below for $^{288}115^*$ and $^{294}118^*$ SHN. The possible branching of α-decay to most probable heavy-cluster decays, the HPR, are discussed in the next sub-subsection.

Table 1 shows the results of our calculated α-decay half lives $T^{\alpha}_{1/2}$, for use of the pocket formula for nuclear proximity potential in PCM(T $\neq 0$) at an average (fixed) E_R of its measured range, with Q_α-value calculated from the binding energies as the sum of liquid drop energy and shell corrections (Sec. 2). PCM(T $\neq 0$) results are compared with PCM(T $=0$) and experimental data. Interestingly, the PCM(T $\neq 0$) calculations agree nicely with the experimental data without any multiplying factor of the

Table 1. PCM(T $\neq 0$) calculated α-decay half-lives for Z = 107-118 SHEs occurring in α-decay chains of SHN $^{288}115^*$ and $^{294}118^*$ synthesized in different heavy-ion reactions via xn emission, compared with PCM(T = 0) results and the experimental data.

SHN	Ref. Expt.	$T_\alpha^{1/2}(expt.)$	PCM(T $\neq$ 0) Ref. 3	PCM(T = 0)x10^4 Ref. 18,44
$^{288}115$	45	171^{+42}_{-28} ms	171 ms	170 ms
$^{284}113$	45	$0.97^{+0.25}_{-0.17}$ s	0.967 s	2.48 s
$^{280}111$	45	$3.6^{+0.9}_{-0.6}$ s	3.69 s	3.52 s
$^{276}109$	45	$0.54^{+0.14}_{-0.09}$ s	0.542 s	0.52 s
$^{272}107$	45	$12.0^{+3.1}_{-2.1}$ s	12.04 s	12.10 s
$^{294}118$	37	$0.69^{+0.64}_{-0.22}$ ms	0.68 ms	0.69 ms
$^{290}116$	37	$8.3^{+3.5}_{-1.9}$ ms	8.3 ms	8.39 ms
$^{286}114$	37	120^{+40}_{-20} ms	120 ms	119 ms

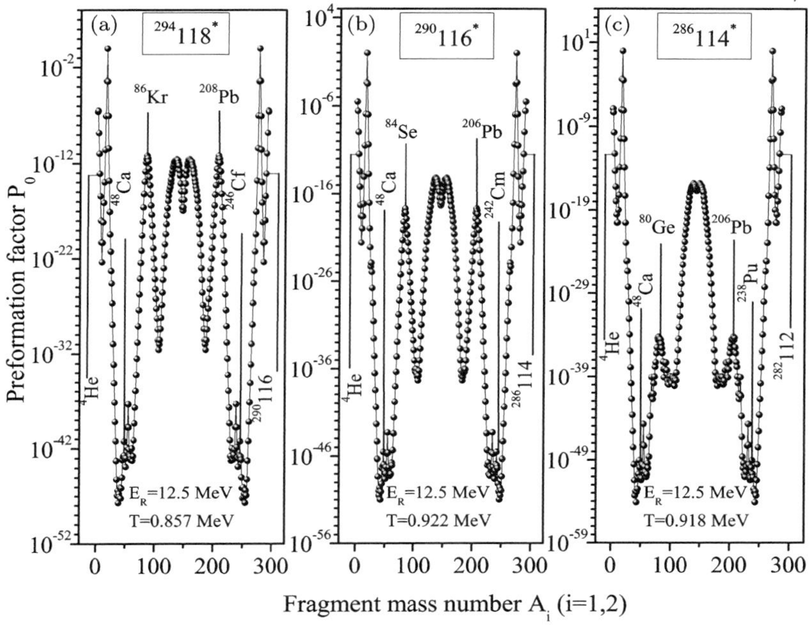

Fig. 7. Preformation factor P_0 for all the parents occurring in α-decay chain of $^{294}118^*$ using PCM(T $\neq$ 0) with pocket formula of nuclear proximity potential at $E_R = 12.5$ MeV, with β_{2i} and "hot" optimum orientations at the respective ΔR fitting α-decay half-life.

type used in PCM(T = 0). The only parameter of model is the neck-length parameter ΔR, which is found to be nearly constant ($\approx 0.95 \pm 0.05$ fm) for all the α-decay chains of various isotopes of Z = 113 − 118 nuclei.

3.2.2. *Heavier-cluster emissions from the parents of α-decay chain*

The competitive heavier-cluster emissions, other than α-particle, is probed here first in terms of preformation factors P_0, illustrated in Fig. 7 for the α-decay chain of SHN $^{294}118^*$ at a T-value equivalent of $E_R = 12.5$ MeV.[18] Note that P_0 gives the relative distribution which means that a slight change in fragmentation potential at any fragment leads to a redistribution of P_0 among all the fragments. Interestingly, we find that, in addition to α-particle, the most prominent clusters are ^{48}Ca and a heavier cluster like ^{80}Ge, ^{84}Se and ^{86}Kr emitted, respectively, from $Z = 114, 116$ and 118 parent. Note that ^{48}Ca is a doubly magic nucleus, and the daughter nucleus of other heavy clusters is a doubly magic ^{208}Pb or its neighboring nucleus.

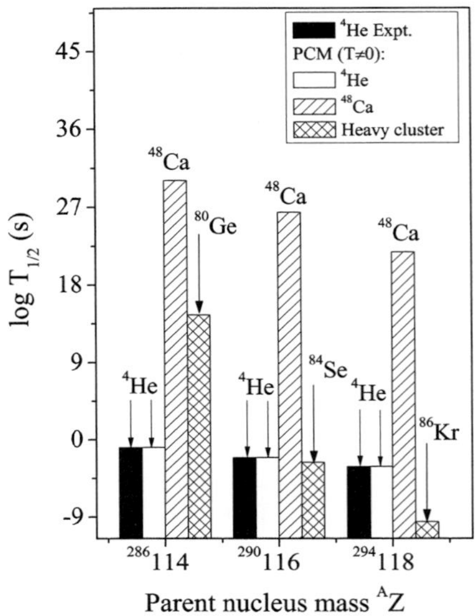

Fig. 8. Logarithm of decay half-lives calculated on PCM ($T \neq 0$) for α and other cluster decays for all parents of the α-decay chain of $^{294}118^*$. Data for α-decay are also shown.[37]

The combined effect of P_0 and P is shown explicitly for the half-life time $T_{1/2}$ in Fig. 8. We notice that our calculated $T^{\alpha}_{1/2}$ agree nicely with the experimental data, which in turn gives a confidence that PCM($T \neq 0$) may impart a reasonable estimate of the most probable emitted heavier clusters. Apparently, this effect is solely due to the combined effect of spherical doubly magic ^{208}Pb and the deformed magic shells around $Z_2 = 36$. For ^{48}Ca cluster, however, though doubly spherical magic nucleus, the predicted decay half-life lies far above the heavy clusters simply because the heavier fragment (^{246}Cf, ^{242}Cm or ^{238}Pu) is a strongly deformed nucleus. Thus, PCM($T \neq 0$) predicts interesting possibility of heavier clusters, in addition to α-decay, in parents of an α-decay chain of SHN.

3.3. *Spontaneous cluster radioactivity*

Spontaneous ($T = 0$) cluster emissions, heavier than α-particle, from trans-lead radioactive nuclei, namely ^{14}C, ^{18,20}O, ^{23}F, ^{22,24,26}Ne, ^{28,30}Mg, ^{32}Si and ^{34}Si have been, respectively, observed or indicated as possible candidates from $^{221-224,226}$Ra, ^{226,228}Th, ^{231}Pa, ^{230,232,234}U, ^{236,238}Pu, ^{238}Pu and ^{242}Cm parents.[46,47] Also, ^{14}C decay from ^{221}Fr and ^{223,225}Ac and ^{34}Si from

^{238}U are observed.[48] It is now well established that these decays find their origin in the closed shell effects of daughter nucleus ^{208}Pb or a neighboring nucleus, and the theoretical models advanced could be classified mainly as the "unified fission models (UFM)" and the "preformed cluster models (PCM)" (see, e.g., a review in Ref. 47). In PCM,[11,49] the α-particle as well as the heavy cluster(s) are assumed to be pre-born with different preformation probabilities P_0 in the ground-state of parent nucleus, before they could penetrate the relevant barrier with the available Q-value(s). On the other hand, in UFM,[50] the cluster radioactivity is considered simply as a barrier penetration phenomenon, in between the fission and the alpha-decay, without invoking the concept of cluster preformation.

Table 2 presents the results of our recent calculations,[51–53] using PCM($T = 0$) with effects of deformations and orientations of "cold" compact configuration included, where "optimum" orientations θ_i^{opt} are given for β_{2i} alone,[10] and "compact" orientations[9] θ_i^c for $\beta_{2i} - \beta_{4i}$ deformations. We notice from Table 2 that deformations and orientations are a must (compare with the spherical case) and that multipole deformations (β_{2i}-$\beta_{4i}, i = 1, 2$) are essential for ^{14}C clusters decays[53] though other measured cluster decays are explained nicely with β_{2i} alone.[51,52] This happens mainly due to P_0.

3.4. *Cluster structure of halo nuclei*

The proton- and neutron-rich light nuclei in the chart of nuclides are characterized by weak binding energies that lead to "exotic" features termed halos. Most of the halo nuclei are confirmed as neutron halos, like 1n-halo nuclei ^{11}Be, ^{14}B, ^{15,17,19}C, ^{22}N, $^{22-24}$O, ^{24,26}F, ^{29,31}Ne and 2n-halo nuclei ^{6,8}He, ^{11}Li, ^{12,14}Be, ^{17,19}B, ^{22}C, ^{23}N, and ^{27,29}F, since the presence of repulsive Coulomb interaction hinders the formation of a proton halo. So far, the 1p-halo structures are established for ^{8}B, ^{11}N and ^{17}F and 2p-halos for only ^{17}Ne, but many more are proposed as the likely candidates. The halo structures of above noted drip-line nuclei at the ground-state ($T = 0$) are investigated here by using our cluster-core model (CCM)[54–58] based on core + valence-nucleons picture. The 'valence-nucleons' here means the cluster of nucleons forming a halo configuration. It is further shown[56] that the halo structure is associated with closed shell effects of the core nucleus.

The CCM is based on the QMFT where the halo nature of a nucleus is studied via the minima in fragmentation potential energy surfaces (PES),

Table 2. PCM(T = 0)-calculated cluster-decay half-life times,[52,53] compared with experimental data, for decays of various parent nuclei to ground states of their respective daughter nuclei, for cases of spherical, β_{2i} alone and $(\beta_{2i} - \beta_{4i})$ deformed nuclei, with appropriate cold "compact" orientations $\theta_i^{opt.}$ or θ_i^c. $Q_{M.N.}$ refers to Q-value calculated by using binding energies of Möller *et al.*[39]

Decay	$R_a (= R_t + \Delta R)$ (fm)	$Q_{M.N.}$ (MeV)	Half-lives $log_{10}T_{1/2}(s)$: PCM: *Sph.*	PCM: $(\beta_{2i}, \theta_i^{opt.})$	PCM: $(\beta_{2i} - \beta_{4i}, \theta_i^c)$	Expt.
$^{221}Fr \rightarrow {}^{14}C + {}^{207}Tl$	$R_t + 1.371$	31.09	20.21	18.44	14.51	14.52
$^{221}Ra \rightarrow {}^{14}C + {}^{207}Pb$	$R_t + 1.410$	32.13	19.43	18.32	13.39	13.39
$^{222}Ra \rightarrow {}^{14}C + {}^{208}Pb$	$R_t + 1.383$	32.47	17.60	16.43	11.01	11.01
$^{223}Ra \rightarrow {}^{14}C + {}^{209}Pb$	$R_t + 1.381$	31.89	20.44	19.42	15.07	15.06
$^{224}Ra \rightarrow {}^{14}C + {}^{210}Pb$	$R_t + 1.414$	31.16	20.44	19.70	15.86	15.86
$^{226}Ra \rightarrow {}^{14}C + {}^{212}Pb$	$R_t + 1.342$	28.61	27.70	26.59	21.19	21.19
$^{223}Ac \rightarrow {}^{14}C + {}^{209}Bi$	$R_t + 1.511$	32.89	17.71	16.75	12.61	12.60
$^{225}Ac \rightarrow {}^{14}C + {}^{211}Bi$	$R_t + 1.484$	31.08	21.95	20.93	17.15	17.16
$^{226}Th \rightarrow {}^{18}O + {}^{208}Pb$	$R_t + 0.65$	47.55	20.10	18.35		> 15.3
$^{228}Th \rightarrow {}^{20}O + {}^{208}Pb$	$R_t + 0.5$	45.91	21.72	20.40		20.87
$^{230}Th \rightarrow {}^{24}Ne + {}^{206}Hg$	$R_t + 0.5$	58.57	24.06	24.19		24.61
$^{231}Pa \rightarrow {}^{23}F + {}^{208}Pb$	$R_t + 0.25$	50.81	28.73	27.38	9.28	26.02
$^{231}Pa \rightarrow {}^{24}Ne + {}^{207}Tl$	$R_t + 0.25$	60.59	25.86	23.07		23.23
$^{230}U \rightarrow {}^{22}Ne + {}^{208}Pb$	R_t	61.69	26.15	19.41	1.27	19.57
$^{230}U \rightarrow {}^{24}Ne + {}^{206}Pb$	R_t	61.81	25.35	23.07		> 18.2
$^{232}U \rightarrow {}^{24}Ne + {}^{208}Pb$	R_t	62.03	24.04	22.88	11.19	21.05
$^{232}U \rightarrow {}^{28}Mg + {}^{204}Hg$	R_t	74.06	28.78	20.31	21.23	> 22.26
$^{233}U \rightarrow {}^{24}Ne + {}^{209}Pb$	$R_t + 0.5$	60.55	24.06	24.57		24.84
$^{233}U \rightarrow {}^{28}Mg + {}^{205}Hg$	$R_t + 0.5$	74.42	24.06	19.22	15.02	> 27.59
$^{234}U \rightarrow {}^{24}Ne + {}^{210}Pb$	$R_t + 0.5$	59.30	24.06	25.99		25.92
$^{234}U \rightarrow {}^{25}Ne + {}^{209}Pb$	$R_t + 0.5$	56.67	24.06	30.38		25.92
$^{234}U \rightarrow {}^{26}Ne + {}^{208}Pb$	$R_t + 0.5$	58.65	28.00	26.49	16.15	25.92
$^{234}U \rightarrow {}^{28}Mg + {}^{206}Hg$	$R_t + 0.5$	74.22	26.72	18.44	15.02	25.54
$^{235}U \rightarrow {}^{28}Mg + {}^{207}Hg$	R_t	72.62	30.08	26.19		> 28.09
$^{236}U \rightarrow {}^{28}Mg + {}^{208}Hg$	R_t	71.33	31.53	25.55		27.58
$^{238}U \rightarrow {}^{34}Si + {}^{204}Pt$	$R_t + 0.215$	85.82	30.37	29.04		29.04
$^{237}Np \rightarrow {}^{30}Mg + {}^{207}Tl$	R_t	75.18	28.12	26.44		> 26.93
$^{236}Pu \rightarrow {}^{28}Mg + {}^{208}Pb$	$R_t - 0.4$	78.75	28.7	21.57		21.67
$^{238}Pu \rightarrow {}^{28}Mg + {}^{210}Pb$	$R_t - 0.4$	75.83	33.7	25.40		25.70
$^{238}Pu \rightarrow {}^{30}Mg + {}^{208}Pb$	$R_t + 0.5$	76.82	24.77	21.87		25.70
$^{238}Pu \rightarrow {}^{32}Si + {}^{206}Hg$	$R_t + 0.5$	90.03		25.16		25.27
$^{241}Am \rightarrow {}^{34}Si + {}^{207}Tl$	R_t	93.62	25.23	22.83		> 22.71
$^{242}Cm \rightarrow {}^{34}Si + {}^{208}Pb$	R_t	95.78	23.59	21.47		23.24

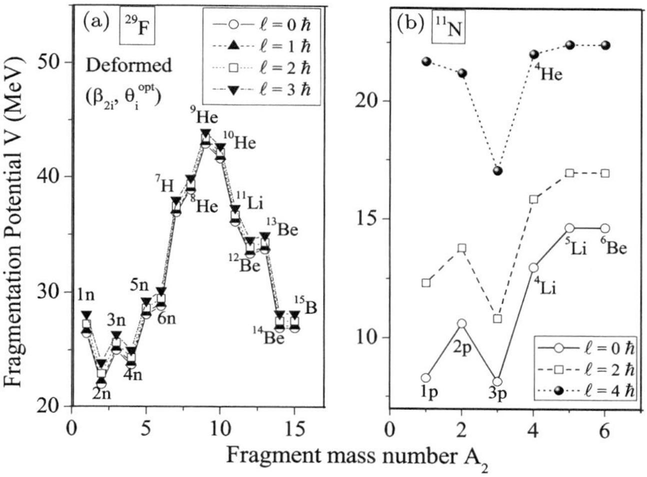

Fig. 9. Fragmentation potential for (a) ^{29}F and (b) ^{11}N, taking the two fragments as β_{2i} deformed, and at some arbitrary ℓ values.[57,58]

which in turn corresponds to the most probable halo-configuration. In Fig. 9, we illustrate our calculated PES, with effects of deformations and angular momentum ℓ included, for 2n-halo nucleus ^{29}F, and 1p-halo nucleus ^{11}N. In Fig. 9(a), we notice that 2n + core minima is almost as deep as 4n + core configuration, though ^{29}F is a 2n-halo nucleus from the point of view of the separation energy. Similarly, Fig. 9(b) illustrates the PES for ^{11}N nucleus, with 1p + core minimum almost as deep as the 3p + core configuration at $\ell = 0$, which gradually changes to 3p-halo structure with the increase in ℓ value. This result may correspond to mixed angular momentum (and parity) states for the ground state configuration.[59] Also, S_{3p} ($= -2619.2$ keV) comes out to be comparable with S_{1p} ($= -2046.3$ keV). Apparently, further experiments and calculations are necessary for halo status of ^{11}N.

4. Conclusion

Collective clusterization phenomenon is established in terms of a Dynamical cluster-decay model (DCM), worked out by introducing collective coordinates of mass (and charge) fragmentation $\eta(\eta_Z)$ together with the relative separation coordinate R, deformations $\beta_{\lambda i}$ ($\lambda = 2, 3, 4$, $i = 1, 2$) and orientations (θ_i, Φ); θ_i for co-planar, and θ_i and Φ for non-coplanar nuclei.

The DCM is developed for the decay of a hot and rotating (T and ℓ effects included) CN, which contains the spontaneous (T = 0, $\ell = 0$) ground-state decay (cluster radioactivity) and decay of non-rotating ($\ell = 0$) recoiled superheavy nuclei as its special cases. The DCM uses the concept of preformed clusters (thereby including nuclear structure effects explicitly) penetrating the confining nucleus-nucleus interaction barrier, first advanced as a preformed cluster model [PCM(T = 0)] for exotic cluster radioactivity observed at T = 0, $\ell = 0$, and more recently applied as PCM(T $\neq$ 0) to the observed α-decay chains and for predicting heavy particle/cluster radioactivity (HPR) of superheavy nuclei (SHN) carrying a recoil energy E_R without ℓ effects ($\ell = 0$). Clustering is also shown to be present in exotic light-halo nuclei, treated within a cluster-core model (CCM) based on cluster + valence-nucleons picture at T = 0.

Acknowledgments

The author is thankful to his students, whose published work provided the source for this review. The work is supported by Department of Science and Technology, Govt. of India, New Delhi.

References

1. A. Kaur, S. Chopra and R. K. Gupta, *Phys. Rev. C* **90**, 024619 (2014).
2. S. Chopra, A. Kaur and R. K. Gupta, *Phys. Rev. C* **91**, 034613 (2015).
3. Niyti, G. Sawhney, M. K. Sharma and R. K. Gupta, *Phys. Rev. C* **91**, 054606 (2015).
4. R. K. Gupta, S. Kumar, R. Kumar, M. Balasubramaniam and W. Schied, *J. Phy. G: Nucl. Part. Phys.* **28**, 2875 (2002).
5. S. Kumar, M. Balasubramaniam, R. K. Gupta, G. Münzenberg and W. Schied, *J. Phy. G: Nucl. Part. Phys.* **29**, 625 (2003).
6. D. N. Poenaru, R. A. Gherghescu and W. Greiner, *Phys. Rev. Lett.* **107**, 062503 (2011).
7. D. N. Poenaru, R. A. Gherghescu and W. Greiner, *Phys. Rev. C* **85**, 034615 (2012).
8. G. Sawhney, K. Sandhu, M. K. Sharma and R. K. Gupta, *Eur. Phys. J. A* **50**, 175 (2014).
9. R. K. Gupta, M. Manhas and W. Greiner, *Phys. Rev. C* **73**, 054307 (2006).
10. R. K. Gupta, M. Balasubramaniam, R. Kumar, N. Singh, M. Manhas and W. Greiner, *J. Phys. G: Nucl. Part. Phys.* **31**, 631 (2005).
11. S. S. Malik and R. K. Gupta, *Phys. Rev. C* **39**, 1992 (1989).
12. S. Kumar and R. K. Gupta, *Phys. Rev. C* **55**, 218 (1997).

13. R. K. Gupta, in *Heavy Elements and Related New Phenomena*, edited by W. Greiner and R. K. Gupta, Vol. **II** (World Scientific, Singapore, 1999), Chapt. 18, p. 730.
14. J. Maruhn and W. Greiner, *Phys. Rev. Lett.* **32**, 548 (1974).
15. R. K. Gupta, W. Scheid and W. Greiner, *Phys. Rev. Lett.* **35**, 353 (1975).
16. R. K. Gupta, in Lecture Notes in Physics, 818, Vol. **1**, *Clusters in Nuclei*, edited by C. Beck (Springer-Verlag, Berlin, Heidelberg, 2010), pp. 223–264.
17. S. Chopra, M. Bansal, M. K. Sharma and R. K. Gupta, *Phys. Rev. C* **88**, 014615 (2013).
18. G. Sawhney, A. Kaur, M. K. Sharma and R. K. Gupta, *Phys. Rev. C* **92**, 064303 (2015).
19. H. Kröger and W. Scheid, *J. Phy. G: Nucl. Part. Phys.* **6**, L85 (1980).
20. N. J. Davidson, S. S. Hsiao, J. Markram, H. G. Miller and Y. Tzeng, *Nucl. Phys. A* **570**, 61c (1994).
21. W. D. Myers and W. J. Swiatecki, *Nucl. Phys. A* **81**, 1 (1966).
22. R. K. Gupta, N. Singh and M. Manhas, *Phys. Rev. C* **70**, 034608 (2004).
23. J. Blocki, J. Randrup, W. J. Swiatecki and C.F. Tsang, *Ann. Phys. (N.Y.)* **105** 427 (1977).
24. W. D. Myers and W. J. Swiatecki *Phys. Rev. C* **62**, 044610 (2000).
25. R. Kumar and M. K. Sharma, *Phys. Rev. C* **85**, 054612 (2012).
26. K. Sandhu, M. K. Sharma, A. Kaur and R. K. Gupta, *Phys. Rev. C* **90**, 034610 (2014).
27. M. Brack, C. Guet and H. -B. Hakansson, *Phys. Rep.* **123**, 275 (1985).
28. J. Friedrich and P.-G. Reinhardt, *Phys. Rev. C* **33**, 335 (1986).
29. B. K. Agrawal, S. K. Dhiman and R. Kumar, *Phys. Rev. C* **73**, 034319 (2006).
30. G. Royer and J. Mignen, *J. Phys. G: Nucl. Part. Phys.* **18**, 1781 (1992).
31. H. S. Khosla, S. S. Malik and R. K. Gupta, *Nucl. Phys. A* **513**, 115 (1990).
32. R. K. Gupta, S. Kumar and W. Scheid, *Int. J. Mod. Phys. E* **6**, 259 (1997).
33. T. Matsuse, C. Beck, R. Nouicer and D. Mahboub, *Phys. Rev. C* **55**, 1380 (1997).
34. S. J. Sanders, *Phys. Rev. C* **44**, 2676 (1991); S. J. Sanders, D.G. Kovar, B.B. Back, C. Beck, D. J. Henderson, R. V. F. Janssens, T. F. Wang and B. D. Wilkins, *Phys. Rev. C* **40**, 2091 (1989).
35. S. K. Arun, R. Kumar and R.K. Gupta, *J. Phys. G: Nucl. Part. Phys.* **36**, 085105 (2009).
36. M. Bansal, S. Chopra, R. K. Gupta, R. Kumar and M. K. Sharma, *Phys. Rev. C* **86**, 034604 (2012)
37. Yu. Ts. Oganessian, *et al.*, *Phys. Rev. Lett.* **109**, 162501 (2012).
38. G. Audi, A. H. Wapstra and C. Thibault, *Nucl. Phys. A* **729**, 337 (2003).
39. P. Möller, J. R. Nix, W. D. Myers and W. J. Swiatecki, *At. Data Nucl. Data Tables* **59**, 185 (1995).
40. S. Chopra, Hemdeep, A. Kaur and R. K. Gupta, *Phys. Rev. C* **93**, 024603 (2016).

41. T. Ahmad, I. A. Rizvi, A. Agarwal, R. Kumar, K. S. Golda and A. K. Chaubey, *Int. J. Mod Phys. E* **20**, 645 (2011).
42. S. Chopra, A. Kaur and R. K. Gupta, *Phys. Rev. C* **91**, 014602 (2015).
43. R. K. Gupta and M. Bansal, *Int. Rev. Phys. (IREPHY)* **5**, 74 (2011).
44. R. Kumar, K. Sandhu, M. K. Sharma and R. K. Gupta, *Phys. Rev. C* **87**, 054610 (2013).
45. Yu. Ts. Oganessian, *et al.*, *Phys. Rev. C* **87**, 014302 (2013).
46. R. Bonetti and A. Guglielmetti, in *Heavy Elements and Related New Phenomena*, Vol. II, Edited by W. Greiner and R.K. Gupta, (World Scientific, Singapore) (1999), p. 643.
47. R. K. Gupta and W. Greiner, *Int. J. Mod. Phys. E* **3**, 335 (1994, Suppl.).
48. R. Bonetti and A. Guglielmetti, *Romanian Reports in Phys.* **59**, 301 (2007) and Private Communication to R.K.G.
49. R. K. Gupta, in *Proc. 5th Int. Conf. on Nuclear Reaction Mechanisms*, Varenna, edited by E. Gadioli (Ricerca Scientifica ed Educazione Permanente, Milano), p. 416 (1988).
50. A. Săndulescu, D. N. Poenaru and W. Greiner, *Sov. J. Part. Nucl.* **11**, 528 (1980).
51. S. K. Arun, R. K. Gupta, B. B. Singh, S. Kanwar and M. K. Sharma, *Phys. Rev. C* **79**, 064616 (2009).
52. S. K. Arun, R. K. Gupta, S. Kanwar, B. B. Singh and M. K. Sharma, *Phys. Rev. C* **80**, 034317 (2009).
53. G. Sawhney, M. K. Sharma and R. K. Gupta, *Phys. Rev. C* **83**, 064610 (2011).
54. R. K. Gupta, *et al.*, *J. Phys. G: Nucl. Part. Phys.* **26**, L23 (2000).
55. R. K. Gupta, *et al.*, *J. Phys. G: Nucl. Part. Phys.* **28**, 699 (2002).
56. R. K. Gupta, *et al.*, *J. Phys. G: Nucl. Part. Phys.* **32**, 565 (2006).
57. G. Sawhney, *et al.*, *J. Phys. G: Nucl. Part. Phys.* **41**, 055101 (2014).
58. G. Sawhney, R. K. Gupta, M. K. Sharma, XXXIV Mazurian Lakes Conference on Physics, Piaski, Poland, September 6–13, 2015; Published in *Acta Physica Polonica B* **47**, 959 (2016).
59. H. Simon, *et al.*, *Phys. Rev. Lett.* **83**, 496 (1999).

CHAPTER 19

MULTIFRAGMENTATION IN THE PERSPECTIVES OF VARIOUS CLUSTERIZATION ALGORITHMS

Rohit Kumar, Arun Sharma*, Samiksha Sood,
and Rajeev K. Puri†

Department of Physics, Panjab University,
Chandigarh -160 014, INDIA

We present a comparative study of fragmentation using isospin-dependent and independent versions of quantum molecular dynamics model. In addition, outcome of different clusterization algorithms is also explored. We will show that the algorithm based on the energy minimization explains experimental data quite nicely.

1. Introduction

The phenomenon of breaking of composite system formed during a heavy-ion collision (at excitation energies comparable with the binding energy of the nucleus) accompanied by the emission of different mass fragments is dubbed as "multifragmentation".[1–4] With the advancements in the 4π-detector arrays around the world, it has become easier to detect the charged as well as uncharged fragments/particles produced in a heavy-ion collision.[4] As a result, a rich experimental information has accumulated on fragmentation. Unfortunately, all experimental measurements are done when matter is cold, therefore, demand for a reliable theoretical approach/model that can reveal the physics behind fragmentation. In this context, one can see a variety of models in the literature describing fragmentation.[5–10]

*Permanent address: Department of Physics, University of Jammu, Jammu -180 006, India.

†Email: rkpuri@pu.ac.in

Generally, One can divide various theoretical attempts into two categories: (i) statistical[9,10] and (ii) dynamical approaches.[5–8] The statistical approaches assume that the fragments are formed from a thermalized composite system. In contrast, dynamical models follow the reaction dynamics from the start where both projectile and target are well separated to the final state where matter is cold and fragmented. These models, therefore, provide rich information about the dynamics behind fragmentation. Among dynamical approaches, models based on the molecular dynamics picture have an edge over other approaches due to the preservation of the correlations among nucleons. Unfortunately, all dynamical models follow the evolution of single nucleons only, therefore, after certain time, clusterization algorithms are employed to identify the fragments. Interestingly, large number of clusterization algorithms has been proposed in the literature. Some of them are: minimum spanning tree (MST) method and its variants,[5,11–14] GEMINI,[10] restructured aggregation model (RAM)[15] etc. More complex secondary algorithms such as early cluster recognition algorithm (ECRA),[16] simulated annealing clusterization algorithm (SACA),[17] fragment recognition in general application (FRIGA)[18] have also been advocated. These secondary algorithms have undergone tremendous success in explaining the physics behind fragmentation.

In this present article, we would like to discuss the formation of fragments within quantum molecular dynamics model using different secondary algorithms. We will show that molecular dynamics approach when coupled with proper secondary algorithm explains the process of fragmentation nicely. In Sec. 2, we will discuss the model in brief and detailed analysis of theoretical calculations is presented in Sec. 3. We summarize our results in Sec. 4.

2. The formalism

The quantum molecular dynamics (QMD) model[5] is based on the molecular dynamics approach where quantum features are also inserted into it. Here, model simulates heavy-ion collision on an event-by-event basis by taking all the fluctuations and correlations into account. In this model, each hadron is represented by a Gaussian in the position and momentum space. The centroid of each Gaussian propagates within the Hamilton's equations of motion of the form:

$$\frac{d\mathbf{r_i}}{dt} = \frac{d\langle H\rangle}{d\mathbf{p_i}}; \quad \frac{d\mathbf{p_i}}{dt} = -\frac{d\langle H\rangle}{d\mathbf{r_i}}, \tag{1}$$

with

$$\begin{aligned}\langle H\rangle &= \langle T\rangle + \langle V\rangle \\ &= \sum_i \frac{p_i^2}{2m_i} + \sum_i \sum_{j>i} \int f_i(\mathbf{r}_i, \mathbf{p}_i, t) V_{ij}(\mathbf{r}_i, \mathbf{r}_j) \\ &\quad \times f_j(\mathbf{r}_j, \mathbf{p}_j, t) d\mathbf{r}_i d\mathbf{r}_j d\mathbf{p}_i d\mathbf{p}_j. \end{aligned} \tag{2}$$

Here, V_{ij} is the baryonic potential that has essential components like Skyrme, Yukawa and Coulomb potentials. Depending on the version one is using, additional large number of variants can also been introduced. Here, we will use quantum molecular dynamics (QMD) model and its isospin variant isospin-dependent quantum molecular dynamics (IQMD) model for the present discussion. The details of the model is described in Refs. 6 and 7.

3. Results and discussion

Here, in the first part, IQMD model will be robust against experimental data. After a comparison of the results obtained using QMD and IQMD models, we will discuss outcome in terms of different secondary algorithms.

Among various observations regarding fragmentation, the rise and fall in the production of intermediate mass fragments (IMFs) was most prominent feature.[2,4,19,20] It has been reported by various experimental and theoretical groups that the multiplicity of IMFs should increase with the incident energy in lower energy regime[4] whereas, the possibility of survival of IMFs is extremely weak at higher incident energies. Therefore, one expects the production of IMFs to be peaked at some intermediate energy (labeled as $E_{c.m.}^{max}$).[2,4,19–22] The peak incident energy (where such maximum production occurs) and corresponding number of IMFs ($\langle N_{IMF}\rangle^{max}$) have been reported to depend on the composite mass of the colliding partners.[20,22] In a latest experimental measurement, it was found that $E_{c.m.}^{max}$ and corresponding intermediate mass fragment production ($\langle N_{IMF}\rangle^{max}$) at $E_{c.m.}^{max}$ exhibits a power law dependence.[20] The percolation model, however, failed to reproduce the trends.[20] Our calculations using IQMD model showed a very close agreement with experimental measurements both for maximal peak energy as well as for the maximum fragment production at the peak incident energy (see Fig. 1). Here, the yield was calculated using soft equation of state with 0.8 σ_{nn} (where σ_{nn} is the total nucleon-nucleon cross-section). This combination of equation of state and nn cross-section

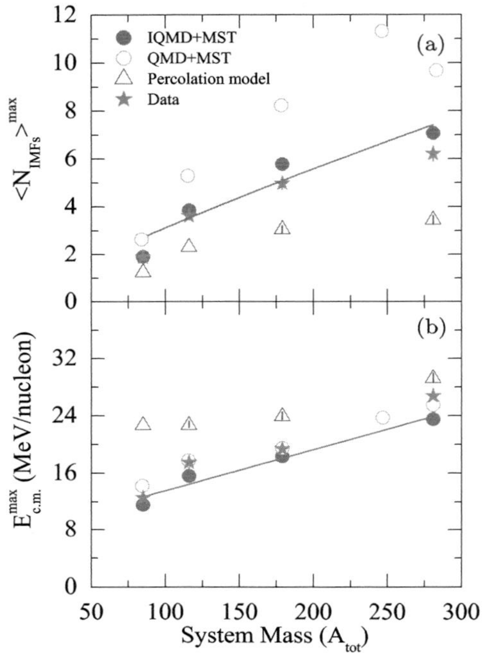

Fig. 1. (Color online) $\langle N_{IMFs}\rangle^{max}$ (upper panel) and $E^{max}_{c.m.}$ (lower panel) as a function of composite mass of the system (A_{tot}). Comparison of IQMD model calculations (filled circles) is made with experimental data (stars). The percolation model calculations (open triangles) as well as QMD model calculations (open circles) are also shown in the figure. Various lines are explained in the text (see Refs. 21, and 22 for original discussion).

was reported to reproduce experimental data very accurately.[7] For this analysis, we had simulated the reactions of ^{40}Ar+^{45}Sc (E_{lab} = 35–115 MeV/nucleon), ^{58}Ni+^{58}Ni (E_{lab} = 35–105 MeV/nucleon), ^{86}Kr+^{93}Nb (E_{lab} = 35–95 MeV/nucleon) and ^{84}Kr+^{197}Au (E_{lab}= 35–400 MeV/nucleon).

In Fig. 1(a), we display the peak multiplicity of IMFs ($\langle N_{IMF}\rangle^{max}$) as a function of total system mass.[21] The corresponding incident energies ($E^{max}_{c.m.}$) are displayed in Fig. 1(b). The solid stars represent experimental data, whereas, solid circles correspond to our theoretical calculations using IQMD model. We also display the results of percolation model calculations (marked as open triangles).[20] From the figure, one noticed that $E^{max}_{c.m.}$ and $\langle N_{IMFs}\rangle^{max}$ increase with system size. The line in the upper (lower) panel of the figure represents power-law (linear) fit to the theoretical points. It is evident that our calculations using IQMD model coupled with MST method (see solid circles) are able to reproduce experimental results reasonably well. On the other hand, the isospin independent version i.e., QMD+MST method (see open circles) is not successful in explaining the experimental measurements and therefore, need of refined method for clusterization was felt in Ref. 22. Interestingly, previously reported Percolation model based on the Boltzmann-Uheling-Uhlenbeck (BUU) model also failed to reproduce the experimental trends.[20]

It is worth mentioning that couple of decades ago, it was pointed out that molecular dynamics model are not able to reproduce experimental trends in asymmetric reactions where matter is not highly compressed, and thermodynamical properties lead the reaction dynamics.[2,23] In the following, we revisited the problem and tried to find the cause of such discrepancy if any.[23] This study was done using both QMD and IQMD variants. In Fig. 2(a), we show the multiplicities of intermediate mass fragments ($<N_{IMFs}>$) against incident energy for the central reaction of ^{84}Kr+^{197}Au.[7] The shaded area represents the IMFs mass window between [$5 \leq A_f \leq 43$] and [$6 \leq A_f \leq 43$] and is guided by the experimental measurements.[2] We have also displayed the experimental data (see stars) along with the previously computed results using QMD model with Pauli-potential (see solid lines). It has been observed that though the QMD model with Pauli-potential is not able to reproduce the multiplicity of IMFs, our current IQMD model (with MST method) reproduces experimental data nicely below 100 MeV/nucleon. This ability of the IQMD model to reproduce the experimental findings is because of higher Fermi-momentum in the IQMD model compared to that in the QMD model. A higher Fermi-momentum may, though yield unbound nucleon close to the surface in the initial evolution.[6] The higher Fermi-momentum in the IQMD model increases the yield of IMFs and thus, overcomes the inadequacy of the QMD model used in Ref. 2 at lower incident energies. To support our view point, we simulated the reaction of ^{84}Kr+^{197}Au with reduced Fermi-momentum (by 40%) that now matches closely with the one used in the QMD model. Our findings are displayed in Fig. 2(b) and are labeled as RFM (see dark shaded region). This reduction in the Fermi-momentum slows down the IMFs production at lower incident energies, though, does not affect production at higher incident energies. Another important ingredient that leads to different results with IQMD and QMD models is the symmetry potential in the former one. Obviously, this ingredient in the IQMD model will increase the yield of fragments at lower beam energies due to its repulsive nature. To

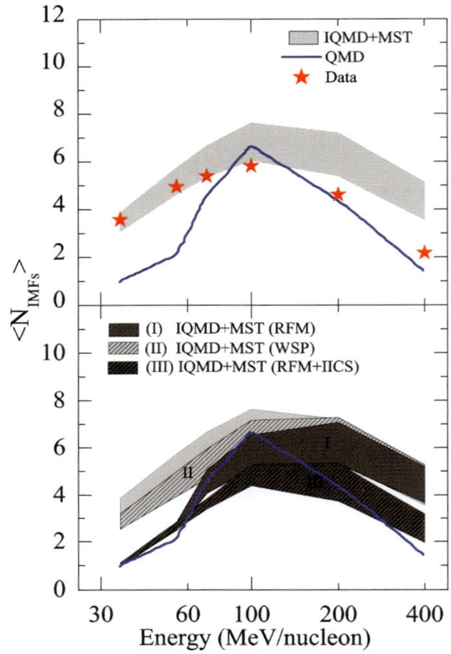

Fig. 2. The multiplicities of intermediate mass fragments ($< N_{IMFs} >$) obtained in the central reaction of ^{84}Kr+^{197}Au as a function of beam energy. Stars represent the experimental data taken from Ref. 2. The light shaded region represents our present calculation. The dark shaded, light crossed and dark crossed regions represent our present calculations (using IQMD model) with reduced Fermi-momentum (RFM), without symmetry potential (WSP) and isospin independent nn cross-section along with reduced Fermi-momentum (RFM+IICS), respectively. The previously calculated results of QMD model (solid line) (unfiltered) are also displayed for comparison (see Ref. 7 for original discussion).

verify this behavior of symmetry potential, we also perform IQMD model simulations by neglecting the symmetry potential (labelled as WSP). Our analysis reports decrease in the multiplicity of IMFs (by $\sim 15\%$) at lower beam energies without having any effect at higher incident energies. Therefore, one can conclude that the repulsive character of symmetry potential as well as higher Fermi-momentum alters the production of intermediate mass fragments at lower beam energies, therefore, yield different results using IQMD model compared to QMD model. As noted above, no effect is visible at higher incident energies. Further, we also perform IQMD model calculations by taking reduced Fermi-momentum along with isospin independent nn cross-section and corresponding results are represented by dark crossed area (labelled as RFM+IICS). We now find that this combination reduces the IMFs multiplicity at higher beam energies by 30–40%. This happens because the isospin independent nn cross-section along with reduced Fermi-momentum forbids massive destruction of correlations among nucleons. This reduction further bridges the difference between QMD and IQMD models.[7]

At the same time, many cases are reported in the literature where minimum spanning tree (MST) method when coupled with QMD model did not yield proper results as found in the experiments.[1] At this point of time, methods based on the energy minimization of the system like ECRA[16] was put forward. Later on, Puri, Hartnack and Aichelin developed SACA method[17] which overcame the difficulties of ECRA method. Recently, SACA has been upgraded to fragment recognition in general application (FRIGA) model.[18] In the following paragraphs, we will first give a brief introduction of the SACA method (which is very successful in reproducing experimental results) and then will discuss results of QMD model coupled with SACA method.

The simulated annealing clusterization algorithm (SACA)[17] is based on the principle of global energy minimization of the structure. Here, we start with any random configuration consists of nucleons and fragments of different size. One can also choose configuration identified by the MST method as preliminary configuration. Nucleons or fragments are then transferred to each other using well accepted Monte-Carlo procedure and using simulated annealing technique.[24] Finally, most bound configuration is accepted as most probable one. At this stage, we make sure that each fragment, identified by the above procedure is properly bound. Accordingly, we demand each fragment to fulfil the following binding energy criteria;

$$\zeta = \sum_{i=1}^{A_f} \left[\sqrt{\left(\mathbf{p}_i - \mathbf{p}_{A_f}^{cm}\right)^2 + m_i^2} - m_i + \frac{1}{2}\sum_{j \neq i}^{A_f} V_{ij}(\mathbf{x}_i, \mathbf{x}_j) \right] < E_{bind} \times A_f, \quad (3)$$

where $E_{bind} = -4.0$ MeV for $A_f \geq 3$ and 0 otherwise. In this equation, A_f is the size of the fragment and $\mathbf{p}_{A_f}^{cm}$ is the center-of-mass momentum of the fragment. It is worth mentioning here that the influence of conditions of different binding energies in the eq. 3 has negligible role to play on the final fragment configuration.[25]

Here, we plan to robust QMD+SACA against experimental measurements at low incident energies; where it is well known that the excitation energy stored in the composite system is small. In particular, we will

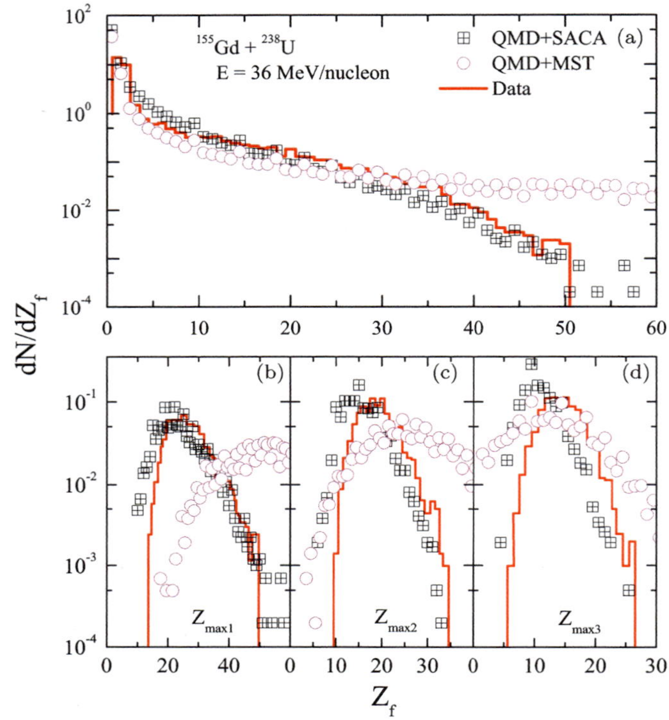

Fig. 3. The total charge distribution (upper panel) and the charge distribution of largest, second largest and third largest fragments (lower panels) for the central reaction of ^{155}Gd+^{238}U at incident energy of 36 MeV/nucleon. Crossed squares and open circles represent the calculations with SACA and MST approaches, respectively. The red lines represent the experimental data taken from Ref. 26.

compare our calculated results for the central reactions of ^{155}Gd+^{238}U at incident energy of 36 MeV/nucleon.[26]

In Fig. 3, we display the fragments yield as well as largest, second largest and third largest fragment. From the figure, we observe that QMD+MST method (open circles) fails badly to reproduce the charge distribution (upper panel) as well as event-by-event distributions of the largest, second largest and third largest fragments (lower panels). We also observe that QMD+MST method even can not reproduce trends as well as peak of the charge distributions for largest, second largest and third largest fragments properly. This happens because of the lesser energy stored in the system due to which nucleons accumulate close to each other in coordinate space and MST method by definition can't recognize fragments which are close in the space. On the other hand, QMD+SACA method (crossed squares) reproduces the experimental data to a great extent. This happens due to the capability of SACA method to single out the fragments even if they are overlapping in coordinate space. This shows that the QMD+SACA method not only reproduces the average fragment properties of the reaction, but also stores the correlations among the fragments within each event. Further, the results of Fig. 3 illustrates that checking only spatial correlations among nucleons is not sufficient to explain the experimental results where excitation energy stored in the system is low and hence, algorithm based on the energy minimization can act as a key source to deduce proper physics of fragmentation.

Next in Fig. 4, we display the QMD+SACA calculations for the central reactions of ^{40}Ar+^{45}Sc in the Fermi-energy region of 15–35 MeV/nucleon.

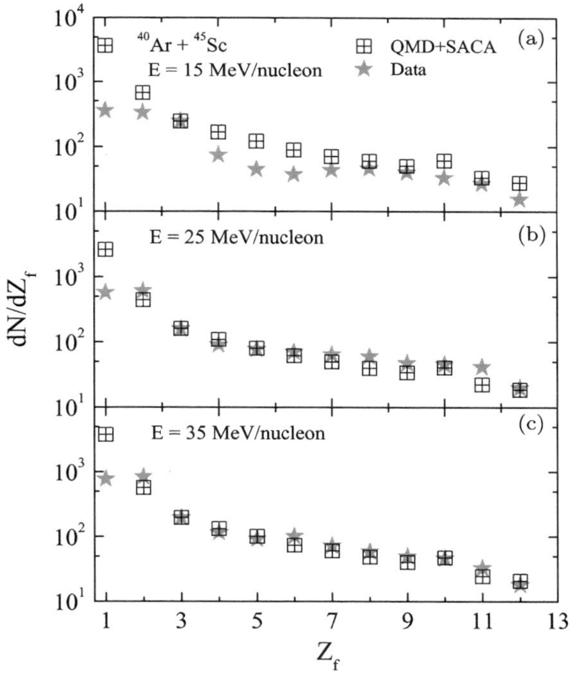

Fig. 4. (Color online) The normalized charge distribution for the central reactions of ^{40}Ar+^{45}Sc at incident energy range of 15–35 MeV/nucleon. Crossed squares represent the calculations with SACA and stars represent measurements extracted from Ref. 27.

From the figure, it is very clear that the slope of the charge distribution steepens with increase in incident energy. This happens due to increase in the deposited excitation energy in the composite system that breaks the system violently. Again, QMD+SACA method calculations give consistent attribution to the experimental observations of Ref. 27. Our calculated results (in Figs. 3 and 4) show the suitability of QMD+SACA combination for investigating the dynamics of fragmentation observables. It is worth mentioning that the advantage of SACA method is not only in its capability to give consistent explanation of the experimental results, but it also realizes the stable fragment structures as early as $\sim$ 60–90 fm/c, when occupancy of phase space is still high (or system is still in the compressed stage).

4. Summary

In this article, we presented our recent calculations involving isospin-dependent and independent versions of the quantum molecular dynamics model for different observables associated with the multifragmentation phenomena at intermediate energy. In particular, we studied the role of different clusterization algorithms and tried to map the efficiency and applicability of various combinations of primary dynamical models with secondary clusterization algorithms. We concluded that the proper choice of clusterization algorithms can greatly affect the outcome of the reaction observables of heavy-ion collisions.

Acknowledgements

This work is supported by a grant from the University Grant Commission (UGC), Government of India vide Grant No. 42-811/2013 (SR).

References

1. A. Schüttauf *et al.*, *Nucl. Phys. A* **607**, 457 (1996).
2. G. F. Peaslee *et al.*, *Phys. Rev. C* **49**, R2271 (1994).
3. Z. Y. Sun *et al.*, *Phys. Rev. C* **82**, 051603(R) (2010).
4. B. Borderie and M. F. Rivet, *Prog. Part. Nucl. Phys.* **61**, 551 (2008).
5. J. Aichelin, *Phys. Rep.* **202**, 233 (1991).
6. C. Hartnack *et al.*, *Eur. Phys. J. A* **1**, 151 (1998).
7. A. Sharma, A. Bharti, S. Gautam and R. K. Puri, *Nucl. Phys. A* **945**, 95 (2016).
8. W. Bauer *et al.*, *Phys. Rev. C* **34**, 2127 (1986).

9. J. P. Bondorf *et al., Nucl. Phys. A* **443**, 321 (1985); D. H. E. Gross, *Rep. Prog. Phys.* **53**, 605 (1990) and references therin; H. W. Barz *et al., Nucl. Phys. A* **561**, 466 (1993).
10. R. J. Charity *et al., Nucl. Phys. A* **483**, 371 (1988); L .Wang *et al., Nucl. Phys. A* **920**, 1 (2013).
11. S. Kumar and R. K. Puri, *Phys. Rev. C* **58**, 320 (1998); J. Singh and R. K. Puri, *J. Phys. G: Nucl. Part. Phys.* **27**, 2091 (2001).
12. S. Kumar and R. K. Puri, *Phys. Rev. C* **58**, 2858 (1998); S. Goyal and R. K. Puri, *Phys. Rev. C* **83**, 047601 (2011).
13. R. Kumar, S. Gautam and R. K. Puri, *Phys. Rev. C* **89**, 064608 (2014).
14. S. Pratt, C. Montoya and F. Romming, *Phys. Lett. B* **349**, 261 (1995); J. Pan and S. D. Gupta, *Phys. Rev. C* **51**, 1384 (1995).
15. C. Ngõ *et al., Nucl. Phys. A* **499**, 148 (1989); S. Leray *et al., Nucl. Phys.* A **531**, 177 (1991).
16. C. Dorso and J. Randrup, *Phys. Lett. B* **301**, 328 (1993).
17. R. K. Puri, Ch. Hartnack and J. Aichelin, *Phys. Rev C* **54** R28 (1996); R. K. Puri and J. Aichelin, *J. Comput. Phys.* **162**, 245 (2000); Y. K. Vermani *et al., J. Phys. G: Nucl. Part. Phys.* **37**, 015105 (2010); Y. K. Vermani and R. K. Puri, *Eur. Phys. Lett.* **85**, 62001 (2009).
18. A. Le Fevre, Y. Leifels, J. Aichelin, C. Hartnack, V. Kireyev and E. Bratkovskaya, *J. Phys.: Conf. Ser.* **668**, 012021 (2016).
19. N. T. B. Stone *et al.*, *Phys. Rev. Lett.* **78**, 2084 (1997), and references therein.
20. D. Sisan *et al.*, *Phys. Rev. C* **63**, 027602 (2001).
21. S. Kaur and R. K. Puri, *Phys. Rev. C* **87**, 014620 (2013).
22. Y. K. Vermani and R. K. Puri, *J. Phys. G: Nucl. Part. Phys.* **36**, 105103 (2009); S. Kaur and A. D. Sood, *Phys. Rev. C* **82**, 054611 (2010).
23. R. Donangelo and S. R. Souza, *Phys. Rev. C* **52**, 326 (1995).
24. P. J. M. Laarhoven and E. H. L. Aarts, *Simulated Annealing: Theory and Applications* (Reidel, Dordrecht, 1987).
25. R. Kumar and R. K. Puri, *Phys. Rev. C* (to be submitted).
26. Al. H. Raduta and Ad. R. Raduta, *Phys. Rev. C* **65**, 054610 (2002).
27. T. Li *et al., Phys. Rev. C* **49**, 1630 (1994).

CHAPTER 20

USING FAST PROCESSES TO INVESTIGATE CLUSTER STATES AND NUCLEAR CORRELATIONS IN MEDIUM-HEAVY NUCLEI: SPECIFIC TOOLS AND NEW OPPORTUNITIES WITH RADIOACTIVE ION BEAMS

T. Marchi

KU Leuven, Dep. of Physics and Astronomy,
Instituut voor Kern- en Stralingsfysica, Leuven, Belgium.
tommaso.marchi@kuleuven.be

F. Gramegna

INFN Laboratori Nazionali di Legnaro,
Legnaro (PD), Italy.

L. Morelli

University of Bologna, Dep. of Physics and Astronomy and
INFN Sezione di Bologna, Bologna, Italy

S. Barlini

University of Florence, Dep. of Physics and Astronomy and
INFN Sezione di Firenze, Firenze, Italy

D. Fabris

INFN Sezione di Padova, Padova, Italy

V.L. Kravchuk

National Research Centre "Kurchatov Institute",
Moscow, Russia

O.V. Fotina

D.V. Skobeltsyn Institue of Nuclear Physics,
Moscow State Univ., Russia

Clustering effects show up in several ways along the Nuclear Chart. While for light nuclei (A≤40) the existence of clusters has been directly linked to nuclear structure and reaction dynamics observations, this is less the case when moving towards medium-mass systems. Pre-equilibrium emission is a nice example of mechanism capable to fill this gap, linking structure with dynamics and, eventually, to clustering in medium-mass nuclei. For central collisions, cluster pre-formation probabilities can be derived by studying light particle production cross sections and multiplicities. Several theories have been developed in the framework of exciton models to describe pre-equilibrium processes and try to explain how these are influenced by the structure of the reacting nuclei. The forthcoming availability of intense radioactive ion beams will open new challenging scenarios in this field of study, allowing the comparison of new systems where projectiles exhibit extreme neutron to proton ratios, bringing to structural configurations where clusters may show up in a more evident way.

1. Introduction

The interplay between nuclear structure and reaction dynamics is a topic of renewed interest in today's Nuclear Physics. Among others, processes such as the ones connected to the formation, transfer and emission of clusters play a crucial role in the explanation of several experimental results. Even if clustering effects show up in several ways along the Nuclear Chart, very different phenomena can be related to either their dynamical formation or their structural presence (pre-formation). While for light-mass nuclei (A≤40) several links between cluster (d, t, ^{3}He, ^{4}He) emission and its connection with nuclear structure have been pointed out,[1–3] this is less the case when moving towards medium-mass ($40 \leq A \leq 200$) systems. On the other side, studies on heavy nuclei ($A \geq 200$) have been carried on for many years, since their decay through α-chain or heavier cluster has been evidenced (e.g., the ground state α emission of Actinides).[4]

A typical example of a light nucleus where an α-cluster structure has been observed is ^{12}C.[5] The debate about the nature of its second excited state at 7.65 MeV, known as the Hoyle state, is still ongoing.[6–8] The ^{12}C nucleus can be considered a cluster itself but, due to its possibility of arrangement in a triangular configuration,[9,10] it is a so-called "non-inert" cluster.[1] On the contrary, the ^{16}O nucleus (described as formed by four α-particles[11,12]) can adjust its structure in such a way to be an inert cluster, meaning that it could be pre-formed and exchanged within heavier nuclei. In medium-mass nuclei the landscape is less clear but evidences of such formations, mainly close to the nuclear surface,[13] have been claimed in the past.[14] The interplay between surface and volume emission from an excited nuclear system can be characterized by the measurement of pre-equilibrium particles. These are fast, forward focused, light charged particles and/or neutrons emitted during the very early stages of the collision, before the reacting nuclei complete their merging and form a thermally equilibrated compound system. Pre-equilibrium emission, considered as an intermediate process between direct reactions and the most dissipative collisions, depends on reaction parameters like the projectile's velocity, entrance channel mass asymmetry and the initial structure of the reacting partners. Given the fast nature of the process, an obvious question arises when cluster particles are emitted: do they exist as pre-formed objects in the reacting nuclei? Are they, on the contrary, dynamically formed during the collision?

It is thus important to study central collisions focusing on the competition between evaporation and fast emission of light particles and clusters as functions of different entrance channel parameters. A quantitative comparison of the experimental results with theoretical models[15–18] capable to describe both the evaporative and the pre-equilibrium emissions simultaneously is then needed.

2. Nuclear reaction mechanisms

Nuclear reactions between a projectile and a target nucleus can proceed according to different mechanisms, which depend typically on the so called "entrance channel" parameters like: the mass (A_p, A_t), the charge (Z_p, Z_t) and structure of the projectile and target nuclei; the impact parameter of the projectile; the projectile's kinetic energy. Reactions induced by light-ions can be grouped as: *direct, compound* and *pre-equilibrium* processes. The difference can be basically linked to the typical time scale of such processes, that is related to the number of intranuclear collisions taking place before any nucleon emission occurs.[19] With the growing mass of the projectile, other mechanisms will start to be effective like Deep Inelastic Collisions (DIC), where a large dissipation of energy and angular momentum is expected.[20]

Typical times for *direct reactions*, which are normally one-step processes, are of the order of 10^{-22} s (compatible with the time needed by the projectile to cross the target nucleus). On the other side, *compound nucleus reactions* proceed through a long series of intranuclear collisions that lead to the energy equilibration of the fused object. Typical time scales of these processes are therefore much longer ($10^{-18} - 10^{-16}$ s for low energy reactions). The equilibrated compound nucleus might have accumulated

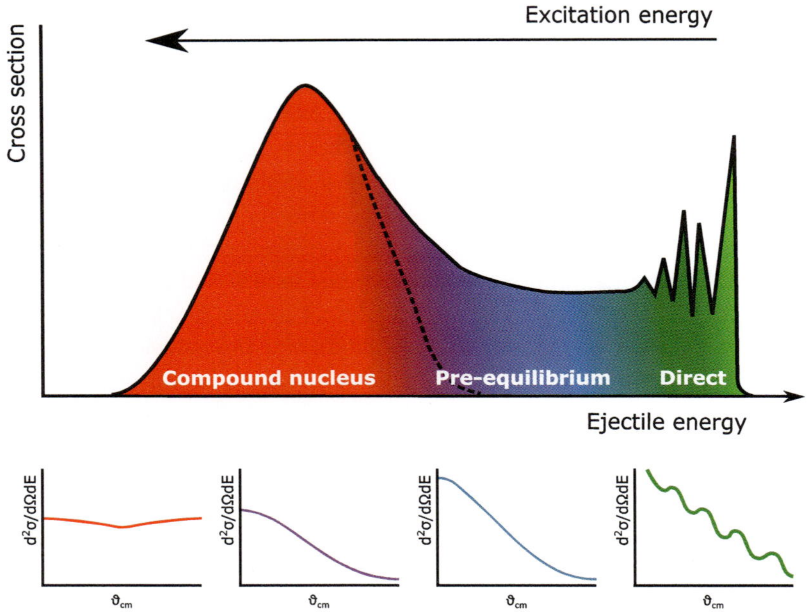

Fig. 1. Energy spectrum of a reaction between nuclei A(a,b)B and associated angular distributions, which show a gradual transition from forward peaked oscillatory behavior to isotropy for decreasing energy in the output channel. Figure adapted from Ref. 19.

some excitation energy with respect to its ground state and emits (*evaporates*) statistically either light particles or γ-rays in order to settle to a stable condition.[21]

The direct reaction against compound nucleus model is clearly an oversimplification. In fact, between these two extreme possibilities, a continuous spectrum of processes has to be expected (as shown in Figure 1) as a function of energy dissipation, among the *pre-equilibrium* mechanisms. This classification of the reactions is supported by the observation of the typical angular distributions. Forward peaked angular distributions are observed in the case of direct reactions where the oscillatory behavior of the cross-section is characteristic of the structure of the residual nucleus. Symmetric angular distributions are associated to compound nucleus reactions where the thermalization process had caused a complete loss of information about the entrance channel.

In the intermediate region, energetic light particles and clusters were observed in experiments, especially in reactions induced by light projectiles at energies larger than 8-10 MeV/u. The observed emission was initially associated to the so-called "projectile break up" reactions.[22] However, when coincidence experiments started to be performed and more exclusive observable were registered it became clear that, even in those reactions where the major part of the projectile were absorbed by the target nucleus (incomplete fusion reactions), an important contribution of fast light-particles, namely pre-equilibrium particles, were present. The topic is almost as old as Nuclear Physics and both semi-classical and quantum-mechanical theories have been developed, starting from the 1960s where the pioneering work of Griffin was performed[15,16] on nucleon induced reactions. Since an analogous situation was seen in heavy-ion induced reactions, hybrid models have been updated to describe pre-compound decays.[23] As we will discuss in the following paragraphs, the statistical decay code PACE2[24] has been recently modified to include pre-equilibrium processes at high excitation energies using the *Moscow Pre-equilibrium Model.*[25]

3. Pre-equilibrium models

In order to describe the pre-equilibrium decay, a simplified picture was initially proposed by J.J. Griffin in the so called *Exciton model.*[15,16] The model was devoted to the description of the formation and decay of a compound nucleus in which a weak two-body residual interaction was present, causing transitions among the eigenstates of an independent particle Hamiltonian. These eigenstates were lying in a region of energy dE near the compound nucleus excitation energy E. The main ingredient of the *Exciton model* is the level density $\omega_n(E) = \omega_{ph}(E)$ of a nucleus in a state characterized by an excitation energy E and a number $n = p + h$ of excited quasi-particles (*excitons*). The subscripts p and h indicate respectively the number of excited particles and holes. Using the *detailed balance principle*, the quantity $\omega_{ph}(E)$ enters directly in the calculation of the partial decay rates for particle emission, where the excitation energy of the nucleus is shared between the number $n = p+h$ of excitons. The $\omega_{ph}(E)$ density then becomes proportional to the number of possible states accommodating p particles and h holes on the single particle level. Only energy conserving transitions which change the number of excitons by 0 or $\pm$ 2 are allowed (all transitions are assumed to proceed by binary processes). The decay is assumed to occur in a very short time towards a state with an outgoing particle of energy E_0 and where a residual nucleus of energy U remains. This happens every time a nucleus undergoes a transition to an independent-particle state in which one exciton has the energy E_0 in the continuum and the remaining excitons will share the energy $U = E-(E_0+B)$, with B the binding energy of the emitted particle. The decay is predicted to happen in this way in the case of nucleon induced reaction on an even-even target. For more complex configurations the residual interaction is expected to cause successive transitions to 3-, 5-, 7- exciton states. These decays, which are a small number at this stage, are considered pre-compound decays and describe the high energy tail of the particle's energy spectra.

The density $\omega_{ph}(E)$ is strongly dependent on the single particle level density $g(\epsilon)$, which can be described having an almost linear dependence on the single particle energy ϵ in the vicinity of the Fermi Energy. Increasing the excitation energy, it presents an increase of the partial contribution to ω_n(E) of highly excited single particle states and, in some cases, some excitons can even be located in the continuum. To determine the importance of a correct treatment of a coupling to the continuum, dedicated calculations have been performed.[26] At the end, after a certain number of two body interactions, an equilibrium is reached so that the normal decay from an equilibrated source can be taken into account, from a nucleus where a residual excitation energy U is left.

The main prediction of the *Exciton model* that could be tested by comparison to experimental data was that all targets bombarded by nucleons should give the same initial configuration characterized by the initial exciton parameter $n_0 = p+h$, that depends essentially on the projectile structure, with one unit uncertainty due to pairing effects.

After the first simplified model by J.J. Griffin, many other considerations have been made in order to give a reasonable description of the pre-equilibrium emission: among the others it is worth mentioning the one developed by Harp, Miller and Berne in 1968.[27] Their model is based on a Boltzmann Master Equation (BME) approach, that is a nuclear transport model based on nucleon-nucleon collision processes in the nuclear potential; this model describes the statistical evolution of a sample of systems, which are evolving from an initial state far from statistical equilibrium to an equilibrium state, through a sequence of two-body interactions and through the emission of unbound particles to the continuum. As an intermediate picture, it has to be mentioned the work by M. Blann, who suggested the so-called *Hybrid model.*[17]The formulation suggested by Blann was based on the calculation of the excited particle populations during equilibration, starting from the partial state densities, as in the simplified exciton model, while the intranuclear transition rates of the excited particles were determined by the calculations of the mean free paths of the nucleons in nuclear matter, like in the case of the BME master equation formalism. A first formulation of the mean free paths was performed based on the free nucleon-nucleon scattering cross sections corrected by Pauli principle, while in a second time values obtained by imaginary optical potential were also considered. A further version, the *Geometry-Dependent Hybrid model* (GDH model) was later developed, in which the effects of interactions in the diffuse nuclear surface were also taken into account.[29] In other words, the importance of the nuclear density distribution on pre-equilibrium decay was evaluated.

The Milano group has generalized the model,[14] further developing and refining the approach[13,30] multiplying by an empirically determined constant the calculated values of the nucleon mean free path, so that the calculated and experimental cross section were in agreement at one excitation energy.

When heavy ion induced reactions are considered, the Boltzmann Master Equation (BME) approach can be used, provided some initial condition are fulfilled (low variation of the states of bound nucleons in a potential well during the time evolution of the di-nuclear system in case of incompressible nuclear matter and volume-conserving deformations). In this way, the time evolution of the energy distribution of the nucleons can be evaluated as a result from the two-body nucleon-nucleon interactions and from the emission of particles in the continuum.[31] The BME theory unifies the description of nucleon induced and heavy ion induced reactions, even if this latter are much more complex processes. However, only average multiplicities of the emitted particles can be reproduced and their average energy and angular distributions, not permitting so far to compare results from exclusive experiments where the coincidences between emitted particles are registered.

The use of a Monte Carlo calculation has been therefore suggested,[32] where the differential multiplicities were interpreted as the probability of emission of the considered particle in the considered time interval. The energy and emission angle of the particle are calculated in the center of mass system of the initial composite system and then transformed in the laboratory system, together with the evaluation of the momenta of the recoiling nucleus, always in the laboratory system. The fast emission is stopped when the ejection of fast particles becomes negligible,

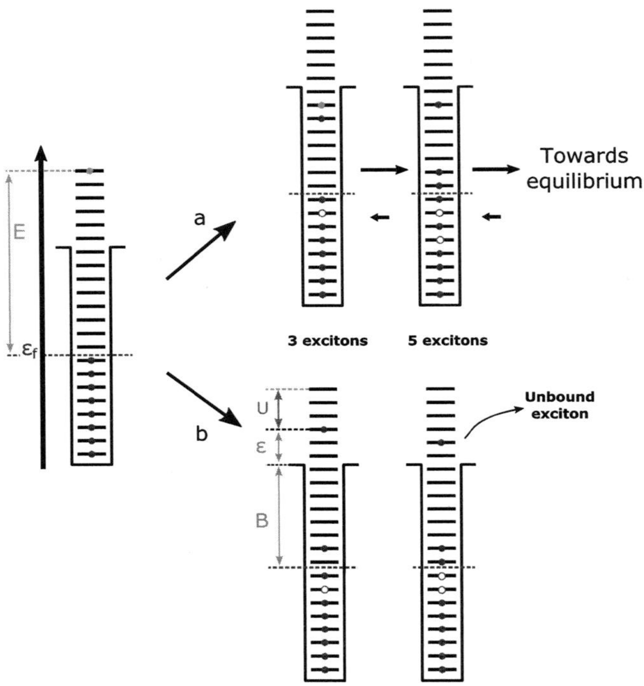

Fig. 2. Schematic representation of the first few stages of a reaction in the pre-equilibrium model. Particles are shown as solid circles, holes as empty circles and horizontal lines are the equally spaced levels in a potential well. The blue dashed line indicates the Fermi Energy of the system and the incoming nucleon is indicated by the red dot. In case (a) the thermalization process proceeds through the formation of exciton-hole couples following the big arrows. A small probability of returning to the previous exciton-hole configuration is shown by the small arrows. In case (b) the exciton is unbound and can be emitted as pre-equilibrium particle. Figure adapted from Ref. 28.

which normally happens about 50–100 fm/c after the two nuclei come in contact. At this time, the equilibration phase is taken into account and a statistical decay is considered. The particle identification, its kinetic energy and emission angle measurement are performed event by event together with the excitation energy, recoil angle and velocity of the residual nucleus. Comparison with experimental data has been performed successfully. The main problem, however, for calculating the pre-equilibrium emission in heavy ion collisions is the determination of the initial configuration of excitons. This has to be performed, essentially, by evaluating the overlap of the colliding nuclei in their momenta space.[18,33]

Another important parameter to be determined is the average squared transition matrix element of the residual interaction $|M|^2$. In principle, the matrix element $|M|^2$ can be calculated based on our knowledge on nucleon-nucleon scattering cross section and nuclear matter, but this does not permit to reproduce the observed cross-sections. Therefore a parametrization is often used:

$$|M|^2 = K' A^{-3} \epsilon^{-1} \tag{1}$$

where K' is a parameter derived from the fit to the data and $\epsilon = E/n$ is the average excitation energy per exciton, with E total available excitation energy.

Of course, dealing with complex systems is different with respect to a simple description of the initial configuration, like the one used with nucleon induced reactions: influence from either target or projectile structure, possible Coulomb interaction between projectiles and targets, coupling of angular momenta etc. can complicate the scenario.

4. Pre-equilibrium models and α-clustering

The formalism needed to describe formation probabilities and number of levels per MeV for particles other than nucleons (i.e. clusters like α, d, t, ^{3}He), together with transition probabilities, involves difficulties due to the complex mechanisms related to the cluster formation itself. Different approaches have been discussed, one related to a generalization of the exciton model with a naive physical meaning, which was, however, not able to reproduce the composite particle spectra.[34] Further methods were based on pre-formation hypothesis, in which the emitted α-particle is pre-formed in the ground state of the target nucleus with a certain probability and it could therefore interact with the incident nucleus, being emitted through a knock out mechanism.[35–37] However, the pre-formation probability needed to reproduce the data is often too high to be physically acceptable.

The main approaches used are:

(a) The cluster can be assumed as pre-formed in the nucleus and it may interact with the same kinematics of nucleons. It is then treated like an exciton in an exciton model formulation, just replacing the exciton number of the residual nucleus (p-1,h) by (p-p_x,h). In this approach it is assumed that the cluster is formed by p_x of the total p excited particles;

(b) Some fraction of all the energy partitions is assumed to have the right number of nucleons and associated momenta to permit a cluster emission;
(c) A quasi-free scattering approach is used for complex particles in the exit and entrance channels.

As previously said, it is a common hypothesis, verified by many observations, that nucleons in nuclei spend part of their time in a cluster structure, especially when they are close to the nuclear surface. This is why the link between pre-equilibrium phenomena and cluster structure is considered very strong. Normally a nucleus is described as composed by nucleons in a constant motion, each occupying a well defined shell-model orbital. Sometimes, however, nucleons can condense to produce a cluster that is a well defined spatially located sub-system, composed by strongly correlated nucleons: the strongest cluster, which in terms of binding energy is the one formed by two neutrons and two protons, that is the α-particle. A transient cluster can escape the nucleus (cluster decay) if it is close to the surface and if its kinetic energy is enough to overcome the binding energy. Sometimes this escape is caused by the knock out reaction induced by an incoming nucleon or by a more massive projectile, in other cases the cluster can be picked up. However, even if the constituents of a cluster are present in the nucleus, since they are surrounded by other nucleons, they cannot have the same properties as if they were in a free state.

For cluster emission in pre-equilibrium reactions, mainly two diametrically opposed models were proposed within the pre-equilibrium formalism in the 70's: on one side the idea that pre-formed α-particles were structurally present in the nucleus was supposed,[14,30,38] based on the fact that the α-particle is a very strongly coupled object, on the other side different coalescence models have been developed, where the formation of a cluster is happening in the course of the reaction in a dynamical way.[34]

The two pictures are quite different in the sense that, while in the first case the α-particle is considered so compact to be described as an exciton state, on the other case nucleons are all equal and the creation of a cluster by nuclear coalescence is a specific process, which can result in different kind of clusters, from weakly bound to strongly coupled ones (d, t ^{3}He, α-particles and heavier fragments etc.). Among the coalescence models, the most justified is the one by Iwamoto-Harada[39]: its basic idea is that both the excited nucleons and those from unexcited levels (lower than the Fermi level) can group together and contribute to form a cluster, with a particular form of pick-up reaction. Iwamoto-Harada suggested a general coalescence model, suitable for different kinds of emitted clusters, where even the spin dependence has been introduced. A better description of both the energy spectra shape and of their absolute values was reached for nucleon and α-induced reactions. The extension to heavy-ion collisions was asking for a description of the initial configuration depending not only on the number of excitons (n_0), but also on the allowed spin distribution of the composite non-equilibrium system,[40,41] which plays a very important role.

A generalization and indubitable improvement of the coalescence model was suggested by Bislinghoff,[42] where it was taken into account that not all the nucleons are available for cluster formation, but only those close to the Fermi energy, determining a kind of *availability band* on the bases of the binding energies of nucleons within the cluster. This fact has, as a consequence, the effect that the results become close to the pre-formed models, in the sense that strongly bound entities like α-particles have a large available space for pick up in this approach, while the pick up possibility for loosely coupled objects like the deuteron becomes very small, but are nevertheless considered. As in the evaporation theory, the coalescence of nucleons to form the cluster (their momentum correlation, angular momentum coupling, etc.) is governed by constraints given by the residual level density accessible. This is very different from the hypothesis of pre-formation or limited coalescence models, in which the key ingredient is the probability of forming a cluster prior to emission.

Further updates on the model were performed in order to keep the consistency between the Compound Nucleus (Weisskopf -Ewing) formulas and the *Exciton model* not only for nucleons, but also for the cluster emission .[40,41,43] Many other empirical formulations which have more predictive power depending on several adjustable parameters have been suggested,[44] while the previous ones can be considered almost parameter-less.

5. Experimental results and interpretation

Huge experimental efforts have been devoted to measure pre-equilibrium processes. Early works were limited to reactions induced by light beams (A $\leq$ 2) on stable targets (e.g. West[45]). Light charged particles were detected using single ΔE-E telescopes, which were moved around the beam axis. Light particles emitted with larger energies than what is averagely expected for central collisions, generally focused in the forward direction, were observed. These particles were attributed as being direct products of the early interaction of the incident projectile with the target nucleus, namely pre-equilibrium particles. Effects of the order of 10% were observed in the high-energy parts of the particle's spectra with respect to statistical model predictions. Among the others, the different types of *Hybrid Exciton* models discussed in the previous paragraphs were used to infer about possible α-particle pre-formation in the target nuclei by comparing simulations to experimental results from many (n,α), (d,α), (^{3}He,α) measured reactions (see Milazzo-Colli,[14,30] Chevarier[46]). A very comprehensive summary can be found in the review by Hodgson and Běták.[3]

With the advent of heavy-ion accelerators, fusion reactions involving two heavy partners became available. This gave the possibility of studying angular momentum effects in central collisions, properly varying the projectile energy and mass with respect to the target mass. In the comparison between particle's yield it became very important to account for quantities like projectile velocity, angular momentum, compound nucleus excitation energy and mass asymmetry of the reaction partners. In particular, the measurement of many reactions forming the same compound nucleus under different conditions still represents a good tool to disentangle between the different effects. The entrance channels are normally chosen in a way to form the same composite system, varying the mass asymmetry and the beam energy to keep constant either the projectile velocity or the compound nucleus excitation energy. For a detailed example, see Bisplinghoff's working Ref. 42 and 47.

A systematic organization of the pre-equilibrium neutron and LCP multiplicities obtained in several experiments has been performed by Cabrera and collaborators.[48] In their work data has been reported as a function of the incoming energy per nucleon (ranging from 5 to 30 MeV/u). A rapid growth of the neutron pre-equilibrium multiplicities and a slower increase of the LCP pre-equilibrium multiplicities has been observed for increasing energy, with the LCP pre-equilibrium process becoming evident around 10 MeV/u.

Similar observations have been obtained from two experimental campaigns carried out by the NUCL-EX collaboration.[a] The measurements

[a]T. Marchi (KU Leuven); F. Gramegna, M. Cinausero (LNL); D. Fabris (INFN Pd); M. Bruno, M. D'Agostino, L. Morelli (Univ and INFN Bo); G. Baiocco (Univ Pv); S. Barlini, M. Bini, G. Casini, A. Olmi, G. Pasquali, S. Piantelli, S. Valdrè (Univ and INFN Fi), E. Vardaci, G. La Rana (Univ Na)

Table 1. Summary of the reactions measured with the GARFIELD setup and discussed in the text.

Projectile			Target	Compound Nucleus			
Ion	E_{lab} (MeV/u)	v_{lab} (cm/ns)	Isotope	Isotope	Mass Asymm.	E* (MeV)	Lab. Vel. (cm/ns)
^{64}Ni	4.7	3.01	^{68}Zn	^{132}Ce	0.0009	100	1.46
^{64}Ni	6.3	3.48	^{68}Zn	^{132}Ce	0.0009	151	1.69
^{64}Ni	7.8	3.89	^{68}Zn	^{132}Ce	0.0009	203	1.88
^{16}O	8.1	3.96	^{116}Sn	^{132}Ce	0.76	100	0.48
^{16}O	12.0	4.82	^{116}Sn	^{132}Ce	0.76	155	0.58
^{16}O	15.6	5.49	^{116}Sn	^{132}Ce	0.76	206	0.67
^{16}O	16.0	5.56	^{65}Cu	^{81}Rb	0.60	209	1.10
^{19}F	16.0	5.56	^{62}Ni	^{81}Rb	0.53	240	1.30

have been performed at the Tandem-Alpi complex of the *Laboratori Nazionali di Legnaro*. The GARFIELD array was employed for light charged particles and fragments detection over a wide angual range ($29° \leq \theta \leq 151°$).[51] In some cases the backward GARFIELD chamber was replaced by large volumes of BaF_2 crystals for high-energy γ-ray detection.[50] The Evaporation Residues (ER) were detected at forward angles ($5° \leq \theta \leq 18°$) using either Multiwire Avalanche Counters[49] (MWAC) or the Ring Counter Array[51] (RCo). These detection configurations allowed us to reconstruct (almost) complete events in charge and to study particle-particle correlations as reported in the paper by Morelli and collaborators..[7,8]

In the two campaigns summarized in Table 1, the decay properties of ^{132}Ce* and ^{81}Rb* nuclei have been measured through the detection of light-charged particles in coincidence with evaporation residues. The excited nuclei were populated in fusion reactions at different energies. The link between the observed pre-equilibrium emission yields and cluster effects in the colliding nuclei was also investigated.

5.1. *Reactions populating* $^{132}Ce^{\star}$

The first campaign was aimed at measuring the decay from the ^{132}Ce* excited nucleus formed through two different mass entrance channels, each of which measured at three excitation energies. The aim of the experiment was to determine and compare the amount of pre-equilibrium emission in the two reactions having different entrance channel mass asymmetries, in order to determine the effective temperature of the source of the high energy γ-rays emitted from the Giant Dipole Resonance (GDR) .[52,53] In

Fig. 3. Pictorial view of the geometrical configuration used for the first pre-equilibrium campaign with the GARFIELD apparatus. On the left the 8 BaF_2 of the HECTOR array are shown for γ-ray detection, in the center the GARFIELD Drift chamber, on the very right the two couple of MWACs for Evaporation Residue tagging.

this experiment, the ER produced in the complete or incomplete fusion reactions have been detected using two couples of MWACs, working in anti-coincidence to disentangle ER from elastic scattering and fast process events. Figure 3 shows a pictorial view of the used experimental configuration. At the beginning, in order to describe the experimental results, the *moving source analysis technique* was used. This global fit allows the consideration of different sources of the emitted particles. The resulting parameters have been obtained considering the whole available angular distribution, cross-checked on the total used detectors (192). A unique evaporative source fit was, in general, not sufficient to describe the experimental spectra in the case of the asymmetric mass entrance channel. A second pre-equilibrium source with a velocity intermediate between the recoiling compound nucleus and the beam had to be added. Eventually two *thermal* sources have been considered. The first source is based on the surface emission from the fully thermalized composite system and associated with the statistical particle distribution (*evaporative source*). The evaporative particle spectrum in the emitting source frame is represented by a Maxwellian distribution:

$$\frac{d^2N_1}{d\Omega dE} = \frac{N_1}{4\pi {T_1}^2}\,(E - E_{C1})\,e^{-\frac{E-E_{C1}}{T_1}}\;[1 + a_2 P_2(cos\theta_{\mathrm{SRF}})], \qquad (2)$$

where E is the energy of the particle, E_{C1} is the Coulomb energy parameter, T_1 is the apparent nuclear temperature of the emitting evaporative source, a_2 is the angular coefficient, P_2 is the Legendre polynomial, θ_{SRF} is the emission angle in the source reference frame and N_1 is the total number of particles emitted from this source.

The second is a *fictitious thermal source* based on a volume emission and associated to the *pre-equilibrium source* (Watt distribution):

$$\frac{d^2N_2}{d\Omega dE} = \frac{N_2}{2(\pi T_2)^{\frac{3}{2}}}\,(E - E_{C2})^{\frac{1}{2}}\, e^{-\frac{(E-E_{C2})}{T_2}}, \tag{3}$$

where E_{C2}, T_2, and N_2 are the Coulomb Energy parameter, the apparent nuclear temperature and the total number of particles emitted from the pre-equilibrium source, respectively. The total experimental yield is described as an incoherent sum of the distributions from two sources. The fit was performed in the laboratory system using the following reference frame transformation for each source:

$$\left.\frac{d^2N}{d\Omega dE}\right|_{\mathrm{LAB}} = \frac{E_{\mathrm{LAB}}}{E'}\left.\frac{d^2N}{d\Omega dE}\right|_{E=E'} \tag{4}$$

with E' energy of the particle in the source reference frame. The fitting procedure gives 8 total output parameters: N_1, T_1, E_{C1}, a_2, N_2, T_2, E_{C2}, and v_2, with v_2 pre-equilibrium source velocity ($v_{CN} \leq v_2 \leq v_{beam}$). The results of the moving source fit are shown in Tables 2 and 3, where the $\frac{N_2}{N_1}$ is the ratio of the total yield of particles from the pre-equilibrium source to the total yield of particles from the evaporative source. The accounted error for the ratio $\frac{N_2}{N_1}$ is 10%. The results of the fit show a visible deviation from pure evaporative behavior for α-particles ($\frac{N_2}{N_1}$=0.21) in the mass asymmetric case even at the lower energy (8.1 MeV/u), while in the mass symmetric reaction, at almost the same beam velocity (7.8 MeV/u), only a feeble deviation can be observed ($\frac{N_2}{N_1}\leq$0.1). At the highest measured energy (15.6 MeV/u), a large fraction of non-equilibrium particles (α-particles and protons) are present in the ^{16}O+^{116}Sn reaction case: in particular, the ratio (N_2/N_1) is 1.3 for α-particles and 0.3 for protons.

To better explain the behavior observed in the ^{16}O+^{116}Sn case, a comparison to a calculation using the *Moscow Pre-equilibrium Model*[25] (MPM) was performed. The model, based on the *Griffin Exciton model*, relies on the statistical code PACE2: the main modification is related to the insertion of a non-equilibrium stage prior to the fusion reaction. The relaxation of

Table 2. Moving source fit parameters for the reaction $^{64}Ni+^{68}Zn$ at 4.7, 6.3 and 7.8 MeV/u. The parameters indicated with * are fixed (no fit procedure has been performed for the 7.8 MeV/u ^{64}Ni + ^{68}Zn reaction, and the parameters were derived by assuming/fixing a pre-equilibrium α-particle distribution capable of describing the small extra yield).

E(MeV/u)	LCP	$\frac{N_2}{N_1}$	E_{C1}(MeV)	T_1(MeV)	E_{C2}(MeV)	T_2(MeV)	$\frac{v_2}{v_{CN}}$
4.7	p	0.0	3.42 ± 0.56	1.93 ± 0.17	—	—	—
4.7	α	0.0	11.6 ± 0.75	2.71 ± 0.26	—	—	—
6.3	p	0.0	3.71 ± 0.98	2.89 ± 0.25	—	—	—
6.3	α	0.0	11.2 ± 1.1	3.93 ± 0.59	—	—	—
7.8	α	≤ 0.1	12.1 ± 0.5	4.24 ± 0.56	12.4*	7.68*	1.60*

Table 3. Moving source fit parameters for the reaction $^{16}O+^{116}Sn$ at 8.1 and 15.6 MeV/u.

E(MeV/u)	LCP	$\frac{N_2}{N_1}$	E_{C1}(MeV)	T_1(MeV)	E_{C2}(MeV)	T_2(MeV)	$\frac{v_2}{v_{CN}}$
8.1	p	0.00	2.77 ± 0.46	3.02 ± 0.16	—	—	—
8.1	α	0.21	11.0 ± 1.50	3.20 ± 0.35	12.4 ± 1.70	3.80 ± 0.75	3.83
15.6	p	0.30	3.16 ± 0.44	3.55 ± 0.47	4.49 ± 1.53	4.68 ± 0.83	5.49
15.6	α	1.30	12.3 ± 0.20	4.38 ± 0.28	12.2 ± 0.90	7.10 ± 0.60	2.85

the compound system to equilibrium is described by the Master Equation:

$$\frac{dq\,(n,t)}{dt} = \sum_{m=n-2}^{m=n+2} \lambda_{m\rightarrow n} q\,(m,t) - q\,(n,t) \left(\omega\,(n) + \sum_{m=n-2}^{m=n+2} \lambda_{n\rightarrow m} \right) \tag{5}$$

where $q(n,t)$ is the occupation probability for the composite nucleus state n, $\omega(n)$ is the emission rate of light particles, and $\lambda_{m\rightarrow n}$ is the internal transition rate. The positive terms at the right-hand side of the equation describe the feeding to the state n from all possible states m, and the negative terms account for the losses of the system due to emission and transitions to other exciton states. The internal transition rates, $\lambda_{m\rightarrow n}$, are determined by the matrix element of the transitions $|M|^2$ and the density of the exciton states to which the transition occurs $\omega_{o,\pm}(n,E)$. The parameter K', which is connected to the transition matrix element according to Eq. 1, was varied within the range of 200 to 800 MeV^3. In the model the density of the exciton states are determined by the single particle level density, which is related to the level density parameter a in the Fermi-gas model. The same values of the evaporative case were used for these parameters. The variation of the parameter a was based on the Fermi-gas model and the level density on the phenomenological model with the Grudzevich parameterization.[55]The details of the transition rate calculations used in the approach can be found in the work by Eremenko *et al.*.[54] The initial exciton configuration $n_0 = p + h$, from which the equilibration process starts, is the free parameter of the model. According to the estimate of E. Běták,[40] the initial number of excitons is derived from the equation $E^\star/n_0 = 6.8 + 0.54(E_{CM} - V_C)/A_p$. In the calculation, the following initial exciton configuration was used: n_0= 17(16p,1h) for the reaction ^{16}O+^{116}Sn. Major details on the calculations can be found in the paper by O. Fotina.[25] Different implementations of the model have been considered. **MPM A** is based on the Exciton model where the angular dependence of the pre-equilibrium emission are derived from the Optical model, from which we determine the angular momentum of the emitting particles and using the associated Legendre functions we select the particles ejection angle. **MPM B** is based on the *Hybrid model.* In this framework, the angular dependence was introduced as $q(n,\Omega,t)$, where Ω is the direction of the fast particle with respect to the incident direction. The states of the system are then characterized by (n,Ω). During the dissipation the memory of the direction is lost. In each act of dissipation the exciton scattering changes the direction from Ω to $\Omega\,'$. A generalized Master Equation is used:

$$\frac{dq\left(n,\Omega,t\right)}{dt} = \sum_{m=n-2}^{m=n+2} \lambda_{m\to n} \int d\Omega' G\left(\Omega,\Omega'\right) q\left(m,\Omega',t\right) + \\ -q\left(n,\Omega t\right)\left(\omega\left(n\right) + \sum_{m=n-2}^{m=n+2} \lambda_{n\to m}\right) \tag{6}$$

where $G(\Omega,\Omega')$ is determined by the free differential nucleon-nucleon scattering cross section.

Assuming the free differential nucleon-nucleon cross section to be isotropic in the center of mass of the nucleons and assumed that the last one is isotropic in the center-of-mass system of the nucleus, the initial condition for the Eq. 6 is

$$q\left(n,\Omega,t=0\right) = N\delta_{n,n_0}\pi^{-1}\cos(\beta\theta_{lab})H\left(\frac{\pi}{2} - \beta\theta_{lab}\right)\frac{d\Omega_{lab}}{d\Omega} \tag{7}$$

where N is a normalization coefficient, H is the Heavyside function, β=$kR/8$, where k is the incident nucleon momentum and R is the nuclear radius. The factor β takes into account the finite nuclear size and leads to significant enhancement of the angular dependence of the double differential spectrum.

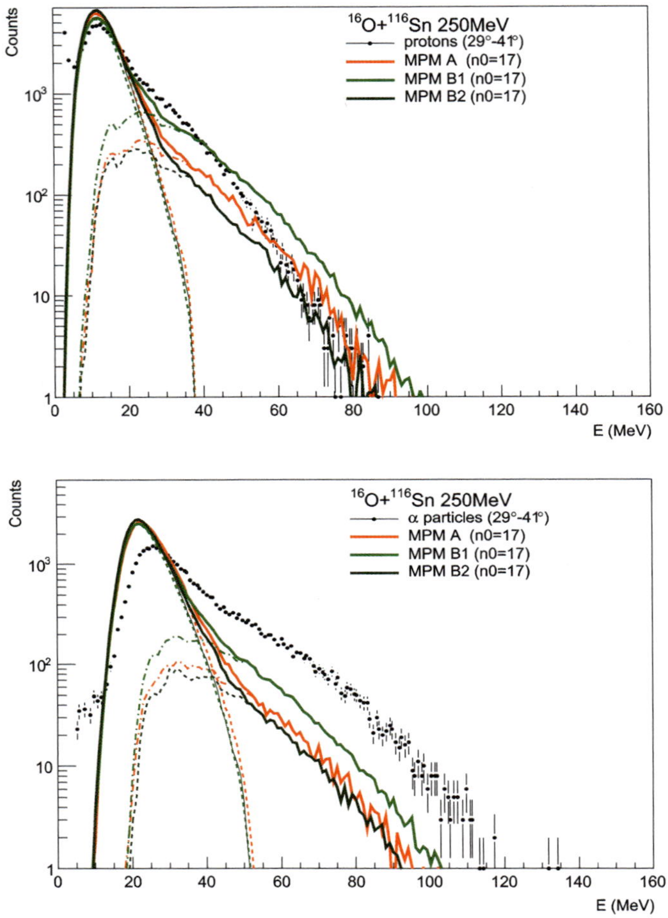

Fig. 4. Relative yield for protons (top panel) and α-particles (bottom panel) emitted in the reaction ^{16}O+^{116}Sn at 250 MeV. The displayed angular range is $29° \leq \theta \leq 41°$. See text for explanation on different simulations. Simulations (MPM A, MPM B1, MPM B2) and data (full circles) are normalized to the area to be compared.

Figure 4 shows the comparison of the experimental distribution of protons and α-particles measured in coincidence with ER and emitted in the angular range $29° \leq \theta \leq 41°$ with different model predictions. For the case B, the parameters which have been used in the calculations are: $\beta = 1.8$, CN pre-formation factor $= 0.5$ for the case B1 and $\beta = -1$, CN pre-formation factor = 0.95 (maximum value allowed in the model) in the case B2. At all measured energies, over the full angular range ($29^0 \leq \theta \leq 85^0$), a slight overestimation was observed for the proton distributions. For α-particles, a quite good agreement persists between theory and experiment at large angles ($\geq 60°$) while a significant underestimation has been observed in the forward direction. In fact, as shown in Figure 4 for the higher energy and the most forward measured angle, all the three model predictions are not in agreement with the experimental α-particle spectra, which still show an extra production. In general, it can be observed that when the model reproduces/overestimates the proton data, with the same input parameters, it strongly underestimates the fast α emission. The analysis of these results led to the conclusion that the pre-equilibrium emission contribution, whatever the implemented recipe, is not sufficient to explain the increased production of α-particles, even considering a value of 0.95 that is the maximum allowed CN pre-formation factor (case B2). It is, therefore, necessary to consider additional effects in the calculation, for example the clustering structure of the projectile, which for the ^{16}O projectile is a possible explanation.

The method for estimating the formation probability of the cluster, together with the algorithm used in the Monte-Carlo procedure to determine the exciton energies for cluster/light ion induced reactions, were presented in the work of M. Blann and M.B. Chadwick.[56]

The expression $e = E(1-(1-x)^{\frac{1}{(n_0-1)}})$ was used to estimate the energy e of the cluster consisting of n_0 initial excitons. Here E is the initial projectile energy and x is a random number with uniform distribution in (0,1). In our case, according to the law of energy conservation, if e_4 is the α-cluster energy emitted by the Oxygen projectile, the remaining Carbon nuclei should have energy $e_{12} = E_{16O} - B_\alpha - e_4$.

The effect of clustering structure is shown in Figure 5, where the experimental data are compared with the model predictions with 10%, 50% and without (0%) *α-clustering probability* in the projectile nucleus. From this comparison it is clear that while the proton distribution seems to be described with no-clustering (0%), this is not the case for the α-particles. Introducing a large clustering probability improves the description of the fast emitted particles at forward angles but the agreement is still not satisfactory. Unfortunately in this campaign the most forward angles ($\theta \leq 30^0$) were covered by MWAC detectors with low efficiency and bad

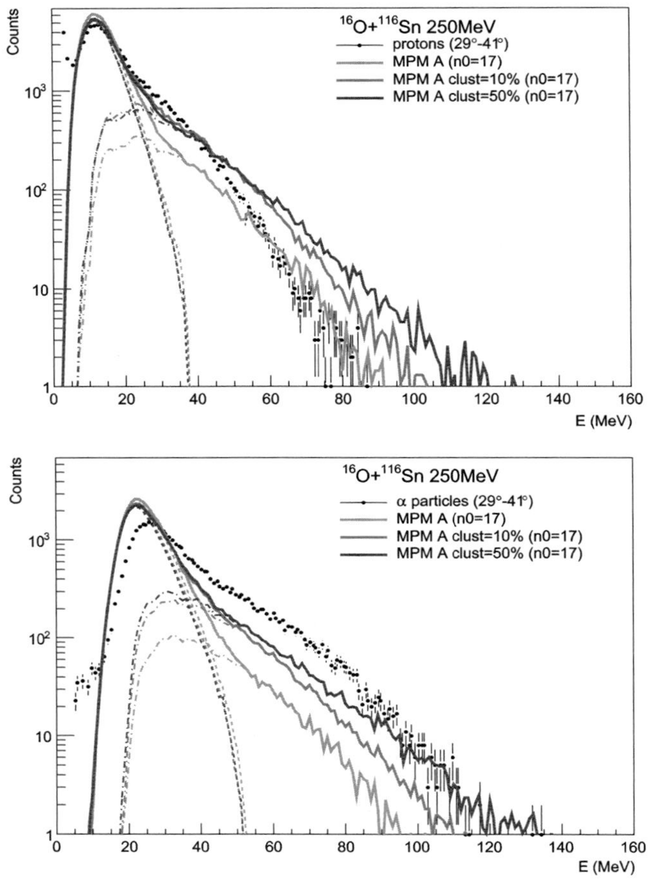

Fig. 5. Relative yield for protons (top panel) and α-particles (bottom panel) emitted in the reaction $^{16}O+^{116}Sn$ at 250 MeV. The displayed angular range is $29^0 \leq \theta \leq 41^0$. See text for explanatin on different simulations. Simulations (MPM A, MPM A clust 10%, MPM A clust 50%) and data (full circles) are normalized to the area to be compared.

energy resolution for light charged particles, so that no further useful information could be derived.

5.2. *Reactions populating $^{81}Rb^{\star}$*

In the second campaign, two reactions have been used to populate the ^{81}Rb nucleus in order to obtain more information about the α-cluster structure of the projectiles. Different mass asymmetric entrance channels have been used to obtain experimental evidences in a model independent way. Moreover, the pre-equilibrium spectra have been studied with the help of the MPM pre-equilibrium model.

The double magic (^{16}O) and the non-magic projectile (^{19}F) impinged respectively on ^{65}Cu and ^{62}Ni targets. In both cases the ^{81}Rb* compound nucleus was populated. In both reactions the beam energy was 16 MeV/u in order to directly compare the light charged particle emission spectra and multiplicities. The same projectile velocity was chosen because the pre-equilibrium emission is expected to depend mainly on this parameter.[48] Indeed, even with an expected small difference in the evaporative part due to the excitation energy (see Table 1), in this situation the fast emission process is predicted to be almost the same for both reactions. Any observed difference and overproduction of fast α-particles between the two cases would suggest, in a model independent way, a possible influence of the projectile α-structure effect. The experimental setup used was an upgraded version of the one employed for the O+Sn reactions. In this case the full GARFIELD setup was used, doubling the angular coverage with respect to the previously mentioned experiment ($29^0 \leq \theta \leq 151^0$). Moreover, the Ring Counter (RCo) array was used at the most forward angles ($5^0 \leq \theta \leq 18^0$), permitting to measure with high resolution the light charged particle spectra also in this angular range. Even if only preliminary results are publisehd to date,[57] from the preliminary analysis the double differential energy spectra obtained in coincidence with ERs for α-particles and protons have been produced.

The two reactions show very similar proton spectra on the whole angular range, except for a small difference at the most forward angle, which can be ascribed to the slightly larger excitation energy of the ^{19}F induced reaction. A much larger difference is, on the contrary, observed in the case of the α-particle emission spectra, especially at the most forward detection angles. The calculated emission spectra performed with an evaporative code like GEMINI,[58] which mainly takes into account the difference in the excitation energy of the compound nucleus, confirms a small difference for the two systems and compatible to that one experimentally observed in the proton decay channel. However, when compared to the data, they also support the idea that a second fast emission source is needed for both systems, to better describe mainly the α-particle decay channel. A first evaluation of the expected amount of fast emission in the two cases was performed by means of the prediction of the **(MPM B)** calculation. A preliminary calculation was performed considering only the GARFIELD angular range, for which the calibration procedure was completely established. An initial exciton number of n_0=17(16p+1h) was used in the case of $^{16}O+^{65}Cu$. With this configuration the calculation seems to reasonably describe the shape of the α-particle spectra, except for an underestimation at the most forward angle in GARFIELD. On the contrary, with the same initial parameters, it strongly overestimates the proton pre-equilibrium emission. In Figures 6 and 7 the comparison of the calculation is shown for protons and α-particles

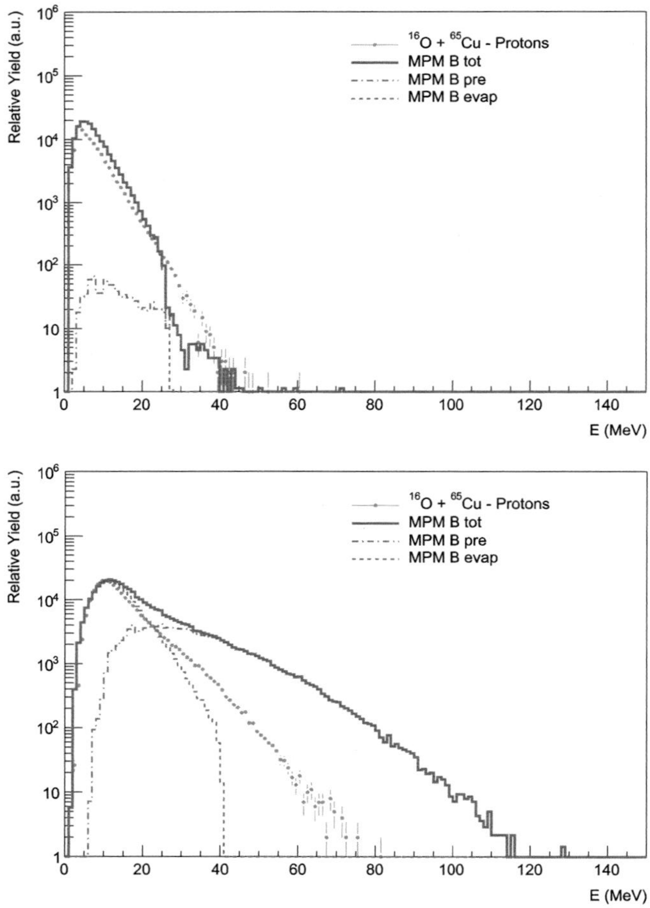

Fig. 6. Relative yield for protons emitted in the reaction $^{16}O+^{65}Cu$. The displayed angular range is $139^0 \leq \theta \leq 151^0$ (top panel) and $29^0 \leq \theta \leq 41^0$ (bottom panel) respectively. Comparison to Hybrid model calculation (B) with n_0=17 (16p+1h) initial configuration.

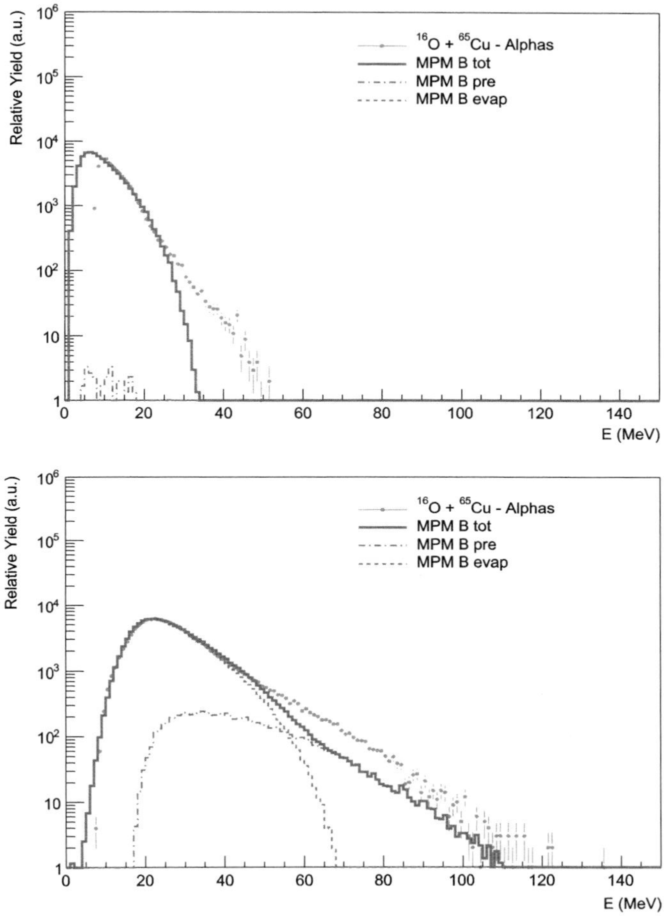

Fig. 7. Relative yield for α-particles emitted in the reaction $^{16}O+^{65}Cu$. The displayed angular range is $139^0 \leq \theta \leq 151^0$ (top panel) and $29^0 \leq \theta \leq 41^0$ (bottom panel) respectively. Comparison to Hybrid model calculation (B) with $n_0=17$ (16p+1h) initial configuration.

in the $^{16}O+^{65}Cu$ system at the most backward (upper panels) and most forward (lower panels) GARFIELD angular range, respectively. Performing the same comparison in the case of the $^{19}F+^{62}Ni$ reaction, where an initial exciton configuration of $n_0=20(19p+1h)$ was used, a quite similar result is obtained, as shown in the Figures 8 and 9, displayed for the same angular range: the experimental fast α-particle overproduction is even larger than in

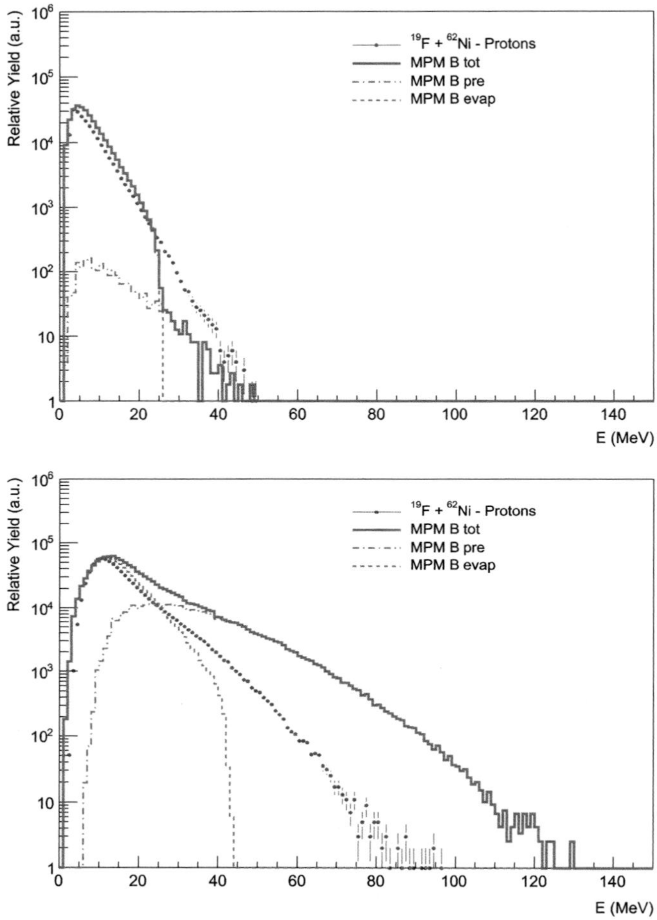

Fig. 8. Relative yield for protons emitted in the reaction $^{19}F+^{62}Ni$. The displayed angular range is $139^0 \leq \theta \leq 151^0$ (top panel) and $29^0 \leq \theta \leq 41^0$ (bottom panel) respectively. Comparison to Hybrid model calculation (B) with n_0=20 (19p+1h) initial configuration.

the ^{16}O induced reaction, while the fast protons are, again, largely overestimated. By changing the initial configuration parameter (i.e. diminishing the exciton number) the description of the α-particle slightly improves, but, on the contrary, this worsen the proton description. A tentative and preliminary explanation for the observed experimental difference between the α-particle decay in the two systems may be the lower energy needed to

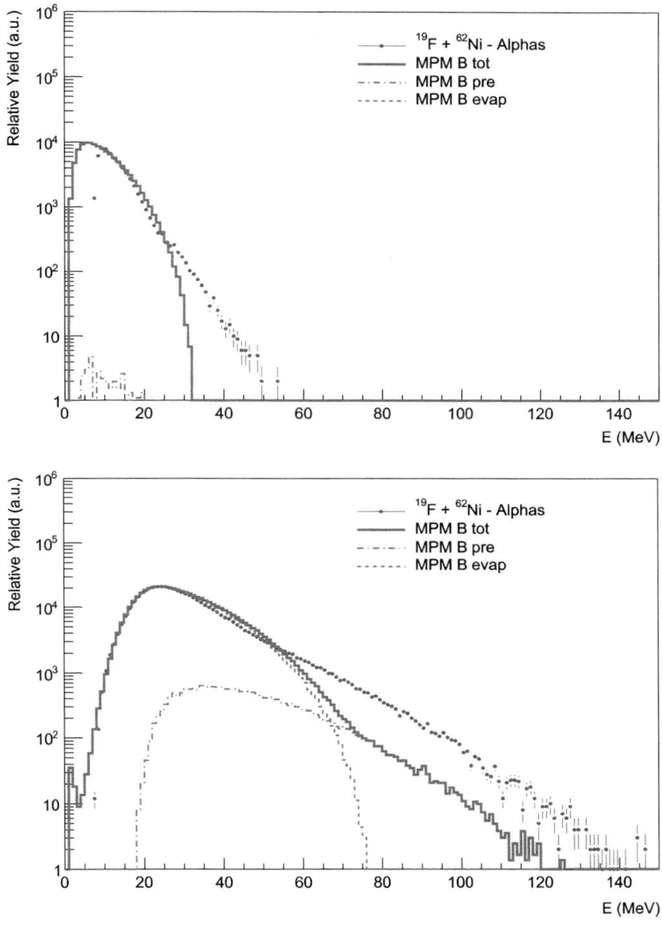

Fig. 9. Relative yield for α-particles emitted in the reaction ^{19}F+^{62}Ni. The displayed angular range is $139^0 \leq \theta \leq 151^0$ (top panel) and $29^0 \leq \theta \leq 41^0$ (bottom panel) respectively. Comparison to Hybrid model calculation (B) with n_0=20 (19p+1h) initial configuration.

break up the ^{19}F nucleus into α+^{15}N (4.01 MeV) with respect to the ^{16}O to be divided into α+^{12}C (7.2 MeV). Once again, like it was observed in the ^{16}O+^{116}Sn case, a unique initial parameter in the Hybrid Exciton model is not able to describe both the proton and α-particle channels suggesting that some more attention has to be paid to the clustering structure effects. It has to be stressed that in the used calculation no weak clusters (deuteron, triton, ^{3}He) emission is considered, which may strongly influence the relative isotopic decay probability. Moreover, a correct filtering of the simulated Evaporation Residue distribution has yet to be introduced. The analysis

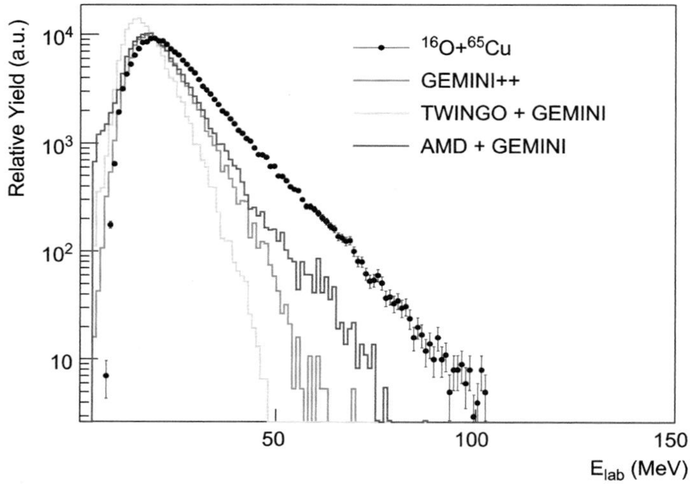

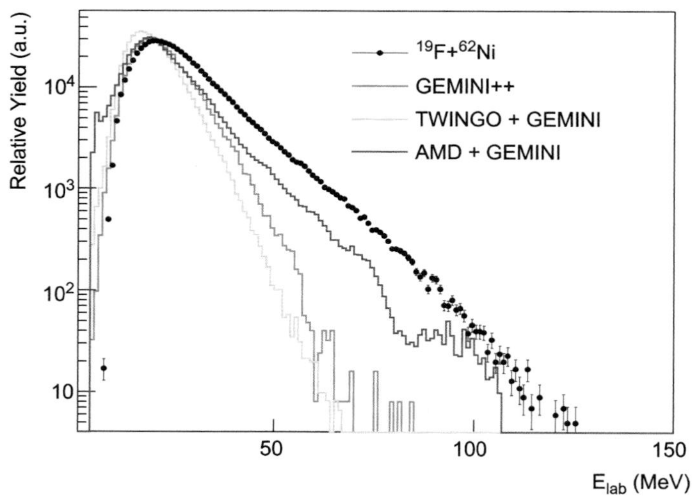

Fig. 10. Relative yield for α-particles emitted in the reactions $^{16}O+^{65}Cu$ (top) and $^{19}F+^{62}Ni$ (bottom) in the $29^0 \leq \theta \leq 41^0$ angular range. Experimental data are compared to GEMINI++ (green), SMF+GEMINI (red) and AMD+GEMINI (blue) models prediction.

of these data is still undergoing together with a new implementation of the MPM model. In particular, the forward detection angles can now be compared to the simulations in the energy range between $5^0 \leq \theta \leq 18^0$. Moreover, different statistical and dynamical calculations are undergoing: GEMINI^{++}, SMF (Stochastic Mean Field)+GEMINI ,[59] AMD (Antisymmetrized Molecular Dynamics)+GEMINI.[60] Preliminary results seem to indicate that approaches like the one predicted by AMD+GEMINI calculations, which consider a certain percentage of clusterization of the projectile, are more appropriate than SMF, as shown in Figure 10.

6. New possibilities using exotic ions

At the new generation radioactive ion beam facilities (like SPES [b] at LNL), exotic beams with energies up to 10-15 MeV/u will soon become available. New systems will be studied in an energy window where the onset of pre-equilibrium processes is expected. Medium-mass radioactive projectiles will exhibit extreme neutron to proton ratios and cluster configurations might show up. Reactions in which a double magic projectile (e.g. ^{132}Sn) interacts either with an N=Z target or with a similar N≠Z target ($Z = \pm 1$) (e.g. ^{28}Si, ^{27}Al and ^{30}Si, respectively), may provide different pre-equilibrium yields, which may be eventually associated to molecular states, possibly enhanced by the valence neutrons [c].

Cluster configurations where one of the clusters is a double-magic nucleus can be also evidenced when studying resonant elastic scattering reactions. Indications of α-cluster states in heavy nuclei have been found in ^{212}Po. Since cluster model calculations predict the existence of these states also in ^{136}Te, the use of the Resonant Scattering Method applied to the reaction ^{132}Sn+^{4}He will be of interest. Exploiting the newly developed *Active Target* technology[61] will allow the measurement of the equivalent of an *infinite target.* In this approach the exotic beam is sent into an extended gas target at such a pressure to be fully stopped. The gas acts at the same time as target and as beam energy degrader. Since interactions between the projectiles and the target nuclei occur continuously at different positions of the target, different energies can be measured at the same time. By Tracking the emitted particles it is possible to reconstruct the interaction vertex position with good precision, overcoming the thick target limitations

---

[b]https://web.infn.it/spes/
[c]T. Marchi *et al*, Letter of Intent to the SPES Scientific Advisory Committee

and allowing the measurement of a scattering excitation function in a given energy range with a single beam energy.

The measurement of the elastic scattering excitation function at backward angles in the center of mass will give information on the intermediate compound nucleus states, which then have a large probability to decay back to the same channel as the entrance one. Performing elastic scattering on α-particles, the study of states which have large α partial-widths as expected for α-cluster states can be carried out. At backward angles, nuclear effects are more important and the resonant scattering is maximized over the Coulomb scattering, so that the visibility of resonances is enhanced.

The importance of studying pre-equilibrium processes and their connection with clustering in medium-mass nuclei has been outlined in the present review. Theories and experiments have been described starting from the early works on nucleon-induced reactions and reaching the most recent works done with heavy ion stable beams and sophisticated experimental setups. Despite the huge amount of data collected and the massive theoretical efforts done to interpret the available information, an ultimate understanding of the mechanism driving the fast emission processes is still missing. It has been also shown that clustering effects play a crucial role in this context, enhancing the interest in studying the effects of the nuclear structure in the dynamical evolution of heavy-ion collisions. The availability of radioactive ion beams and new generation detectors will give the opportunity to explore strongly neutron unbalanced regions of the nuclear chart, where clustering phenomena are expected to play a major role.

References

1. W. von Oertzen *et al.*, *Phys. Rep.* **432**, 43 (2006).
2. M. Freer, *Rep. on Progr. in Phys.* **70**, 2149 (2007).
3. P. Hodgson and E. Běták, *Physics Reports.* **374**, 1 (2003).
4. B. Buck *et al.*, *Journ. of Phys. G* **17**, 1223 (1991).
5. N. Takigawa and A. Arima, *Nuclear Physics A.* **168**, 593 (1971).
6. M. Freer and H. Fynbo, *Progr. in Part. and Nucl. Phys.***78**, 1 (2014).
7. L. Morelli *et al.*, *Journ. of Phys.G: Nucl.&Part.Phys.***41**,075107 (2014).
8. L. Morelli *et al.*, *Journ. of Phys.G: Nucl.&Part.Phys.* **43**,045110 (2016).
9. R. Bijker and F. Iachello, *Annals of Physics.* **298**(2), 334 (2002).
10. M. Gai *et al.*, *Journal of Physics: Conference Series.* **569**, 012011 (2014).
11. D. Brink *et al.*, *Physics Letters B.* **33**, 143 (1970).
12. R. Bijker and F. Iachello, *Phys. Rev. Lett.* **112**, 152501 (2014).

13. D. Brink and J. Castro, *Nuclear Physics A.* **216**, 109 (1973).
14. L. Colli-Milazzo and G. Marcazzan-Braga, *Physics Letters B.* **38**, 155 (1972).
15. J. J. Griffin, *Phys. Rev. Lett.* **17**, 478 (1966).
16. J. J. Griffin, *Phys. Rev. Lett.* **19**, 57 (1967).
17. M. Blann, *Phys. Rev. Lett.* **27**, 337 (1971).
18. N. Cindro *et al.*, *Phys. Rev. Lett.* **66**, 868 (1991).
19. A. J. Koning, J. M. Akkermans *Pre-equilibrium nuclear reactions: An introduction to classical and quantum-mechanical models*, World Scientific (1999).
20. V. Zagrebaev, *Annals of Physics.* **197**, 33 (1990).
21. P. E. Hodgson, *Reports on Progress in Physics.* **50**, 1171 (1987).
22. A. Bonaccorso and D. M. Brink, *Phys. Rev. C.* **43**, 299 (1991).
23. M. Blann, *Nuclear Physics A.* **235**, 211 (1974).
24. A. Gavron, *Phys. Rev. C.* **21**, 230 (1980).
25. O. V. Fotina *et al.*, *Int. Journ. of Mod. Phys. E* **19**, 1134 (2010).
26. S. Shlomo and V. M. Kolomietz, *Rep. on Progr. in Phys.* **68**, 1 (2005).
27. G. D. Harp *et al.*, *Phys. Rev.* **165**, 1166 (1968).
28. C. Cline and M. Blann, *Nuclear Physics A.* **172**, 225 (1971).
29. M. Blann, *Phys. Rev. Lett.* **28**, 757 (1972).
30. L. Milazzo-Colli and G. Braga-Marcazzan, *Nucl. Phys. A.* **210**, 297 (1973).
31. E. Fabrici *et al.*, *Phys. Rev. C.* **40**, 2548 (1989).
32. M. Cavinato *et al.*, *Nucl. Phys. A.* **679**, 753 (2001).
33. N. Cindro, *Fizika B.* **1**, 51 (1992).
34. C. Cline, *Nuclear Physics A.* **193**, 417 (1972).
35. L. M. Colli *et al.*, *La Rivista del Nuovo Cimento (1971–1977).* **3**, 535 (2008).
36. E. Gadioli, E. G. Erba, and J. J. Hogan, *Phys. Rev. C.***16**, 1404 (1977).
37. A. Ferrero *et al.*, *Zeit. für Phys. A Atoms & Nuclei* **293**, 123 (1979).
38. L. Milazzo-Colli and M. Braga-Marcazzan, *Phys. Lett.B* **36**, 447 (1971).
39. A. Iwamoto and K. Harada, *Phys. Rev. C* **26**, 1821 (1982).
40. E. Běták, *Fizika B* **12**, 11 (2003).
41. E. Běták and J. Cseh (2015).
42. J. Bisplinghoff and H. Keuser, *Phys. Rev. C* **35**, 821 (1987).
43. P. Obložinský and I. Ribanský, *Physics Letters B* **74**, 6 (1978).
44. C. Kalbach, *Zeit. für Phys. A Atoms and Nuclei* **283**, 401 (1977).
45. R. W. West, *Phys. Rev.* **141**, 1033 (1966).
46. A. Chevarier *et al.*, *Phys. Rev. C* **11**, 886 (1975).
47. J. Bisplinghoff, *Phys. Rev. C* **50**, 1611 (1994).
48. J. Cabrera *et al.*, *Phys. Rev. C* **68**, 034613 (2003).
49. M. Bini *et al.*, *Nucl Instr and Meth A* **515**, 497 (2003).
50. A. Maj *et al.*, *Nucl Phys A* **571**, 185 (1994).
51. M. Bruno *et al.*, *Eur. Phys. J. A* **49**, 128 (2013).
52. O. Wieland *et al.*, *Journ. of Phys.G: Nucl. & Part.Phys.* **31**, S1973 (2005).
53. A. Corsi *et al.*, *Physics Letters B* **679**, 197 (2009).
54. D. O. Eremenko *et al.*, *Physics of Atomic Nuclei* **65**, 18 (2002).
55. O. Grudzevich *et al.*, *Rad. Prot. Dos.* **126**, 101 (2007).
56. M. Blann and M. B. Chadwick, *Phys. Rev. C* **62**, 034604 (2000).

57. F. Gramegna *et al.*, *Journ. of Phys.: Conf. Series* **580**, 012011 (2015).
58. R. J. Charity, *Phys. Rev. C* **82**, 014610 (2010).
59. M. Colonna *et al.*, *Nuclear Physics A* **642**, 449 (1998).
60. A. Ono, *Phys. Rev. C* **59**, 853 (1999).
61. S. Beceiro-Novo *et al.*, Prog. in Part. and Nucl. Phys. 124 (2015).

CHAPTER 21

"NECKLACE" CLUSTER FRAGMENTATION OF THE DINUCLEAR SYSTEM IN DISSIPATIVE REACTIONS

S. O. Nyibule[*,a], M. J. Quinlan[*], E. Henry[*], H. Singh[*,b], I. Pawelczak[*], J. Tõke[*], W. U. Schröder[*], L. Auditore[†], G. Cardella[‡], M. B. Chatterjee[§], E. De Filippo[‡], G. Lanzalone[†,¶], C. Maiolino[†], A. Pagano[‡], M. Papa[‡], S. Pirrone[‡], G. Politi[‡], F. Rizzo[||], P. Russotto[||], A. Trifiro[‡], G. Verde[‡]

[*] *Departments of Chemistry and Physics, University of Rochester, Rochester, NY 14627, USA*
[†] *INFN-Gr. Coll. Messina and Dip. di Fisica Università, Messina, Italy*
[‡] *INFN Sezione di Catania, Italy*
[§] *Saha Institute for Nuclear Physics, Kolkata, India*
[¶] *INFN LNS, Catania and Università Kore Enna, Italy*
[||] *INFN Laboratori Nazionali del Sud, Catania, Italy*

1. Introduction

Since its beginnings, nuclear science has focused on the macroscopic shapes of atomic nuclei, droplets of a (Fermi) fluid *sui generis*. In spite of the enormous progress in understanding made in the past 75 years, explorations of the characteristic dependencies of the stable nuclear shapes and nuclear disintegration modes, on mass number A, neutron excess $I = (N - Z)/A$, excitation energy E^* and spin angular momentum J, are incomplete and need to continue.[1,2] As expected from macroscopic energetics,[3,4] the most stable nuclei assume compact spherical, or nearly spherical, ground-state shapes that minimize nuclear surface energy. However, intrinsic quantal structure imposes strong deformations on nuclei in certain domains of the Segrè Chart.[5] In turn, internal structure may also constrain[6] nuclear

[a]Now at Department of Physics and Astronomy, Rochester Institute of Technology, Rochester, NY 14623, USA.
[b]Now at Department of Physics, Kurukshetra University, Kurukshetra, India.

shapes from becoming too compact, lending itself to observation and detailed theoretical evaluation.[7–9] Studies of the fine structure of the fission barrier,[10–13] the shape of fission pathways[14] and explorations of the multi-dimensional potential energy landscapes in heavy-ion reactions,[15] e.g., for the production of super-heavy nuclei[16,17] illustrate past utility for progress in nuclear science derived from the connection between macroscopic shape dynamics and intrinsic nuclear structure.

Spontaneous nuclear fission[11,12] and low-energy, fission-like dissipative heavy-ion reactions[15,18] are cases in point. These are essentially binary processes with only two ($M_F = 2$) massive primary fragments in the exit channel, even though Q-values would allow[19] simultaneous emission of multiple massive fragments. Reasons for the inhibition of multi-fragmentation ($M_F > 2$) processes at low and moderate energies can be found[20–22] in the compound "bottle neck" effect faced by system trajectories, which is due to a density of accessible states that is strongly reduced by important exit-channel interaction barriers. Because of intrinsic excitations acquired in the process, the hot primary reaction products de-excite in statistical decay, but on time scales significantly longer than typical collision times. Additional, promptly emitted particles are rare.

Expanding the knowledge horizon, investigations of neck-like structures,[23–29] are of particular interest in this context. Such structures consist of matter bridges that connect temporarily the massive fragments emerging from nuclear fission or from dissipative heavy-ion reactions. In models of low-energy fission,[11,30,31] this "neck" is predicted to have little mass and attain a relatively short maximum length[32,33] before snapping in just one place. Theoretical estimates based on the hydrodynamic capillary stability of a cylindrical fluid column (cylindrical neck) predict[34–36] a "magic" relation of maximum length (ℓ_{N}) to radius (r_{N}) of

$$\ell_N \approx 2\pi \cdot r_N \tag{1}$$

Experimental evidence based on fission fragment multiplicity[37–40] and final fragment kinetic energies[11,12] supports that picture. Significant probabilities for longer necks and/or simultaneous multiple neck ruptures leading to ternary or higher-order nuclear fission are neither predicted by theory nor experimentally observed.[40,41] This fact impedes a more detailed experimental study of necked-in nuclear shapes and their connection to the nuclear equation of state, specifically of the effective tensile strength of finite nuclei subjected to thermal and mechanical stresses.

The above conundrum can be addressed by studying the complex shape evolution in dissipative heavy-ion reactions which, like fission, produce mostly two massive fragments, projectile-like and target-like types. While at asymptotic times the hot primary fragments will have de-excited and regained their compact ground-state shapes, the intermediate, pre-scission state displays the most interesting configurations. Its temporal evolution across the multi-dimensional potential energy landscape, its eventual disintegration mode and associated time scales should reflect the variations of internal nuclear structure and basic fermionic fluid dynamics with the balance between collective and intrinsic (thermal) excitations.

Since excitation energy equilibration times in heavy-ion reactions are very short[15] ($\sim 10^{-22}$ s), the elevated nuclear temperatures attained in dissipative heavy-ion reactions at $E/A \approx (10-50)$ MeV per nucleon are expected to rapidly expand the system and, in particular, soften the nuclear surface. Therefore, in these reactions, pre- and final scission shapes are expected[25,42] to be more complex than those occurring in low-energy fission. Since they would reflect temperature and density dependence of the mechanical, chemical, and thermal stability of nuclear matter, searches for potentially extended neck-like structures produced in heavy-ion reactions are of particular interest. A number of experiments[43–47] have already reported unexpected breakup and splitting phenomena involving the fast-moving projectile- and/or target-like fragments, results of potential relevance to the task.

The present experiment has found and studied such structures and their disintegration in dissipative reactions ^{48}Ca + ^{112}Sn and ^{48}Ca + ^{124}Sn at a bombarding energy of E/A = 45 MeV via the observation of prompt, coaxially aligned binary, ternary, and quaternary fragmentations. The following Sec. 2 summarizes briefly experimental setup and procedures. Section 3 describes some details of the kinematical analysis of ternary and quaternary events. Section 4 provides a qualitative discussion of results. Conclusions and outlook are offered in Sec. 5.

2. Experimental procedures

The experiment was conducted at the Laboratori Nazionali del Sud (LNS) Catania/Italy. Various calibration and production beams were provided by the LNS 15-MV Tandem Van de Graaff accelerator and the K800 superconducting cyclotron, respectively. Production beams of 45 A-MeV ^{48}Ca ions were used to bombard (630–680) μg/cm^2-thick, self-supporting targets of isotopically enriched ^{112}Sn and ^{124}Sn, mounted in the center of the CHIMERA target chamber CICLOPE. The 4π multi-detector array CHIMERA covers the angular range $\{1° \leq \Theta \leq 176°, 0° \lesssim \Phi \lesssim 180°\}$ with 1192 Si/CsI(Tl) charged-particle detector telescopes, arranged in two super structures, forward CONE and central SPHERE. The Si transmission detectors had thicknesses of $\approx$300μm. They provided particle identification in atomic and mass numbers, Z and A, via the Z-dependent specific energy loss and A-dependent time of flight (TOF) information. The residual energy (E) was measured with the (3–12)-cm thick CsI(Tl) scintillators, read out via pin diodes. For light and intermediate mass charged particles, these elements also provided particle identification, utilizing a pulse shape discrimination (PSD) method involving fast and slow scintillator light output components and A-dependent time of flight (TOF) information.

The detectors were serviced by the standard, fast-slow CHIMERA electronics and data acquisition system. In the production runs, events with detected product multiplicity of $M_F > 2$ (minimum bias) were recorded, in addition to a concurrent, 1:20 sampling of elastic scattering events. Further details of CHIMERA detectors and electronics are presented elsewhere.[48,49] All data cited in the following have been tested for instrumental bias, and/or corrected for finite efficiency, using a software replica[50] of the CHIMERA detector and electronics, which has been developed and tested extensively in the past.

An example of raw data obtained for the reaction ^{48}Ca+^{124}Sn is presented in Fig. 1. Here, the response of a Si transmission detector is plotted vs. the corresponding CsI light output in form of a scatter plot. Yield ridges are labeled according to the atomic numbers of the detected particles. The PSD method, applied to the CsI(Tl) light output signals, provides more precise particle identification for the lighter ions ($p, d, t,$ ^{A}He).

In Fig. 2, the Z = 3 (Li) isotopic yield distribution, measured with a forward detector, is plotted vs. a continuous mass pseudo parameter. The resolution is typical for the domain of light and intermediate-mass charged clusters, commonly dubbed *IMF*s. The overall shapes of the isotopic distributions from Li to C are similar to the one depicted in Fig. 2, i.e., slightly asymmetric Gaussians, when plotted vs. actual, discrete mass

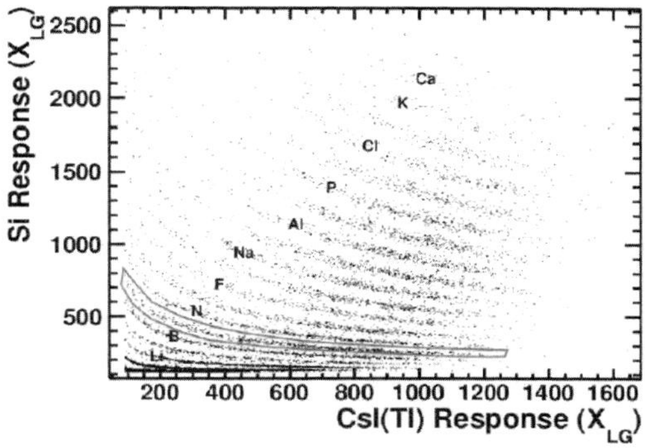

Fig. 1. Events from the reaction ^{48}Ca+ ^{124}Sn at E/A = 45 MeV. Shown is a scatter plot of Si vs. CsI response for a detector telescope at 5.2 degrees. The polygon identifies Z = 6 fragment events.

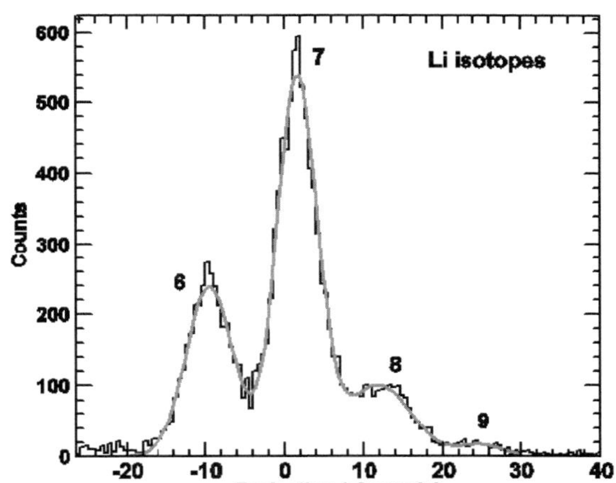

Fig. 2. ^{48}Ca+ ^{124}Sn isotopic distribution for Li ions emitted at forward angles. The yield for Z-3 is plotted vs. a linearized continuous mass pseudo parameter.

number A. As expected, the distributions for the neutron richer target ^{124}Sn are shifted to heavier masses, when compared to those for the n-poorer target ^{112}Sn. In the following discussion, such potentially interesting isotopic differences between the two Ca+Sn reactions are disregarded.

3. Experimental results

3.1. *Overall dissipative reaction features*

First inspection of experimental reaction phenomena[49,50] for both systems $^{48}Ca+^{112,124}Sn$ reveals that, for much of the cross section, reaction phenomena are in qualitative agreement with the gentle, low-energy dissipative mechanism based on the exchange of multiple nucleons between projectile and target. In view of the lack of nuclear matter conditions supporting such mechanism, such behavior is not expected for the elevated bombarding energy of E/A = 45 MeV in the present reactions. In fact, there are also striking differences observed in the data, which turn out to support a different, more abrupt and violent reaction scenario.

Figure 3 depicts an angle-integrated, two-dimensional logarithmic contour diagram, where the atomic number Z is plotted vs. the velocity component in beam direction, as measured for charged particles from the reaction $^{48}Ca+^{124}Sn$ at E/A = 45 MeV. The dashed enclosure defines the distribution of projectile-like fragments (PLF). It corresponds to a measured low event multiplicity of $M_{\text{event}} < 5$. The PLF distribution exhibits an intense peak of quasi-elastic and partially damped events, at $< Z > \approx 20$

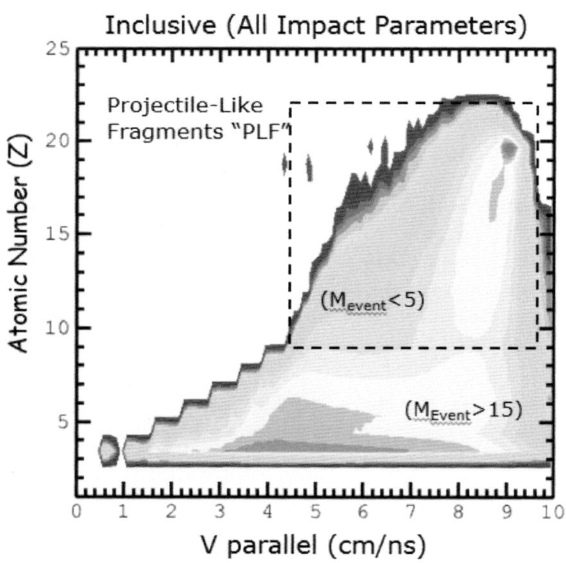

Fig. 3. Atomic-number vs. velocity component in beam direction for $^{48}Ca+^{124}Sn$ reaction events. Dashed square: dissipative reaction domain.

and longitudinal velocity component of $< v_{\|} >= 9\,\mathrm{cm/ns}$, fanning out into a ridge of yield for which reduced velocity correlates with reduced atomic number. The latter correlation is a result that might also be expected from secondary particle evaporation from the hot primary PLFs.

Separated from the above PLF distribution, a domain of light and intermediate IMF clusters with $3 \leq Z < 10$ is observed in the contour plot of Fig. 3. The large number of associated charged particles, "event multiplicities" of $M_{\mathrm{event}} > 15$, identifies this domain with significant levels of energy dissipation ($E_{\mathrm{diss}} \gtrsim 300\,\mathrm{MeV}$). It is noteworthy that these particles have parallel velocity components that range from low values up to those of quasi-elastic PLFs but, interestingly, not beyond. Specific searches of the data did not find any randomly emitting sources for such IMF clusters.

The Wilczyński contour diagrams shown in Fig. 4 represent an effective way to visualize system trajectories in phase space. Here, the lab kinetic energy (Panel a) or lab velocity (Panel b) of PLFs are plotted vs. the corresponding emission angles θ_{PLF}. Actually, just the absolute deflections,

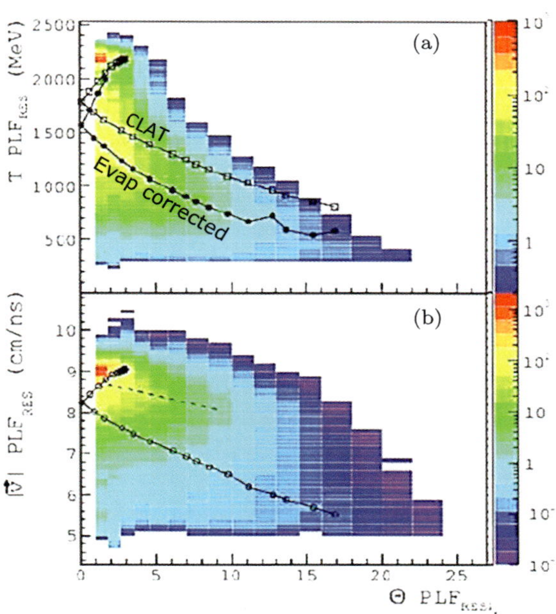

Fig. 4. Wilczynski contour diagram of the cross section (logarithmic scale) for the reaction $^{48}Ca + ^{112}Sn$. Plotted are PLF lab kinetic energy (a) or speed (b) vs. laboratory emission angle $|\theta_{\mathrm{PLF}}|$. Curves represent model calculations for mean values. Dashed line indicates trends.

$|\theta_{\rm PLF}|$ are measured. At first sight, the patterns observed in Fig. 4 seem to identify a dissipative reaction, specifically one characteristic of dissipative orbiting.[15] Here, in a collision between projectile and target nuclei, a dinuclear system is formed transiently, which rotates about its center of gravity, while kinetic energy of relative motion is dissipated.

For large degrees of energy dissipation, the system has rotated past the beam direction ($\theta_{\rm PLF} = 0°$), such that the PLF is emitted at negative angles. In an experiment, where positive angles are not distinguished from negative angles, one observes a superposition of the two parts of a continuous ridge of yield associated with positive and negative angles, respectively.

The symbols and solid lines in Fig. 4 represent averages of semi-classical model predictions obtained with the reaction code CLAT,[51] which is based on the one-body nucleon exchange model.[52] These calculations were complemented by simulations of the sequential statistical fragment decay, employing the code GEMINI.[53] As seen in Fig. 4(a), the model simulations ("Evap corrected") describe the mean PLF energy-angle correlations quite well. Other inclusive observable distributions, such as the PLF mass and Z distributions, are also well reproduced by the above CLAT-plus-GEMINI simulations.

However, this apparent agreement of experimental data for the present Ca+Sn reactions at E/A = 45 MeV with key predictions by a low-energy reaction model is deceptive. An actual discrepancy between data and theory is demonstrated by the Wilczyński plot of Fig. 4(b), which changes the representation of Fig. 4(a) to one in which the PLF velocity, rather than its kinetic energy, is plotted vs. emission angle. Obviously, in this latter rendition (Fig. 4(b)), the theoretical simulation misses the experimental Wilczyński yield ridge by a significant margin, for the larger degrees of damping ("negative-angle" yield ridge). In this domain, the large energy differences relative to the beam, measured for the PLFs, are not caused by reactive, frictional damping of the projectile velocity or by subsequent particle evaporation. Rather, this kinetic energy loss must be due to a prompt loss of mass by the projectile, or by the fast PLF, during the nuclear interaction with the target. To recover the corresponding missing PLF mass, one is led to search the data for relatively fast, intermediate-mass fragments (IMF) that are kinematically correlated with the PLFs or their remnants. The data in Fig. 3 indicate that such IMF particles do indeed exist.

3.2. *Multi-particle correlations in* Ca+Sn *reactions*

In order to test the validity of the above scenario, the ^{48}Ca+^{112}Sn and ^{48}Ca+^{124}Sn data sets were scanned for events in which a Ca-like, PLF or remnant $\mathrm{PLF_R}$, was detected in coincidence with intermediate-mass fragments between lithium and carbon ($3 \leq Z_{IMF} \leq 6$). The particles detected in each event were sorted according to individual Z and A values, velocity vectors, and grouped according to the overall multiplicity M_{lcp} of associated light charged particles (lcp). The latter multiplicity observable can be taken as a measure of the collision impact parameter.[15]

Indeed, for mid-peripheral collisions ($< M_{\mathrm{lcp}} > (\approx M_{event}) \gtrsim 15$) associated with energy losses of several hundred MeV, a significant fraction (~10%) of PLF remnants appear in coincidence with up to 3 IMFs in the scanned ranges for mass, Z, and velocity. As indicated by the atomic number/velocity diagram in Fig. 3, such IMF particles have a range of "mid-velocities," velocities below that of PLFs (or their remnants). This fact renders sequential, statistical evaporation from the fast-moving PLFs an unlikely origin for such IMFs. In the following, mainly two event classes will be discussed. In a *ternary event*, a fast remnant ($\mathrm{PLF_R}$) of the PLF with $\mathrm{Z_{PLFR}} > 10$ is detected in coincidence with an IMF cluster. The associated TLF fragment or its remnant is not directly detected but inferred from kinematics, mass and charge conservation. In *quaternary* events, two IMF clusters are detected in coincidence with a $\mathrm{PLF_R}$.

In Fig. 5, a hypothetical scenario for quaternary events is shown with two IMFs placed in the spatial region between the remnants of PLF and TLF, $\mathrm{PLF_R}$ and $\mathrm{TLF_R}$, respectively. For the purpose of the specific data analysis, the reaction plane has been defined by the beam axis and the vector sum of the velocities of the two heaviest fragments, $\mathrm{PLF_R}$ and $\mathrm{IMF_1}$, as measured in forward direction ("PLF-TLF Reference"). In a ternary scenario, in which these latter two particles result from a dynamic splitting of a primary PLF, with a basically intact TLF, this reference would indicate the reconstructed mean direction of flight of the PLF prior to splitting.

How well the reaction plane is defined by this procedure can be seen from Fig. 6 showing the out-of-plane angular distribution $dP/d\beta$ of the normalized emission probability (cross section σ) $P \propto \sigma$, where β is the out-of-plane angle of the relative velocity vector $\vec{v}_{rel} = \vec{v}_{PLF_R} - \vec{v}_{IMF_1}$ for $\mathrm{PLF_R}$-$\mathrm{IMF_1}$ pairs of fragments. The resulting distribution (cf. Fig. 6) turns out to be unimodal, symmetric, and peaked about $\beta = 0^0$, demonstrating the absence of any strong bias in the analysis procedure. A significant

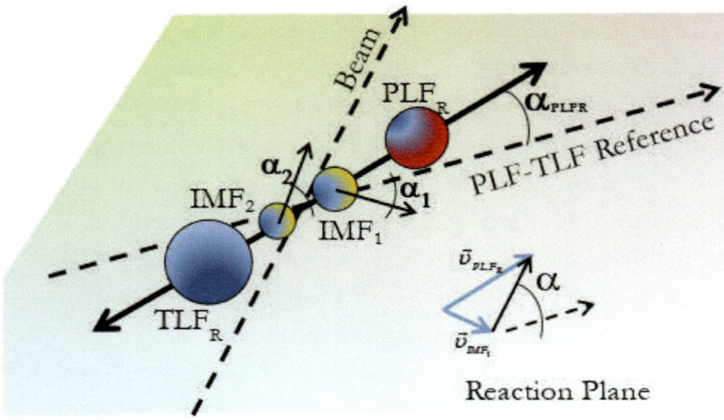

Fig. 5. Schematics of reaction scenario with remnants of primary PLF and TLF and two neck-IMFs in the neck region between them. The inset defines angle α between relative PLF$_R$-IMF$_1$ velocity vector $\vec{v}_{rel}=\vec{v}_{PLF_R} - \vec{v}_{PLF_R}$ and reference direction.

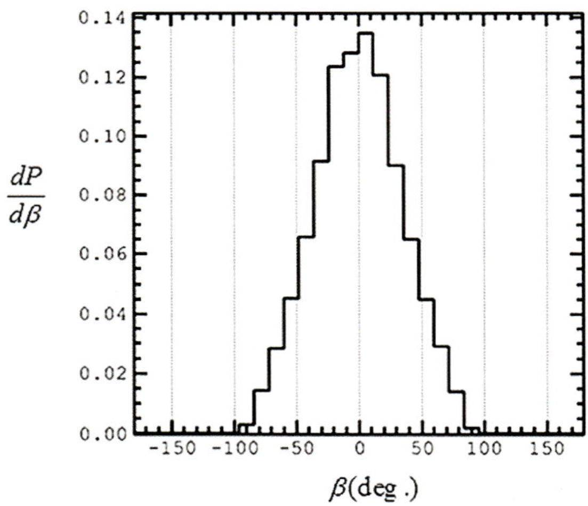

Fig. 6. Out-of-plane angular correlations between PLF remnant and IMF.

width of the out-of-plane distribution is expected from sequential evaporative deexcitation of hot primary reaction products.

The orientation of the PLF remnant/IMF pair relative to the reference direction (cf. Fig. 5) in the reaction plane is illustrated in Fig. 7, where the in-plane angular distribution of the coincidence yield, $d\sigma/d\Omega_\alpha$, is plotted vs.

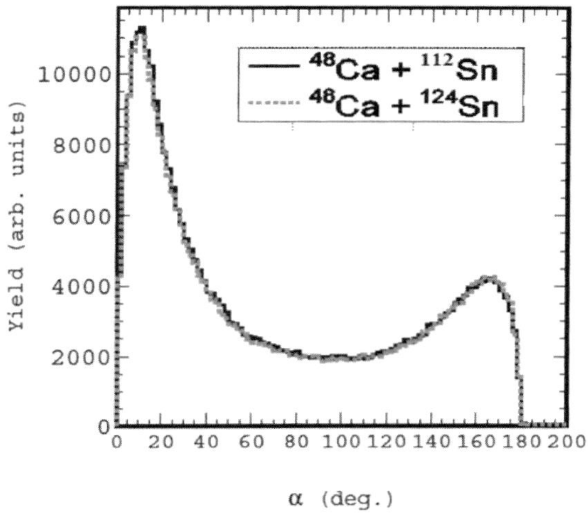

Fig. 7. In-plane angular distributions $d\sigma/d\Omega_\alpha$ for light IMF-heavy PLF remnant pair, taken relative to the heavy PLF remnant, for the systems $^{48}Ca+^{112}Sn$ (solid-black (online) histogram) and $^{48}Ca+^{124}Sn$ (dotted-red (online) histogram).

the angle α between the relative velocity vector $\vec{v}_{rel}$ and the hypothetical PLF-TLF Reference direction. A relative angle of $\alpha < 90°$ indicates forward emission of the PLF_R and backward emission of the IMF (toward the TLF), in the relative frame of reference. For $\alpha > 90°$, the intermediate mass fragment is emitted forward of the PLF_R.

Obviously, the angular distribution in Fig. 7 is strongly anisotropic and not symmetric about $\alpha = 90°$. It peaks strongly at forward angles and, to a somewhat lesser extent, also at backward angles. This anisotropy reflects strong alignment of the PLF_R-IMF_1 axis with respect to the PLF-TLF reference direction (cf. Fig. 5). It indicates a dominantly dynamical, non-statistical fragmentation mechanism producing the measured aligned PLF_R-IMF pairs. In contrast, a statistical, fission-like process would lead to a forward/backward symmetric distribution $d\sigma/d\Omega_\alpha$, and an isotropic angular distribution, $d\sigma/d\alpha \approx$ const. Maximum contribution by a possible statistical component can be estimated from the backward peak in $d\sigma/d\Omega_\alpha$ amounting to a fraction of <25% of the considered class of events.

A dominantly dynamic mechanism for the observed PLF_R-IMF correlations is corroborated by the measured distribution of relative velocities between the PLF remnant and the associated IMF presented in Fig. 8. Here,

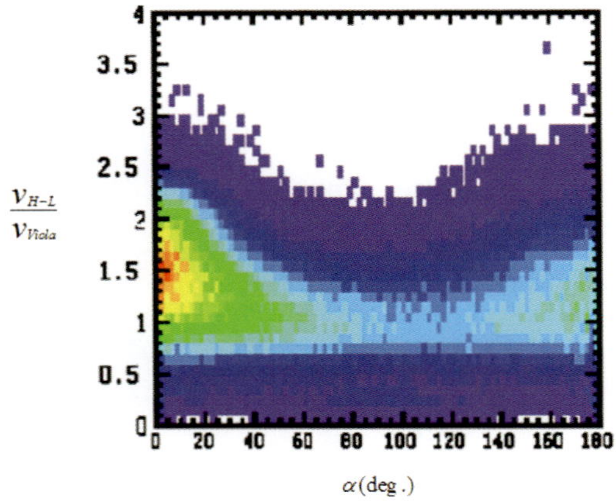

Fig. 8. Logarithmic contour diagram of the probability plotted vs. IMF-PLF remnant relative velocity and the in-plane orientation angle α of the fragmentation axis.

the yield is plotted in terms of a logarithmic contour diagram vs. orientation angle α and relative velocity, v_{rel}/v_{Viola}, normalized to the corresponding Coulomb velocity,[54]

$$v_{Viola} = \sqrt{\frac{2}{\mu}\left(0.755\frac{Z_{PLFR}Z_{IMF}}{A_{\mathrm{PLFR}}^{1/3} + A_{IMF}^{1/3}} + 7.3\right)} \tag{2}$$

The above Eq. 2 predicts the relative velocity expected from just the mutual Coulomb repulsion between $\mathrm{PLF_R}$ and $\mathrm{IMF_1}$ in a pair. The formula is based on experimental "Viola" fission systematics.[54] The quantity μ is the reduced mass of the pair; Z and A notations are obvious.

As expected, the intensity pattern seen in the contour plot of Fig. 8 recalls the non-statistical angular distribution apparent already in Fig. 7. However, the fact that forward emission of the $\mathrm{PLF_R}$ is associated with high relative $\mathrm{PLF_R}$-$\mathrm{IMF_1}$ velocities, exceeding those due to Coulomb repulsion by 50% or more, lends additional support to a dynamical production mechanism for these fragment pairs.

For events corresponding to emission of the $\mathrm{PLF_R}$ backward of the IMF, i.e., in the direction toward the TLF remnant, the distribution indicates a smaller mean excess velocity. This latter, smaller excess relative velocity, which is still 10% above the Coulomb systematics, could perhaps be explained by centrifugal effects adding to Coulomb repulsion. Collective rotation of the primary PLF corresponding to a mean angular momentum of $J \approx 30\hbar$ could, in fact, explain the excess relative velocities observed for backward emission events in terms of a statistical disintegration of the primary PLF. Angular momentum transfer of this magnitude is consistent with CLAT simulations performed for the systems of interest here.

The above observations regarding the anisotropic angular distributions of yields and excess relative PLF_R-IMF velocities have been made consistently for both Ca+Sn reactions. They lead one to conclude that such fragment pairs do not result from sequential decay of hypothetical primary PLFs occurring outside the nuclear field of the other massive TLF reaction partner. Notably, the observed excess relative velocities require a dynamic "pull" by reaction partners left behind by the fast PLFs.

Further elucidation of the complex reaction mechanism for the Ca+Sn reactions of interest here comes from extending the analysis to quaternary and more complex events, in which more than one IMF in the range $3 \leq Z_{IMF} \leq 6$ is detected in coincidence with a fast-moving PLF remnant. However, the probability $P(M_{IMF})$ of the multiplicity (M_{IMF}) of IMFs *coincident* with PLF_Rs decreases by roughly one order of magnitude per M_{IMF} unit. Although events with up to 3 of such IMFs have been identified in the Ca+Sn data sets, the low statistics for the higher multiplicities prevents detailed correlation analysis for $M_{IMF} \geq 3$.

Nevertheless, IMF-IMF pairs {Li-Li, Li-Be, ..., C-C} coincident with PLF_R's have been detected with reasonable statistics, as shown in Fig. 9. Here, examples are presented for such mid-velocity, di-IMF data

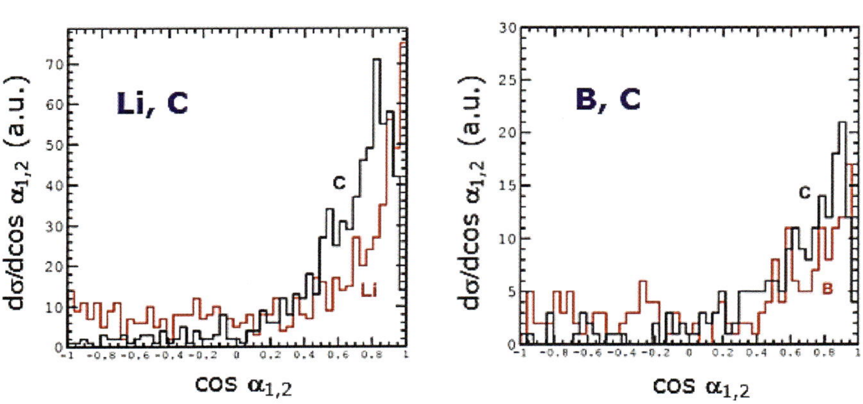

Fig. 9. Correlated pairs Li-C (left) and B-C (right) of IMF pairs measured in coincidence with PLF_R are analyzed for the reaction $^{48}Ca+^{124}Sn$. Angles are measured relative to the PLF-TLF reference (cf. Fig. 5).

for ^{48}Ca+^{124}Sn. In these graphs, angular distributions $d\sigma_i/d\cos\alpha_i$ are illustrated for the individual IMF_i $(i = 1, 2)$ in a pair of IMFs coincident with a fast-moving $\mathrm{PLF_R}$, where α_i is the emission angle of $\mathrm{IMF_i}$, measured relative to the PLF-TLF reference (cf. Fig. 5). Obviously, the IMF angular distributions are all anisotropic, strongly forward peaked ($\cos\alpha_\mathrm{i} > 0$), eliminating a statistical origin of these particles. The emission scenario of an IMF pair in coincidence with a $\mathrm{PLF_R}$ corresponds to that anticipated in Fig. 5, where both IMFs are lined up, trailing behind the $\mathrm{PLF_R}$. The following discussion is exemplary but concentrates on the statistically most prominent Li-Li correlations.

In order to quantify the angular correlation of pairs of IMFs trailing a $\mathrm{PLF_R}$, consider the opening angle $\Theta = \alpha_{IMF_1} - \alpha_{IMF_2}$ defined by their velocity vectors $\vec{v}_{IMF_i}(i = 1, 2)$. For random relative orientations of the two IMF velocity vectors, the relative probability for an actual opening angle to range below the angle Θ is given by the associated solid angle,

$$\Delta\Omega = 2\pi(1 - \cos(\Theta/2)) \tag{3}$$

The corresponding un-normalized experimental probability distribution $P(\Theta)$ is plotted in Fig. 10. Obviously, this distribution is decidedly not random, since it is strongly peaked at $\Delta\Omega \approx 0$. This fact implies an approximately collinear, strongly aligned configuration of the fragment triplets $\mathrm{PLF_R}$-$\mathrm{IMF_1}$-$\mathrm{IMF_2}$. To appreciate the degree of alignment of the $\mathrm{PLF_R}$-IMF triplet, assume a mean $\Delta\Omega = (0.2–0.3)$ and recall the definition of the PLF-TLF Reference (Fig. 5) in terms of $\mathrm{PLF_R}$ and $\mathrm{IMF_1}$. Then, the velocity vector of the second IMF is measured to be misaligned with the $\mathrm{PLF_R}/\mathrm{IMF_1}$ direction of flight by less than $\pm(15^\circ - 20^\circ)$.

An alternative and popular method of analyzing particle-particle correlations in terms of a velocity correlation function $R(v_{rel})$[55–58] has also been applied to the present data for correlated Li-Li pairs. Here, the coincidence yield $Y_{12}(v_{rel})$ of correlated Li pairs, for any given relative velocity v_{rel}, is normalized to an uncorrelated yield $[Y_1(v_1) \cdot Y_2(v_2)]_{v_{rel}}$ constructed at random out of the same data set,

$$1 + R(v_{rel}) := \sum_{12} Y_{12}(v_{rel}) \Bigg/ \left((1/2) \sum_{1,2} Y_1(v_1) \cdot Y_2(v_2) \right)_{v_{rel}} \tag{4}$$

Since correlated actual, and the constructed randomized events have been measured simultaneously and under identical conditions, this procedure is largely insensitive to experimental insufficiencies or artifacts.

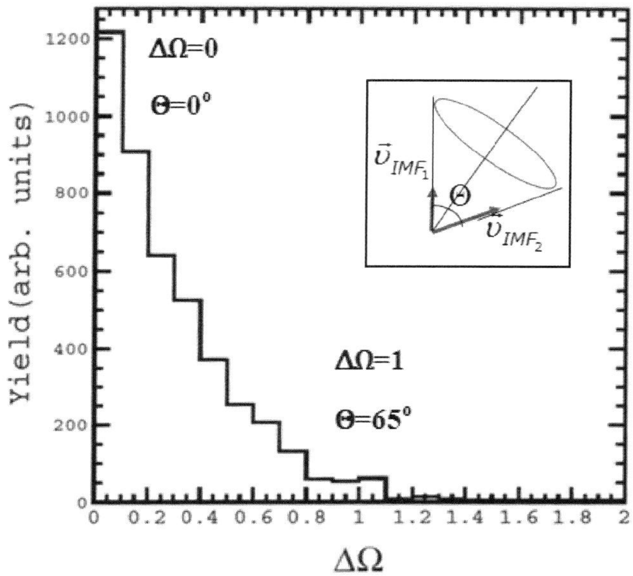

Fig. 10. Correlation between emission angles of two correlated Li IMFs, measured relative to the PLF-TLF reference. The inset illustrates the angular correlation between the two IMFs, where $\Theta = \alpha_{IMF_1} - \alpha_{IMF_2}$.

The resulting correlation yield in Fig. 11 shows the depression for low relative IMF-IMF velocities v_{rel} expected from Coulomb repulsion and a peak positive correlation at $v_{rel} = v_{IMF-IMF} \approx (2.2 \pm 0.2)$ns/cm. This latter feature is characteristic of an attractive interaction, for example, one associated with an intermediate system (resonance) decaying into two IMFs with the above relative velocity. It is interesting that this velocity is also by 10% higher than that given by Coulomb systematics, but application of the corresponding formula (Eq. 1) is not well justified for such light nuclei.

In order to evaluate the possibility that the two Li IMFs could have originated from the statistical decay of an excited ^{12}C intermediate,

$$^{12}C^* \xrightarrow{???} {}^6Li + {}^6Li \tag{5}$$

statistical simulations were performed with the code GEMINI.[59] From associated particle multiplicities for such hypothetical events, a maximum excitation of a ^{12}C-IMF of $E^*(^{12}\mathrm{C}) \leq (24\text{–}36)$ MeV is expected. Carbon nuclei at such low excitations decay mostly by alpha particles and nucleons. A statistical, fission-like splitting of a carbon nucleus into two Li fragments

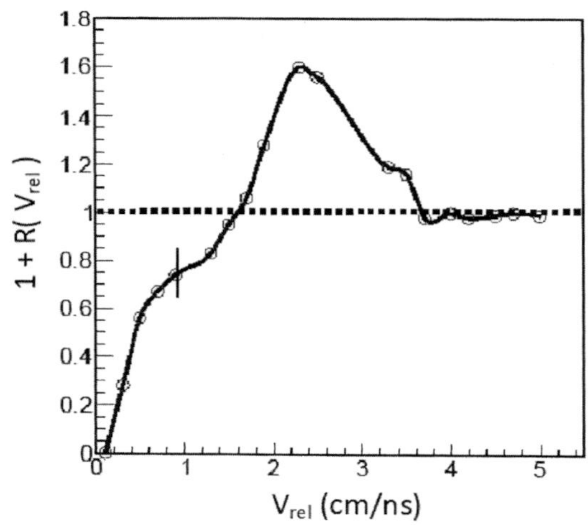

Fig. 11. IMF-IMF correlation function vs. relative velocity for the two co-incident IMF fragments each with Z = 3. Data are for ^{48}Ca + ^{112}Sn.

should be by 3–4 orders of magnitude slower and less likely than alpha emission. Furthermore, in such a slow process, the two Li fragments would be directionally anti-correlated in their c.m. system, but with a scission axis randomly oriented with respect to the PLF-TLF reference. The lifetime of the C* intermediate at low excitations would be so much longer than typical nuclear interaction times[15] for the Ca+Sn reaction, such that directional correlations with the PLF would have disappeared. Such a scenario can therefore not be invoked for an explanation of the strongly aligned emission processes leading to ternary and quaternary events observed in the data.

These considerations further corroborate conclusions derived previously, according to which correlated IMF pairs are directly produced in the Ca+Sn heavy-ion reaction proper at mid-velocity and in the neck region. Alternative scenarios are defeated by counter arguments given above. Although a detailed analysis for IMF triplets in coincidence with PLF remnants has not been conducted for the $^{48}Ca+^{112,124}Sn$ reactions, such events have been detected in kinematical regions consistent with an extended scenario, where a significant matter bridge connecting PLF and TLF in a dissipative reaction breaks in multiple locations.

4. Neck geometry and dynamics

The previous sections have provided compelling evidence for the formation and the subsequent decay of a massive neck region, occurring in energetic Ca+Sn reactions at mid-peripheral collision impact parameters. A thorough theoretical analysis is required to quantitatively assess conditions for formation or the influences of potential energy landscapes and viscous forces on the eventual disintegration, of these neck-like structures. Since this is an ongoing project and because of space limitations, results of such an analysis are relegated to a forthcoming paper.[49] However, most reaction phenomena described in this article are quite plausible and can, at least qualitatively, be understood in some, but not all, reaction scenarios, which are based on different sets of conservative and dissipative nuclear interaction forces. Similar variations in scenarios have been considered[24,31,32] also for the low-energy fission process.

Given the finite range of the nucleus-nucleus interactions, e.g., as represented by either proximity or folding forces, a force-free region must establish itself between the surfaces of projectile and target nuclei at relative distances smaller than approximately $s \approx 2$ fm between their half-density points. During the very brief peripheral interactions between target and projectile, moving at speeds more than twice the mean Fermi velocity, this space is accessed but cannot be completely filled in nucleon transfer processes. Hence, dissipative reactions should prevail for large collision impact parameters, i.e., peripheral reactions.

The situation changes for a range of mid-peripheral collisions. Here, substantial interpenetration of projectile and target nuclei may produce overlap regions of supra-normal densities of hot nuclear matter, which will subsequently expand rapidly and decay into the available space. Dwelling times at the turning point for the radial motion may also become long enough for nucleonic exchange currents to fill more of the force-free space between the surfaces of emerging PLF and TLF fragments. The resulting column of hot, dilute nuclear matter is the natural realm of eventual cluster formation. However, because of the short nucleonic mean free path in hot matter, and since relative motion is still fast in most phases of such mid-peripheral collisions, an adiabatic, mean-field evolution of the neck region cannot be presumed. A theoretical evaluation of stability and decay modes of the neck column is complicated by the time-dependent mechanical stresses imposed on the neck by the accelerating PLF and TLF fragments, to be considered in addition to nuclear surface tension and compressibility. The latter properties are also time dependent and significantly reduced in magnitude, in the course of energy dissipation and nuclear heating.

The experimental observations of approximately collinear emission for three or more reaction fragments reported above implies the transient existence of an elongated, necklace-like spatial pattern of interacting IMF clusters, extending between the two main fragments. If isolated, such a system would be laterally unstable against sideways, off-axis excursions. It would also be longitudinally unstable against fragment reabsorption, unless sustained by longitudinal external forces.

Such forces are provided by motion of the emerging PLF and TLF fragments, which are asymptotically observable as remnants, PLF_R and TLF_R. This scenario explains naturally the collinearity and preference, observed in ternary events, for emission of the heavier PLF_R forward of the intermediate, lighter IMF neck cluster (cf. Fig. 7). Calculations of proximity-plus-Coulomb potential energy surfaces (PES) for aligned ternary events produced in Ca+Sn reactions show[49,50] that such asymmetric "heavy-forward" configurations are energetically preferred.

External forces exerted on the neck region by the moving main PLF and TLF fragments produce a velocity field for particles in the neck region, which depends on the retarding viscous interactions, in addition to the conservative Coulomb, nuclear and inertial forces. Preliminary dynamical simulations for ternary Ca+Sn events suggest that the observed augmented relative IMF-PLF_R velocities (cf. Fig. 8) are caused mainly by the nuclear viscosity, rather than by conservative nuclear or centrifugal forces. Given the rather elongated nuclear shapes of interest here and the short interaction times for Ca+Sn at $E/A = 45$ MeV, one expects from theoretical model calculations,[31] that the dissipative forces are most likely of the two-body viscosity type. Although one-body, mean-field dynamics can also lead to ternary fission, for an energetic heavy-ion reaction, the required shape evolution times are prohibitively long, even for the still rather compact associated scission shapes.

The final topic in this qualitative discussion concerns the A/Z ratios of the IMF clusters remaining from the disintegration of the neck. One expects that these ratios reflect an average of the compositions of mainly the juxtaposed surfaces, which are both neutron rich. The two reaction systems measured in this work use the same, neutron rich projectile, ^{48}Ca, and differ only with respect to target isotope, ^{112}Sn vs. ^{124}Sn. Since the neutron skin thickness for stable nuclei has only a weak dependence on their neutron number,[60] the chemical composition of the neck and, hence, of its fragments should be similar for both reactions.

5. Conclusions

A complex, "necklace" pattern has been observed in the single or multi-IMF decay of neck-like structures formed transiently in ^{48}Ca + ^{112,124}Sn reactions at E/A = 45 MeV. The process occurs for mid-peripheral collisions, as indicated by the multiplicities of associated charged particles, and terminates the normal pattern of dissipative reactions, which still prevail in peripheral collisions. Likely, in peripheral collisions, interaction times are too short, and projectile-target interpenetration too shallow, for massive necks to develop between the nuclei. In more violent, central collisions, hot matter in chaotic motion in the overlap region may not be able to reorganize into a metastable macroscopic neck.

Properties measured in this experiment pertain to masses and charges of several IMFs in a collinear configuration, aligned dynamically between PLF and TLF remnants. The geometry estimated for ternary events suggests a neck radius of $r_N \approx 2.5 fm$ and a total length of $\ell_N \approx (3-4) \cdot r_N$, before it snaps at two locations. For quaternary events observed in this work, with two similar IMF clusters produced in triple neck snapping, the relation of neck length to radius is closer to that (cf. Eq. 1) predicted by theory.[32] In neither case, random simple neck snapping is observed.

In evaluating these discrepancies, one has to keep in mind that predictions of neck geometry and decay have so far been made only for quasi-static, homogeneous nuclear matter near its ground-state. In contrast, the present experiments involve hot matter, at temperatures of approximately (2–3) MeV, which is subjected to significant external stresses.

It is interesting that both variables can, to some extent, be varied independently in such reactions. More extensive measurements of the evolution of "necklace" decay patterns with excitation energy and isospin may allow one to investigate the ability of various proposed approximations of the effective equation of state of finite nuclei on their thermo-mechanical stability.

Acknowledgements

The authors wish to express their gratitude to the LNS Catania accelerator operating staff for the excellent quality of tandem and cyclotron beams provided during the experiments. The University of Rochester group has appreciated the kind hospitality extended to them at the LNS.

This work was supported by the US DOE grant DE-FG02-88ER40414.

References

1. W. Greiner, *Nuclear Physics: Present and Future.* (Springer Verlag, 2015).
2. *75 Years of Nuclear Fission* Present Status and Future Perspectives, (D.C. Biswas, K. Mahata, and V.M. Datar, Editors). (Indian Academy of Sciences, 2015).
3. W. D. Myers, *Droplet Model of Atomic Nuclei.* (IFI/Plenum, 1977).
4. S. Cohen, F. Plasil and W. J. Swiatecki, The Nuclear Rotating Liquid Drop Model. *Ann. Phys.* **82**, 557 (1974).
5. P. Møller, J. R. Nix, W. D. Myers and W. J. Swiatecki, Nuclear Ground-State Masses and Deformations. *Atomic Data and Nuclear Data Tables* **59**, 185–381 (1995).
6. J. Tõke and W. U. Schröder, Liquid drop model and quantum pressure resisting noncompact nuclear geometries. *Phys. Rev. C* **65**, 044319 (2002).
7. A. Bohr and B. R. Mottelson, *Nuclear Structure I.* (Benjamin, 1969).
8. A. Bohr and B. R. Mottelson, *Nuclear Structure II.* (Benjamin, 1975).
9. S. G. Nilsson and Ingemar Ragnarsson, *Shapes and Shells in Nuclear Structure.* (Cambridge University Press, 1995).
10. V. M. Strutinsky, Shell effects in nuclear masses and deformation energies. *Nucl. Phys. A* **95**, 420 (1967).
11. R. Vandenbosch and J. R. Huizenga, *Nuclear Fission.* (Academic Press, New York and London, 1973).
12. C. Wagemans, *The Nuclear Fission Process.* (CRC Press, 1991).
13. H. J. Krappe and K. Pomorski, *Theory of Nuclear Fission.* (Springer Verl., 2012).
14. M. Brack, Funny Hills: The Shell correction approach to nuclear Shell Effects and its applications to the fission process. *Rev. Mod. Phys.* **44**, 320 (1972).
15. W. U. Schröder and J. R. Huizenga, Damped Nuclear Reactions, in *Treatise on Heavy-Ion Science, Vol. 2* (ed. Bromley, D. A.) 113 (Plenum Press, 1984).
16. P. Walker and G. Dracoulis, Energy traps in atomic nuclei. *Nature* **399**, 35–40 (1999).
17. Z. Patyk, A. Sobiczewski, P. Armbruster and K. H. Schmidt, Shell effects in the properties of the heaviest nuclei. *Nucl. Physics A* **491**, 267–280 (1989).
18. L. Tassan-Got and C. Stephan, A way to dissipate energy and angular momentum for reactions in the Fermi energy domain. *Nucl. Phys. A* **524**, 121–140 (1991).
19. W. J. Swiatecki, Second UN Conf. Peaceful Uses of Atomic Energy. in 651 (1958).

20. J. Tõke and W. U. Schröder, A New Type of Shape Instability of Hot Nuclei and Nuclear Fragmentation. *Phys. Rev. Lett.* **82**, 5008 (1999).
21. J. Tõke, Revisiting theoretical concepts in statistical multi-fragmentation. *Nucl. Phys.* **A681**, 374c (2001).
22. J. Tõke, J. Lu and W. U. Schröder, Surface entropy in statistical emission of massive fragments from equilibrated nuclear systems. *Phys. Rev. C* **67**, 034609 (2003).
23. U. Brosa, S. Grossmann and A. Müller, Nuclear Scission. *Nucl. Phys. A* **502**, 423c (1989).
24. H. Diehl and W. Greiner, Theory of ternary fission in the liquid drop model. *Nucl. Phys. A* **229**, 29–46 (1974).
25. V. Baran, M. Colonna and M. Di Toro, Neck fragmentation reaction mechanism. *Nucl. Phys. A* **730**, 329–354 (2004).
26. J. Tõke *et al.*, Intermediate-mass fragment decay of the neck zone formed in peripheral ^{209}Bi+^{136}Xe collisions at Elab/A=28 MeV. *Phys. Rev. Lett.* **75**, 2920–2923 (1995).
27. C. P. Montoya *et al.*, Fragmentation of necklike structures. *Phys. Rev. Lett.* **73**, 3070–3073 (1994).
28. G. Verde, A. Chbihi, R. Ghetti and J. Helgesson, Correlations and characterization of emitting sources. *Eur. Phys. J. A* **30**, 81–108 (2006).
29. R. Kurniadi, Y. S. Perkasa and A. Waris, Neck curve polynomials in neck rupture model., *Proc. 3rd Int. Conf. Adv. Nucl. Sci. Engineering, AIP Conf. Proc.* 1448, 291–296 (2012). doi:10.1063/1.4725467
30. P. Møller and J. Nix, Calculation Of Fission Barriers With The Droplet Model And Folded Yukawa Single-Particle Potential. *Nucl. Phys. A* **229**, 269 (1974).
31. N. Cârjan, A. J. Sierk and J. R. Nix, Effect of dissipation on ternary fission in very heavy nuclear systems. *Nucl. Physics, Sect. A* **452**, 381–397 (1986).
32. U. Brosa, S. Grossmann, A. Müller and E. Becker, Nuclear scission. *Phys. Rep.* **197**, 167–262 (1990).
33. U. Brosa and S. Grossmann, Random neck rupture in deep inelastic collisions. *J. Phys. G. Nucl. Part. Phys.* **10**, 933 (1984).
34. G. F. Bertsch and D. Mundinger, The cohesion of nuclear matter, *Phys. Rev. C* **17**, 1646 (1978).
35. A. Müller and S. Grossmann, Surface excitations of a compressible cylindrical liquid. *Z. Naturforsch.* **40a**, 968–975 (1985).
36. G. F. Bertsch and S. Das Gupta, A guide to microscopic models for intermediate energy heavy ion collisions. *Phys. Rep.* **160**, 189–233 (1988).
37. R. Vandenbosch and J. R. Huizenga, *Nuclear Fission.* (Academic Press, 1973).
38. F. Gönnenwein, M. Mutterer and Y. Kopatch, Ternary and quaternary fission. *Europhys. News* **36**, 11–15 (2005).
39. R. Brandt, Ternary Fission, *Angew. Chemie* **24**, 043 (1971).
40. D. N. Poenaru, R. A. Gherghescu and W. Greiner, Complex fission phenomena. *Nucl. Phys. A* **747**, 182–205 (2005).

41. S. Vermote *et al.*, Ternary α and triton emission in the spontaneous fission of ^{244}Cm, ^{246}Cm and ^{248}Cm and in the neutron induced fission of ^{243}Cm, ^{245}Cm and ^{247}Cm. *Nucl. Phys. A* **806**, 1–14 (2008).
42. M. Di Toro, A. Olmi and R. Roy, Neck dynamics. *Eur. Phys. J. A* **30**, 65–70 (2006).
43. E. De Filippo *et al.*, Dynamical fission in Sn124+Ni64 collision at 35A MeV. *Phys. Rev. C - Nucl. Phys.* **71**, (2005).
44. I. Skwira-Chalot *et al.*, Fast ternary and quaternary breakup of the ^{197}Au + ^{197}Au system in collisions at 15MeV/nucleon. *Phys. Rev. Lett.* **101**, 1–5 (2008).
45. J. Wilczyński *et al.*, How to Calibrate the Time Scale of Emission of Intermediate Mass Fragments. *Int. Jour. Mod. Phys.* **14**, 353–357 (2005).
46. S. Hudan *et al.*, Interplay of initial deformation and Coulomb proximity on nuclear decay, *Phys. Rev. C* **70**, 031601(R) (2004).
47. J. Colin *et al.*, Dynamical effects in multifragmentation at intermediate energies. *Phys. Rev. C* **67**, 064603 (2003).
48. A. Pagano *et al.*, Fragmentation studies with the CHIMERA detector at LNS in Catania: Recent progress, *Nucl. Phys. A* **734**(5), 504–511 (2004).
49. S. O. Nyibule, Prompt aligned binary, ternary, and quaternary fragmentation in ^{48}Ca + ^{112}Sn and ^{48}Ca + ^{124}Sn reactions at a bombarding energy of 45 MeV/A., PhD Thesis, unpublished (University of Rochester, 2016).
50. M. J. Quinlan, Non-equilibrium splits of the projectile and the mechanism of Ca+Sn reactions at 45 A MeV., PhD Thesis, unpublished (University of Rochester, 2011).
51. W. U. Schröder, R. T. De Souza, J. R. Huizenga and L. M. Schmieder, in *Nuclear Fission and Heavy-Ion-Induced Reactions* (ed. W. U. Schröder) 255–291 (Harwood Academic Publishers, 1987).
52. J. Randrup, Some aspects of nuclear dynamics, *Nucl. Inst. Meth.* **146**, 213 (1977).
53. R. J. Charity, Systematic description of evaporation spectra for light and heavy compound nuclei. *Phys. Rev. C* **82**, 014610 (2010).
54. V. E. Viola and K. Kwiatkowski, Systematics of fission fragment total kinetic energy release. *Phys. Rev. C* **31**, 1550–1552 (1985).
55. S. E. Koonin, Proton pictures of high-energy nuclear collisions. *Phys. Lett. B* **70**, 43–47 (1977).
56. F. B. Yano and S. E. Koonin, Determining pion source parameters in relativistic heavy-ion collisions. *Phys. Lett. B* **78**, 556–559 (1978).
57. J. Knoll and J. Randrup, Proton-proton correlations in high-energy nuclear collisions. *Phys. Lett. B* **103**, 264 (1981).
58. S. Pratt and M. B. Tsang, Viewing the liquid-gas phase transition by measuring proton correlations. *Phys. Rev. C* **36**, 2390–2395 (1987).
59. R. J. Charity *et al.*, Systematics of complex fragment emission in niobium induced reactions. *Nucl. Phys* **483**, 371 (1988).
60. F. De-Qing *et al.*, Isospin effect of fragmentation reaction on the isotopic distribution produced in intermediate energy heavy ion collisions, *Chin. Phys. Lett.* **17**, 267 (2000).

INDEX